THE AIR EFFICIENCY AWARD 1942-2005

Air Vice-Marshal William Charles Coleman Gell, CB, DSO*, MC, TD, AE, DL

AOC Balloon Command, from 1 February 1944 to 13 February 1945, late Auxiliary Air Force and Royal Warwickshire Regiment; the highest ranking (at promulgation) British recipient of the Air Efficiency Award between 1942 and 2005.

His full medal entitlement is: CB, DSO & Bar, MC, 1914-15 Star, British War Medal, Victory Medal with Mention in Despatches emblem, Defence Medal, 1939-45 War Medal, 1935 GVR Jubilee, 1937 GVIR Coronation, Territorial Decoration, Air Efficiency Award, Special Constabulary Long Service & Good Conduct Medal, *Al Valore Militare* medal (Italy).

THE AIR EFFICIENCY AWARD
1942-2005

A History and Nominal Rolls of British and Commonwealth Recipients

Christopher Brooks

To David, Best wishes,

Christopher Brooks

24/1/08

**The Orders and Medals
Research Society**

Published by the Orders and Medals Research Society (OMRS)
PO Box 1904
SOUTHAM
CV47 2ZX

First Published 2006

© Christopher Brooks 2005

ISBN 0-9539207-3-9

Printed in Great Britain
by
Print Solutions Partnership

In honour of the part-time volunteer members of the British and Commonwealth Auxiliary Air Forces, who across the years have given dedicated service – and in some cases their lives – so that others may enjoy the privileges of freedom.

To Zita, Haydn and Elliot

CONTENTS

ILLUSTRATIONS

– the medals of:

Flt Lt H.F. Grubb, AE, RAFVR
Sqn Ldr R.A. Kings, AE, RAF
Gp Capt J.S. Kennedy, DFC*, AE, RAF
Sqn Ldr C. Haw, DFC, DFM, AE, RAF
Fg Offr C.F. Rawnsley, DSO, DFC, DFM*, AE, RAuxAF
Wg Cdr D.E. Kingaby, DSO, AFC, DFM**, AE, RAF
Wg Cdr E.B.R. Lockwood, DSO, MBE, AE**, RAuxAF
WO B.D. Davies, RAuxAF

FOREWORD

The story of the Air Efficiency Award (AEA) begins in August 1942, when His Majesty King George VI sanctioned the Award to all ranks of the pre-war Royal Air Force Reserve and Auxiliary Forces of the United Kingdom and the Commonwealth, the majority of whom were members of the Auxiliary Air Force (formed in 1924), Women's Auxiliary Air Force (WAAF), the Royal Air Force Volunteer Reserve (RAFVR) and those of the Commonwealth Auxiliary Air Forces (Australia, Canada and New Zealand).

The Air Efficiency Award ceased to be awarded in 2000 (although residual awards are still being awarded in 2005) following the government's decision to replace it with the Volunteer Reserves Service Medal (VRSM), a tri-service award to all ranks.

Chris Brooks brings two special qualities to bear on his in-depth study of the Air Efficiency Award. First, he has invested a great deal of time and effort over many years collecting, researching and documenting the origins, history, characteristics and, most importantly, compiling the nominal roll of the recipients of this distinctive Award.

The second factor is his dedication to this task and his enthusiasm for his subject, which was enhanced by his brief service with No 600 (City of London) Squadron, Royal Auxiliary Air Force. It was during his service that I met Chris and found we had a mutual interest in the history of the Air Efficiency Award.

This work reminds us that our culture is steeped in voluntary military service and sacrifice, which members of the Auxiliary Air Forces of the United Kingdom and Commonwealth, the Royal Air Force Volunteer Reserve and the Women's Auxiliary Air Force have given in peace and war.

The current members of the Royal Auxiliary Air Force (RAuxAF) and the Royal Air Force Reserve (RAFR) continue the fine traditions of voluntary service by supporting the Royal Air Force in its deployments and air operations world-wide – a fact of which I am sure the Second World War recipients of the Award would be proud.

I had the privilege to have served as a commissioned officer in the RAuxAF for nearly 30 years and was honoured by the award of the Air Efficiency Award and Clasp (the latter relinquished in favour of the VRSM and Clasp) and the Queens Volunteer Reserves Medal (QVRM) for my service.

I was delighted to be asked to write this Foreword and the members of the Orders and Medals Research Society (OMRS), as well as military historians, owe Chris Brooks a tremendous debt of gratitude for documenting, for the past, present and future members of the Air Force's reserves, the history and medal roll of the Air Efficiency Award.

Bruce Blanche, QVRM, AE,
Squadron Leader, RAuxAF (Retired)

Official Historian & Archivist, Royal Auxiliary Air Force

Squadron Leader Blanche

ACKNOWLEDGEMENTS

I owe grateful thanks to the Orders and Medals Research Society, publisher of this book, and for the guidance received from its Publications Team.

I am also indebted to a number of people, who in their capacity as employees of the Armed Forces of Great Britain and the Commonwealth countries, civilian organisations, or as individuals, have assisted in any small way to the collation of the material in this book.

They include, in no particular order:

Mr John Tamplin, MBE, TD, for granting me access to his collection of *Commonwealth* and *Hong Kong Gazette* supplements, and for his wise counsel;

the staff of the Imperial War Museum Reading Room;

the staff in the Department of Records and Information Services at the RAF Museum, Hendon, and the unidentified Friends of the Museum for assisting in the location and photocopying of so many pages of Air Ministry Orders;

Squadron Leader Bruce Blanche, QVRM, AE, RAuxAF (Retired), Official Historian and Archivist of the Royal Auxiliary Air Force, who was able to supply wise words on possible sources of names for the roll, and copies of the relevant RAF Regulations reproduced herein;

Wing Commander Jim Routledge, RAF (Retired), for his assistance with RAF statutes;

Messrs Bulpit, Twynholm and Visick – fellow medal collectors who along the months have provided encouragement in this enterprise;

Flying Officer Arna Barnao, RNZAF, who provided very great assistance to the New Zealand section of the roll;

Mrs Bridget Chapman, for allowing me to use the portrait of her father, AVM W.C.C. Gell, CB, DSO*, MC, TD, AE, DL.

To anyone who has assisted materially or otherwise and whose name is not mentioned in these notes, I offer my sincere apologies for the oversight.

I acknowledge unreservedly any copyrights from published sources past and present. Any errors and omissions in this work are solely mine and I would welcome any corrections to these that readers may be able to offer.

Finally and most importantly, I would like to thank my family – my sons Haydn and Elliot, but most especially my wife, Zita – for their support and encouragement, and for letting me spend endless hours uninterrupted on my computer, with my nose in books, or talking to fellow enthusiasts about medals, when I should have been spending the time with them.

INTRODUCTION

The history of the Air Efficiency Award (AEA) is inextricably linked with the gallant and self-sacrificing voluntary service of the men and women of the British and Commonwealth reserve and auxiliary air forces, from the establishment of the Air Force (Constitution) Act of 1917 through to the present day – a period of 88 years.

During that time, the personnel establishment of the Royal Air Force (RAF) reserve and auxiliary forces has naturally fluctuated according to the prevailing conditions – from tens of thousands of air and ground crews during the Second World War, down to a few hundred during the 1960s and 1970s, to the couple of thousand in today's forces.

However, it is deliberately intended that this study should not detail the service histories of the RAF reserve and auxiliary forces that have been extensively chronicled in other excellent publications. Instead, the focus is on the AEA – the only air force medal awarded to aircrew and ground trades, officers and other ranks alike, for which operational service was not a qualification – and the single common award that united all members of the various RAF reserve and auxiliary forces and their service components.

Broadly, but with some caveats around war and aircrew service, explored in later sections, the AEA was awarded for ten years' continuous voluntary service, comprising regular weekly training attendance and an annual two-week Summer camp. An additional Bar to the AEA was also available for each supplementary ten-year period completed. The first awards were promulgated in Air Ministry Orders (AMOs) of 3 December 1942, recognising the pre-war service of many recipients.

Further awards were made on a regular basis throughout the war years and beyond. However, in keeping with Second World War campaign medals, all thoughts of the actual design and production of the AEA as a physical entity were put on hold until the serious job of winning the war had been accomplished.

DESIGN AND APPROVAL

Instituted by Royal Warrant on 17 August 1942, the establishment of the AEA itself was promoted by the desire of King George VI and the RAF 'top brass' to ensure that the volunteer service of airmen and airwomen was recognised by the award of a long service medal, such as had existed for many years for volunteer members of the Royal Navy and the Army.

The ribbon design preceded the medal by three years and was produced by Group Captain E.H. Hooper, CBE, expert on ceremonial regalia working in the Air Ministry. In the book *The Royal Air Force Volunteer Reserve – Memories*[1], it is recorded that:

> Group Captain Hooper wanted something in green, a particular shade which he saw clearly in his mind's eye. It had to sit well with light blue, for the service. But how to mix them, how to arrange them … where to find the precise matching shades? Then his eye fell upon the cover of a popular magazine of the day. It was a favourite with servicemen everywhere, and it was titled *Lilliput*. The solution was there in front of him. A multicoloured illustration on its front cover contained the exact combination of colours he had toiled to create. Scissors in hand, he trimmed out a section, pasted it onto cardboard and placed it in the out-tray. The word came down. This was it.

Records from the National Archives show that the first activities within the Air Ministry toward the actual design and production of the medal itself commenced in July 1945, at the same time that the designs for wartime campaign stars had been produced for the King's approval. Air Ministry minutes[2] from the time note that:

> … it would be desirable to settle and obtain the King's approval for the design of the AEA so that manufacture can proceed as soon as labour and materials are available.

It is also recorded by Air Ministry civil servants that 'a design rather superior to the ordinary medal' be achieved since it was to be 'common to officers and airmen'. As part of the deliberations, consideration was given to the designs of the existing Efficiency Decoration and Efficiency Medal as a comparator for the style and visual impact that the AEA was expected to achieve.

---

[1] Dickson, A. (Gp Capt, RAFR), 1997, *The Royal Air Force Volunteer Reserve – Memories*, published by the MOD
[2] Minute from C.G. White (Air Ministry) to DGPI dated 17 July 1945, NA file ref AIR2/6892

Early suggestions were also put forward from within the Air Ministry[3] that the AEA could bear a resemblance to the uniform badge given to Americans who voluntarily entered into service with the RAF in the early days of the war. The mooted design, shown in the following diagram, was for a round skeletal badge incorporating an eagle with wings outspread, this surmounted by the royal crown and the whole surrounded by an oak laurel wreath.

An early design suggestion

Perhaps sadly, this rather attractive design did not gain much credence. It was recognised by the Air Ministry that suggestions for the design should be sought from the Royal Mint in common with the majority of other British medallic designs, subject to the opportunity for senior personnel at

---

[3] Minute from DGP3 to CAS and Sec of State dated 24 September 1945, NA file ref AIR2/6892

the Air Ministry to comment and critique the Royal Mint design prior to submission to the King for approval.

Accordingly, on 19 July 1945, a letter enclosing a copy of the Royal Warrant and a specimen of the ribbon, was sent from the Air Ministry to the Royal Mint. The letter highlighted the fact that the AEA was to be manufactured in silver, and noted the initial ideas from within the Air Ministry.

An amusing exchange of minutes[4] then followed between the two august institutions to settle the question of the 'type of bird' contained within the RAF badge – was it an eagle or an albatross? I am sure that many serving and former members of the RAF were able to breath a sigh of relief when the Air Ministry was able to categorically state that it should, of course, be an eagle as evidenced in the College of Heralds' description of the RAF badge as a 'circlet bearing the motto PER ARDUA AD ASTRA with an eagle superimposed, the whole surmounted by a Royal crown'.

Having settled on the correct type of bird to be used, the Royal Mint commissioned the noted artist and designer, Mr Percy Metcalfe, CVO, RDI, to work on the designs.

Metcalfe was already established as one of the noted numismatic and medallic designers of the era, having beaten 65 other design entries to be chosen as designer of the coinage for the Irish Free State in 1928. Among other notable achievements, he also designed the Mount Everest Flight Expedition Medal, 1933, that had been awarded to Air Commodore P.F.M. Fellowes, DSO. However, arguably his most famous design up to that point had also been in the medallic field – The George Cross – instituted in 1940.

Designs for the AEA were swiftly delivered by Metcalfe and, on 27 August 1945, the Royal Mint was able to submit three sketches of designs to the Air Ministry for consideration. At the same time, the Air Ministry was invited to participate in the September meeting of the Royal Mint Advisory Committee on Designs, to be held at the National Gallery in London, in order to discuss and agree the design for the AEA[5].

---

[4] Minute from G. Tucker (Air Ministry) to J.H.McC. Craig (Royal Mint) dated 19 July 1945 and reply dated 25 July 1945, NA file ref AIR2/6892

[5] Minute from Craig (Royal Mint) to Tucker (Air Ministry) dated 27 August 1945, NA file ref AIR2/6892

The following drawings show the obverse and reverse of each of the three submissions, which comprised:

Design 'A' – an oval-shaped medal with a suspender in the form of an eagle with wings outspread, but plain on the reverse. The obverse bore the effigy of King George VI with the royal cipher, and on the reverse, the words AIR EFFICIENCY AWARD.

Obverse of design 'A'   *Reverse of design 'A'*

Design 'B' – a round-shaped medal with straight plain suspender. The obverse bore the effigy of the King with the royal cipher, and on the reverse an eagle with wings outspread, this surmounted by the royal crown and a spray of oak leaves under each wing.

Design 'C' – a variant on design 'A' with a round instead of oval medal.

Obverse of design 'B'

Reverse of design 'B'

Obverse of design 'C'

Reverse of design 'C'

The Advisory Committee on Designs meeting was duly held on 4 September 1945, with Mr C.G. White in attendance on behalf of the Air Ministry. From the meeting, there is no documentary evidence of a firm recommendation on any of the three designs by the Royal Mint, save that it was recorded that Mr Metcalfe was 'inclined to deprecate the addition of the elaborate bar on design 'B'', because from a design perspective, a medal with an eagle on the obverse would clash all to readily with an eagle suspension bar[6].

As can be seen from the illustrations, Design 'A' – the oval-shaped medal – with soon-to-be-familiar eagle suspender was a clear contender from the start of the design process. The minutes do however record a debate on which version of the King's effigy – crowned or uncrowned – should be used, but no final agreement on the design of the AEA itself was minuted. Instead, the Air Ministry was requested to go away and make a provisional selection from the three designs submitted, for further refinement by Mr Metcalfe.

No evidence of a formal minute from the Air Ministry has been found within the National or Royal Mint Archives, recording the selection of Design 'A' as the 'winning design', although this is known to have been the case.

Further communications in September 1945 between the Royal Mint and the Air Ministry did settle that the King's effigy on the obverse of the AEA should be the uncrowned version, as on the existing Air Force and Distinguished Flying Medals. Additionally, the Air Ministry requested that the eagle should be shown in relief on both sides of the medal mount and not just on the obverse, with a plain reverse[7].

Incorporating the Air Ministry request on the mount, Mr Metcalfe produced plaster casts of the proposed final design and they were duly submitted to the Air Ministry in November 1945. It was also at this time that the question of the ribbon width was raised by the Royal Mint, which noted that the one and a half-inch wide ribbon as specified in the Royal Warrant was one quarter-inch wider than the universal one and a quarter-inch ribbon width that was the prevailing standard for most other contemporary medals[8].

In the interim, the Air Ministry pressed ahead with the task in hand and, on 8 December 1945, the Right Honourable Viscount Stansgate, DSO, DFC, Secretary of State for Air, submitted Design 'A' to Buckingham Palace[9]

---

[6] Minutes of the 119th meeting of the Honours & Awards Committee held on 4 September 1945, NA file ref AIR2/6892
[7] Letter C.G. White (Air Ministry) to J.H.McC. Craig (Royal Mint) dated 29 September 1945 and reply 3 October 1945, NA file ref AIR2/6892
[8] Letter Craig to White dated 29 November 1945 and reply 15 December 1945, NA file ref AIR2/6892
[9] Letter Stansgate to Rt Hon Sir A. Lascelles, KCB, KCVO, CMG, MC, dated 8 December 1945, NA file ref AIR2/6892

for provisional Royal approval. Stansgate noted in his submission that 'fourteen hundred officers and two hundred and seventy airmen have so far been authorised to wear the ribbon of the AEA'.

It is also quite apparent that Stansgate did not appreciate the work done on the medal by either his staff or the Royal Mint, as he recorded that 'I do not feel any particular enthusiasm for the design but it seems to compare reasonably well with that of the Efficiency Medal'. However, even in the absence of a ringing endorsement from the Secretary of State, Buckingham Palace gave their informal approval for the design in a letter to Stansgate dated 11 December 1945, pending a formal submission to be made at a later date.

Thus, the Air Ministry was able to report the informal Royal approval to the Royal Mint and request final specimens, in plaster or metal, to be used in the final submission to the Palace[10]. At the same time, the Ministry addressed the ribbon width issue, noting as previously implied, that in order to compare favourably with the existing Efficiency Decoration the ribbon should be wider than standard as per the Royal Warrant. This was on the basis that the personnel awarded the AEA would 'naturally compare the design with that of the Efficiency Decoration'. Clearly, Air Ministry personnel were of the view that service personnel were preoccupied with ribbon width and that the bigger the width of ribbon, the higher the esteem in which the award would be held by the recipient!

Possibly because of the lack of availability of suitable labour and materials prevailing at the time, the first actual silver cast specimen of the AEA was not delivered to the Air Ministry until April 1946, some four months after the informal Royal approval. Upon receipt at the Air Ministry, the silver specimen was swiftly forwarded to the Palace under the cover of another formal submission signed by Viscount Stansgate[11], most humbly requesting 'that Your Majesty may be graciously pleased to approve the design of the Air Efficiency Award'.

As shown in the following illustration, on 14 April 1946, in a letter from Windsor Castle to Wing Commander Gordon Sinclair, DFC, of the Air Ministry, it was duly recorded that the King had signed the submission that day and that, in a gesture that would impress all medal collectors, had 'retained the sample for his own collection'!

---

[10] Letter White to Craig dated 15 December 1945, NA file ref AIR2/6892
[11] Submission papers April 1946, NA file ref AIR2/6892

WINDSOR CASTLE

14th April, 1946.

My dear Sinclair,

    The King has today signed
the Submission of the Secretary
of State for Air with regard to
the design of the Air Efficiency
Award, and has retained the sample
which you sent for his own collection.

       Yours sincerely,

Wing-Commander
  Gordon Sinclair, DFC.,
  Air Ministry.

Letter from King George VI

With considered tact, the Air Ministry overlooked the collecting habits of the King and took delivery of a second (presumably silver) specimen from the Royal Mint as a replacement[12].

And so the Air Efficiency Award had received Royal approval, enabling the production process to begin, on the agreed basis that the Air Ministry would meet the cost of the silver metal used in the medal and the Royal Mint all of the other associated production costs.

I have not been able to find an officially documented date for the first actual issuance of the medal to recipients, but based upon contemporary letters from RAF Records to recipients and photographs from the period, I would estimate that the first medals were issued during the late 1947 and early 1948 period.

---

[12] Letter to F.S. Yuill (Air Ministry) from the Royal Mint dated 27 April 1946, NA file ref AIR2/6892

THE ROYAL WARRANT, AND REGULATIONS GOVERNING THE AWARD OF THE AEA

The Royal Warrant was published in the *London Gazette* on 17 August 1942. A facsimile of that notice is shown below, plus copies of the 1942, 1951, and 1996 RAF regulations governing the award of the AEA.

THIRD SUPPLEMENT
TO
The London Gazette

Of TUESDAY, the 8th of SEPTEMBER, 1942

Published by Authority

Registered as a newspaper

FRIDAY, 11 SEPTEMBER, 1942

Air Ministry. 11th September. 1942.

THE AIR EFFICIENCY AWARD.

Royal Warrant.

GEORGE R.I.

GEORGE THE SIXTH, by the Grace of God, of Great Britain, Ireland and the British Dominions beyond the Seas, King, Defender of the Faith, Emperor of India; To all to whom these Presents shall come. Greeting.

Whereas We are desirous of signifying Our appreciation of long and meritorious service in Our Auxiliary and Volunteer Air Forces;

We do hereby for Us, Our Heirs and Successors, institute and create a medal to be available for officers and airmen and to be designated the "Air Efficiency Award";

And We do hereby direct that the following regulations shall govern the said award:—

1. *Description.*—The Air Efficiency Award shall be in silver.

2. *Ribbon.*—The award shall be worn on the left breast pendent from a ribbon of one inch and a half in width which shall be in colour green with two central stripes of pale blue, one-eighth of an inch in width.

3. *Miniatures.*—Reproductions in miniature, which may be worn on certain occasions by those on whom this award is conferred, shall be approximately half the size of the award and a sealed pattern of the miniature award shall be kept in the Central Chancery of Our Orders of Knighthood.

4. *Eligibility (Royalty).*—The award may be worn by Us, Our Heirs and Successors, Kings and Queens Regnant of Great Britain, Ireland and the British Dominions beyond the Seas, and it shall be competent for Us, Our Heirs and Successors, to confer at Our pleasure the award upon any Princes of the blood Royal.

5. *Eligibility (general).*—The award may be conferred on officers and airmen of any Auxiliary or Volunteer Air Force raised in Our United Kingdom of Great Britain and Northern Ireland, Our Indian Empire, Burma, any of Our Colonies or a territory under Our protection; or within any other part of Our Dominions Our Government whereof shall so desire or within any territory under Our protection administered by Us in such Government.

6. *Service required.*—The period of service requisite for the award shall be ten years' qualifying service (as defined in the regulations hereinafter mentioned) and shall include not less than five years' actual service (as defined in the regulations hereinafter mentioned) in an Auxiliary or Volunteer Air Force.

7. *Publication and registration.*—The names of those upon whom the award is conferred shall be published and a register kept in the manner prescribed in the regulations hereinafter mentioned.

8. *Forfeiture and restoration.*—It shall be competent for Our Air Council in regard to Our Auxiliary Air Force and Our Royal Air Force Volunteer Reserve, or for the Governors-General, Governors or Officers administering the Government, as the case may be, in regard to the Auxiliary or Volunteer Air Forces of Our respective Dominions, Colonies or territories under Our protection, or for our Viceroy in regard to the Auxiliary or Volunteer Air Forces in India, or for the Governor in regard to the Auxiliary or Volunteer Air Forces in Burma, to cancel and annul the conferment of the award on any person and also to restore a forfeited award under the conditions laid down in the regulations hereinafter mentioned.

9. *Further regulations.*—The award shall be conferred under such regulations as to grant, forfeiture, restoration and other matters, in amplification of these Our regulations as may, with Our approval be issued from time to time by Our Air Council in regard to Our Auxiliary Air Force and Our Royal Air Force Volunteer Reserve, or by the Governors-General, Governors or Officers administering the Government, as the case may be, in regard to the Auxiliary or Volunteer Reserve Air Forces of Our respective Dominions, Colonies or territories under Our protection, or by Our Viceroy in regard to the Auxiliary or Volunteer Reserve Air Forces in India or by the Governor in regard to the Auxiliary or Volunteer Reserve Air Forces in Burma.

10. *Annulment, etc., of regulations.*—We reserve to Ourself, Our Heirs and Successors, full power of annulling, altering, abrogating, augmenting, interpreting, or dispensing with these regulations or any part thereof by a notification under Our Sign Manual.

Given at Our Court at St. James's this seventeenth day of August, 1942, in the sixth year of Our Reign.

By His Majesty's Command.
Archibald Sinclair.

Royal Warrant from the London Gazette

In addition to Great Britain, several Commonwealth countries published the Warrant and subsequently issued the medal. These countries include: Australia, Canada, New Zealand and Hong Kong, for which nominal rolls have been produced herein.

It was always intended that South African airmen would also be eligible for the medal. However, diplomatic wrangles about the wording of clauses to the Royal Warrant, and a protracted correspondence between the Air Ministry and the South African Government during the period 1943 to 1946, regarding repeated South African requests for a dual English/Afrikaans version of the medal, effectively forestalled any attempts to issue the medal in South Africa, and in the event none was ever issued.

With the benefit of hindsight it is hard to see what the possible objections to a South African version could have been, particularly given the existence of South African versions of other medals, such as the Efficiency Medal.

Had it come to pass, the South African version would have had a dual English/Afrikaans reverse, worded 'Air Efficiency Award – Toekenning Vir Bekwaamheid (Lugmag)' with a suspender bar bearing the words 'Union of South Africa – Unie Van Suid-Afrika'[13].

For those readers not wishing to trudge through the detail of Royal Warrant or Regulations, they are preceded by a high-level summary of the original (and subsequently amended) award criteria as shown in the following table:

---

[13] Various inter-governmental items of correspondence refer, NA file ref T333/44

	1942	1951	1996
	Initial regulations underpinning the 1942 Royal Warrant	Amended regulations to reflect the amended Royal Warrant particularly WW2 service & the issue of Clasps	Additional Defence Council regulations & expression of preference for presentation of the award
Summary			
Eligibility	Service commenced pre 4/9/39	Same plus additional eligibility of post 18/8/46	All Officers, airmen and airwomen of the RAuxAF and RAFVR
	Completed required training and certified as deserving	Same	Same
	Completed requisite service	Same	Same
Service	10 years	Same	Same
	At least 5 in AAF or RAFVR (but not Reserves)	Same; AAF amended to RAuxAF and WAAF added	Removed
Qualifying service	AAF & RAFVR aircrew = time and a half	Same plus: a) Pre 3/9/39 engagements and WW2 aircrew service = treble time b) Post 18/8/46 aircrew engagement = time and a half	Amended to: a) WW2 aircrew service = treble time b) Post WW2 aircrew engagement = double time
	AAF & RAFVR other = single time	Same plus: a) Pre 3/9/39 engagements and WW2 other service = double time b) Post 18/8/46 other engagement = single time	Amended to: a) WW2 other service = double time b) Post WW2 other engagement = time and a half
	n/a	All other WW2 service = single time	Same
	n/a	WAAF service = single time	Same
	AAF & RAFVR aircrew Reserve Lists = three quarter time	Same for pre 3/9/39 engagements	Same
	AAF & RAFVR other Reserve Lists = half time	Same for pre 3/9/39 engagements	Same
	WW1 regular force service counts = single time	Same	Amended to WW1 or WW2 regular force service counts = single time
	WW1 non-regular force service counts = double time	Same	Amended to WW1 or WW2 peacetime service or non-regular force service counts = double time
	Service in W Africa pre 30/9/37 = double time	Removed	n/a
	Service in W Africa post 30/9/37 = single time	Removed	n/a
	Service in other non-regular forces = single time	Same	Same
	n/a	n/a	Service since 10/1/55 on the RAuxAF General List = single time
Already reckoned	Periods of service for which an efficiency medal has already been granted does not count as qualifying service	Same	Same
Continuity	Qualifying service must be continuous unless directed otherwise	Same plus service during and after WW2 can count provided the break in service is not more than 6 months from 18/8/46 or day of release (whichever is later) or 12 months for persons from an overseas force	Same but date amended to 8/11/46 Additionally National Service 1948 - 1955, although not qualifying time, would not be regarded as breaking the continuity of qualifying service
Clasp	None noted	For further 10 year service subject to existing conditions	Same but amended to award for each 10 year period of additional service plus note of a rosette on the ribbon for each clasp
Forfeiture	None noted	As laid down for RAF LS&GC medal	Same
Restoration	None noted	At the discretion of the Air Council	Same but read Defence Council for Air Council

13

1942 AIR MINISTRY ORDER A969 FOR THE AEA

AIR MINISTRY
10th September 1942

The following Orders are hereby promulgated for information and guidance and necessary action.

By Command of the Air Council.

PERSONNEL

A.969. – Air Efficiency Award
(A.186572/41/S.10(a), - 10.9.42.)

1.   The King has been graciously pleased to institute a medal, entitled the "Air Efficiency Award", in recognition of long and meritorious service in the auxiliary and volunteer air forces of the Empire.   The warrant is being published in the London Gazette and, together with the regulations made by the Air Council in regard to the conferment of the Award on members of the Auxiliary Air Force and the R.A.F. Volunteer Reserve, will be included in due course in, K.R. & A.C.I.

2.   A summary of the provisions of the warrant and of the regulations is contained in appendix A to this order.   Appendix A indicates the conditions applicable to members of the Auxiliary Air Force and the R.A.F. Volunteer Reserve whose service commenced before 4th September, 1939; the conditions applicable to those whose service commenced on or after that date and who continue in the post-war non-regular air forces will be determined after the conclusion of hostilities.

3.   Recommendations for the Award are to be submitted in the form shown in appendix B, which is to be prepared by units and forwarded to the Air Ministry (S.10(a)), for officers, and to the Air Officer i/c Records, for airmen.   Qualified persons will be authorised to wear the ribbon, but the Award itself will not be issued during the war.

4.   Recommendations for airmen from overseas commands holding service documents are to be accompanied by Forms 200, prepared from Forms 1433 or 1996, as appropriate. When forwarding to the Air Ministry recommendations from home units and overseas units not holding service documents, the Air Officer i/c Records will attach Forms 1433 or 1996, as appropriate. The Air Officer i/c Records will also compile and forward

Source	Service No	Force	Notes	Surname
N1007 1947	72366	RAFVR		POLLOCK
N174 1952	840795	RAuxAF		POWELL
N174 1952	91244	RAuxAF		PRICKETT
N802 1961	801392	AAF	wef 22/11/42	PRITCHARD
N1060 1947	72035	RAFVR		RAYNER
N693 1952	2681596	RAuxAF	Formerly 800448	READ
N1007 1947	840048	AAF		RILETT
N1060 1947	101781	RAFVR		RILEY
N693 1952	188106	RAFVR		ROBERTSON
N693 1952	2682522	RAuxAF	Formerly 1253612	ROBINS
N1007 1947	84724	RAFVR		ROBINSON
N100 1963	800282	AAF	wef 1/4/44	ROLFE
N693 1952	106020	RAFVR		ROXBURGH
N802 1961	801153	AAF	wef 13/6/44	SAUNDERS
N174 1952	743582	RAFVR		SAXTON
N174 1952	701484	RAFVR		SCOTT
N693 1952	136060	RAuxAF		SCOTT
N693 1952	2689525	RAuxAF	Formerly 1233462	SEAMER
N1007 1947	752345	RAFVR		SEYMOUR
N174 1952	770613	RAFVR		SHAWYER
N1007 1947	856016	AAF		SMITH
N693 1952	2686004	RAuxAF	1st Clasp; Formerly 808081	SMITH
N100 1963	2664235	WRAuxAF	wef 1/12/62	SMITH
N693 1952	863353	RAuxAF		SPILLER
N1060 1947	845291	AAF		STARK
N100 1963	2692172	RAFVR	wef 16/12/62	STEVENS
N802 1961	803246	AAF	wef 10/1/37	STEWART
N1060 1947	812136	AAF		STITT
N1060 1947	747154	RAFVR		TANNER
N1060 1947	206	WAAF		TAUDEVIN
N693 1952	104109	RAFVR		TAYLOR
N802 1961	870921	AAF	wef 12/7/44	THOMAS
LG 2006	90100	RAuxAF		TOLLEMACHE
LG 2006	90100	RAuxAF	1st Clasp	TOLLEMACHE
N1060 1947	801555	AAF		TOWNSHEND
N1007 1947	840789	AAF		TRINDER
N1007 1947	846780	AAF		TRIPTREE
N1007 1947	840518	AAF		TURNER
N693 1952	186653	RAFVR		WARD
N1007 1947	72968	RAFVR		WICKHAM
N693 1952	2681644	RAuxAF	Formerly 815174	WILD
N100 1963	2672185	WRAuxAF	wef 26/5/62	WILKINS
N1007 1947	816073	AAF		WILKINSON
N693 1952	799943	RAFVR		WILSON
N1007 1947	85959	RAFVR		WOOD
N1007 1947	85958	RAFVR		WOOD
N693 1952	868541	RAuxAF		WOODWARD
N1007 1947	841119	AAF		YORKE

Surname	Initials	Rank	Status	Date
YULE	R C	FO		06/03/1952

The figures in the 'Total' column to the right of pages 421 and 423 are incorrect; the year from the left column has been added-in, creating excessive totals.

Excluding the above 143 additional names, the correct totals are:

Year	Total		Year	Total
1942	155		1958	205
1943	371		1959	269
1944	490		1960	271
1945	790		1961	151
1946	754		1962	109
1947	1212		1963	110
1948	1093		1964	115
1949	517		1965	90
1950	721		1966	60
1951	399		1967	14
1952	269		1968	33
1953	310		1969	22
1954	303		1970	28
1955	259		1971	33
1956	194		1972	32
1957	249		1973	30

Source	Service No	Force	Notes		Surname
N174 1952	187294	RAuxAF	DFC		YULE

Year	Total	Year	Total
1974	32	1990	53
1975	18	1991	23
1976	6	1992	35
1977	10	1993	62
1978	9	1994	57
1979	11	1995	77
1980	11	1996	49
1981	12	1997	84
1982	13	1998	64
1983	15	1999	39
1984	27	2000	24
1985	26	2001	17
1986	23	2002	7
1987	25	2003	11
1988	21	2004	18
1989	49	2005	10
		No date	9
		Total	**10605**

The Air Efficiency Award 1942-2005

Surname	Initials			Rank	Status	Date
POLLOCK	C	C	A	SQD LDR		04/12/1947
POWELL	W	L		SGT		06/03/1952
PRICKETT	L	A		FO		06/03/1952
PRITCHARD	A	B		SGT	Temporary	18/10/1961
RAYNER	E	B		WING COMM	Acting	18/12/1947
READ	R	H		SGT		11/09/1952
RILETT	C	B		WO		04/12/1947
RILEY	B	T		SQD LDR	Acting	18/12/1947
ROBERTSON	R	I		FL		11/09/1952
ROBINS	R	V	F	CPL		11/09/1952
ROBINSON	E	L		SQD LDR	Acting	04/12/1947
ROLFE	A	G		SGT	Temporary	06/02/1963
ROXBURGH	T			FO		11/09/1952
SAUNDERS	G	H		CPL	Temporary	18/10/1961
SAXTON	J	W		LAC		06/03/1952
SCOTT	T			LAC		06/03/1952
SCOTT	J	H		FO		11/09/1952
SEAMER	A	R		SGT		11/09/1952
SEYMOUR	A	E		LAC		04/12/1947
SHAWYER	S	G		LAC		06/03/1952
SMITH	G	T		AC1		04/12/1947
SMITH	L	V		SGT		11/09/1952
SMITH	J	A	S	SGT		06/02/1963
SPILLER	W	C	J	CPL		11/09/1952
STARK	G	T		SGT		18/12/1947
STEVENS	A	G		CPL		06/02/1963
STEWART	A	L		CPL	Acting	18/10/1961
STITT	J	L		SGT		18/12/1947
TANNER	S	V		CPL		18/12/1947
TAUDEVIN	N	I	de C	SQD OFF	Acting	18/12/1947
TAYLOR	M	O		FO		11/09/1952
THOMAS	G	W		LAC		18/10/1961
TOLLEMACHE	A	H	H	FL		07/03/2006
TOLLEMACHE	A	H	H	FL		07/03/2006
TOWNSHEND	T	W		WO		18/12/1947
TRINDER	G	A		CPL		04/12/1947
TRIPTREE	F	J		CPL		04/12/1947
TURNER	L	J		CPL		04/12/1947
WARD	S	W		FL		11/09/1952
WICKHAM	N	W		FL		04/12/1947
WILD	F	A		SGT		11/09/1952
WILKINS	M	M		SACW		06/02/1963
WILKINSON	W			AC2		04/12/1947
WILSON	W			SGT		11/09/1952
WOOD	N	W		FL		04/12/1947
WOOD	A	G		FO		04/12/1947
WOODWARD	H			CPL		11/09/1952
YORKE	C	H		CPL		04/12/1947

Source	Service No	Force	Notes	Surname
N693 1952	873761	RAuxAF		ADAM
N1007 1947	804194	AAF		ANDREWS
N693 1952	2614018	RAFVR	Formerly 752821	ANLEY
N693 1952	73594	RAFVR		AUSTIN-SMITH
N1007 1947	860299	AAF		BACON
N1007 1947	846974	AAF		BALMAN
N1007 1947	840400	AAF		BAMPTON
N1060 1947	90703	AAF		BANWELL
N693 1952	150194	RAuxAF		BAXTER
N1060 1947	76254	RAFVR		BLACKMON
N1007 1947	746601	RAFVR		BLATHERWICK
N1007 1947	172708	RAFVR	DFM	BLOOD
N693 1952	196457	RAFVR		BOTSFORD
N174 1952	858266	RAuxAF		BOWERS
N1060 1947	73092	RAFVR		BOYCOTT
N693 1952	2602502	RAFVR	Formerly 1625122	CASH
N1007 1947	840752	AAF		CHEW
N1007 1947	860454	AAF		CHIDLEY
N174 1952	845574	RAuxAF		CLARKE
N693 1952	846510	RAuxAF		CLIFFORD
N1060 1947	869494	AAF		COCKTON
N802 1961	805495	AAF	wef 19/5/44	COLBURN
N87 1948	90229	RAuxAF		COLTART
N693 1952	812217	RAuxAF		COOPER
N100 1963	2650962	RAuxAF	wef 25/11/62	CORBETT
N1007 1947	869492	AAF		COX
N693 1952	2686002	RAuxAF	1st Clasp; Formerly 808176	CROSBY
N100 1963	855876	AAF	wef 5/6/44	CULLEN
N802 1961	196976	RAFVR	wef 14/5/61	CULSHAW
N1060 1947	743802	RAFVR		DOUGALL
N1060 1947	853837	AAF		DRURY
N693 1952	107220	RAFVR		DUNLOP
N693 1952	184171	RAFVR		DUNSTER
N1007 1947	801561	AAF		EATON
N802 1961	818177	AAF	wef 24/6/44	EDWARDS
N693 1952	63760	RAFVR		ELWIN
N693 1952	85281	RAFVR	DFC,AFC	FARRELL
LG 2006	2601521	RAFVR		FISHER
N1060 1947	812159	AAF		FLETCHER
N1060 1947	112114	RAFVR		FOAD
N693 1952	171443	RAFVR	DFC	FROST
N693 1952	843157	RAuxAF		GEORGE
N693 1952	138438	RAFVR		GORDON
N174 1952	180194	RAuxAF		GRANT
N1007 1947	110405	RAFVR		GUERRIER
N1007 1947	72407	RAFVR		HAMBLIN

The Air Efficiency Award 1942-2005

Surname	Initials			Rank	Status	Date
HARRISON	J			SGT		18/12/1947
HAVILAND	J	K		FL		04/12/1947
HAYES	E			FLT OFF	Acting	18/12/1947
HEATH	E	C		SQD LDR	Acting	18/12/1947
HEATON	K			FL		11/09/1952
HEMFREY	H	S		SGT		11/09/1952
HIDSON	T	G	P	SQD LDR		04/12/1947
HINKS	J	F		SQD LDR		11/09/1952
HITCHCOCK	R	F		FL		07/03/2006
HODGSON	J	W		CPL		11/09/1952
HOPKINS	A	T		CPL		04/12/1947
HURDMAN	A	R	N	FL	Acting	18/12/1947
IZARD	P			FL		18/10/1961
JACKSON	J	G		SGT		06/03/1952
JAMIESON	J	A		SGT		11/09/1952
JONES	F			LAC		18/10/1961
KANE	J			SGT		04/12/1947
KAYES	I	A		FL		04/12/1947
KNIGHT	A	G	F	CPL		04/12/1947
KNOX	G			SGT		06/03/1952
LAWFORD	D	N		FL		18/12/1947
LAWSON	C	H		SQD LDR	Acting	11/09/1952
LENSON	D	J		SGT		04/12/1947
LLOYD	G	K		LAC		18/10/1961
LOCKE	W	L	D	CPL		11/09/1952
LOFTUS	M	C		WING OFF		11/09/1952
LONGMAN	L	E		SGT		04/12/1947
LUNT	H	L		CPL		11/09/1952
MARSBURY	C	S		LAC		04/12/1947
MEEHAN	K	P		FLT SGT		11/09/1952
MERRETT	E	M		FLT OFF		04/12/1947
MIDGLEY	W	S		FO		11/09/1952
MILLARD	S	J		LAC		11/09/1952
MILLS	B			LAC		18/12/1947
MILLS	A	F		SGT	Temporary	06/02/1963
MILNER	A	J	M	FL		11/09/1952
MORRIS	E	G		LAC		18/10/1961
NICOLL	J	R		FL		11/09/1952
NIXON	J			SGT		06/03/1952
NOBLE	R			FO		06/03/1952
OXBY	J	E		CPL		18/12/1947
PAINE	H	S		FL		04/12/1947
PARAMOR	W	A		CPL		18/12/1947
PEARCE	J	A		LAC		18/10/1961
PETITT	A	G	L	CPL		18/12/1947
PHILLIPS	T	R		CPL	Temporary	18/10/1961
PILKINGTON	M			SGT		18/12/1947
POCOCK	L	J		FL		18/12/1947

Source	Service No	Force	Notes	Surname
N1060 1947	857416	AAF		HARRISON
N1007 1947	82690	RAFVR	DFC	HAVILAND
N1060 1947	6414	WAAF		HAYES
N1060 1947	123781	RAFVR		HEATH
N693 1952	115750	RAFVR		HEATON
N693 1952	749070	RAFVR		HEMFREY
N1007 1947	72266	RAFVR		HIDSON
N693 1952	72426	RAFVR		HINKS
LG 2006	137430	RAFVR		HITCHCOCK
N693 1952	858184	RAuxAF		HODGSON
N1007 1947	846584	AAF		HOPKINS
N1060 1947	69512	RAFVR		HURDMAN
N802 1961	2607450	RAFVR	wef 10/1/61	IZARD
N174 1952	799764	RAFVR		JACKSON
N693 1952	2611375	RAuxAF	Formerly 807273	JAMIESON
N802 1961	818120	AAF	wef 27/12/45	JONES
N1007 1947	816020	AAF		KANE
N1007 1947	85283	RAFVR		KAYES
N1007 1947	845314	AAF		KNIGHT
N174 1952	803534	RAuxAF		KNOX
N1060 1947	113266	RAFVR		LAWFORD
N693 1952	132068	RAFVR		LAWSON
N1007 1947	800647	AAF		LENSON
N802 1961	810120	AAF	wef 13/11/43	LLOYD
N693 1952	865107	RAuxAF		LOCKE
N693 1952	229	WAAF		LOFTUS
N1007 1947	801540	AAF		LONGMAN
N693 1952	810167	RAuxAF		LUNT
N1007 1947	840406	AAF		MARSBURY
N693 1952	2680003	RAuxAF	Formerly 1458321	MEEHAN
N1007 1947	1554	WAAF		MERRETT
N693 1952	144672	RAFVR		MIDGLEY
N693 1952	2606774	RAuxAF	Formerly 844695	MILLARD
N1060 1947	759084	RAFVR		MILLS
N100 1963	746553	RAFVR	wef 14/7/44	MILLS
N693 1952	153164	RAuxAF		MILNER
N802 1961	843991	AAF	wef 30/3/44	MORRIS
N693 1952	126060	RAFVR		NICOLL
N174 1952	874061	RAuxAF		NIXON
N174 1952	155451	RAuxAF		NOBLE
N1060 1947	814155	AAF		OXBY
N1007 1947	72391	RAFVR		PAINE
N1060 1947	845354	AAF		PARAMOR
N802 1961	860380	AAF	wef 13/7/44	PEARCE
N1060 1947	849972	AAF		PETITT
N802 1961	818234	AAF	wef 4/7/44	PHILLIPS
N1060 1947	857594	AAF		PILKINGTON
N1060 1947	158132	RAFVR		POCOCK

The following recipients were revealed after the book went to Press:

Surname	Initials			Rank	Status	Date
ADAM	A			LAC		11/09/1952
ANDREWS	J	J	T	FLT SGT		04/12/1947
ANLEY	R	S		SGT		11/09/1952
AUSTIN-SMITH	G	H		FL		11/09/1952
BACON	L			CPL		04/12/1947
BALMAN	C	J		FLT SGT		04/12/1947
BAMPTON	B	A		CPL		04/12/1947
BANWELL	J	A		SQD LDR		18/12/1947
BAXTER	J			FL		11/09/1952
BLACKMON	F	A		FL		18/12/1947
BLATHERWICK	H			FLT SGT		04/12/1947
BLOOD	L	J	B	FL		04/12/1947
BOTSFORD	R	H		FL		11/09/1952
BOWERS	N			CPL		06/03/1952
BOYCOTT	A	G		WING COMM	Acting	18/12/1947
CASH	D			SGT		11/09/1952
CHEW	G	T		LAC		04/12/1947
CHIDLEY	R	L		SGT		04/12/1947
CLARKE	C	G	S	CPL		06/03/1952
CLIFFORD	W	A		CPL		11/09/1952
COCKTON	J	H		SGT		18/12/1947
COLBURN	G			CPL	Temporary	18/10/1961
COLTART	W	D		WING COMM		05/02/1948
COOPER	L	A		CPL		11/09/1952
CORBETT	H			SGT		06/02/1963
COX	J	W		WO		04/12/1947
CROSBY	J	C	R	CPL		11/09/1952
CULLEN	G			CPL	Temporary	06/02/1963
CULSHAW	I			FL		18/10/1961
DOUGALL	D			CPL		18/12/1947
DRURY	W			FLT SGT	Acting	18/12/1947
DUNLOP	J	P		FL		11/09/1952
DUNSTER	R	A		FL		11/09/1952
EATON	W	H		CPL		04/12/1947
EDWARDS	H	C		LAC		18/10/1961
ELWIN	R	E	O	FL		11/09/1952
FARRELL	A	W		FL		11/09/1952
FISHER	T	A		SGT		07/03/2006
FLETCHER	N	P		FLT SGT		18/12/1947
FOAD	L	J		FL		18/12/1947
FROST	N	H		FL		11/09/1952
GEORGE	S	R		CPL		11/09/1952
GORDON	R	A		FL		11/09/1952
GRANT	A	C		FO		06/03/1952
GUERRIER	E	G	J	FL		04/12/1947
HAMBLIN	F	W		SQD LDR		04/12/1947

Forms 200 for all service rendered by personnel prior to joining the Auxiliary Air Force or the R.A.F. Volunteer Reserve.

5.   Adjutants are to bring to the notice of C.O.'s. the name of every officer and airman who becomes eligible for the Award, but the fact that the requisite service has been completed confers no right to receive the Award. C.O.'s. are to recommend only such persons as are "efficient and deserving of the Award".

APPENDIX A

AIR EFFICIENCY AWARD – OUTLINE OF CONDITIONS FOR THE AUXILIARY AIR FORCE AND THE R.A.F VOLUNTEER RESERVE

1.   *Description –* The Award will be of silver and worn on the left breast from a green ribbon one inch and a half in width with two central stripes of pale blue, one eighth of an inch in width.

2.   *Eligibility –* The Award may be conferred on officers and airmen of the Auxiliary Air Force or the R.A.F. Volunteer Reserve provided that –
(1)   their service commenced before 4th September, 1939;
(2)   they have completed the required periods of training and are certified by the responsible air force authorities as efficient and deserving of the Award; and
(3)   they have completed the requisite service.

3.   *Service required –* The service requisite for the Award is 10 years qualifying service (as defined below), including not less than five years' service in –
(1)   the Auxiliary Air Force (excluding the Auxiliary Air Force General List, the Auxiliary Air Force Reserve of Officers and the Auxiliary Air Force Reserve of Airmen);
(2)   the R.A.F. Volunteer Reserve (including classes of the R.AF. Reserve which were the predecessors of the R.A.F. Volunteer Reserve).

4.   *Qualifying service –* Qualifying service is reckoned as follows:
(1)   Service in an air crew category (including service under training as air crew) in the Auxiliary Air Force or the R.A.F. Volunteer Reserve counts as time and a half.
(2)   Other service in the Auxiliary Air Force or the R.A.F. Volunteer Reserve counts as single time.

(3) Service in an air crew category (including service under training as air crew) on the Auxiliary Air Force General List, in the Auxiliary Air Force Reserve of Officers or in the Auxiliary Air Force Reserve of Airmen counts as three-quarter time, provided the specified minimum amount of flying per annum has been carried out.

(4) Other service on the Auxiliary Air Force General List, in the Auxiliary Air Force Reserve of Officers or in the Auxiliary Air Force Reserve of Airmen counts as half time, provided there is a liability for annual training.

(5) Service in the Great War in a regular force counts as single time.

(6) Service in the Great War in a non-regular force counts double, provided it can be counted double for the efficiency award of the force concerned.

(7) Service in West Africa prior to 30th September, 1937, counts double. Any period spent on leave from West Africa and any service rendered after 30th September, 1937 counts as single time (unless reckonable as time and a half under (i) above).

(8) Service in other non-regular forces by personnel of the Auxiliary Air Force or the R.A.F. Volunteer Reserve counts as single time or to such less extent as it would have counted towards the efficiency award of the force in question. Service in the classes of the R.A.F. Reserve which were the predecessors of the R.A.F. Volunteer Reserve (i.e., in classes "AA", "BB", "E (ii) (b)" and "F") counts as if it were service in the R.A.F. Volunteer Reserve.

5.   *Service without training liability* – Service in a reserve or auxiliary force involving & liability for service only and no liability for training in peace cannot count as qualifying service for the Award. (This does not, however, debar the counting of service in respect of which a training liability was not enforced.).

6.   *Service already reckoned* – No period of service for which an efficiency decoration or medal or a long service and good conduct medal has already been awarded can count as qualifying service for the Award.

7.   *Continuity* – Qualifying service must be continuous unless the Air Council, by regulation or in special circumstances, shall otherwise direct.

APPENDIX B
ROYAL AIR FORCE
RECOMMENDATION FOR AIR EFFICIENCY AWARD

No............... Rank................   Name (in full) ....

Notes

1. Enter service in chronological order.
2. Show in column 1 whether Auxiliary Air Force, Auxiliary Air Force Reserve of Officers, R.A.F. Volunteer Reserve, etc.
3. Use separate lines whenever service counts differently under A.M.O. A.969/42, appendix A, para. 4.
4. Show in column 6 how the service counts. Put "1" for single time, "1½" for time and a half, etc. Calculate the qualifying service for column 7 accordingly.
5. Exclude service carrying no liability for training and service already counted towards an efficiency award or long service and good conduct medal.

A.   *Service Details*

1	2	3	4		5		6	7	
Type of service	Class or branch	Whether air crew or non air crew	Dates of service		Period of service		How counted	Qualifying service	
			From	To	Years	Days		Years	Days
				Total			Total		

B.   *Previous service.* – If previous naval or military service is included above, give identification particulars, i.e. official number, rank, type of reserve, regiment, etc.

C.   *Certificate from C.O.* – In my opinion, the above-named officer (airman) is efficient and deserving of the Air Efficiency Award.

Station ....   Signature and rank ....

Date...................   Unit ....

D.   Certificate from Air Officer i/c Records (for airmen). – The particulars shown have been verified from the airman's permanent record.

Date...................   Signature ....

1951 AIR COUNCIL REVISED REGULATIONS A244
FOR THE AEA

1. PERSONNEL – R.A.F. AND W.R.A.F.

A.244. – AIR EFFICIENCY AWARD – REVISED REGULATIONS
[A.80171/51/S.7(e) – 3rd May, 1951.]

1. The King has been graciously pleased to approve a revised Royal Warrant and Air Council Regulations governing the Air Efficiency Award, to provide –

   (a) for the award of a clasp to those members of the Royal Auxiliary Air Force and R.A.F. Volunteer Reserve who, after qualifying for the award, complete & further ten years approved service;
   (b) for the counting of war service, under certain specified conditions, by members of the reconstituted Auxiliary and Reserve Air Forces, whose service in those Forces commenced after 18th August, 1946.

2. A summary of the provisions of the Warrant and Regulations is published at Appendix A to this order, which indicates the conditions applicable both to members of the Royal Auxiliary Air Force and the R.A.F. Volunteer Reserve whose service began before 3rd September, 1939, and to those whose service began after that date.

3. Recommendations for the award or clasp are to be submitted in the form shown at Appendix B to this order, which is to be prepared by units and forwarded to the Air Ministry [S.7(e)] (for officers), or to the Commanding Officer, R.A.F. Record Office, Gloucester (for airmen and airwomen). Qualified persons will be authorized to wear the ribbon; the award itself, or the clasp, will be issued in due course. The names of those granted the award or clasp will be promulgated, from time to time, in Air Ministry Orders, Class N.

4. Responsible administrative officers are to bring to the notice of C.O.s the name of every officer, airman, and airwoman, who becomes eligible for the award or clasp, but the fact that the requisite service has been completed confers no right to receive the award. C.O.s are to recommend only those who are "efficient and deserving of the Award".

(A.M.O. AM969 and amending order A.54;57 cancelled.)

Appendix A

OUTLINE OF CONDITIONS OF GRANT OF THE AIR EFFICIENCY AWARD FOR THE ROYAL AUXILIARY AIR FORCE AND R.A.F.VOLUNTEER RESERVE

1. *Description* – The Award shall be in silver, and worn on the left breast from a green ribbon 1½ inches wide, with two central stripes of pale blue ⅛ inch wide. The clasp shall be in silver, bearing on the obverse an eagle with out-stretched wings surmounted by a crown. When the ribbon alone is worn, the award of a clasp shall be denoted by the wearing of a silver rose emblem of approved pattern on the centre of the ribbon.

2. *Eligibility* – The award may be conferred on members of the Royal Auxiliary Air Force and the R.A.F. Volunteer Reserve provided that –

    (a) at the date of qualification they were serving on an engagement which, began before 3rd September, 1939, or after 18th August, 1946;
    (b) they have completed the required periods of training, and are certified by the responsible air force authorities as efficient and in every way deserving of the award; and
    (c) they have completed the requisite service.

3. *Service required* – The service required for the award is ten years qualifying service (as defined below), including not less than five years actual service on a peacetime engagement in —

    (a) the Royal Auxiliary Air Force (excluding the R.Aux.A.F. General List, the R.Aux.A.F. Reserve of Officers, and the R.Aux.A.F. Reserve of Airmen);
    (b) the R.A.F. Volunteer Reserve [including those classes of the R.A.F. Reserve which were the predecessors of the R.A.F. Volunteer Reserve, i.e., classes "AA", "BB", "E2(b)", and "F", provided the invitation to transfer to the R.A.F. Volunteer Reserve on its formation was accepted]; or
    (c) the Women's Auxiliary Air Force.

4. *Qualifying service* – Qualifying service is reckoned as follows: –

(a) *Service on an engagement which began before 3rd September, 1939:* –

(i) Service in a flying duties category in the Royal Auxiliary Air Force or the R.A.F. Volunteer Reserve ordinarily counts as time and a half, but embodied or mobilized service in such a category during and immediately after the Second World War, 1939-1945, counts as treble time.

(ii) Other service in the Royal Auxiliary Air Force, R.A.F. Volunteer Reserve, or Women's Auxiliary' Air Force, ordinarily counts as single time, but embodied or mobilized Service during and immediately after the Second World War, 1939-1945, counts as double time.

(b) *Service on an engagement which began after 18th August, 1946;* –

(i) Service in a flying duties category in the Royal Auxiliary Air Force or the R.A.F. Volunteer Reserve counts as time and a half.

(ii) Other service in the Royal Auxiliary Air Force or the R.A.F. Volunteer Reserve counts as single time.

(c) *Other reckonable service:* –

(i) Except as provided in sub-para. (a) above, embodied or mobilized service in the Royal Auxiliary Air Force, the R.A.F. Volunteer Reserve, or the Women's Auxiliary Air Force, during and immediately after the Second World War, 1939-1945, counts as single time.

(ii) Service in the Women's Auxiliary Air Force/R. List or the Women's Auxiliary Air Force/V.R. List counts as single time.

(iii) Service in a flying duties category on the R.Aux.A.F. General List, in the R.Aux.A.F. Reserve of Officers, or in the R.Aux.A.F. Reserve of Airmen, on an engagement which began before 3rd September, 1939, counts as three-quarters time, provided the specified minimum amount of flying per annum has been carried out.

(iv) Other service on the R.Aux.A.F. General List, in the R.Aux.A.F. Reserve of Officers, or in the R.Aux.A.F. Reserve of Airmen, on an engagement which began before 3rd September, 1939, counts as half-time, provided that there is a liability for annual training.

(v) Service in the First World War, 1914-1919, in a regular Force, counts as single time.

(vi) Service in the First World War, 1914-1919, in a non-regular Force, counts as double time, provided that it can be counted as double time for the efficiency award of the Force in question.

(vii) Service in other non-regular Forces, by personnel of the Royal Auxiliary Air Force, the R.A.F. Volunteer Reserve, or the Women's Auxiliary Air Force, on an engagement which began before 3rd September, 1939, counts as single time or to such less extent as it would have counted towards the efficiency award of the Force in question. Service in the classes of the R.A.F. Reserve which were the predecessors of the R.A.F. Volunteer Reserve counts as if it were service in the R.A.F. Volunteer Reserve [but see para. 3, sub para. (b) above].

5.   *Service without training liability.* – Service in an auxiliary or reserve Force, involving a liability for service only and no liability for training in peacetime, does not count as qualifying service for the award. (This does not, however, debar the counting of service in respect of which a training liability was not enforced).

6.   *Service already reckoned.* – No period of service for which an efficiency decoration or medal or a long service and good conduct medal has already been awarded can count as qualifying service for the award.

7.   *Continuity of service.* – Qualifying service must be continuous, except that a period of service during and immediately after the Second World War, 1939-1945, may be allowed to count towards the period of qualifying service, provided the break in service before rejoining the Royal Auxiliary Air Force or the R.A.F. Volunteer Reserve is not more than six months from. 18th August, 1946, or six months from the last day of release leave, whichever is the later, or twelve months for persons from an overseas Force.

8.   *Clasp.* – A clasp to the award may be conferred on members of the Royal Auxiliary Air Force and the R.A.F. Volunteer Reserve who, subsequent to the grant of the award, complete a further ten years

qualifying service as defined and computed in accordance with para. 4 of this Appendix.

9. *Forfeiture.* – The conditions governing forfeiture of the award by an officer, airman, or airwoman, are the same as those laid down for the Long Service and Good Conduct Medal in K.R. 295 (1) and (2). Forfeiture of the Award in any circumstances will entail forfeiture of the clasp, if any.

10. *Restoration* – An award and clasp, if any, forfeited in accordance with para. 8 of this Appendix, may be restored at the discretion of the Air Council.

Appendix B

ROYAL AIR FORCE
RECOMMENDATION FOR AIR EFFICIENCY AWARD OR CLASP

The same notes and form as that used in the 1942 orders.

1996 DEFENCE COUNCIL REGULATIONS FOR THE AEA

THE AIR EFFICIENCY AWARD – RAUXAF

PART I – General.

1.   Regulations governing the Air Efficiency Award were approved by the late King George VI by virtue of a Royal Warrant dated 12 April 1951. They are published at Part II of this annex.

2.   In pursuance of Regulation 10 of those regulations, the Defence Council made further regulations, which are published at Part III of this annex.

3.   Recommendations for the award or clasps are to be submitted in the form shown at Appendix I. They are to be prepared by units and forwarded to HQ PTC, P Man 4a(3)(b)(RAF) (through DCRF(RAF))who should, on approval, return a copy to the originating unit and the Inspector RAuxAF.

4.   Responsible administrative officers are to bring to the notice of COs the name of every officer, airman and airwoman who completes the requisite service for the award or clasp. The fact that such service has been completed, however, confers no right to receive the award or clasp, and COs are to recommend only those who are efficient and in every way deserving of the award (See Regulation I at Part III of this annex).

5.   Qualified personnel will be given the opportunity either:

    a.   To express a preference for presentation of the award by:
        (i) The Lord Lieutenant of the County or his representative, or
        (ii) The AOCinC or his representative, or
    b.   To elect to have the award sent to them by registered post

6.   The RAF PMC will ascertain the wish of each officer or airman concerned and:
    a.   If presentation by the Lord Lieutenant or his representative is desired, will notify and forward the medal to the appropriate TAVRA for them to arrange the ceremony with the recipient's CO, or
    b.   If presentation by the AOCinC or his representative is desired, will notify and forward the medal to the appropriate command for them to arrange the presentation with the recipient's CO, or
    c.   Will despatch the medal by registered post.

7. Recipients who elect to be presented with the award in accordance with paragraph 5(a) above will not be eligible for the reimbursement of any travelling expenses which they may incur.

PART II – Extract From Royal Warrant.

1. Description. The Air Efficiency Award shall be in silver and oval shaped, bearing on the obverse the Effigy of the Sovereign and on the reverse the words "Air Efficiency Award".

2. Clasp. The clasp shall be in silver, bearing on the obverse an Eagle with outstretched wings surmounted by a Crown.

3. Ribbon. The award shall be worn on the left breast pendent from a ribbon of one inch and a half in width which shall be in colour green with two central stripes of pale blue, one eighth of an inch in width.

4. Miniatures. Reproduction in miniature, which may be worn on certain occasions by those on whom this award is conferred, shall be approximately half the size of the award and a scaled pattern of the miniature award shall be kept in the Central Chancery of Our Orders of Knighthood.

5. Eligibility (Royalty). The award may be worn by Us, Our Heirs and Successors, Kings and Queens Regnant of Great Britain, Ireland and the British Dominions beyond the Seas, and it shall be competent for Us, Our Heirs and Successors, to confer at Our Pleasure the award upon any Princes or Princesses of the blood Royal.

6. Eligibility (General). The award and clasp may be conferred on officers, airmen and airwomen of any Auxiliary or Voluntary Air Force raised in Our United Kingdom of Great Britain and Northern Ireland, any of Our Colonies or a territory under Our protection, or within any other part of Our Dominions, Our Government whereof shall so desire, or within any territory under Our protection administered by Us in such Government.

7. Service Required. The period of service requisite for the award of clasp shall be ten years' qualifying service (as defined in the regulations hereinafter mentioned).

8.   Publication and Registration.
a.   The names of those upon whom the award is conferred shall be published and a register of those names shall be kept in the manner prescribed in the regulations hereinafter mentioned.
b.   Commissioned officers who have received this award shall be entitled to the addition after their names of the letters 'AE'.

9.   Forfeiture and Restoration.   It shall be competent for Our Defence Council in regard to Our Royal Auxiliary Air Force and Our Royal Air Force Volunteer Reserve, or for the Governors or Officers administering the Government, as the case may be, in regard to the Auxiliary or Volunteer Air Forces of Our respective Colonies or territories under Our protection, or the appropriate Minister of State in regard to the Auxiliary or Volunteer Air Forces in any of Our Dominions, Our Government whereof shall so desire, or within any territory under Our protection administered by Us in such Government, to cancel and annul the conferment of the award or clasp on any person and also to restore a forfeited award or clasp under the conditions published in the regulations hereinafter mentioned.

10.   Further Regulations.   The award and clasp shall be conferred under such regulations as to grant, forfeiture, restoration and other matters, in amplification of these Our regulations as may, with Our approval, be issued from time to time by Our Defence Council in regard to Our Royal Auxiliary Air Force and Our Royal Air Force Volunteer Reserve, or by the Governors or Officers administering the Government, as the case may be, in regard to the Auxiliary or Volunteer Air Forces of Our respective Colonies or territories under Our protection, or the appropriate Minister of State in regard to the Auxiliary or Volunteer Air Forces in any of Our Dominions, Our Government whereof shall so desire, or within any territory under Our protection administered by Us in such Government.

11.   Annulment, etc of Regulations.   We reserve to Ourself, Our Heirs and Successors, full power of annulling, altering, abrogating, augmenting, interpreting, or dispensing with these regulations or any part thereof by a notification under Our Sign Manual.

PART III – Further Regulations By The Defence Council Regarding The Royal Auxiliary Air Force And The Royal Air Force Volunteer Reserve.

1.   Eligibility.   Officers, airmen and airwomen of the Royal Auxiliary Air Force and the Royal Air Force Volunteer Reserve shall be eligible for the Air Efficiency Award, provided they are qualified by service calculated under the terms of Regulation 2 below and they have completed the

required periods of training and are certified by the responsible Air Force authorities as efficient and in every way deserving of the award.

2. Qualifying Service. For the purposes of Regulation 7 of the Royal Warrant "qualifying service" shall be:

    a. Service on an engagement which commenced before the 3 September 1939, at the date of qualification:

        (i) Service in a flying duties category in the Royal Auxiliary Air Force or the Royal Air Force Volunteer Reserve shall ordinarily count as time and a half, but embodied or mobilized service in such category during the Second World War shall count as treble time.

        (ii) Other service in the Royal Auxiliary Air Force, the Royal Air Force Volunteer Reserve or the Women's Auxiliary Air Force shall ordinarily count as single time, but embodied or mobilized service during the Second World War shall count as double time.

    b. Service on an engagement which commenced after the Second World War, at the date of qualification:

        (i) Service in a flying duties category in the Royal Auxiliary Air Force or the Royal Air Force Volunteer Reserve shall ordinarily count as time and a half, but periods of service or training under the Reserve and Auxiliary Forces (Training) Act 1951, shall count as double time.

        (ii) Other service in the Royal Auxiliary Air Force or the Royal Air Force Volunteer Reserve shall ordinarily count as single time, but periods of service or training under the Reserve and Auxiliary Forces (Training) Act 1951, shall count as time and a half.

    c. Other reckonable service:

        (i) Except as provided in paragraph (a) above, embodied or mobilized service in the Royal Auxiliary Air Force, the Royal Air Force Volunteer Reserve or the Women's Auxiliary Air Force during the Second World War shall count as single time.

        (ii) Service in the Women's Auxiliary Air Force Reserve List or the Women's Auxiliary Air Force Volunteer Reserve List shall count as single time.

        (iii) Service in a flying duties category on the Royal Auxiliary Air Force General List or in the Royal Auxiliary Air Force Reserve (including the Royal Auxiliary Air Force Reserve of Officers) on an engagement which commenced before the 3 September 1939, shall count as three-quarter time.

        (iv) Other service on the Royal Auxiliary Air Force General List or in the Royal Auxiliary Air Force Reserve (including the Royal Auxiliary Air Force Reserve of Officers) on an engagement which commenced before the 3 September 1939, shall count as half-time, provided there is a liability for annual training.

(v) Service since the 10 January 1955 on the Royal Auxiliary Air Force General List shall count as single time.

(vi) Service in a regular force in the First World War or in the Second World War shall count as single time.

(vii) Embodied or mobilized service on a peace-time engagement in other non-regular forces in the First World War or in the Second World War shall count as double time or to such less extent as it would have counted for the efficiency award of the force in question.

(viii) Service in other non-regular forces (not being service to which sub-paragraph (vii) of this paragraph applies) shall count as single time or to such less extent as it would have counted for the efficiency award of the force in question.

(iv) Service in the classes of the Royal Air Force Reserve which were the predecessors of the Royal Air Force Volunteer Reserve, ie, classes "AA", "BB", "E2(b)" and "F", shall count as if it were service in the Royal Air Force Volunteer Reserve, provided the invitation to transfer to the Royal Air Force Volunteer Reserve on its formation was accepted.

3.   Service Without Training Liability. Service in an auxiliary or reserve force involving a liability for service only and no liability for training in peace shall not reckon as qualifying service for the Air Efficiency Award or clasp.

4.   Service Already Reckoned.   A period of service for which an efficiency decoration or medal or a long service and good conduct medal has already been awarded shall not reckon as qualifying service for the Air Efficiency Award or clasp.

5.   Continuity.
   a.   Except as provided in paragraphs (b), (c) and (d) of this Regulation, qualifying service shall be continuous unless the Defence Council by regulation or in special circumstances shall otherwise direct.
   b.   A break between a period of qualifying service as defined in Regulation 2 above during the Second World War and the date of subsequently joining or rejoining the Royal Auxiliary Air Force or the Royal Air Force Volunteer Reserve will not be regarded as breaking the continuity or qualifying service provided the date of joining or rejoining the Royal Auxiliary Air Force or the Royal Air Force Volunteer Reserve is not later than six months (twelve months in the

case of a person from an overseas force) after the 8 November 1946, or such later date when recruiting actually began in the area in which the individual resided, or the last day of release leave, whichever is the latest.

c. A period of National Service under the National Service Acts, 1948 to 1955 though inadmissible as qualifying service for the Air Efficiency Award or clasp, will not be regarded as breaking the continuity of qualifying service.

d. A break between two periods of qualifying service, as defined in Regulation 2 above, after the Second World War will not be regarded as breaking the continuity of qualifying service provided that the break is due to circumstances outside the individual's control and does not exceed six months.

6. Registration. A register of those of whom the Air Efficiency Award and clasps thereto have been conferred will be maintained at the RAF PMC by PMan 4a(3)(b)(RAF).

7. Forfeiture. Forfeiture by an officer, airman or airwoman of the Air Efficiency Award or clasp will be determined by the same conditions as are published for the Long Service and Good Conduct Medal in Appendix 28C of the Queen's Regulations for the Royal Air Force (Fourth Edition).

8. Restoration. An Air Efficiency Award or clasp which has been forfeited may be restored at the discretion of the Defence Council.

9. Clasp.

a. Officers, airmen and airwomen of the Royal Auxiliary Air Force and the Royal Air Force Volunteer Reserve who, subsequent to the award of the medal, complete a further ten years' qualifying service as defined and computed in accordance with Regulation 2 above, may be awarded a clasp to the medal.

b. A further clasp may be awarded on the completion of each additional ten years' qualifying service.

c. A rose emblem denoting the award of each clasp shall be worn on the ribbon when the ribbon only is worn.

SUPPLEMENTARY NOTE

The regulations regarding the award and subsequent clasps to the AEA were documented in the various editions/amendments of Air Publication (AP) AP 968 for Auxiliary/Royal Auxiliary Air Force, RAuxAF Reserve (including the RAuxAF Reserve of Officers) and Air Publication AP938 with respect to the Royal Air Force Reserve. These APs were superseded by Air Publication AP 400B – 001 / AP 3392 in 1996 following the amalgamation of the RAFVR and the RAuxAF at RAF Cranwell on the 5 April 1997 via the provisions of the Reserve Forces Act 1996, governing members of the Air Force Reserves.

MEDAL AND RIBBON

The AEA is made of silver. It is an oval-shaped medal (38mm high at its maximum by 32mm wide and 2.5mm thick at its rim) with a ribbon mount in the form of an eagle with wings outspread. The detailed relief of the eagle is visible from obverse and reverse views. The obverse of the AEA bears the monarch's effigy, and the reverse, the words AIR EFFICIENCY AWARD.

Additional award bars are also made of silver and are identical to those used on the RAF Long Service and Good Conduct Medal. The bars are rectangular in shape (40mm long by 7mm high) with recessed corners. The face of the bar bears the form of an eagle with wings outspread, surmounted by the royal crown, framed by an inset border running parallel to the edge of the bar. The bar is designed to be loose on the ribbon and is not attached to the medal or ribbon suspension bar.

The ribbon is one and a half inches (38mm) wide and has two vertical light blue stripes running along its length against a green background, in approximate dimensions (from left to right) of 14mm green, 3mm light blue, 3mm green, 3mm light blue, 14mm green.

The following illustration shows the obverse and reverse views of the medal (George VI first type) and clasp:

Both faces of the medal, and the clasp

MEDAL NAMING

In keeping with the tradition of other medals and awards to the RAF, the British medal has always been issued named to the recipient. Medals to officer ranks included details in the order of: rank, initials, surname, and service, whilst naming to other ranks would also include the airman's service number. In accordance with other RAF medal issues, existing post-nominal details were not added to the medal naming.

Medals to all ranks of Australian, Canadian and New Zealand recipients differ from the British awards as they include the recipient's service number irrespective of rank. Additionally, New Zealand awards include the service after the surname.

Recognising that RAF ranks have evolved over time, including the scrapping of a separate structure for females, the following table provides a list of the abbreviations observed in AEA medal naming. The table is not meant to be a definitive list of official RAF rank abbreviations and will differ from the published abbreviation lists attributable to certain periods in history.

Air Force Rank (Male, all ranks)	Abbreviation on AEA
Wing Commander	WG CDR
Squadron Leader	SQN LDR
Flight Lieutenant	FLT LT
Flying Officer	FG OFF
Pilot Officer	PLT OFF
Warrant Officer	WO
Flight Sergeant	F/SGT
Sergeant	SGT
Corporal	CPL
Senior Aircraftman	SAC
Leading Aircraftman	LAC
Aircraftman Class 1	AC1
Aircraftman Class 2	AC2

Air Force Rank (Female, all ranks)	Abbreviation on AEA
Wing Officer	WG OFF
Squadron Officer	SQD OFF
Flight Officer	FLT OFF
Section Officer	SEC OFF
Assistant Section Officer	ASS SEC OFF

Additionally, due to the various wartime movements between ranks, it was quite usual to find an airman/woman posted into a rank higher than his/her substantive rank. Thus where an AEA has been awarded to someone in an acting rank, the abbreviation ACT is included in the rank naming. For example, a medal to an Acting Flight Lieutenant would be named as ACT FLT LT. An uncommon occurrence appears in the example of late issue (impressed naming) awards with the prefix TEMP for temporary before an airman's rank, and only one example of this has been seen by this author.

There are two main types of medal naming – engraved (in common with RAF-issue gallantry medals) covering the period 1942 to (approximately) 1960, and impressed, for the period 1960 onwards.

With the exception of NZ awards, all AEAs have the recipient's details on the rim of the medal. NZ awards are almost exclusively named in small capitals on the reverse of the medal, although it is reported that examples of rim naming do exist.

In view of the number of late claims and awards, the impressed naming styles can vary quite considerably in font and size, as can the sequence of details for other ranks. Impressed examples have been seen with service numbers placed at the beginning and at the end of the recipient's details, sometimes with or without brackets around the service number, but not sufficient to place them within a set time period.

The following illustrations show examples of both engraved and impressed naming.

Naming styles

APPLICATION FOR, AND AWARD OF, THE MEDAL

Award of the AEA was not automatic, but was made following application by serving personnel using the requisite RAF Form as appended to standing regulations. The completed form, including details of applicable service and relevant dates, would then be countersigned by the local senior officer and submitted to RAF central records for validation. Once verified and approved, the medal issue process could commence. The application process for former airmen would simply be via a letter to the Air Ministry or RAF Records, providing details of service, which would be checked, queried if necessary, and then approved or rejected.

Originally, award of the medal would be notified to the recipient by letter, and the award itself would be despatched once it had been named. If the recipient was still serving, the letter and award would usually be sent to the Squadron for presentation. Where the recipient was no longer serving, they would be sent direct to the home address.

The 1996 regulations introduced an option for a recipient to elect for presentation of the award by the Lord Lieutenant or the Air Officer Commanding (AOC), or their relevant representatives. Alternatively, the publicity-shy airman could elect to have the award posted direct to his home address!

The AEA was usually delivered in the standard medal-issue white cardboard box, bearing the name of the medal (and for later issues, whether it was a GVI or ERII type) plus the recipient's name, all of which was either printed onto a label or hand-written on the box. Additional bars to the medals were also claimed and issued in the same way.

The medal itself would be issued swing-mounted on a single pin for immediate presentation on parade. An example of the package components, including covering letter and award box, is shown in the illustration overleaf:

Delivery package and covering letter

During the period 1957 to 1963, some 924 of the British awards were formally recorded in the AMOs with the date of qualification for the actual award plus an earlier 'effective from' date, most probably due to the efficiency of the civil service clerk assigned to compile the lists during this period rather than any official policy to record such details.

There are several examples in the roll of quite large time gaps between the date of qualification for the award and the 'effective from' date, including for example, an award on 19 October 1960 under AMO N775 to Sergeant F.D. Pearse, with an effective date of 1 December 1943, some 17 years earlier!

The high number of awards for this period with separate award and 'effective' dates is clear evidence of an 'application-only' as opposed to an 'automatic-award' system.

POST-NOMINAL 'AE'

Under the statute changes of 1975, past and present recipients of the AE who held commissioned rank were also awarded the privilege of using the post-nominal 'AE' after their name, as specified in AMO T36/75:

THE AIR EFFICIENCY AWARD

The Air Efficiency Award was introduced in 1942 with the object of recognizing meritorious service rendered by officers, airmen and airwomen of the Royal Auxiliary Air Force and the Royal Air Force Volunteer Reserve. Hitherto there has been no provision for the use by an individual of post-nominal letters to denote that he holds the award, but recently Her Majesty the Queen has approved an amendment to the Royal Warrant permitting officers, including retired officers, who have received the Award to add the letters "AE" after their names. These officers are thus now accorded the same privilege as officers who have rendered similar service in the Reserve Forces of the Royal Navy and the Army that qualifies them to use the post-nominal letters "RD", "VRD" or "TD" as appropriate.

POSTHUMOUS AWARDS

No provision was made for posthumous awards of the AEA under the terms of the Royal Warrant, but despite the official position there are documented examples contained within the AMOs of the medal being formally awarded to 15 deceased personnel.

That 14 of these 15 awards all occur within the period February 1948 to March 1949, and include several gallantry award winners (including a Victoria Cross), would lead one to surmise that this short run of awards was the result of a determined effort by an individual or individuals within the Air Ministry and RAF to recognise the volunteer service of those airmen, and that their efforts were halted when they were moved-on or replaced by others who took a more arbitrary view of the terms and conditions for award of the medal.

Beyond this supposition, one can see no other reason for the awards, comprising both Bomber and Fighter Command aircrew who were killed on active duty across the years 1940 to 1945.

There is also one award documented in the *Commonwealth Gazette*, to a deceased member of the Royal Australian Air Force – Citizens Air Force Reserve (RAAF – CAFR) that was granted following a written submission by his widow. However her subsequent request for the grant of an additional award bar was refused because he did not have the necessary reckonable service for qualification.

Doubtless there will also be cases of other posthumous awards that are not officially recorded, but one would expect the number of such cases to be very limited, given the unambiguous wording of the Royal Warrant and various Service regulations governing the conditions for the grant of the award.

MONARCH

Following replacement of the AEA with the Volunteer Reserves Service Medal (VRSM), the AEA will only ever have been awarded with one of two monarch's effigies on the obverse – King George VI or Queen Elizabeth II – for which there are three variants and applicable time periods.

Note that the time periods noted below are those that have actually been observed on real examples of the AEA. These time periods do not accord with other publications that record different, more generalised time periods for the variations listed, nor does it include the Elizabeth II coinage profile BRITT:OMN for the period 1953-1954, and it is doubtful that this obverse was ever used on the AEA.

Therefore, it is believed that the only available obverse variations on the AEA are:

a) George VI coinage profile IND:IMP for awards in the period 1942 to March 1953,
b) George VI coinage profile FID:DEF for awards in the period April 1953 to October 1953,
c) Elizabeth II coinage profile DEI GRATIA issued from October 1953 onwards.

Note that late claim George VI awards were usually issued in the IND:IMP variant.

The following illustrations show the three different monarch profiles:

George VI first type obverse, IND:IMP legend

George VI second type obverse, FID:DEF legend

Elizabeth II obverse

From the availability dates it can be seen that the George VI FID:DEF type is the rarest of the three obverse types. The author has only physically seen two examples, one in his collection and one in the collection of the Royal Northumberland Fusiliers Museum[14], at Alnwick Castle.

It is by tracing the issue dates of these two awards on the Nominal Roll that it has been possible to formulate the approximate start and end dates as noted above, but there is a fair margin of error in this method. In any event, it is estimated that there are no more than 125 such examples in existence.

---

[14] The AEA awarded to AVM W.C.C. Gell, whose portrait appears as the frontispiece of this book

FAKES AND FORGERIES

As a general note, the author has yet to come across any deliberate fakes of the AEA, but there are certainly unnamed copy medals available covering both GVI and QEII periods. Some of these are of a poor quality, often reflected in the asking price, but others are cast in silver and much more akin to the real item.

Anyone buying an AEA should always satisfy himself that he is getting the original item and that the naming is correct for the period or for a verified late award. This is especially important in relation to RAF medal groups where the value can be significantly enhanced by the presence of a named medal to an individual, or to single AEAs named to high-ranking or famous airmen.

On the positive side, when considering the acquisition of an AEA for one's medal collection, it should be remembered that unlike instances arising with other medals, such as the General Service Medal, 1918-1962, and the Imperial Service Medal, the recipient's post-nominal details, should there be an entitlement, are not included on the AEA. Thus what appears to be an uninteresting example could in fact be to a gallantry award winner.

As ever in the world of medal collecting, 'research, research, research' (and a little luck) is the key to capturing an interesting medal.

DEMISE OF THE AEA

Sadly, in common with other single-service long service awards for reservists, and no doubt prompted by government cost-cutting at the time, the AEA was formally replaced by the Volunteer Reserves Service Medal (VRSM) in 1999, thus ending a 57-year period of availability to Air Force personnel.

However, the issuance of the AEA does not end quite so neatly, as serving officers who had completed at least five years of their ten-year eligibility prior to the 1999 end-date were still entitled to claim an AEA, should they so desire, in lieu of the VRSM. Clearly this anomaly should have officially ended at the end of 2004, but the odd 'late award' can still be found in current *London Gazette* Supplements for the Ministry of Defence.

It should also be noted that in common with recipients of territorial or auxiliary medals in the other services, recipients of the AEA and bar/s can and often do indulge in the present-day phenomenon of 'officially trading' subsequent bars and reckonable service time for the award of the new VRSM and/or additional bars. It is not unknown for holders of the AEA and bar to officially trade that bar for a VRSM, thus ending up with two medals instead of one.

ERRORS & OMISSIONS IN THE NOMINAL ROLL

Every effort has been made to ensure the completeness and accuracy of the roll by cross checking different reference sources for the names of AEA recipients. These sources include: Air Ministry Orders; *London Gazettes*; *Commonwealth Gazettes*; *Royal Auxiliary Air Force Yearbooks*; files at the National Archives and other published sources; files from Commonwealth Archives and Government Departments, and input from fellow researchers.

Additionally, manual error-checking was undertaken during compilation of the roll itself and entries that are blatantly wrong have been amended, for example where a recipient's initials may have been transposed. Missing information has been added where available, for example officer service numbers from the relevant *Air Force Lists*.

However, it is inevitable that a combination of original publishing errors within source material and the manual transcription of the AEA roll into electronic format for publishing may have let some errors and omissions creep in. If any errors are found, please do provide corrected information to the author via the Publisher. This should ensure that any future updates are corrected accordingly.

Equally, if you are the proud recipient or current custodian of an AEA that is not listed in the roll, please accept the author's apologies. It could simply be an error on his part, or it is quite possible that it was a post-1963 late issue award, for which, as yet, it has not been possible to locate an official source for a record of such awards, if indeed one actually exists.

NOTES TO BE USED IN CONJUNCTION WITH THE NOMINAL ROLLS

The nominal rolls for each country are formatted into the following generic columns, although the content in each column may differ by country:

- **Surname** – recipient's family name.
- **Initials** – recipient's initials, up to a maximum of four.
- **Rank** – recipient's rank as officially recorded at the date of qualification for the award. Note that RAF rank abbreviations have changed across the wartime, post-Second World War, and modern time periods. For example, abbreviations for Flight Lieutenant have changed from FL to F/L and then to Flt Lt. Therefore the abbreviations used in the rolls are not true for any specific time period but are the author's generic choice, based upon those more commonly recognised. For a full list of all ranks used in the rolls, please see the Statistics table showing awards by rank.
- **Status** – if rank is non-substantive, e.g., acting or temporary.
- **Date** – publication date of the data source, e.g., for Air Ministry Orders, this is the date the AMO was published and is deemed the official date of qualification for the award.
- **Source** – data source. Examples include:

    GB Listings:
    - *N611 1957* – Royal Air Force Air Ministry Orders 'N' Series and year. This is the official Air Ministry roll of GB awards promulgated for the period 1942 to 1963.
    - *MOD 1964* – MOD official memorandum and papers held at the UK National Archive in 'AIR' class files. The ad hoc papers contain various notations and lists of awards, and copies of correspondence between claimants and the MOD concerning claims, refusals and re-instatements of awards. The papers cover the years 1964 to 1968.
    - *YB 1999 – Royal Auxiliary Air Force Yearbook* and year. The only published source from which recent awards can be derived for the period 1999 to 2004.
    - *LG 2003 – London Gazette* and year. The last few awards of the AEA started to appear in the *London Gazette* in March 2004.

    Australia Listings:
    - *N61 – Commonwealth Gazette* notice for the date given.

Hong Kong Listings:
- o *N141 – Hong Kong Government Gazette* notice for the date given.

New Zealand Listings:
- o All of the awards of the AEA to New Zealand personnel were extracted either from the excellent reference work *By Such Deeds – Honours & Awards in the New Zealand Air Force 1923 – 1999* by Group Captain C.M. Hanson, OBE, RNZAF (Retd) which lists most of the awards to officers, or direct from New Zealand Defence Force records for awards to other ranks.

- **Service No** – recipient's service number, either as recorded in the official source or inserted from other material, e.g., the GB *Air Force List* for officers. There are a few instances where no service number was located.

- **Force** – The service within which the recipient was serving at the time of qualification for the award. Note that recipients no longer serving at the time of the award are shown with the prefix 'EX' to the force name.

GB Air Forces:

AAF	- Auxiliary Air Force
AAFRO	- Auxiliary Air Force Reserve of Officers
RAFRO	- Royal Air Force Reserve of Officers
RAFVR	- Royal Air Force Volunteer Reserve
RAFVR(T)	- Royal Air Force Volunteer Reserve (Training)[15]
RAuxAF	- Royal Auxiliary Air Force
WAAF	- Women's Auxiliary Air Force
WRAAF	- Women's Royal Auxiliary Air Force
WRAVR	- Women's Royal Air Force Volunteer Reserve

GB Other:

AER	- Army Emergency Reserve
TA	- Territorial Army

---

[15] Official sources list nine such awards to the RAFVR(T). These recipients may well have served at some point in time in reserve forces eligible for the AEA, however it is fairly certain that the medals would be named to the RAFVR, on the basis that members of the RAFVR(T) were not mobilised reserves and were, in any case, eligible for the Cadet Forces Medal for their service.

Australian Air Forces:
AFR　　　- Air Force Reserve
CAF-R　　- Citizen's Air Force - Reserve
PAF　　　- Permanent Air Force
RETIRED - Retired
WRAAF　 - Women's Royal Australian Air Force

Canadian Air Forces:
RCAF　　 *- Royal Canadian Air Force*

Hong Kong Air Forces:
GFS　　　- Government Flying Service
HKAAF　 - HK Royal Auxiliary Air Force
HKWAAF - HK Women's Royal Auxiliary Air Force

- **Notes** – additional information noted from the data source which may include some or all of the following combinations:

  o *Civil & Military Awards held at the time of the publication date of the AEA award, e.g., DFC for Distinguished Flying Cross. It is very important to note that the awards, or lack of them, against an individual name should **not** be taken as a summary of awards made during their entire civil or Air Force career, but only of awards officially recorded up to that point in time. For example, the December 1942 listing for Acting Wing Commander E.B.R. Lockwood does not record his 1945 award of the Distinguished Service Order (DSO), but it does record his October 1942 appointment as a Member of the Order of the British Empire (MBE), in the Military Division.*
  Additionally, for any awards and orders noted, only the first award is recorded – bars for subsequent awards are not. For example, the 1946 listing for Wing Commander D.E. Kingaby records him as DFC, DFM, but does not record the two bars already awarded to his DFM.
  o *Other post-nominals, e.g., medical qualifications.*
  o *Title, e.g., 'The Duke of ... '*
  o *Clasp – whether it is a 1ˢᵗ, 2ⁿᵈ or 3ʳᵈ clasp award.*
  o *Effective date – some GB awards were back-dated.*
  o *Ad hoc, e.g., deceased recipient.*

o *Branch / Trade / Squadron – the Australian and NZ rolls include this data against some recipients. The following abbreviations will be noted:*

Abbreviation used in roll	Branch
ACCT	Accountant
ADMIN	Administration
ARM	Armourer
ATC	Air Traffic Control
CHAP	Chaplain
ENGIN	Engineering
EQUIP	Equipment
GD	General Duties (denotes aircrew)
LEGAL	Legal
MEDIC	Medical
SD	Special Duties
SEC	Secretarial
SIG	Signals
SUP	Supply
TECH	Technical

Abbreviation used in roll	Trade
AG/NAV	Air Gunner/ Navigator
AIRASST	Air Assistant (catch-all trade for General Service hands)
AIR SIG	Air Signaller (aircrew)
ARM	Armourer
ARTIST	Official Artist
FITTER	Fitter
MTDRVR	Driver
MUSN	Musician
NAV	Navigator
PILOT	Pilot
WO/AG	Wireless Operator / Air Gunner

NOMINAL ROLL FOR GREAT BRITAIN

Surname	Initials			Rank	Status	Date
ABBISS	J	B		CPL		01/07/1948
ABBOTT	A	F		FO		10/02/1949
ABBOTT	K			FL		19/10/1960
ABBOTT	W	J		LAC		07/04/1955
ABEL	A	J		WO		24/11/1949
ABEL	R	F		FO		06/03/1947
ABERCROMBIE	J	G		FL		20/11/1957
ABRAHAM	L	F		FL	Acting	04/11/1948
ABRAHAMS	D	S		CPL		19/09/1996
ABRAHAMS	H	J		FL		03/06/1959
ACKERLEY	J			SAC		21/10/1959
ACKERS	H			SGT		22/05/1952
ACKLAM	W	G		LAC		05/06/1947
ACLAND	E	F	D	FL		06/03/1947
ACONS	F	N		FL		20/05/1948
ADAM	A	W		FL		05/01/1950
ADAMS	A	F		WO	Acting	08/02/1945
ADAMS	C	J		SAC		19/02/1999
ADAMS	C	M		FL		06/03/1947
ADAMS	D	A		FL		06/05/1943
ADAMS	D	C		FL		24/01/1946
ADAMS	D	V	G	CPL		30/09/1948
ADAMS	E			LAC		22/05/1952
ADAMS	F	J	F	FL		05/01/1950
ADAMS	F	W	G	FO		28/02/1946
ADAMS	G	W		WING COMM	Acting	15/07/1948
ADAMS	G	W		FL	Acting	06/09/1951
ADAMS	H			WING COMM	Acting	23/05/1946
ADAMS	J	C		FL		30/03/1944
ADAMS	J	F	W	FL		11/07/1946
ADAMS	J			SGT		05/06/1947
ADAMS	J			FL	Acting	04/09/1947
ADAMS	L	C		CPL		14/12/1994
ADAMS	R	D	M	SGT		04/09/1957
ADAMS	R	E	R	FL		13/12/1945
ADAMS	R	G	C	FL		30/09/1948
ADAMS	R			SGT		14/10/1948
ADAMS	W			LAC		01/10/1953
ADAMSON	G	D	W	SQD LDR		15/01/1991
ADAMSON	J			FL		25/05/1960
ADAMSON	T	L		SQD LDR		25/05/1960
ADCOCK	H	C		SGT		16/10/1947
ADDICOTT	H	F		LAC		04/09/1947
ADDINGTON	F	F		FL		11/07/1946
ADDISON	A			LAC		24/11/1949
ADDISON	H			PO		13/01/1949
ADDISON	W			CPL	Temporary	01/05/1957
ADDY	R	G		FL		11/11/1943
ADDYMAN	G	V		CPL		15/07/1948
ADDYMAN	J			CPL		23/02/1950
ADKINS	W	R		SQD LDR	Acting	27/03/1947
ADLAM	B			CPL		04/09/1947
AGATE	L	G	L	LAC		20/05/1948
AGATE	L	G	L	SGT	Acting	20/03/1957
AGATE	W	L		FLT SGT		20/05/1948
AGATE	W	L		FLT SGT		26/04/1956
AGNEW	J	B		FL		13/12/1945
AGNEW	P	G		WING COMM	Acting	04/10/1945
AGNEW	R	M		LAC		15/11/1961
AHEARN	A	S		SQD LDR		07/08/1989

Source	Service No	Force	Notes	Surname
N527 1948	850580	RAuxAF		ABBISS
N123 1949	130719	RAFVR	Deceased	ABBOTT
N775 1960	162279	RAuxAF	wef 21/7/60	ABBOTT
N277 1955	846895	RAuxAF		ABBOTT
N1193 1949	880861	WAAF		ABEL
N187 1947	191245	RAFVR		ABEL
N835 1957	152667	RAFVR	wef 14/6/56	ABERCROMBIE
N934 1948	63227	RAFVR		ABRAHAM
YB 1999	n/a	RAuxAF		ABRAHAMS
N368 1959	183761	RAFVR	wef 26/2/54	ABRAHAMS
N708 1959	2688637	RAuxAF	wef 21/6/59	ACKERLEY
N402 1952	2688527	RAuxAF		ACKERS
N460 1947	871534	AAF		ACKLAM
N187 1947	73059	RAFVR		ACLAND
N395 1948	139931	RAFVR		ACONS
N1 1950	123997	RAFVR		ADAM
N139 1945	812014	AAF		ADAMS
YB 2000	n/a	RAuxAF		ADAMS
N187 1947	118908	RAFVR		ADAMS
N526 1943	90537	AAF		ADAMS
N82 1946	88018	RAFVR		ADAMS
N805 1948	746167	RAFVR		ADAMS
N402 1952	814130	RAuxAF		ADAMS
N1 1950	145498	RAFVR	DFM	ADAMS
N197 1946	186159	RAFVR		ADAMS
N573 1948	73057	RAFVR	OBE	ADAMS
N913 1951	104681	RAFVR		ADAMS
N468 1946	90099	AAF	OBE	ADAMS
N281 1944	78978	RAFVR		ADAMS
N594 1946	138683	RAFVR	DFC,AFC	ADAMS
N460 1947	743808	RAFVR		ADAMS
N740 1947	178748	RAFVR		ADAMS
YB 1996	n/a	RAuxAF		ADAMS
N611 1957	2687011	RAuxAF	wef 13/7/57	ADAMS
N1355 1945	85656	RAFVR		ADAMS
N805 1948	74045	RAFVR		ADAMS
N856 1948	804268	RAuxAF		ADAMS
N796 1953	850540	RAuxAF		ADAMS
YB 1996	91426	RAuxAF		ADAMSON
N375 1960	115660	RAuxAF	wef 9/8/52	ADAMSON
N375 1960	171428	RAuxAF	wef 12/4/60	ADAMSON
N846 1947	801536	AAF		ADCOCK
N740 1947	840798	AAF		ADDICOTT
N594 1946	78442	RAFVR	DFC	ADDINGTON
N1193 1949	817046	RAuxAF		ADDISON
N30 1949	199903	RAFVR		ADDISON
N277 1957	799888	RAFVR	wef 7/6/44	ADDISON
N1177 1943	84699	RAFVR	AFC	ADDY
N573 1948	799937	RAFVR		ADDYMAN
N179 1950	747059	RAFVR		ADDYMAN
N250 1947	72596	RAFVR		ADKINS
N740 1947	868585	AAF		ADLAM
N395 1948	819121	RAuxAF		AGATE
N198 1957	2689511	RAuxAF	1st Clasp; wef 14/5/53	AGATE
N395 1948	819005	RAuxAF		AGATE
N318 1956	2689500	RAuxAF	BEM; 1st Clasp	AGATE
N1355 1945	133806	RAFVR		AGNEW
N1089 1945	72531	RAFVR		AGNEW
N902 1961	2650925	RAuxAF	wef 9/10/61	AGNEW
YB 1998	2625171	RAuxAF		AHEARN

Surname	Initials			Rank	Status	Date
AHEARN	A	S		SQD LDR		07/08/1999
AHERN	J	C	S	WING COMM		27/11/1952
AILES	M	K	G	FL		08/11/1989
AINSLEY	J	G		SGT		10/06/1948
AINSLIE	A			FL		18/04/1946
AINSWORTH	C	W		FL		25/05/1960
AINSWORTH	D	B		FL		13/12/1945
AINSWORTH	H	B		SQD LDR	Acting	24/01/1946
AINSWORTH	W	R		FLT SGT		27/03/1963
AIREY	C	F		FO		17/04/1947
AIREY	C	R	W	CPL	Temporary	01/05/1957
AITKEN	J			WING COMM	Acting	06/03/1947
AITKEN	M			WING COMM		03/12/1942
AITON	J	M		WING COMM		30/11/1944
AKEHURST	J			WO		02/08/1951
AKERS	E			FL		18/04/1946
ALABASTER	R	C		WING COMM	Acting	30/09/1948
ALBERTINI	A	V		FL	Acting	28/11/1946
ALBRECHT	A	R		SQD LDR	Acting	01/08/1946
ALBURY	T	H		FLT SGT		07/08/1947
ALCOCK	A			CPL		05/06/1947
ALCOCK	H			FLT SGT		04/11/1948
ALCOCK	R			WING COMM	Acting	05/06/1947
ALDAY	D			SGT		22/05/1998
ALDER	A	W	C	FL		30/10/1957
ALDERSON	R	H		FL		09/12/1959
ALDERTON	P			SQD LDR		12/06/1947
ALDOUS	G	H		FL	Acting	18/10/1951
ALDRICK	M	J		WING COMM	Acting	17/04/1947
ALDRIDGE	A			SQD LDR	Acting	23/05/1946
ALDRIDGE	C	F	G	LAC		20/05/1948
ALDRIDGE	E	G		FL		09/12/1959
ALDRIDGE	J	S		FL		30/03/1944
ALDRIDGE	W	W		FL		28/02/1946
ALDWINCKLE	A	J	M	FL		30/03/1944
ALEXANDER	A	E		FL	Acting	10/06/1948
ALEXANDER	F	B		WO		22/04/1943
ALEXANDER	J			WING COMM	Acting	15/06/1944
ALFORD	G	F	R	FL		18/05/1944
ALLAN	A	B		CPL		27/07/1944
ALLAN	C			SGT	Acting	22/02/1961
ALLAN	D			CPL		27/04/1950
ALLAN	J	C	H	WING COMM		03/12/1942
ALLAN	J	W		WING COMM		12/04/1945
ALLAN	J	W		FL		06/03/1947
ALLAN	M	B		FL		05/01/1950
ALLAN	P	L		SAC		26/06/1997
ALLAN	W	D		CPL		04/09/1957
ALLAN	W			CPL		05/06/1947
ALLDIS	C	A		WING COMM	Acting	21/03/1946
ALLELY	T	D		LAC		26/08/1948
ALLEN	A	J		CPL		07/07/1949
ALLEN	A	L		WING COMM	Acting	20/05/1948
ALLEN	C			CPL		15/09/1949
ALLEN	D	L		FL		28/10/1986
ALLEN	D	L		FL		28/10/1996
ALLEN	G	A		FL	Acting	27/03/1947
ALLEN	J	W		FL		18/04/1946
ALLEN	K	M		FL		09/05/1946
ALLEN	L	A		PO		21/10/1959

Source	Service No	Force	Notes	Surname
YB 2000	2625171	RAuxAF	1st Clasp	AHEARN
N915 1952	72597	RAFVR		AHERN
YB 1997	210250	RAuxAF		AILES
N456 1948	808337	RAuxAF		AINSLEY
N358 1946	113437	RAFVR		AINSLIE
N375 1960	196394	RAFVR	wef 18/1/57	AINSWORTH
N1355 1945	81031	RAFVR		AINSWORTH
N82 1946	78974	RAFVR		AINSWORTH
N239 1963	2676406	RAuxAF	wef 20/8/62	AINSWORTH
N289 1947	142050	RAFVR		AIREY
N277 1957	812286	RAuxAF	wef 27/6/44	AIREY
N187 1947	90551	AAF		AITKEN
N1638 1942	90128	AAF	DSO,DFC; The Honorable	AITKEN
N1225 1944	90584	AAF		AITON
N777 1951	751666	RAFVR	DFM	AKEHURST
N358 1946	123093	RAFVR	DFC	AKERS
N805 1948	81065	RAFVR	DSO,DFC	ALABASTER
N1004 1946	119844	RAFVR		ALBERTINI
N652 1946	86218	RAFVR		ALBRECHT
N662 1947	750048	RAFVR		ALBURY
N460 1947	811215	AAF		ALCOCK
N934 1948	847682	RAuxAF		ALCOCK
N460 1947	73562	RAFVR		ALCOCK
YB 1999	n/a	RAuxAF		ALDAY
N775 1957	188085	RAFVR	wef 2/12/52	ALDER
N843 1959	107529	RAFVR	wef 31/8/44	ALDERSON
N478 1947	73732	RAFVR		ALDERTON
N1077 1951	88255	RAFVR		ALDOUS
N289 1947	72122	RAFVR		ALDRICK
N468 1946	84716	RAFVR		ALDRIDGE
N395 1948	841744	RAuxAF		ALDRIDGE
N843 1959	178384	RAFVR	wef 21/7/59	ALDRIDGE
N281 1944	84669	RAFVR		ALDRIDGE
N197 1946	129202	RAFVR		ALDRIDGE
N281 1944	83288	RAFVR		ALDWINCKLE
N456 1948	68953	RAFVR		ALEXANDER
N475 1943	741120	RAFVR		ALEXANDER
N596 1944	90384	AAF	DFC	ALEXANDER
N473 1944	108852	RAFVR		ALFORD
N763 1944	803380	AAF		ALLAN
N153 1961	2649557	RAuxAF	wef 1/10/60	ALLAN
N406 1950	870111	RAuxAF		ALLAN
N1638 1942	90173	AAF	MB,ChB	ALLAN
N381 1945	89617	RAFVR	DSO,DFC	ALLAN
N187 1947	88046	RAFVR		ALLAN
N1 1950	91196	RAuxAF		ALLAN
YB 1998	n/a	RAuxAF		ALLAN
N611 1957	2683527	RAuxAF	wef 22/6/57	ALLAN
N460 1947	811033	AAF		ALLAN
N263 1946	74657	RAFVR	DFC	ALLDIS
N694 1948	816174	RAuxAF		ALLELY
N673 1949	770504	RAFVR		ALLEN
N395 1948	72287	RAFVR		ALLEN
N953 1949	857569	RAuxAF		ALLEN
YB 1997	2591644	RAuxAF		ALLEN
YB 1997	2591644	RAuxAF	1st Clasp	ALLEN
N250 1947	87076	RAFVR		ALLEN
N358 1946	135866	RAFVR		ALLEN
N415 1946	146710	RAFVR		ALLEN
N708 1959	2692539	RAuxAF	wef 29/8/59	ALLEN

Surname	Initials			Rank	Status	Date
ALLEN	L	W		SQD LDR		05/01/1950
ALLEN	L			SGT		05/02/1948
ALLEN	P	F		SQD LDR	Acting	18/04/1946
ALLEN	R	W	J	FL		20/11/1957
ALLEN	R			FO		08/11/1945
ALLEN	T			WO		28/11/1956
ALLEN	W	J		CPL		05/06/1947
ALLEN	W			CPL		07/12/1950
ALLEN-JONES	M	A		FLT OFF		05/06/1947
ALLETSON	J	C		CPL		05/06/1947
ALLIES	E	M		SQD LDR	Acting	11/07/1946
ALLINGTON	K	L	H	FL		15/09/1949
ALLINSON	C	C		FL		07/08/1947
ALLISON	T			SGT		13/12/1951
ALLSEBROOK	A	E		WO	Acting	31/10/1956
ALLSOP	A	J		CPL		07/07/1949
ALLWAY	F	E		LAC		24/11/1949
ALMOND	M			WING COMM		03/10/2000
ALSOP	E	R		FL		21/06/1945
ALTHAM	R	J	L	FL		22/10/1958
AMBERTON	M	Y		FLT OFF		06/03/1947
AMBLER	F	T		AC2		22/09/1955
AMBLER	G	H		AIR COMMD	Acting	03/12/1942
AMBROSE	C	M		SQD OFF		17/04/1947
AMES	C	F		SGT		13/01/1949
AMIES	E	J		FL		08/07/1959
AMLOT	W	J		SQD LDR		13/01/1949
AMOS	C	E		SAC		20/07/1960
ANDERS	E	N		SGT		09/12/1992
ANDERS	G			WO		04/08/1996
ANDERSON	A	H		FL		20/05/1948
ANDERSON	A			SGT		30/09/1948
ANDERSON	B	C		CPL		13/01/1949
ANDERSON	D	M		FL		24/07/1952
ANDERSON	F	A		WO		24/12/1942
ANDERSON	F			LAC		04/09/1947
ANDERSON	G	F		SQD LDR		03/12/1942
ANDERSON	I			FLT SGT		15/09/1949
ANDERSON	J	D		FL		21/03/1946
ANDERSON	J	G	B	LAC		03/07/1957
ANDERSON	J	R		SQD LDR		30/03/1960
ANDERSON	J	S		FL		24/01/1946
ANDERSON	J			WO		04/09/1947
ANDERSON	K	M	G	SQD LDR		29/06/1944
ANDERSON	K	S		FL	Acting	17/04/1947
ANDERSON	L	G		SQD LDR		30/10/1957
ANDERSON	M	F		WING COMM	Acting	10/12/1942
ANDERSON	M	W		WING COMM	Acting	05/10/1944
ANDERSON	O	S		SGT		27/03/1947
ANDERSON	P			FO		16/10/1947
ANDERSON	R			CPL		23/02/1950
ANDERSON	T	W		FL	Acting	24/11/1949
ANDERTON	E			SQD LDR	Acting	18/11/1948
ANDERTON	J	L		FL		10/05/1945
ANDREWS	A	E	C	FL		26/01/1950
ANDREWS	C	A		CPL		30/09/1948
ANDREWS	D	H		SGT		05/02/1948
ANDREWS	D	O		FL		24/01/1946
ANDREWS	F	W	T	FL		13/12/1945
ANDREWS	F			CPL	Temporary	15/01/1958

Source	Service No	Force	Notes	Surname
N1 1950	72772	RAFVR	MC	ALLEN
N87 1948	770114	RAFVR		ALLEN
N358 1946	102973	RAFVR	DFC	ALLEN
N835 1957	191719	RAFVR	DFM; wef 27/8/56	ALLEN
N1232 1945	168831	RAFVR	AFC	ALLEN
N833 1956	2676800	RAuxAF		ALLEN
N460 1947	811184	AAF		ALLEN
N1210 1950	849245	RAuxAF		ALLEN
N460 1947	2427	WAAF		ALLEN-JONES
N460 1947	853857	AAF		ALLETSON
N594 1946	77375	RAFVR	DFC	ALLIES
N953 1949	124210	RAFVR		ALLINGTON
N662 1947	72773	RAFVR		ALLINSON
N1271 1951	807224	RAuxAF		ALLISON
N761 1956	2677203	RAuxAF		ALLSEBROOK
N673 1949	860417	RAuxAF		ALLSOP
N1193 1949	847042	RAuxAF		ALLWAY
YB 2001	5206994	RAuxAF		ALMOND
N671 1945	120872	RAFVR		ALSOP
N716 1958	181105	RAFVR	wef 23/3/52	ALTHAM
N187 1947	253	WAAF		AMBERTON
N708 1955	865110	RAuxAF		AMBLER
N1638 1942	90296	AAF	CBE,AFC	AMBLER
N289 1947	255	WAAF		AMBROSE
N30 1949	744535	RAFVR		AMES
N454 1959	199624	RAuxAF	wef 10/5/59	AMIES
N30 1949	91017	RAuxAF		AMLOT
N531 1960	2692073	RAuxAF	wef 2/3/60	AMOS
YB 1996	n/a	RAuxAF		ANDERS
YB 1999	n/a	RAuxAF		ANDERS
N395 1948	112587	RAFVR		ANDERSON
N805 1948	807221	RAuxAF		ANDERSON
N30 1949	799763	RAFVR		ANDERSON
N569 1952	91267	RAuxAF		ANDERSON
N1758 1942	807011	AAF		ANDERSON
N740 1947	808359	AAF		ANDERSON
N1638 1942	90110	AAF		ANDERSON
N953 1949	743833	RAFVR		ANDERSON
N263 1946	131473	RAFVR		ANDERSON
N456 1957	803621	AAF	wef 15/5/44	ANDERSON
N229 1960	202537	RAuxAF	MB,ChB; wef 7/7/59	ANDERSON
N82 1946	142002	RAFVR		ANDERSON
N740 1947	802556	AAF		ANDERSON
N652 1944	70014	RAFVR		ANDERSON
N289 1947	191705	RAFVR		ANDERSON
N775 1957	72090	RAFVR	wef 25/11/43	ANDERSON
N1680 1942	90206	AAF	DFC	ANDERSON
N1007 1944	72079	RAFVR		ANDERSON
N250 1947	752793	RAFVR		ANDERSON
N846 1947	176292	RAFVR		ANDERSON
N179 1950	853602	RAuxAF		ANDERSON
N1193 1949	149419	RAFVR		ANDERSON
N984 1948	77200	RAFVR		ANDERTON
N485 1945	122994	RAFVR	AFC	ANDERTON
N94 1950	72774	RAFVR		ANDREWS
N805 1948	812165	RAuxAF		ANDREWS
N87 1948	809046	RAuxAF		ANDREWS
N82 1946	117585	RAFVR	DFC	ANDREWS
N1355 1945	122990	RAFVR	DFC	ANDREWS
N40 1958	858252	AAF	wef 5/7/44	ANDREWS

Surname	Initials			Rank	Status	Date
ANDREWS	F			SAC		26/11/1958
ANDREWS	G	E		FO		10/06/1948
ANDREWS	H	W	G	SQD LDR		05/01/1950
ANDREWS	J	M		SQD OFF	Acting	12/06/1947
ANDREWS	P	M		FL		30/03/1944
ANDREWS	S	D		FO		09/05/1946
ANDREWS	T	F		LAC		09/01/1963
ANDREWS	V	H		SQD LDR	Acting	08/07/1959
ANGIER	G	M		SQD LDR		01/09/1944
ANGROVE	H	C		SAC		05/04/1997
ANGUS	I	G		FLT SGT		14/01/1954
ANGUS	J	M		LAC		24/11/1949
ANGUS	J	M		CPL		28/11/1956
ANGUS	K	F		GRP CAP	Acting	27/04/1944
ANGUS	R	S		SQD LDR	Acting	14/10/1948
ANGWIN	A	E		SGT		18/07/1956
ANKIN	E	T		CPL		06/03/1947
ANLEY	R	S		SGT		20/09/1961
ANNAN	R	H		WING COMM	Acting	06/03/1947
ANNAND	A	C		FO		08/03/1956
ANNANDALE	J			SQD LDR	Acting	08/11/1945
ANNING	J	E		FL		01/01/2003
ANNING	W	F		SQD LDR	Acting	08/05/1947
ANNING	W	F		FL		01/10/1953
ANNISS	G	T		FL		20/05/1948
ANSTEY	J	L		FL	Acting	22/09/1955
ANSTEY	J	M		SGT		13/11/1947
ANSTEY	J	P		FL		22/02/1961
ANSTIE	J	A		FL		30/12/1943
ANTRAM	P	D		SQD LDR	Acting	13/12/1945
ANYON	N	H		FL		20/03/1957
APPERLEY	D	M		FLT OFF		12/06/1947
APPLEBY	M	J		FL		05/01/1950
APPLEFORD	H	J		FL		09/02/1951
APPLETON	E	P		WING COMM	Acting	06/03/1947
APPLETON	H	C		SQD LDR		20/05/1948
APPS	N	T		SGT	Temporary	28/07/1955
APPS	R	W		FLT SGT		26/08/1948
ARBER	I	K		FO		20/09/1945
ARBUTHNOT	W			FLT SGT		01/07/1948
ARCHDALE	G	E		FL		26/10/1944
ARCHER	D	P	D	FL		17/04/1947
ARCHER	E	L		WING COMM	Acting	08/11/1945
ARCHER	R	A		FL		01/07/1948
ARCHER	R	D	J	SQD LDR	Acting	27/03/1947
ARCHER	R	G		FL		16/01/1957
ARCHER	R	G		FL		31/08/1963
ARCHER	S			FLT SGT		22/05/1952
ARCHIBALD	E	L		CPL		07/02/1962
ARCHIBALD	G	D		LAC		13/01/1949
ARDLEY	B			FO		25/06/1953
ARDLEY	B			FL		20/03/1957
ARGENT	W	C		CPL		26/08/1948
ARIES	E	W		FL		20/09/1945
ARIES	E	W		FL		27/03/1952
ARIES	M	R		FL		15/06/1944
ARKELL	M	F		SQD OFF		27/03/1947
ARKLESS	R	S		SQD LDR		16/10/1947
ARMITAGE	D	L		SQD LDR	Acting	13/12/1945
ARMITAGE	G			FLT SGT		14/09/1950

Source	Service No	Force	Notes	Surname
N807 1958	2604569	RAFVR	wef 28/3/56	ANDREWS
N456 1948	189461	RAFVR		ANDREWS
N1 1950	85639	RAFVR	DFC	ANDREWS
N478 1947	1726	WAAF		ANDREWS
N281 1944	83729	RAFVR		ANDREWS
N415 1946	179156	RAFVR	DFC	ANDREWS
N22 1963	856023	AAF	wef 21/6/44	ANDREWS
N454 1959	194302	RAuxAF	wef 10/5/59	ANDREWS
MOD 1964	75428	RAFVR		ANGIER
YB 1998	n/a	RAuxAF		ANGROVE
N18 1954	2601811	RAFVR		ANGUS
N1193 1949	817090	RAuxAF		ANGUS
N833 1956	2684011	RAuxAF	1st Clasp	ANGUS
N396 1944	90589	AAF	OBE,MC,TD	ANGUS
N856 1948	90473	RAuxAF	MBE	ANGUS
N513 1956	2697500	RAuxAF		ANGWIN
N187 1947	865552	AAF		ANKIN
N720 1961	2614018	RAFVR	1st Clasp; wef 30/6/61	ANLEY
N187 1947	60282	RAFVR	DSO	ANNAN
N187 1956	62720	RAF		ANNAND
N1232 1945	78458	RAFVR	DFC	ANNANDALE
YB 2003	213025	RAuxAF		ANNING
N358 1947	72599	RAFVR		ANNING
N796 1953	72599	RAFVR	1st Clasp	ANNING
N395 1948	89216	RAFVR		ANNISS
N708 1955	106070	RAFVR		ANSTEY
N933 1947	840588	AAF		ANSTEY
N153 1961	114195	RAuxAF	wef 22/1/61	ANSTEY
N1350 1943	68150	RAFVR		ANSTIE
N1355 1945	113438	RAFVR		ANTRAM
N198 1957	130949	RAuxAF	wef 10/12/56	ANYON
N478 1947	1302	WAAF		APPERLEY
N1 1950	90962	RAuxAF		APPLEBY
N145 1951	117962	RAFVR		APPLEFORD
N187 1947	62451	RAFVR		APPLETON
N395 1948	72180	RAFVR		APPLETON
N565 1955	756290	RAFVR		APPS
N694 1948	801194	RAuxAF	BEM	APPS
N1036 1945	156944	RAFVR	AFC	ARBER
N527 1948	850264	RAuxAF		ARBUTHNOT
N1094 1944	128976	RAFVR		ARCHDALE
N289 1947	135430	RAFVR	DFC,DFM	ARCHER
N1232 1945	78735	RAFVR	DSO,AFC	ARCHER
N527 1948	133012	RAFVR		ARCHER
N250 1947	77907	RAFVR		ARCHER
N24 1957	168246	RAFVR		ARCHER
MOD 1966	168246	RAFVR	1st Clasp	ARCHER
N402 1952	803473	RAuxAF		ARCHER
N102 1962	2659471	WRAAF	wef 9/12/61	ARCHIBALD
N30 1949	873944	RAuxAF		ARCHIBALD
N527 1953	144076	RAFVR		ARDLEY
N198 1957	144076	RAuxAF	1st Clasp; wef 31/3/54	ARDLEY
N694 1948	746441	RAFVR		ARGENT
N1036 1945	79555	RAFVR	AFC	ARIES
N224 1952	79555	RAuxAF	AFC; 1st Clasp	ARIES
N596 1944	87645	RAFVR		ARIES
N250 1947	937	WAAF		ARKELL
N846 1947	90801	AAF		ARKLESS
N1355 1945	76573	RAFVR	DFC	ARMITAGE
N926 1950	771019	RAFVR		ARMITAGE

Surname	Initials			Rank	Status	Date
ARMITAGE	J	W		SQD LDR	Acting	15/07/1948
ARMSTRONG	E	S		FL	Acting	17/04/1947
ARMSTRONG	G	M		FO		13/10/1955
ARMSTRONG	H	G		SQD LDR	Acting	30/09/1948
ARMSTRONG	J			SGT	Temporary	15/11/1961
ARMSTRONG	M	W		SGT		20/08/1958
ARMSTRONG	M			FO		20/07/1960
ARMSTRONG	R			FL		05/09/1946
ARMSTRONG	V			FO		19/08/1959
ARMSTRONG	W			FO		04/11/1943
ARNOLD	A	H		FLT SGT		16/10/1947
ARNOLD	C			WING COMM	Acting	13/11/1947
ARNOLD	J	F		FO		21/09/1944
ARNOLD	J	F		SQD LDR	Acting	17/04/1947
ARNOLD	S	E		FO		01/05/1957
ARNOTT	A	R		SQD OFF		23/02/1950
ARNOTT	G			CPL		01/01/1999
ARNOTT	R	B		FLT SGT		10/02/1949
ARSCOTT	J	E	J	FL		30/03/1960
ARUNDEL	D	T		FL		28/11/1946
ASCOUGH	J	S		FL		20/01/1960
ASH	L	S		FL		04/11/1943
ASH	O			PO		15/04/1943
ASHBEE	D	J		FL		20/07/1997
ASHBUREY	W	F		SQD LDR	Acting	13/01/1949
ASHBY	E	J		SQD LDR		03/10/1946
ASHBY	F	E		CPL		07/05/1953
ASHBY	H	M	P	SQD LDR	Acting	11/01/1945
ASHBY	M	S		FL		13/11/1947
ASHER	C	C		FO		17/06/1954
ASHFIELD	P	J		LAC		09/11/1950
ASHFORD	R	C		CPL		15/07/1948
ASHLEY	R	Y		SQD LDR	Acting	09/11/1950
ASHLEY-COOPER	L	M	M	FLT OFF		16/10/1947
ASHMAN	L	E	W	SGT		15/09/1949
ASHMORE	R	C		CPL		11/07/1946
ASHTON	H	M		SQD LDR	Acting	24/07/1952
ASHTON	J	H		WING COMM	Acting	13/12/1945
ASHTON	T	W		SQD LDR		14/10/1943
ASHTON	W			LAC		29/11/1951
ASHWORTH	A	H		FL	Acting	30/09/1948
ASKE	R	E		SAC		20/07/1960
ASKINS	K	T	B	FL		20/03/1957
ASLETT	A	T	R	FL		21/03/1946
ASLIN	D	J		FL		13/12/1945
ASPINALL	P	P		CPL		27/04/1950
ASPLAND	E			AC1		13/07/1950
ASPRAY	A	J		FL		25/05/1960
ASSINDER	G	E		FL		01/10/1953
ASSITER	M	F	G	CPL		26/08/1948
ASTALL	J			LAC		07/05/1953
ASTBURY	J	G		FL		29/06/1944
ASTON	W	E	R	SGT		29/06/1944
ATKIN	T	P		CPL		21/06/1951
ATKINS	A	H		WO		19/08/1959
ATKINS	E	G		WO		07/08/1947
ATKINS	E			FL		13/02/1957
ATKINS	G	W		FL		05/02/1948
ATKINS	K	R		CPL		15/09/1949
ATKINSON	A			LAC		13/02/1957

Source	Service No	Force	Notes	Surname
N573 1948	90843	RAuxAF		ARMITAGE
N289 1947	131715	RAFVR		ARMSTRONG
N762 1955	110392	RAFVR		ARMSTRONG
N805 1948	117194	RAFVR		ARMSTRONG
N902 1961	857340	AAF	wef 30/5/44	ARMSTRONG
N559 1958	2685517	RAFVR	wef 23/7/57	ARMSTRONG
N531 1960	2663657	WRAAF	wef 5/8/58	ARMSTRONG
N751 1946	129928	RAFVR	DFM	ARMSTRONG
N552 1959	2663750	WRAAF	wef 10/5/59	ARMSTRONG
N1144 1943	70023	RAFVR	AFC	ARMSTRONG
N846 1947	864857	AAF		ARNOLD
N933 1947	90786	AAF		ARNOLD
N961 1944	158318	RAFVR		ARNOLD
N289 1947	73733	RAFVR		ARNOLD
N277 1957	2600564	RAuxAF	wef 22/5/55	ARNOLD
N179 1950	215	WAAF		ARNOTT
YB 2000	n/a	RAuxAF		ARNOTT
N123 1949	844169	RAuxAF		ARNOTT
N229 1960	131776	RAFVR	wef 14/12/55	ARSCOTT
N1004 1946	86653	RAFVR		ARUNDEL
N33 1960	168146	RAFVR	wef 4/5/53	ASCOUGH
N1144 1943	70026	RAFVR		ASH
N454 1943	116029	RAFVR		ASH
YB 1998	1939555	RAuxAF		ASHBEE
N30 1949	90738	RAuxAF		ASHBUREY
N837 1946	73068	RAFVR		ASHBY
N395 1953	816205	RAuxAF		ASHBY
N14 1945	85943	RAFVR		ASHBY
N933 1947	79761	RAFVR		ASHBY
N471 1954	2603784	RAFVR		ASHER
N1113 1950	746589	RAFVR		ASHFIELD
N573 1948	801595	RAuxAF		ASHFORD
N1113 1950	79744	RAFVR	DFC	ASHLEY
N846 1947	2063	WAAF	Lady	ASHLEY-COOPER
N953 1949	748921	RAFVR		ASHMAN
N594 1946	805377	AAF		ASHMORE
N569 1952	75998	RAFVR	DFC	ASHTON
N1355 1945	77456	RAFVR	DFC	ASHTON
N1065 1943	70027	RAFVR		ASHTON
N1219 1951	858064	RAuxAF		ASHTON
N805 1948	141942	RAFVR		ASHWORTH
N531 1960	2692080	RAuxAF	wef 8/5/60	ASKE
N198 1957	55006	RAuxAF	DFC; wef 21/10/52	ASKINS
N263 1946	170758	RAFVR		ASLETT
N1355 1945	102097	RAFVR		ASLIN
N406 1950	745642	RAFVR		ASPINALL
N675 1950	868608	RAuxAF		ASPLAND
N375 1960	2601049	RAFVR	wef 6/11/58	ASPRAY
N796 1953	87466	RAFVR		ASSINDER
N694 1948	846526	RAuxAF		ASSITER
N395 1953	856122	RAuxAF		ASTALL
N652 1944	66000	RAFVR	AFM	ASTBURY
N652 1944	805098	EX-AAF		ASTON
N621 1951	855135	RAuxAF		ATKIN
N552 1959	2676410	RAFVR	wef 19/7/58	ATKINS
N662 1947	864218	AAF		ATKINS
N104 1957	157599	RAFVR	wef 6/1/56	ATKINS
N87 1948	112515	RAFVR	DFC	ATKINS
N953 1949	865696	RAuxAF		ATKINS
N104 1957	869400	RAuxAF	wef 2/7/44	ATKINSON

Surname	Initials			Rank	Status	Date
ATKINSON	C	L		WING COMM	Acting	17/04/1947
ATKINSON	D			SAC		07/12/1960
ATKINSON	H	K		FL		27/04/1944
ATKINSON	J	D		FLT SGT		28/07/1955
ATKINSON	J	D		FL		18/05/1963
ATKINSON	S	B		FL		30/03/1943
ATKINSON	T	St	B	FO		25/06/1953
ATKINSON	W	R		SGT		26/08/1948
ATTREE	E	E		SGT		26/08/1948
ATTREE	F			WING COMM	Acting	16/10/1947
ATTWOOD	J			WING COMM		07/08/1947
ATTWOOD	T	M		SQD LDR	Acting	30/08/1945
AUCH	J	R		WO		30/03/1943
AUCHINVOLE	D	B		SQD LDR	Acting	05/06/1947
AULD	A			FL		09/05/1946
AULD	D	R		FL		27/04/1950
AULD	J			LAC		15/07/1948
AULD	J			SAC		05/09/1962
AUSTIN	A	T		FL		05/01/1950
AUSTIN	A	W		LAC		05/01/1950
AUSTIN	F	J		SQD LDR		21/06/1951
AUSTIN	G	W	B	SQD LDR		03/12/1942
AUSTIN	J	B		SQD LDR	Acting	15/07/1948
AUSTIN	L	J		FO		04/11/1948
AUSTIN	P	D		FL		01/01/2003
AUSTIN	S	W		WO	Acting	15/07/1948
AUSTIN	T	A		FL		08/03/1956
AUSTIN	T	A		FL		02/12/1962
AVENS	A	R		SQD LDR	Acting	26/01/1950
AVENT	M	T		WING COMM		12/08/1943
AXCELL	V	C		SGT		26/04/1961
AXON	R	P		FL		10/02/1949
AYERS	R	H		WO		11/11/1943
AYERS	R	H		FL		17/01/1952
AYERS	W	T		CPL	Temporary	22/09/1955
AYLING	D			FLT SGT		03/07/1957
AYLOTT	R	W		FL		22/10/1958
AYRE	H	W		FL		10/08/1944
AYRES	A	E	H	FL		07/12/1960
AYRES	A			FLT SGT		21/06/1951
AYRES	A			FLT SGT		10/04/1952
AYRES	G	A		SGT		08/05/1947
AYRES	K	A		FO		21/09/1944
AYSCOUGH	S	G	H	FL		06/01/1955
BABBAGE	C	F		SQD LDR	Acting	12/07/1945
BACKHOUSE	E			SGT		15/07/1948
BACKHOUSE	J			SEC OFF		23/02/1950
BACKLER	G	E		CPL		07/02/1962
BACON	F	J	H	FL	Acting	06/03/1947
BADCOCK	J	K		FL		06/03/1947
BADCOCK	W	J	A	SAC		03/07/1957
BADEN	S	R		WING COMM		13/11/1947
BADEN	S	R		FL		26/04/1956
BADENOCH	A	W		WING COMM		26/01/1950
BADLEY	J	W		FLT SGT		07/08/1947
BADSEY	H	H		CPL		07/07/1949
BAGLEY	J	V		FO		06/10/2000
BAGLEY	K	E		SQD OFF		16/10/1947
BAGLEY	K	E		FLT OFF		26/11/1953
BAGLEY	R	G	F	FL		22/10/1958

Source	Service No	Force	Notes	Surname
N289 1947	73697	RAFVR		ATKINSON
N933 1960	2650828	RAuxAF	wef 23/5/60	ATKINSON
N396 1944	81664	RAFVR		ATKINSON
N565 1955	2604004	RAFVR		ATKINSON
MOD 1966	2604004	RAFVR	1st Clasp	ATKINSON
N1014 1943	70031	RAFVR		ATKINSON
N527 1953	121837	RAFVR		ATKINSON
N694 1948	800491	RAuxAF		ATKINSON
N694 1948	805023	RAuxAF		ATTREE
N846 1947	72152	RAFVR		ATTREE
N662 1947	90103	AAF		ATTWOOD
N961 1945	121468	RAFVR		ATTWOOD
N1014 1943	815006	AAF		AUCH
N460 1947	139779	RAFVR		AUCHINVOLE
N415 1946	139935	RAFVR		AULD
N406 1950	82760	RAFVR		AULD
N573 1948	743533	RAFVR		AULD
N655 1962	2602030	RAFVR	1st Clasp; wef 25/11/58	AULD
N1 1950	129122	RAFVR		AUSTIN
N1 1950	752108	RAFVR		AUSTIN
N621 1951	111933	RAFVR		AUSTIN
N1638 1942	90258	AAF		AUSTIN
N573 1948	104448	RAFVR	DFC	AUSTIN
N934 1948	161170	RAFVR		AUSTIN
YB 2003	213464	RAuxAF		AUSTIN
N573 1948	753557	RAFVR		AUSTIN
N187 1956	171386	RAFVR	DFC	AUSTIN
MOD 1966	171386	RAFVR	DFC; 1st Clasp	AUSTIN
N94 1950	91032	RAuxAF		AVENS
N850 1943	90242	AAF		AVENT
N339 1961	2684642	RAFVR	wef 22/10/60	AXCELL
N123 1949	157981	RAFVR		AXON
N1177 1943	741030	RAFVR		AYERS
N35 1952	168832	RAFVR	1st Clasp	AYERS
N708 1955	860352	RAuxAF		AYERS
N456 1957	819031	AAF	wef 17/12/43	AYLING
N716 1958	183370	RAFVR	wef 21/10/52	AYLOTT
N812 1944	120739	RAFVR	AFC	AYRE
N933 1960	157075	RAFVR	wef 15/12/56	AYRES
N621 1951	808172	RAuxAF		AYRES
N278 1952	2686005	RAuxAF	1st Clasp	AYRES
N358 1947	746559	RAFVR		AYRES
N961 1944	134509	RAFVR	AFC	AYRES
N1 1955	62368	RAFVR		AYSCOUGH
N759 1945	89298	RAFVR	DFM	BABBAGE
N573 1948	746031	RAFVR		BACKHOUSE
N179 1950	3958	WAAF		BACKHOUSE
N102 1962	2688683	RAFVR	wef 19/10/60	BACKLER
N187 1947	72602	RAFVR		BACON
N187 1947	74659	RAFVR		BADCOCK
N456 1957	2606502	RAuxAF	wef 4/12/53	BADCOCK
N933 1947	72155	RAFVR		BADEN
N318 1956	72155	RAFVR	1st Clasp	BADEN
N94 1950	72046	RAFVR	MD,ChB,FRCS	BADENOCH
N662 1947	880441	WAAF		BADLEY
N673 1949	857289	RAuxAF		BADSEY
YB 2002	2641909	RAuxAF		BAGLEY
N846 1947	584	WAAF		BAGLEY
N952 1953	584	WRAAF	1st Clasp	BAGLEY
N716 1958	139318	RAFVR	wef 7/6/53	BAGLEY

Surname	Initials			Rank	Status	Date
BAGNALD	E	H		WING COMM	Acting	24/01/1946
BAGNALL-OAKLEY	J	R		FL		06/03/1947
BAIKIE	A	H	G	FO		26/04/1961
BAILEY	A	C		WO		13/12/1945
BAILEY	C	A		FL		09/01/1963
BAILEY	E	A		FO		26/06/1947
BAILEY	F	J		FL		04/11/1943
BAILEY	G	H		FO		13/07/1950
BAILEY	G	J		FL		10/05/1945
BAILEY	H	N	D	SQD LDR	Acting	20/09/1945
BAILEY	H	S		FL		20/05/1948
BAILEY	J	W	C	LAC		01/07/1948
BAILEY	P			FL		08/03/1956
BAILEY	W	F		SGT		07/08/1947
BAILEY	W	H		SGT		30/03/1943
BAILLIE	A	A		FL	Acting	24/01/1946
BAILLIE	A	S	H	FL		30/03/1944
BAILLIE	A			SAC		26/11/1958
BAIN	A			FL		21/03/1946
BAIN	G	S	P	SQD LDR		06/03/1947
BAIN	J			FL		03/07/1952
BAIN	J			FL		03/07/1952
BAIN	T	N		SGT		19/08/1959
BAIN	V	A		WO		12/04/1945
BAIN	W	S		CPL		20/07/1960
BAIN	W			FO		05/10/1944
BAINBRIDGE	F	K		SQD LDR	Acting	27/07/1944
BAINBRIDGE	J	T		FO		03/07/1957
BAINBRIDGE	W			CPL		26/08/1948
BAINES	A	R		FL		30/04/1997
BAIRD	J	B		FL		07/04/1955
BAIRD	J	H		SQD LDR		20/05/1948
BAIRD	W	C		SQD LDR		20/05/1948
BAKER	A	C		FL		04/10/1945
BAKER	A	G		WING COMM	Acting	23/02/1950
BAKER	A	G		SAC		19/08/1959
BAKER	A	P		CPL		01/05/1952
BAKER	A	V		SGT		14/10/1948
BAKER	C			SAC		31/07/1998
BAKER	E	F		FLT SGT	Acting	22/09/1955
BAKER	E	W		SAC		28/02/1997
BAKER	F	A		CPL	Temporary	20/01/1960
BAKER	F	C		CPL		07/07/1949
BAKER	F	G		SGT		15/07/1948
BAKER	F	H		FL		16/10/1952
BAKER	F	J		CPL		15/07/1948
BAKER	G	P		SGT		20/07/1960
BAKER	H	G		SGT		16/03/1950
BAKER	J	M	A	WO		27/03/1947
BAKER	J			FLT SGT		08/11/1945
BAKER	L	C		FL	Acting	30/08/1945
BAKER	L	E		FO		15/07/1948
BAKER	R	E		CPL		23/02/1950
BAKER	R			SQD LDR		10/05/1945
BAKER	R			FL		04/10/1945
BAKER	R			FL		22/10/1958
BAKER	S			SGT		26/08/1948
BAKER	W	A		LAC		07/04/1955
BAKER	W	R	J	LAC		07/08/1947
BALCOMBE	F	J		SQD LDR	Acting	04/09/1947

Source	Service No	Force	Notes	Surname
N82 1946	85020	RAFVR	DFC	BAGNALD
N187 1947	91029	AAF		BAGNALL-OAKLEY
N339 1961	2659389	WRAAF	wef 28/10/60	BAIKIE
N1355 1945	740182	RAFVR		BAILEY
N22 1963	5008555	RAFVR	wef 8/12/62	BAILEY
N519 1947	144407	RAFVR	MBE	BAILEY
N1144 1943	70033	RAFVR		BAILEY
N675 1950	117197	RAFVR		BAILEY
N485 1945	106355	RAFVR		BAILEY
N1036 1945	84957	RAFVR		BAILEY
N395 1948	90929	RAuxAF		BAILEY
N527 1948	841909	RAuxAF		BAILEY
N187 1956	178884	RAuxAF	DFC	BAILEY
N662 1947	863549	AAF		BAILEY
N1014 1943	808006	AAF		BAILEY
N82 1946	177030	RAFVR		BAILLIE
N281 1944	88430	RAFVR		BAILLIE
N807 1958	2677447	RAuxAF	wef 9/10/58	BAILLIE
N263 1946	128968	RAFVR		BAIN
N187 1947	85647	RAFVR		BAIN
N513 1952	102589	RAFVR	DFM	BAIN
N513 1952	102589	RAFVR	DFM; 1st Clasp	BAIN
N552 1959	2699504	RAuxAF	wef 2/6/54	BAIN
N381 1945	741596	RAFVR		BAIN
N531 1960	2605873	RAFVR	wef 1/5/60	BAIN
N1007 1944	156315	RAFVR		BAIN
N763 1944	64285	RAFVR	DFC	BAINBRIDGE
N456 1957	2685599	RAuxAF	wef 13/1/57	BAINBRIDGE
N694 1948	867911	RAuxAF		BAINBRIDGE
YB 1998	408334	RAuxAF		BAINES
N277 1955	202702	RAuxAF		BAIRD
N395 1948	74263	RAFVR		BAIRD
N395 1948	73591	RAFVR		BAIRD
N1089 1945	64892	RAFVR	DFC	BAKER
N179 1950	72603	RAFVR		BAKER
N552 1959	2692534	RAuxAF	wef 20/6/59	BAKER
N336 1952	770625	RAFVR		BAKER
N856 1948	843145	RAuxAF		BAKER
YB 2002	n/a	RAuxAF		BAKER
N708 1955	800463	RAuxAF		BAKER
YB 1998	n/a	RAuxAF		BAKER
N33 1960	818109	AAF	wef 31/5/44	BAKER
N673 1949	842879	RAuxAF		BAKER
N573 1948	750946	RAFVR		BAKER
N798 1952	195778	RAuxAF	DFC	BAKER
N573 1948	846915	RAuxAF		BAKER
N531 1960	2650402	RAuxAF	wef 30/4/60	BAKER
N257 1950	770940	RAFVR		BAKER
N250 1947	755046	RAFVR		BAKER
N1232 1945	804181	AAF		BAKER
N961 1945	139748	RAFVR		BAKER
N573 1948	122774	RAFVR		BAKER
N179 1950	843215	RAuxAF		BAKER
N485 1945	81930	RAFVR	DSO,DFC	BAKER
N1089 1945	82174	RAFVR		BAKER
N716 1958	149623	RAFVR	wef 3/3/52	BAKER
N694 1948	812134	RAuxAF		BAKER
N277 1955	864142	RAuxAF		BAKER
N662 1947	844082	AAF		BAKER
N740 1947	73072	RAFVR		BALCOMBE

Surname	Initials			Rank	Status	Date
BALCOMBE	G			FL		20/01/1960
BALDIE	A	C		SGT		30/03/1944
BALDRY	J	H		FL		12/04/1945
BALDWIN	J	W		AC1		26/04/1951
BALDWIN	K	McG		WO		13/01/1949
BALFOUR	C	J		FL		07/07/1949
BALFOUR	E			FL		09/01/1963
BALL	C	H		FL		07/07/1949
BALL	E	V		CPL		23/02/1950
BALL	H	J		WING COMM		20/05/1948
BALL	H	M		FLT SGT		20/05/1948
BALL	H			FL		16/04/1958
BALL	J	E		LAC		09/01/1963
BALL	J	O		FL	Acting	23/02/1950
BALL	T	C		CPL		30/09/1948
BALL	V			LAC		19/10/1960
BALL	W	G		FL		03/06/1959
BALLARD	A	G		FL	Acting	13/11/1947
BALLINGALL	G			FO		01/12/1955
BALLINGALL	L	A		SQD LDR	Acting	15/07/1948
BALNAVE	D	M		FO		08/02/1945
BAMBERGER	C	S		FL		20/04/1944
BAMFORD	C			FLT SGT	Acting	01/07/1948
BANFIELD	A	E		FL		28/08/1963
BANKIN	E	E		SGT		16/10/1947
BANKS	C	E		WING COMM	Acting	06/03/1947
BANKS	J	H		LAC		26/08/1948
BANKS	J	H		CPL	Acting	03/07/1952
BANNER	G			FO		25/05/1944
BANNISTER	W	J		CPL		06/01/1955
BANNON	P	M		FLT SGT	Acting	07/07/1949
BAPTIE	M			SGT		25/05/1960
BARBER	E	E		FL		06/03/1947
BARBER	G	W		CPL	Temporary	27/03/1963
BARBER	J	A	A	FL		27/04/1950
BARBER	R	B		SGT		19/08/1959
BARBER	R	J		LAC		26/08/1948
BARBOUR	J	A	N	FO		29/04/1954
BARCLAY	A	B	G	FL		04/05/1944
BARCLAY	G	D		SGT		18/02/1954
BARCLAY	J	C		WING COMM	Acting	01/03/1945
BARCLAY	J	O		FO		15/01/1953
BARCLAY	W	E	B	WING COMM		08/05/1947
BARDSLEY	J	H		LAC		07/12/1960
BARDSWELL	C	G		SQD LDR		26/08/1964
BARKER	A	D		FL		22/10/1958
BARKER	C	W		CPL		15/09/1949
BARKER	C			SAC		31/07/1998
BARKER	D			FL		22/09/1955
BARKER	H	G		FL		28/06/1945
BARKER	H	L		FLT SGT	Acting	05/06/1947
BARKER	H			WING COMM		25/05/1950
BARKER	J			WO		11/07/1946
BARKER	M			FL		24/01/1946
BARKER	S			SAC		28/09/1995
BARKER	W	H		FLT SGT	Temporary	25/05/1960
BARLEY	B	G		FL		08/11/1945
BARLEY	G	R		WO		09/05/1946
BARLEY	K			CPL		04/11/1948
BARLOW	A	W		FL		19/08/1959

Source	Service No	Force	Notes	Surname
N33 1960	128967	RAFVR	wef 11/6/57	BALCOMBE
N281 1944	803374	AAF		BALDIE
N381 1945	87311	RAFVR		BALDRY
N402 1951	843028	RAuxAF		BALDWIN
N30 1949	743467	RAFVR		BALDWIN
N673 1949	104905	RAFVR		BALFOUR
N22 1963	188623	RAFVR	wef 15/6/59	BALFOUR
N673 1949	139731	RAFVR		BALL
N179 1950	748993	RAFVR		BALL
N395 1948	78100	RAFVR		BALL
N395 1948	743159	RAFVR	BEM	BALL
N258 1958	136739	RAFVR	wef 4/3/54	BALL
N22 1963	815129	AAF	wef 19/7/43	BALL
N179 1950	102199	RAFVR		BALL
N805 1948	845435	RAuxAF		BALL
N775 1960	856619	AAF	wef 29/6/44	BALL
N368 1959	79856	RAFVR	wef 17/1/59	BALL
N933 1947	122775	RAFVR		BALLARD
N895 1955	2683570	RAuxAF		BALLINGALL
N573 1948	81697	RAFVR		BALLINGALL
N139 1945	126842	RAFVR		BALNAVE
N363 1944	116515	RAFVR	DFC	BAMBERGER
N527 1948	750110	RAFVR		BAMFORD
N621 1963	189740	RAFVR	wef 27/11/60	BANFIELD
N846 1947	813259	AAF		BANKIN
N187 1947	84585	RAFVR	OBE	BANKS
N694 1948	808334	RAuxAF		BANKS
N513 1952	2683316	RAuxAF		BANKS
N496 1944	126142	RAFVR		BANNER
N1 1955	849920	RAuxAF		BANNISTER
N673 1949	810118	RAuxAF		BANNON
N375 1960	2659313	WRAAF	wef 26/1/60	BAPTIE
N187 1947	118670	RAFVR	MBE	BARBER
N239 1963	747234	RAFVR	wef 26/6/44	BARBER
N406 1950	72779	RAFVR		BARBER
N552 1959	2676412	RAFVR	wef 12/8/58	BARBER
N694 1948	846071	RAuxAF		BARBER
N329 1954	142931	RAFVR		BARBOUR
N425 1944	100027	RAFVR		BARCLAY
N137 1954	2677401	RAuxAF		BARCLAY
N225 1945	90046	AAF		BARCLAY
N28 1953	2601754	RAFVR		BARCLAY
N358 1947	90428	AAF	MC	BARCLAY
N933 1960	2691093	RAuxAF	wef 9/10/60	BARDSLEY
MOD 1964	81869	RAFVR		BARDSWELL
N716 1958	166466	RAFVR	wef 25/7/52	BARKER
N953 1949	799885	RAFVR		BARKER
YB 1999	n/a	RAuxAF		BARKER
N708 1955	83616	RAFVR		BARKER
N706 1945	121091	RAFVR	AFC	BARKER
N460 1947	752750	RAFVR		BARKER
N517 1950	72780	RAFVR		BARKER
N594 1946	751030	RAFVR		BARKER
N82 1946	128444	RAFVR		BARKER
YB 1996	n/a	RAuxAF		BARKER
N375 1960	2601667	RAFVR	wef 2/4/60	BARKER
N1232 1945	138202	RAFVR		BARLEY
N415 1946	751076	RAFVR		BARLEY
N934 1948	872520	RAuxAF		BARLEY
N552 1959	205264	RAFVR	wef 22/7/59	BARLOW

Surname	Initials			Rank	Status	Date
BARLOW	F	W		FO		28/07/1955
BARLOW	T	P	E	SQD LDR		07/04/1955
BARNACLE	R	H		FL		19/09/1956
BARNES	D	H		SQD LDR		06/03/1947
BARNES	D			WING OFF	Acting	12/06/1947
BARNES	E	S		WING COMM	Acting	10/05/1945
BARNES	H	G	H	FO		01/05/1952
BARNES	H	J	S	CPL		07/05/1953
BARNES	J	G	C	SQD LDR		11/05/1944
BARNES	J	H		WO	Acting	27/04/1944
BARNES	L	D		FL		31/05/1945
BARNES	L	W		LAC		12/06/1947
BARNES	S	R		CPL		10/02/1949
BARNES	W			FL		06/03/1947
BARNETT	E	A		FLT SGT	Acting	25/06/1953
BARNETT	E	E		SQD LDR	Acting	01/07/1948
BARNETT	G	W	F	FL		07/06/1951
BARNETT	H	J		SGT		07/04/1955
BARNETT	J	D	A	FLT OFF	Acting	06/03/1947
BARNETT	J	G		SQD LDR	Acting	24/01/1946
BARNETT	L	O		FL		20/09/1945
BARNETT	M	H		WING OFF	Acting	27/03/1947
BARNFATHER	R	R		FL		23/06/1955
BARNFIELD	H	H		CPL		03/07/1957
BARON	A			SGT		08/11/1945
BARR	J	G	H	FO		12/06/1947
BARR	S			WING COMM	Acting	27/03/1947
BARR	S			SQD LDR	Acting	16/01/1957
BARRABLE	R	F		FO		01/07/1948
BARRACLOUGH	A	R		CPL		07/08/1947
BARRACLOUGH	R	G	V	SQD LDR	Acting	05/10/1944
BARRALET	A	F		FL		24/01/1946
BARRAS	H			FL		03/10/1946
BARRELL	D	R	J	LAC		15/01/1953
BARRELL	F			FL		28/02/1946
BARRETT	C	A		FL		10/08/1944
BARRETT	C	B		SQD LDR		26/11/1953
BARRETT	C			CPL		10/05/1951
BARRETT	D	A		SGT		31/10/1956
BARRETT	J	C		FL		14/10/1948
BARRETT	M	J		SGT		08/07/1993
BARRETT	N	J		FLT OFF		15/01/1953
BARRETT	W	E		WO		18/04/1946
BARRIE	G	L		SGT		31/10/1956
BARRIE	T			WO		22/04/1943
BARRINGTON	D	G		SGT		25/06/1953
BARRITT	C	C	J	SQD LDR		10/12/1942
BARRON	N	P	C	FL		30/11/1944
BARRON	Y			FLT OFF		16/10/1947
BARROWMAN	G	McC.		FL		28/06/1945
BARROWS	L			WO		21/03/1946
BARRY	J	R		SQD LDR	Acting	24/01/1946
BARSON	H	C		SQD LDR	Acting	29/06/1944
BARTHOLOMEW	A	S		SGT		07/08/1947
BARTLAM	C	D	W	WO		08/11/1945
BARTLETT	E			SGT		05/06/1947
BARTLETT	H	A		WING COMM	Acting	06/03/1947
BARTLETT	L	H		WING COMM		01/08/1946
BARTLETT	M	E		FL		10/02/1949
BARTLETT	R	A		JNR TECH		22/09/1955

Source	Service No	Force	Notes	Surname
N565 1955	2688644	RAuxAF		BARLOW
N277 1955	74662	RAFVR	DFC	BARLOW
N666 1956	157250	RAFVR		BARNACLE
N187 1947	91108	AAF		BARNES
N478 1947	294	WAAF	Lady	BARNES
N485 1945	78523	RAFVR	DFC	BARNES
N336 1952	204320	RAFVR		BARNES
N395 1953	757560	RAFVR		BARNES
N444 1944	90101	AAF		BARNES
N396 1944	803170	AAF		BARNES
N573 1945	60325	RAFVR		BARNES
N478 1947	800687	AAF		BARNES
N123 1949	840311	RAuxAF		BARNES
N187 1947	90294	AAF		BARNES
N527 1953	752033	RAFVR		BARNETT
N527 1948	68952	RAFVR		BARNETT
N560 1951	128979	RAFVR		BARNETT
N277 1955	849750	RAuxAF		BARNETT
N187 1947	1056	WAAF		BARNETT
N82 1946	75999	RAFVR		BARNETT
N1036 1945	67038	RAFVR	DFC	BARNETT
N250 1947	144	WAAF		BARNETT
N473 1955	120211	RAFVR		BARNFATHER
N456 1957	809160	AAF	wef 23/5/44	BARNFIELD
N1232 1945	740505	RAFVR		BARON
N478 1947	188790	RAFVR		BARR
N250 1947	72181	RAFVR		BARR
N24 1957	72181	RAFVR	1st Clasp	BARR
N527 1948	123713	RAFVR		BARRABLE
N662 1947	868784	AAF		BARRACLOUGH
N1007 1944	66487	RAFVR		BARRACLOUGH
N82 1946	85233	RAFVR		BARRALET
N837 1946	120326	RAFVR		BARRAS
N28 1953	755651	RAFVR	DFM	BARRELL
N197 1946	117346	RAFVR		BARRELL
N812 1944	82684	RAFVR		BARRETT
N952 1953	103637	RAuxAF		BARRETT
N466 1951	857336	RAuxAF		BARRETT
N761 1956	2602786	RAFVR		BARRETT
N856 1948	115879	RAFVR		BARRETT
YB 1996	n/a	RAuxAF		BARRETT
N28 1953	5374	WAAF		BARRETT
N358 1946	751810	RAFVR	DFM	BARRETT
N761 1956	2649504	RAuxAF		BARRIE
N475 1943	802010	AAF		BARRIE
N527 1953	844182	RAuxAF		BARRINGTON
N1680 1942	70045	RAFVR		BARRITT
N1225 1944	88649	RAFVR		BARRON
N846 1947	42	WAAF		BARRON
N706 1945	117039	RAFVR		BARROWMAN
N263 1946	743061	RAFVR		BARROWS
N82 1946	79154	RAFVR		BARRY
N652 1944	80835	RAFVR	AFC	BARSON
N662 1947	800474	AAF		BARTHOLOMEW
N1232 1945	740047	RAFVR		BARTLAM
N460 1947	881062	WAAF		BARTLETT
N187 1947	72782	RAFVR		BARTLETT
N652 1946	102959	RAFVR	DSO	BARTLETT
N123 1949	67731	RAFVR		BARTLETT
N708 1955	2682270	RAFVR		BARTLETT

Surname	Initials				Rank	Status	Date
BARTLETT	R	H			SGT		15/07/1948
BARTLETT	S	A	C		SQD LDR		18/03/1948
BARTLEY	J	W			SGT		10/06/1948
BARTLEY	J	W			FLT SGT		04/09/1957
BARTON	D	M			SQD OFF	Acting	04/09/1947
BARTON	E	H			FL		09/12/1959
BARTON	E	L	T		WING COMM	Acting	01/05/1957
BARTON	J				LAC		15/11/1951
BARTON	R	W			SQD LDR		14/10/1943
BARTON	W	A			SQD LDR	Acting	04/09/1947
BARTON	W	F			CPL		18/02/1954
BARTON	W	J			SGT		15/09/1949
BARTON	W	J			CPL	Temporary	30/10/1957
BARTON	W	R			SGT		26/08/1948
BARWELL	E	G			WING COMM	Acting	21/06/1945
BARWOOD	A	W			SQD LDR	Acting	06/03/1947
BASCOMB	R	McK			LAC		23/06/1955
BASEDEN	M	W			SQD LDR	Acting	27/04/1950
BASHER	W	F	H	W	FL		01/07/1948
BASHFORD	H				FL		14/09/1950
BASS	E	W	H		SGT		14/10/1948
BASS	G	B			FL		30/03/1944
BASSETT	C	C	H		SGT		07/08/1947
BASSETT	F	W			FL		08/05/1947
BASSETT	J	F			FL		28/02/1946
BASSON	W	N			SQD LDR	Acting	01/03/1945
BASTARD	F				FL		28/02/1946
BATCHELOR	G	S			FO		26/08/1948
BATCHELOR	L	R			FL		20/04/1944
BATE	H	V			LAC		26/11/1953
BATE	H				WO		26/08/1948
BATE	J				CPL		24/05/2005
BATEMAN	D	F			FL		22/10/1958
BATES	C	A			FL	Acting	14/09/1950
BATES	G	M			FL		08/02/1945
BATES	T	C	L		FLT SGT		01/07/1948
BATES	T				CPL		20/01/1960
BATES	W	A			LAC		23/08/1951
BATESON	W	S			FL		22/10/1958
BATH	H	F			FL		14/01/1954
BATHER	E	C			GRP OFF	Acting	13/11/1947
BATIN	L				FL		26/11/1953
BATT	F	T			SGT		30/09/1948
BATT	J	R			SGT		26/08/1948
BATT	L	G			FO		11/01/1945
BATT	L	G			FL		22/10/1958
BATT	L	G			FL		22/10/1958
BATTISON	J	H			FL		30/03/1944
BATWELL	D	A			GRP CAP		15/07/1948
BATY	R	E			SQD LDR		26/06/1947
BAXENDALE	D	H			FO		15/07/1948
BAXENDALE	E				FO		25/06/1953
BAXENDALE	F				LAC		27/03/1963
BAXTER	E	B			WO		28/11/1946
BAXTER	F	M			FLT OFF		12/06/1947
BAXTER	G	W			FLT SGT		12/06/1947
BAXTER	G	W			CPL		12/10/1950
BAXTER	J	G			FL		22/09/1955
BAXTER	W	E			FLT SGT		16/10/1947
BAYLES	I	N			WING COMM	Acting	28/02/1946

Source	Service No	Force	Notes	Surname
N573 1948	801500	RAuxAF		BARTLETT
N232 1948	73607	RAFVR		BARTLETT
N456 1948	808415	RAuxAF		BARTLEY
N611 1957	2686042	RAuxAF	1st Clasp; wef 17/6/57	BARTLEY
N740 1947	1348	WAAF	OBE	BARTON
N843 1959	197817	RAFVR	wef 27/8/57	BARTON
N277 1957	85191	RAuxAF	OBE,ERD; wef 3/8/55	BARTON
N1185 1951	856727	RAuxAF		BARTON
N1065 1943	70046	RAFVR		BARTON
N740 1947	170153	RAFVR		BARTON
N137 1954	843258	RAuxAF		BARTON
N953 1949	847703	RAuxAF		BARTON
N775 1957	847348	AAF	wef 2/9/44	BARTON
N694 1948	756107	RAFVR		BARTON
N671 1945	77454	RAFVR	DFC	BARWELL
N187 1947	81654	RAFVR	DFC	BARWOOD
N473 1955	843427	RAuxAF		BASCOMB
N406 1950	72783	RAFVR		BASEDEN
N527 1948	147336	RAFVR		BASHER
N926 1950	141156	RAFVR		BASHFORD
N856 1948	804149	RAuxAF		BASS
N281 1944	84978	RAFVR		BASS
N662 1947	842351	AAF		BASSETT
N358 1947	138595	RAFVR		BASSETT
N197 1946	135461	RAFVR	DFC,DFM	BASSETT
N225 1945	78539	RAFVR		BASSON
N197 1946	101012	RAFVR	DFM	BASTARD
N694 1948	174775	RAFVR	DFC,AFC	BATCHELOR
N363 1944	72978	RAFVR		BATCHELOR
N952 1953	856011	RAuxAF		BATE
N694 1948	860355	RAuxAF		BATE
LG 2005	856133	RAFVR		BATE
N716 1958	124383	RAFVR	DFC; wef 6/5/52	BATEMAN
N926 1950	144078	RAFVR		BATES
N139 1945	102059	RAFVR	DFM	BATES
N527 1948	848500	RAuxAF		BATES
N33 1960	2681096	RAuxAF	wef 28/10/59	BATES
N866 1951	849064	RAuxAF		BATES
N716 1958	153085	RAFVR	wef 17/5/54	BATESON
N18 1954	77761	RAFVR		BATH
N933 1947	163	WAAF	OBE	BATHER
N952 1953	194366	RAFVR		BATIN
N805 1948	813116	RAuxAF		BATT
N694 1948	804319	RAuxAF		BATT
N14 1945	145514	RAFVR		BATT
N716 1958	145514	RAFVR	1st Clasp; wef 5/10/49	BATT
N716 1958	145514	RAFVR	2nd Clasp; wef 13/12/57	BATT
N281 1944	61237	RAFVR		BATTISON
N573 1948	72784	RAFVR		BATWELL
N519 1947	90753	AAF	MC	BATY
N573 1948	110393	RAFVR		BAXENDALE
N527 1953	2687513	RAuxAF		BAXENDALE
N239 1963	2688041	RAuxAF	wef 16/10/57	BAXENDALE
N1004 1946	747890	RAFVR		BAXTER
N478 1947	1889	WAAF		BAXTER
N478 1947	845032	AAF		BAXTER
N1018 1950	863307	RAuxAF		BAXTER
N708 1955	152597	RAuxAF		BAXTER
N846 1947	863392	AAF		BAXTER
N197 1946	74327	RAFVR	DFC	BAYLES

Surname	Initials				Rank	Status	Date
BAYLEY	F	H			FL	Acting	18/11/1948
BAYLEY	F				SQD LDR	Acting	30/09/1948
BAYLISS	W	J			SGT		24/12/1942
BAYSTON	C	L			FL		19/08/1954
BAZIN	J	M			WING COMM	Acting	03/12/1942
BEACH	D	O			FO		12/02/1953
BEACH	W				FLT SGT		30/03/1943
BEADLE	W	A			FO		24/01/1946
BEAGLE	F	G			WO		22/04/1943
BEAKE	P	H			SQD LDR	Acting	23/02/1950
BEALE	G	A			CPL		30/09/1948
BEALE	J	A			CPL		04/10/1951
BEALE	V	P			FO		25/05/1960
BEAN	C	J			CPL		11/04/1997
BEAN	F				SGT		26/08/1948
BEANES	A	J			CPL		27/04/1950
BEAR	A	M			FO		10/12/1942
BEARD	A	J			WING COMM		26/01/1950
BEARD	J	H	C		SQD LDR		24/12/1942
BEARD	J	M	B		SQD LDR	Acting	20/04/1944
BEARDSALL	G	J			SQD LDR		01/10/1953
BEARDSLEY	R	A			FL		11/01/1945
BEARDSWORTH	E				FLT OFF		26/04/1951
BEASLEY	H	J	S		SQD LDR		31/05/1945
BEATON	A	J			WING COMM		16/03/2004
BEATTIE	A	G			FL		21/03/1946
BEATTIE	B	S	E		SQD LDR	Acting	04/05/1944
BEATTIE	W	M			SGT		20/11/1957
BEATTY	M	A			FL		18/04/1946
BEAUMONT	C	K	M	H	FLT OFF		25/03/1954
BEAUMONT	G				FL		20/05/1948
BEAUMONT	L	A			SGT		01/07/1948
BEAUMONT	S	G			WING COMM		04/02/1943
BEAUMONT	W	A			FL		07/04/1955
BEAUMONT	W	R	L		FL		07/01/1943
BEAUMONT	W	R	L		FL		26/11/1958
BEAVINGTON	R	B			FL		24/01/1946
BECK	E	S			FL	Acting	26/01/1950
BECKETT	R	A			FO		10/08/1950
BECKETT	R	T			FLT SGT		26/06/1947
BEDDARD	T	E			SQD LDR	Acting	07/08/1947
BEDDHAM	D	W			FL		22/02/1961
BEDDOW	H	D			SQD LDR	Acting	17/02/1944
BEDDOW	W	P			SQD LDR		01/07/1948
BEDFORD	H	M			FL		18/04/1946
BEDNELL	R	D			FL		10/12/1942
BEE	D	G			SQD LDR	Acting	01/12/1955
BEECHAM	J	H			FL		28/11/1946
BEECHEY	A	F			SQD LDR	Acting	24/01/1946
BEECROFT	B	C			GRP OFF	Acting	06/03/1947
BEECROFT	B	D			WO		21/03/1946
BEEL	A	F			FLT SGT	Acting	25/05/1950
BEER	C	L			FLT OFF		07/08/1947
BEERS	D				FL		23/05/1946
BEESLEY	T				CPL	Temporary	21/01/1959
BEESTON	K	H			FO		20/05/1948
BEGBIE	W				CPL		03/06/1959
BEGG	M	G			WING COMM	Acting	19/07/1951
BEGG	W	D			FL		22/10/1958
BELASCO	F	C			SQD LDR	Acting	14/10/1943

Source	Service No	Force	Notes	Surname
N984 1948	137012	RAFVR		BAYLEY
N805 1948	76249	RAFVR		BAYLEY
N1758 1942	805260	AAF		BAYLISS
N672 1954	123543	RAFVR		BAYSTON
N1638 1942	90281	AAF	DFC	BAZIN
N115 1953	116132	RAFVR		BEACH
N1014 1943	808001	AAF		BEACH
N82 1946	173322	RAFVR		BEADLE
N475 1943	804099	AAF		BEAGLE
N179 1950	84923	RAFVR	DFC	BEAKE
N805 1948	842653	RAuxAF		BEALE
N1019 1951	841221	RAuxAF		BEALE
N375 1960	170863	RAFVR	wef 15/3/60	BEALE
YB 1998	n/a	RAuxAF		BEAN
N694 1948	852046	RAuxAF		BEAN
N406 1950	842012	RAuxAF		BEANES
N1680 1942	112443	RAFVR		BEAR
N94 1950	90503	RAuxAF	MC	BEARD
N1758 1942	70052	RAFVR		BEARD
N363 1944	89588	RAFVR	DFM	BEARD
N796 1953	133428	RAuxAF		BEARDSALL
N14 1945	100607	RAFVR	DFC	BEARDSLEY
N402 1951	341	WAAF		BEARDSWORTH
N573 1945	73023	RAFVR	DFC	BEASLEY
LG 2004	4233216M	RAuxAF		BEATON
N263 1946	137576	RAFVR		BEATTIE
N425 1944	83990	RAFVR	DFC	BEATTIE
N835 1957	2650758	RAuxAF	wef 6/4/55	BEATTIE
N358 1946	69455	RAFVR		BEATTY
N249 1954	2658052	WRAAF		BEAUMONT
N395 1948	141153	RAFVR	DFC	BEAUMONT
N527 1948	748848	RAFVR		BEAUMONT
N131 1943	90319	AAF		BEAUMONT
N277 1955	166709	RAuxAF		BEAUMONT
N3 1943	72979	RAFVR		BEAUMONT
N807 1958	72979	RAFVR	DFC; 1st Clasp; wef 9/1/46	BEAUMONT
N82 1946	132333	RAFVR		BEAVINGTON
N94 1950	89420	RAFVR		BECK
N780 1950	178728	RAFVR		BECKETT
N519 1947	743558	RAFVR		BECKETT
N662 1947	90510	AAF		BEDDARD
N153 1961	112396	RAFVR	wef 13/10/43	BEDDHAM
N132 1944	60511	RAFVR		BEDDOW
N527 1948	72352	RAFVR		BEDDOW
N358 1946	130106	RAFVR		BEDFORD
N1680 1942	70054	RAFVR		BEDNELL
N895 1955	83618	RAFVR		BEE
N1004 1946	124208	RAFVR		BEECHAM
N82 1946	113913	RAFVR	DFC	BEECHEY
N187 1947	133	WAAF	OBE	BEECROFT
N263 1946	745643	RAFVR		BEECROFT
N517 1950	770345	RAFVR		BEEL
N662 1947	1312	WAAF		BEER
N468 1946	122408	RAFVR		BEERS
N28 1959	850075	RAFVR	wef 24/6/44	BEESLEY
N395 1948	169273	RAFVR		BEESTON
N368 1959	2692026	RAuxAF	wef 22/3/59	BEGBIE
N724 1951	90430	RAuxAF	MC	BEGG
N716 1958	180612	RAFVR	wef 10/7/52	BEGG
N1065 1943	84327	RAFVR		BELASCO

Surname	Initials			Rank	Status	Date
BELCHER	B	J	E	SQD LDR		26/01/1950
BELCHER	K	A		SGT	Temporary	03/06/1959
BELL	A			CPL		09/08/1945
BELL	A			SGT		20/05/1948
BELL	C	T		SQD LDR		07/07/1949
BELL	E	R		FLT OFF		27/03/1947
BELL	E	V	N	SQD LDR		10/12/1942
BELL	E			FL		22/10/1958
BELL	J	A		FL		25/05/1960
BELL	J	A		FL		07/12/1960
BELL	J	C	G	SQD LDR	Acting	30/03/1944
BELL	J			FL		28/11/1946
BELL	J			WING COMM	Acting	16/10/1947
BELL	J			FLT SGT		26/02/1958
BELL	L	D	C	SQD OFF	Acting	04/11/1948
BELL	S	E	H	FL		20/11/1957
BELL	T	L	I	SQD LDR		30/12/1943
BELL	T	R		CPL		27/04/1944
BELL	T			FL		30/09/1948
BELL	W			CPL		22/02/1951
BELLHOUSE	M	A	H	WING COMM	Acting	04/09/1947
BELLIEU	T	A		SGT		20/05/1948
BELLINGHAM	D	J		FL		18/04/1946
BELLINGHAM	J	E		FL		04/10/1945
BELSEY	A	J		FL		09/05/1946
BELSON	J	R	J	FL		10/08/1944
BELTON	B	J	N	FO		14/01/1954
BENDREY	L	M		FL		28/02/1946
BENDRY	E	T		SGT		26/08/1948
BENFIELD	M	G	H	A	FL		22/10/1958
BENGE	C	F		SGT		15/09/1949
BENHAM	D	I		SQD LDR	Acting	13/12/1945
BENJAMIN	E	A		SQD LDR	Acting	10/08/1944
BENNELL	S	E		WO		20/06/1990
BENNETT	A	L		SGT	Acting	13/12/1951
BENNETT	A			LAC		23/06/1955
BENNETT	C	P		SQD LDR		15/09/1949
BENNETT	D	B		WING COMM		26/11/1953
BENNETT	E	J		CPL		26/08/1948
BENNETT	F	A		CPL		02/07/1958
BENNETT	G			FL		13/12/1945
BENNETT	J	A		SGT		25/06/1953
BENNETT	J	P		SQD LDR		06/03/1947
BENNETT	J	S		FL		23/06/1955
BENNETT	M			SGT		30/08/1945
BENNETT	R	H		FL		22/10/1958
BENNETT	R			FL		03/12/1942
BENNETT	R			LAC		05/06/1947
BENNETT	S			FL		21/03/1946
BENNETT	W	G		FL		19/08/1954
BENSON	J	G		WING COMM	Acting	04/11/1948
BENSON	N	A		SGT		05/04/1997
BENSON	N	D		FO		20/09/1945
BENSON	R	A		SQD LDR	Acting	01/08/1946
BENSON	R	H		SGT		15/09/1949
BENTLEY	C			SGT		25/05/1950
BENTLEY	H	M		SQD LDR		27/03/1947
BENTLEY	J	F		FL		10/05/1945
BENTLEY	J	T		AC2		22/09/1955
BENTLEY	R	R		FL		12/06/1947

Source	Service No	Force	Notes	Surname
N94 1950	72609	RAFVR	DFC	BELCHER
N368 1959	819131	AAF	wef 10/5/44	BELCHER
N874 1945	816055	AAF		BELL
N395 1948	870019	RAuxAF		BELL
N673 1949	84931	RAFVR	DFC	BELL
N250 1947	1313	WAAF		BELL
N1680 1942	90162	AAF		BELL
N716 1958	151368	RAuxAF	wef 22/5/58	BELL
N375 1960	55560	RAFVR	wef 22/1/53	BELL
N933 1960	126487	RAFVR	DFC; wef 1/7/44	BELL
N281 1944	90040	AAF		BELL
N1004 1946	177188	RAFVR	DFC	BELL
N846 1947	73938	RAFVR		BELL
N149 1958	2683530	RAuxAF	BEM; wef 7/12/57	BELL
N934 1948	988	WAAF		BELL
N835 1957	177851	RAFVR	wef 19/5/53	BELL
N1350 1943	70056	RAFVR		BELL
N396 1944	808201	AAF		BELL
N805 1948	110894	RAFVR		BELL
N194 1951	870154	RAuxAF		BELL
N740 1947	90838	AAF		BELLHOUSE
N395 1948	816094	RAuxAF		BELLIEU
N358 1946	130806	RAFVR	DFC,AFC	BELLINGHAM
N1089 1945	77776	RAFVR		BELLINGHAM
N415 1946	127845	RAFVR		BELSEY
N812 1944	121947	RAFVR		BELSON
N18 1954	199220	RAFVR		BELTON
N197 1946	85025	RAFVR		BENDREY
N694 1948	743773	RAFVR		BENDRY
N716 1958	163944	RAFVR	wef 21/3/54	BENFIELD
N953 1949	819000	RAuxAF		BENGE
N1355 1945	104443	RAFVR	DFC,AFC	BENHAM
N812 1944	77777	RAFVR	DFC	BENJAMIN
YB 1997	n/a	RAuxAF		BENNELL
N1271 1951	842655	RAuxAF		BENNETT
N473 1955	845259	RAuxAF		BENNETT
N953 1949	90992	RAuxAF		BENNETT
N952 1953	81663	RAFVR		BENNETT
N694 1948	841539	RAuxAF		BENNETT
N452 1958	812261	AAF	wef 9/6/44	BENNETT
N1355 1945	86674	RAFVR		BENNETT
N527 1953	803441	RAuxAF		BENNETT
N187 1947	72252	RAFVR		BENNETT
N473 1955	150188	RAFVR		BENNETT
N961 1945	802431	AAF		BENNETT
N716 1958	172620	RAFVR	wef 14/6/53	BENNETT
N1638 1942	70058	RAFVR		BENNETT
N460 1947	811208	AAF		BENNETT
N263 1946	126773	RAFVR		BENNETT
N672 1954	81703	RAuxAF		BENNETT
N934 1948	81365	RAFVR	DSO,DFC	BENSON
YB 1998	n/a	RAuxAF		BENSON
N1036 1945	68209	RAFVR		BENSON
N652 1946	124538	RAFVR	DFC	BENSON
N953 1949	866434	RAuxAF		BENSON
N517 1950	815178	RAuxAF		BENTLEY
N250 1947	72787	RAFVR		BENTLEY
N485 1945	89307	RAFVR		BENTLEY
N708 1955	870863	RAuxAF		BENTLEY
N478 1947	73565	RAFVR	MC,AFC	BENTLEY

Surname	Initials			Rank	Status	Date
BENTON	G	T		FL		04/09/1947
BENWELL	E	M		CPL		19/08/1959
BENZIE	D	G	E	SQD LDR		24/01/1946
BENZIES	J			SGT		05/06/1947
BERCOT	P	V		FL		20/07/1960
BERG	V	W		FL		20/04/1944
BERKELEY	O	M		WING COMM	Acting	06/03/1947
BERNARD	E	C		FO		10/05/1945
BERNARD	H	G		FL		06/03/1947
BERNARD	H	G		FL		22/09/1955
BERRIDGE	H	W	W	FL		26/01/1950
BERRIE	J			LAC		26/11/1953
BERRY	J			SGT		05/01/1950
BERRY	R	H		SQD LDR		17/03/1949
BERRY	R	M		SQD LDR	Acting	04/10/1945
BERRY	R			WING COMM	Acting	30/03/1944
BERRY	S	J		WING COMM		20/05/1948
BERRY	T	A		SGT	Temporary	20/01/1960
BERRY	T	F		FLT SGT		30/09/1948
BERRY	W	L	S	PO		20/09/1951
BERRYMAN	R	H		WING COMM		13/11/1947
BERWICK	P	G		FO		10/05/1945
BESLEY	A	P		GRP CAP		27/04/1950
BESSELL	P	A		CPL		07/07/1949
BESSEY	J	C		FL		15/06/1944
BEST	B			FLT SGT	Acting	04/09/1947
BEST	B			FLT OFF	Acting	07/12/1950
BEST	C	N		CPL		02/09/1998
BEST	J			CPL		01/07/1948
BESWICK	S			LAC		01/11/1951
BETTERIDGE	F	C		LAC		12/02/1953
BETTERIDGE	F	J		LAC		17/06/1954
BETTERIDGE	L	G		SQD LDR	Acting	01/08/1946
BETTINSON	L	G		SQD LDR	Acting	21/03/1946
BETTS	A	A		FL		11/01/1945
BETTS	E			FL		19/09/1956
BETTS	J	M		SQD LDR	Acting	28/02/1946
BEVAN	F	H		FO		09/11/1944
BEVERIDGE	A			SQD LDR		07/08/1947
BEVERIDGE	J	W		FL	Acting	10/02/1949
BEVIS	D	R		FL		03/06/1943
BEWLEY	N			CPL		10/08/1950
BICHENO	E	G		LAC		15/09/1949
BICKNELL	N			WING COMM	Acting	11/07/1946
BIDDLECOMBE	F	R		FL		08/07/1959
BIDE	J	A		FL		26/06/1947
BIDEN	R	M		SQD LDR	Acting	09/05/1946
BIGGAR	A	C		FL		22/02/1961
BIGGAR	A	C		FL		28/08/1963
BIGGS	S	J		LAC		07/06/1951
BIGNELL	D	E		FL		07/05/1953
BILES	A	G		FLT SGT	Temporary	30/09/1954
BILLINGHAM	W			SGT		26/11/1953
BILLINGS	R			SQD LDR	Acting	21/06/1945
BILLINGS	R			FL		07/12/1960
BILLSON	A	E		FL		15/07/1948
BILLYEALD	P			SQD LDR	Acting	10/05/1945
BILTON	J	E		FL		26/08/1948
BILTON	W	B		SQD LDR	Acting	04/10/1945
BINGHAM	A	E		FL		21/03/1946

Source	Service No	Force	Notes	Surname
N740 1947	81704	RAFVR		BENTON
N552 1959	2655806	WRAAF	wef 22/3/59	BENWELL
N82 1946	90372	AAF	DFC	BENZIE
N460 1947	845286	AAF		BENZIES
N531 1960	183384	RAFVR	wef 4/5/60	BERCOT
N363 1944	78717	RAFVR		BERG
N187 1947	72319	RAFVR		BERKELEY
N485 1945	178934	RAFVR		BERNARD
N187 1947	129135	RAFVR	DFC	BERNARD
N708 1955	129135	RAFVR	DFC; 1st Clasp	BERNARD
N94 1950	115634	RAFVR	DSO,DFC	BERRIDGE
N952 1953	2695508	RAuxAF		BERRIE
N1 1950	756754	RAFVR		BERRY
N231 1949	72254	RAFVR		BERRY
N1089 1945	62252	RAFVR		BERRY
N281 1944	78538	RAFVR	DSO,DFC	BERRY
N395 1948	72123	RAFVR	OBE	BERRY
N33 1960	814173	AAF	wef 1/7/44	BERRY
N805 1948	857554	RAuxAF		BERRY
N963 1951	746136	RAFVR		BERRY
N933 1947	90443	AAF	OBE	BERRYMAN
N485 1945	161290	RAFVR		BERWICK
N406 1950	90594	RAuxAF	CBE	BESLEY
N673 1949	866508	RAuxAF		BESSELL
N596 1944	116894	RAFVR		BESSEY
N740 1947	843260	AAF		BEST
N1210 1950	3710	WAAF		BEST
YB 1999	n/a	RAuxAF		BEST
N527 1948	749196	RAFVR		BEST
N1126 1951	856717	RAuxAF		BESWICK
N115 1953	859596	RAuxAF		BETTERIDGE
N471 1954	744021	RAFVR		BETTERIDGE
N652 1946	120089	RAFVR	DFC	BETTERIDGE
N263 1946	84727	RAFVR	DFC	BETTINSON
N14 1945	102068	RAFVR	AFC	BETTS
N666 1956	80945	RAFVR		BETTS
N197 1946	85657	RAFVR	DFC,AFC	BETTS
N1150 1944	158693	RAFVR		BEVAN
N662 1947	73080	RAFVR		BEVERIDGE
N123 1949	122773	RAFVR		BEVERIDGE
N609 1943	83984	RAFVR		BEVIS
N780 1950	811049	RAuxAF		BEWLEY
N953 1949	864074	RAuxAF		BICHENO
N594 1946	73041	RAFVR	DSO,DFC	BICKNELL
N454 1959	182298	RAFVR	wef 20/9/56	BIDDLECOMBE
N519 1947	86739	RAFVR		BIDE
N415 1946	78084	RAFVR		BIDEN
N153 1961	195088	RAuxAF	wef 21/10/52	BIGGAR
N621 1963	195088	RAuxAF	1st Clasp; wef 8/7/62	BIGGAR
N560 1951	864064	RAuxAF		BIGGS
N395 1953	159338	RAFVR		BIGNELL
N799 1954	849242	RAuxAF		BILES
N952 1953	744841	RAFVR		BILLINGHAM
N671 1945	77457	RAFVR		BILLINGS
N933 1960	77457	RAuxAF	1st Clasp; wef 16/11/60	BILLINGS
N573 1948	130521	RAFVR	DFC	BILLSON
N485 1945	77779	RAFVR	DFC	BILLYEALD
N694 1948	81552	RAFVR		BILTON
N1089 1945	77780	RAFVR		BILTON
N263 1946	161311	RAFVR	AFC	BINGHAM

Surname	Initials			Rank	Status	Date
BINGHAM	S	H		FL		26/01/1950
BINGLEY	S	R		FLT SGT		21/02/1952
BINKS	W	R		FL		17/06/1954
BINNS	F			CPL	Temporary	30/09/1954
BINSTEAD	C	V		FO		08/05/1947
BIRCH	A	G		SQD LDR		05/06/1947
BIRCHALL	H	G		LAC		25/05/1950
BIRD	B	A		SQD LDR		05/06/1947
BIRD	F	G		SGT		20/09/1945
BIRD	H	A		FL		10/06/1948
BIRD	H	G		WING COMM		25/06/1953
BIRD	J	W		FL		07/12/1960
BIRD	J			FL		22/10/1958
BIRD	L	M		SGT		14/10/1948
BIRD	P	D		SQD LDR	Acting	04/10/1945
BIRD	P	R		FLT SGT		26/08/1948
BIRD	T	R		WING COMM		26/06/1947
BIRDSALL	R	W		LAC		12/06/1947
BIRKBECK	J	W	L	SQD LDR		08/05/1947
BIRKIN	J	M		SQD LDR		21/06/1945
BIRKIN	J	M		GRP CAP		02/04/1953
BIRKIN	J	M		GRP CAP		28/03/1962
BIRLEY	R	N		SQD LDR	Acting	06/03/1947
BIRLEY	R	N		FL		01/11/1951
BIRMINGHAM	J	C		CPL		15/09/1949
BIRRELL	R			FL		08/03/1956
BIRSCHEL	A	D		CPL	Acting	28/09/1950
BISDEE	J	D		GRP CAP		10/05/1945
BISHOP	H	F		LAC		23/02/1950
BISHOP	L			CPL		13/01/1949
BISHOP	R	M		FL		25/07/1962
BITTON	H	W		FL		13/02/1957
BITTON	H	W		FL		05/01/1966
BIZLEY	N	C		WO		08/11/1945
BJORKEGREN	H	W	S	B	FO	20/03/1957
BLACK	A	J		WING COMM	Acting	14/10/1943
BLACK	A	J		SQD LDR		04/10/1951
BLACK	B	G		SGT		17/04/1990
BLACK	B	G		SGT		17/04/2000
BLACKADDER	W	F		WING COMM		22/04/1943
BLACKBURN	L			SGT		12/06/1947
BLACKBURNE	R	A		CPL		30/09/1948
BLACKIE	N	F		SQD LDR		14/10/1943
BLACKING	L	R		FL		15/11/1961
BLACKMORE	F	C		SQD LDR	Acting	10/05/1945
BLACKMORE	J	P		FL		22/10/1958
BLACKSTONE	M	E		WING COMM	Acting	09/08/1945
BLADES	W	R		CPL		16/10/1952
BLAIKIE	W	A		SGT	Acting	01/10/1953
BLAIR	A	N		WO	Acting	12/06/1947
BLAIR	D			WO		16/10/1947
BLAIR	P			LAC		26/04/1961
BLAKE	C	E	P	FL		13/07/1950
BLAKE	C	F		WO		17/02/1944
BLAKE	G			SGT		14/09/1950
BLAKE	J	C		LAC		20/01/1960
BLAKE	L	A		FO		26/01/1950

Source	Service No	Force	Notes	Surname
N94 1950	72613	RAFVR		BINGHAM
N134 1952	847688	RAuxAF		BINGLEY
N471 1954	156672	RAFVR		BINKS
N799 1954	2697000	RAuxAF		BINNS
N358 1947	157637	RAFVR		BINSTEAD
N460 1947	90506	AAF		BIRCH
N517 1950	811204	RAuxAF		BIRCHALL
N460 1947	73801	RAFVR		BIRD
N1036 1945	812010	AAF		BIRD
N456 1948	73082	RAFVR		BIRD
N527 1953	91030	RAuxAF		BIRD
N933 1960	205493	RAFVR	wef 17/8/60	BIRD
N716 1958	187516	RAFVR	DFC; wef 11/7/52	BIRD
N856 1948	882627	WAAF		BIRD
N1089 1945	123832	RAFVR		BIRD
N694 1948	800243	RAuxAF		BIRD
N519 1947	90418	AAF		BIRD
N478 1947	811214	AAF		BIRDSALL
N358 1947	73526	RAFVR	AFC	BIRKBECK
N671 1945	81350	RAFVR	DSO,DFC,AFC	BIRKIN
N298 1953	81350	RAuxAF	DSO,OBE,DFC, AFC; 1st Clasp	BIRKIN
N242 1962	81350	RAuxAF	CB,DSO,OBE,DFC,AFC, ADC,MA; 2nd Clasp; wef 14/1/62	BIRKIN
N187 1947	90356	AAF		BIRLEY
N1126 1951	90356	RAuxAF	1st Clasp	BIRLEY
N953 1949	855132	RAuxAF		BIRMINGHAM
N187 1956	131762	RAFVR	DFC	BIRRELL
N968 1950	810135	RAuxAF		BIRSCHEL
N485 1945	76575	RAFVR	DFC	BISDEE
N179 1950	752688	RAFVR		BISHOP
N30 1949	845087	RAuxAF		BISHOP
N549 1962	2602417	RAFVR	wef 6/10/61	BISHOP
N104 1957	140355	RAFVR	wef 5/1/56	BITTON
MOD 1967	140355	RAFVR	1st Clasp	BITTON
N1232 1945	742472	RAFVR		BIZLEY
N198 1957	184591	RAFVR	wef 24/5/55	BJORKEGREN
N1065 1943	70066	RAFVR		BLACK
N1019 1951	70066	RAFVR	1st Clasp	BLACK
YB 1996	n/a	RAuxAF		BLACK
YB 2000	n/a	RAuxAF	1st Clasp	BLACK
N475 1943	90282	AAF	DSO	BLACKADDER
N478 1947	811023	AAF		BLACKBURN
N805 1948	810183	RAuxAF		BLACKBURNE
N1065 1943	90421	AAF		BLACKIE
N902 1961	2055798	RAFVR	wef 6/10/61	BLACKING
N485 1945	64281	RAFVR	AFC	BLACKMORE
N716 1958	123096	RAFVR	wef 30/6/52	BLACKMORE
N874 1945	72257	RAFVR	DFC	BLACKSTONE
N798 1952	799900	RAFVR		BLADES
N796 1953	803490	RAuxAF		BLAIKIE
N478 1947	811191	AAF		BLAIR
N846 1947	742018	RAFVR		BLAIR
N339 1961	2650879	RAuxAF	wef 13/2/61	BLAIR
N675 1950	85036	RAFVR		BLAKE
N132 1944	740150	RAFVR		BLAKE
N926 1950	749000	RAFVR		BLAKE
N33 1960	2691034	RAuxAF	wef 25/5/59	BLAKE
N94 1950	163077	RAFVR		BLAKE

Surname	Initials			Rank	Status	Date
BLAKE	L	G	D	FL		28/06/1959
BLAKE	P	H		LAC		26/08/1948
BLAKENEY	W	J	J	SGT	Temporary	25/03/1954
BLANCHE	J	B		SQD LDR		01/12/1985
BLANCHE	J	B		SQD LDR		01/12/1995
BLANDE	A			WING COMM	Acting	25/05/1950
BLANDFORD	P			SQD LDR	Acting	20/04/1944
BLANEY	M			SGT		07/08/1947
BLANKS	T	V	G	FL		11/05/1944
BLANN	A	H	P	FL		09/08/1945
BLASHILL	K			FO		15/07/1948
BLASHILL	K			FL		07/04/1955
BLAYDON	B	F		FLT SGT		24/11/1949
BLAYNEY	A	J		FL		04/11/1943
BLISS	R	E	P	SQD LDR	Acting	24/01/1946
BLOCKSIDGE	R	T		FO		08/05/1947
BLOFELD	J	S		SGT	Temporary	21/06/1961
BLOMFIELD	W	J		SQD LDR	Acting	25/01/1951
BLOOMFIELD	A	L		SGT		29/11/1951
BLOOMFIELD	D	F		CPL		07/07/1949
BLORE	G			WO	Acting	15/09/1949
BLORE	J	H		CPL		26/06/1947
BLOTT	W			FL		18/04/1946
BLUNSDON	M	A		SAC		02/12/1992
BLYTH	T	S		FL		26/04/1956
BLYTHE	R			MST NAV		27/03/1963
BOAL	D			CPL	Temporary	23/06/1955
BOARD	G	J		SAC		25/05/1995
BOARDMAN	T	B		LAC		05/06/1947
BOAST	R	S		SQD LDR	Acting	04/10/1945
BOATFIELD	A	E		FO		24/11/1949
BODDINGTON	M	C	B	SQD LDR	Acting	27/07/1944
BODDINGTON	P			FO		21/03/1946
BODDY	J	T		CPL		20/08/1958
BODDY	W	D		FL		25/05/1960
BODFIELD	A	W		FL		12/02/1953
BOEREE	A	R		SQD LDR		04/11/1943
BOGLE	W	J		CPL		19/06/1963
BOLTON	P	R		FO		19/10/1960
BOLTON	P	W		FO		29/04/1954
BOLTON	S	G		CPL		21/06/1951
BOND	E	F		SQD LDR		01/01/2002
BOND	H	St.	G	SQD LDR	Acting	10/05/1945
BOND	K	A		SAC		21/09/1995
BOND	V	C		LAC		26/04/1961
BOND	W	J		SQD LDR		07/12/1960
BONE	A	G	R	CPL		25/03/1954
BONE	C	R	T	FO		08/11/1945
BONIFACE	H	G		CPL		18/10/1951
BONIFACE	W	C	L	SGT		24/06/1943
BONNAR	K	J		FL		20/04/1944
BONNER	G	C		SQD LDR		03/12/1942
BONNETT	H	W		LAC		17/01/1952
BONOME	E	A		LAC		08/05/1947
BONSER	S	H		WING COMM		19/08/1954
BOON	A	E		CPL		20/05/1948
BOON	L	C		CPL		14/10/1948
BOON	P	C		FL		22/10/1958
BOOT	P	V		FL		11/05/1944
BOOTH	F			SGT	Temporary	30/10/1957

Source	Service No	Force	Notes	Surname
MOD 1964	138954	RAFVR		BLAKE
N694 1948	869293	RAuxAF		BLAKE
N249 1954	840966	RAuxAF		BLAKENEY
YB 1996	91410	RAuxAF		BLANCHE
YB 1997	91410	RAuxAF	1st Clasp	BLANCHE
N517 1950	90668	RAuxAF	DSO	BLANDE
N363 1944	80834	RAFVR		BLANDFORD
N662 1947	881945	WAAF		BLANEY
N444 1944	106144	RAFVR		BLANKS
N874 1945	117155	RAFVR		BLANN
N573 1948	134779	RAFVR		BLASHILL
N277 1955	134779	RAuxAF	1st Clasp	BLASHILL
N1193 1949	744401	RAFVR		BLAYDON
N1144 1943	900528	AAF		BLAYNEY
N82 1946	78718	RAFVR		BLISS
N358 1947	190898	RAFVR		BLOCKSIDGE
N478 1961	761248	RAFVR	wef 6/9/44	BLOFELD
N93 1951	115667	RAFVR		BLOMFIELD
N1219 1951	749753	RAFVR		BLOOMFIELD
N673 1949	744746	RAFVR		BLOOMFIELD
N953 1949	762159	RAFVR	BEM	BLORE
N519 1947	815268	AAF		BLORE
N358 1946	115303	RAFVR		BLOTT
YB 1996	n/a	RAuxAF		BLUNSDON
N318 1956	126153	RAFVR	DFC,AFC	BLYTH
N239 1963	755342	RAF	wef 8/7/44	BLYTHE
N473 1955	873809	RAuxAF		BOAL
YB 1996	n/a	RAuxAF		BOARD
N460 1947	811224	AAF		BOARDMAN
N1089 1945	81066	RAFVR	DFC	BOAST
N1193 1949	137691	RAFVR		BOATFIELD
N763 1944	88017	RAFVR	DFC,DFM	BODDINGTON
N263 1946	172580	RAFVR		BODDINGTON
N559 1958	2691505	RAuxAF	wef 23/7/58	BODDY
N375 1960	103658	RAuxAF	wef 26/4/59	BODDY
N115 1953	83852	RAFVR		BODFIELD
N1144 1943	70071	RAFVR		BOEREE
N434 1963	2650973	RAuxAF	wef 16/4/63	BOGLE
N775 1960	204867	RAFVR	wef 23/6/53	BOLTON
N329 1954	175731	RAuxAF		BOLTON
N621 1951	841112	RAuxAF		BOLTON
YB 2003	2643631E	RAuxAF		BOND
N485 1945	76569	RAFVR		BOND
YB 1996	n/a	RAuxAF		BOND
N339 1961	2650885	RAuxAF	wef 27/2/61	BOND
N933 1960	91136	AAF	wef 17/7/44	BOND
N249 1954	260527	RAFVR		BONE
N1232 1945	161680	RAFVR		BONE
N1077 1951	864039	RAuxAF		BONIFACE
N677 1943	813023	AAF		BONIFACE
N363 1944	69437	RAFVR	AFC	BONNAR
N1638 1942	90108	AAF		BONNER
N35 1952	855955	RAuxAF		BONNETT
N358 1947	842831	AAF		BONOME
N672 1954	77590	RAFVR	MBE	BONSER
N395 1948	841046	RAuxAF		BOON
N856 1948	846427	RAuxAF		BOON
N716 1958	196461	RAFVR	wef 14/6/53	BOON
N444 1944	76455	RAFVR	DFC	BOOT
N775 1957	744822	RAFVR	wef 28/8/44	BOOTH

Surname	Initials			Rank	Status	Date
BOOTH	J	G	F	FL		22/10/1958
BOOTH	J	J		FO		11/07/1946
BOOTH	J	S		FL		28/11/1956
BOOTH	K	R		FL		05/09/1946
BOOTH	L	E		FO		13/01/1949
BOOTH	L	W	N	SQD LDR	Acting	28/02/1946
BOREHAM	G	B		FL		23/05/1946
BORGMAN	P	S		FL		09/10/2001
BORLAND	A	C		AC1		07/06/1951
BORTHWICK	A	B		LAC		25/03/1954
BOSLEY	G	F		LAC		07/08/1947
BOSLEY	J	A		FL		02/07/1958
BOSLEY	J	E		FL		10/06/1948
BOSWORTH	E	C	C	SQD LDR	Acting	28/11/1956
BOTHAM	C	W		SQD LDR	Acting	26/06/1947
BOTT	H	A		WO		26/01/1950
BOUCHER	F	H		SGT		23/02/1950
BOUGH	A	H		CPL	Temporary	20/01/1960
BOUGHTON-THOMAS	K	A		SQD LDR		19/08/1954
BOULTER	K	C	J	FL		06/03/1947
BOULTON	H			SGT		26/08/1948
BOULTON	L	A		FO		20/05/1996
BOULTON	S	H		SGT		19/08/1954
BOURKE	C			LAC		02/07/1958
BOURKE	C			LAC		27/03/1963
BOURNE	D	J		FL		07/07/1949
BOURNE	J	H		FL		21/09/1944
BOURNE	S	N		FL		08/05/1947
BOURNE	W	J		FL		20/05/1948
BOUSFIELD	S	P	A	WING COMM	Acting	06/03/1947
BOVILL	M	W		WING COMM		05/02/1948
BOWD	J	E		FL		19/08/1954
BOWDEN	L	M		FO		31/07/1998
BOWDEN	S			SAC		20/07/1997
BOWDLER	J	D		SQD LDR	Acting	26/08/1948
BOWELL	L			FLT SGT		11/01/1945
BOWEN	H	W	J	SGT		20/05/1948
BOWEN	T	A		FL		09/08/1945
BOWERS	G	S		FL		09/05/1946
BOWES	R	L		SQD LDR		04/02/1943
BOWES	R	L		WING COMM		23/08/1951
BOWIE	G	J		SQD LDR	Acting	15/07/1948
BOWLER	J			FLT SGT	Acting	14/10/1948
BOWLES	C	B	J	CPL		16/03/1950
BOWLES	I	S	J	FL		08/07/1959
BOWLES	L	G		CPL		07/07/1994
BOWLES	M			WING COMM	Acting	07/12/1950
BOWMAN	W	B	F	SGT		16/03/1950
BOWN	H	W		LAC		13/01/1949
BOWN	L	G		FLT SGT		13/01/1949
BOWNAS	H	C		SQD LDR	Acting	09/05/1946
BOWRING	B	H		SQD LDR		21/06/1945
BOX	E	D	D	FLT OFF		08/05/1947
BOYCE	E	A		FL		21/03/1946
BOYCE	F	A		SQD LDR		07/06/1951
BOYCE	K	D		FL		08/11/1945
BOYD	A	D	McN.	WING COMM	Acting	20/09/1945
BOYD	R	F		WING COMM		04/02/1943
BOYD	T	H		SGT		20/05/1948
BOYLAND	T	D	McA.	WING COMM	Acting	27/03/1947

Source	Service No	Force	Notes	Surname
N716 1958	61288	RAFVR	wef 10/7/52	BOOTH
N594 1946	171689	RAFVR		BOOTH
N833 1956	150659	RAFVR		BOOTH
N751 1946	133926	RAFVR	DFC,DFM	BOOTH
N30 1949	199592	RAFVR	Deceased	BOOTH
N197 1946	145299	RAFVR	DFC	BOOTH
N468 1946	120069	RAFVR	DFC	BOREHAM
YB 2002	2642043	RAuxAF		BORGMAN
N560 1951	743149	RAFVR		BORLAND
N249 1954	873961	RAuxAF		BORTHWICK
N662 1947	841235	AAF		BOSLEY
N452 1958	188889	RAFVR	wef 24/9/52	BOSLEY
N456 1948	89623	RAFVR		BOSLEY
N833 1956	136095	RAFVR		BOSWORTH
N519 1947	90904	AAF		BOTHAM
N94 1950	746813	RAFVR		BOTT
N179 1950	813160	RAuxAF		BOUCHER
N33 1960	850304	AAF	wef 22/7/44	BOUGH
N672 1954	77520	RAuxAF		BOUGHTON-THOMAS
N187 1947	130459	RAFVR		BOULTER
N694 1948	869254	RAuxAF		BOULTON
YB 1997	2635949	RAuxAF		BOULTON
N672 1954	2695054	RAuxAF		BOULTON
N452 1958	2692146	RAuxAF	wef 7/4/44	BOURKE
N239 1963	2692146	RAuxAF	1st Clasp; wef 28/7/62	BOURKE
N673 1949	137590	RAFVR		BOURNE
N961 1944	72020	RAFVR		BOURNE
N358 1947	72790	RAFVR		BOURNE
N395 1948	73029	RAFVR		BOURNE
N187 1947	72791	RAFVR		BOUSFIELD
N87 1948	73609	RAFVR		BOVILL
N672 1954	109548	RAuxAF		BOWD
YB 1999	2646924	RAuxAF		BOWDEN
YB 1998	n/a	RAuxAF		BOWDEN
N694 1948	85536	RAFVR		BOWDLER
N14 1945	808121	AAF		BOWELL
N395 1948	746443	RAFVR		BOWEN
N874 1945	85658	RAFVR		BOWEN
N415 1946	67598	RAFVR	DFC	BOWERS
N131 1943	70072	RAFVR		BOWES
N866 1951	70072	RAFVR	DFC; 1st Clasp	BOWES
N573 1948	157460	RAFVR		BOWIE
N856 1948	856536	RAuxAF		BOWLER
N257 1950	756283	RAFVR		BOWLES
N454 1959	202413	RAuxAF	wef 25/1/59	BOWLES
YB 1996	n/a	RAuxAF		BOWLES
N1210 1950	72285	RAFVR		BOWLES
N257 1950	752104	RAFVR		BOWMAN
N30 1949	860453	RAuxAF		BOWN
N30 1949	860460	RAuxAF		BOWN
N415 1946	67044	RAFVR		BOWNAS
N671 1945	90105	AAF		BOWRING
N358 1947	2000	WAAF		BOX
N263 1946	122057	RAFVR	DFC	BOYCE
N560 1951	72792	RAFVR		BOYCE
N1232 1945	134529	RAFVR	DFM	BOYCE
N1036 1945	72461	RAFVR	DSO,DFC	BOYD
N131 1943	90165	AAF	DSO,DFC	BOYD
N395 1948	816083	RAuxAF		BOYD
N250 1947	72210	RAFVR		BOYLAND

Surname	Initials			Rank	Status	Date
BOYLE	C	R		FL	Acting	23/02/1950
BOYLE	J			SGT		05/06/1947
BOYLE	T	M		SQD LDR		21/01/1959
BOYLE-BANTOFT	F	N		WING COMM	Acting	22/02/1961
BOYNE	A	M		SQD LDR		15/07/1948
BOYNTON	N	D		FL		15/07/1948
BOYS	D	L		FL		06/03/1947
BRACEGIRDLE	J			CPL		30/09/1954
BRACEY	A	H		FO		10/06/1948
BRACEY	M	A		SGT	Acting	26/01/1950
BRACKSTON	H	J		LAC		07/06/1951
BRADBURY	G	E		SGT		12/06/1947
BRADFIELD	C	A	W	CPL		15/01/1953
BRADFORD	J	A		FL		12/04/1945
BRADFORD	J	R	T	WING COMM	Acting	24/06/1943
BRADFORD	L			FL		24/07/1952
BRADLEY	C	S		FL		27/11/1952
BRADLEY	R			FL		07/04/1955
BRADSHAW	H	W		WO		28/02/1946
BRADSHAW	W	H	A	FL		09/05/1946
BRADY	D	G		FL		04/10/1945
BRAITHWAITE	D	A		SQD LDR	Acting	21/03/1946
BRAMLEY	D			FLT SGT		09/12/1959
BRAMLEY	R	C		FL		28/02/1946
BRAMSON	A	E		FL		21/06/2005
BRANAGAN	J			FLT SGT		15/07/1948
BRAND	E			LAC		26/01/1950
BRAND	G			FO		27/03/1947
BRAND	H	C	S	SQD LDR	Acting	03/12/1942
BRANDER	F	E		FL		30/03/1944
BRANN	J			SGT		25/03/1954
BRANSKI	P	B		FL		14/03/1998
BRANTON	P	M		FL		24/01/1946
BRATLEY	J	L		FO		26/11/1958
BRATTON	E	G	R	FL		24/10/1989
BRATTON	E	G	R	SQD LDR		24/10/1999
BRAUN	R	P		SQD LDR		09/11/1944
BRAY	D			FO		26/06/1947
BRAY	H	L		FL		18/02/1954
BRAY	R	B		LAC		26/01/1950
BRAYLEY	F	J		FL		24/01/1946
BRAYTON	W	A	O'M	FL		24/11/1949
BREAKSPEAR	H	F		SQD LDR	Acting	30/03/1944
BREARLY	W	D		FO		31/05/1945
BREEDON	M	G		SQD OFF	Acting	17/04/1947
BREINGAN	R	A		SGT		12/06/1947
BRENTNALL	D	I		SGT		26/04/1951
BRENTNALL	G	C		CPL		21/06/1951
BRETT	L			FO		19/08/1954
BRETT	R	H		WO		21/06/1945
BREW	S			WING COMM		01/07/1948
BREWER	A	E		FO		20/05/1948
BREWER	D	W		LAC		21/01/1959
BREWER	J	W		CPL		01/10/1953
BREWER	R	W		SGT		18/07/1956
BRICE	D	A		FL		27/07/1944
BRICE	G	P		SGT		26/08/1948
BRICE	J	L	A	LAC		21/02/1952
BRICKWOOD	R	R		SQD LDR		19/01/1956
BRIDEN	H			WO		15/07/1948

Source	Service No	Force	Notes	Surname
N179 1950	134781	RAFVR		BOYLE
N460 1947	811174	AAF		BOYLE
N28 1959	90329	AAF	MB,ChB,FRCS,LRCP; wef 23/12/42	BOYLE
N153 1961	44534	RAuxAF	TD; wef 29/11/60	BOYLE-BANTOFT
N573 1948	90915	RAuxAF		BOYNE
N573 1948	110360	RAFVR		BOYNTON
N187 1947	126849	RAFVR		BOYS
N799 1954	2650010	RAuxAF		BRACEGIRDLE
N456 1948	120152	RAFVR		BRACEY
N94 1950	813131	RAuxAF		BRACEY
N560 1951	840297	RAuxAF		BRACKSTON
N478 1947	809103	AAF		BRADBURY
N28 1953	841507	RAuxAF		BRADFIELD
N381 1945	77458	RAFVR		BRADFORD
N677 1943	90029	AAF	OBE	BRADFORD
N569 1952	151122	RAFVR		BRADFORD
N915 1952	115670	RAuxAF		BRADLEY
N277 1955	175017	RAuxAF		BRADLEY
N197 1946	751058	RAFVR		BRADSHAW
N415 1946	115313	RAFVR	DFC	BRADSHAW
N1089 1945	125942	RAFVR		BRADY
N263 1946	91060	AAF	DFC	BRAITHWAITE
N843 1959	2685646	RAuxAF	wef 22/5/54	BRAMLEY
N197 1946	169145	RAFVR		BRAMLEY
LG2005	150694	RAFVR		BRAMSON
N573 1948	746796	RAFVR		BRANAGAN
N94 1950	841320	RAuxAF		BRAND
N250 1947	187385	RAFVR		BRAND
N1638 1942	70079	RAFVR		BRAND
N281 1944	88853	RAFVR		BRANDER
N249 1954	2694500	RAuxAF		BRANN
YB 1999	n/a	RAuxAF		BRANSKI
N82 1946	126155	RAFVR	DFC	BRANTON
N807 1958	2657140	WRAAF	wef 30/9/58	BRATLEY
YB 1996	208695	RAuxAF		BRATTON
YB 2000	208695	RAuxAF	1st Clasp	BRATTON
N1150 1944	90114	AAF		BRAUN
N519 1947	167590	RAFVR		BRAY
N137 1954	75766	RAFVR		BRAY
N94 1950	859642	RAuxAF		BRAY
N82 1946	78261	RAFVR		BRAYLEY
N1193 1949	81542	RAFVR	DFC	BRAYTON
N281 1944	80824	RAFVR		BREAKSPEAR
N573 1945	173515	RAFVR	AFC	BREARLY
N289 1947	347	WAAF	MBE	BREEDON
N478 1947	811096	AAF		BREINGAN
N402 1951	880531	WAAF		BRENTNALL
N621 1951	852851	RAuxAF		BRENTNALL
N672 1954	184322	RAuxAF		BRETT
N671 1945	741919	RAFVR		BRETT
N527 1948	73097	RAFVR		BREW
N395 1948	162679	RAFVR		BREWER
N28 1959	2692514	RAuxAF	wef 4/10/58	BREWER
N796 1953	869317	RAuxAF		BREWER
N513 1956	750660	RAFVR		BREWER
N763 1944	86484	RAFVR		BRICE
N694 1948	818079	RAuxAF		BRICE
N134 1952	2689018	RAuxAF		BRICE
N43 1956	90699	RAuxAF	Sir	BRICKWOOD
N573 1948	755163	RAFVR		BRIDEN

Surname	Initials			Rank	Status	Date
BRIDGE	A	M		SGT		06/12/1995
BRIDGE	H	M	P	CPL		20/05/1948
BRIDGE	L	H	E	FL		21/03/1946
BRIDGE	P	G		SGT		26/11/1953
BRIDGE	S	L		LAC		15/09/1949
BRIDGER	J	C		FL		08/11/1945
BRIERLY	F			SQD LDR	Acting	27/07/1944
BRIGGS	A	G		FO		26/08/1948
BRIGGS	J	J		SQD LDR		05/01/1950
BRIGGS	L	R		GRP CAP	Acting	20/04/1944
BRIGGS	P	I		SQD LDR	Acting	26/04/1956
BRIGGS	P	S	E	FL		28/02/1946
BRIGHT	A	T	F	FLT SGT		10/06/1948
BRIGHT	F	G		SGT	Temporary	29/04/1954
BRIGHT	H			CPL		07/07/1949
BRIGHTMORE	A	G	P	WING COMM		12/07/1945
BRIGINSHAW	W			WO		15/07/1948
BRIGSTOCKE	L	M		SQD OFF	Acting	17/04/1947
BRILL	C	C		FL		25/01/1951
BRIND	E	J		LAC		30/09/1948
BRINKHURST	W	H		WO		15/09/1949
BRINTON	R	H		FO		24/12/1942
BRISCALL	A	G		FLT SGT		30/09/1948
BRISLEY	G	S		SGT		25/03/1959
BRISTOW	A	N		FL		13/12/1945
BRISTOW	H	M		FLT OFF		17/04/1947
BRISTOW	P	M		FL		03/12/1942
BRISTOW	R	G	W	FL		12/04/1945
BRITCHFORD	M	L	W	FL		22/10/1958
BRITTAIN	H	G		SQD LDR	Acting	12/07/1945
BRITTON	E			SGT		04/11/1948
BROAD	C	T		FL	Acting	30/11/1944
BROADBENT	C	J		SGT		08/05/1947
BROADBENT	L			FLT SGT	Acting	19/01/1956
BROADBENT	R			SQD LDR	Acting	08/05/1947
BROADHEAD	S			SQD LDR		16/04/1958
BROADRIBB	P	D		SAC		23/03/1997
BROCK	A	T		SQD LDR	Acting	10/12/1942
BROCK	A	T		FL		30/09/1954
BRODIE	E	J	C	FLT OFF		13/07/1950
BRODIE	J			PO		01/12/1955
BROMAGE	C	C		CPL		10/06/1948
BRONKS	N			SGT		18/03/1948
BROOK	B	L		CPL		09/02/1951
BROOK	L	T		FO		04/02/1943
BROOK	S	A		CPL		08/05/1947
BROOKE	S			CPL		29/10/1990
BROOKES	H	H		SGT		17/01/1952
BROOKES	H			LAC		29/11/1951
BROOKES	J	R		CPL		18/03/1948
BROOKIN	W	J		SGT		10/08/1950
BROOKING	M			FL		06/01/1955
BROOKS	C			FL		28/11/1946
BROOKS	E	H		SGT		23/02/1950
BROOKS	F	W		CPL		27/02/1963
BROOKS	S	K		SAC		31/07/1998
BROOME	G	E		LAC		05/02/1948
BROOMFIELD	D	J	P	FL		21/06/1945
BROUGH	S			LAC		05/01/1950
BROUGH	T			SGT		19/07/1951

Source	Service No	Force	Notes	Surname
YB 2000	n/a	RAuxAF		BRIDGE
N395 1948	869498	RAuxAF		BRIDGE
N263 1946	64294	RAFVR		BRIDGE
N952 1953	2603502	RAFVR		BRIDGE
N953 1949	846536	RAuxAF		BRIDGE
N1232 1945	103516	RAFVR	DFC	BRIDGER
N763 1944	120738	RAFVR	DFM	BRIERLY
N694 1948	127559	RAFVR		BRIGGS
N1 1950	90513	RAuxAF		BRIGGS
N363 1944	90031	AAF		BRIGGS
N318 1956	152596	RAuxAF	DFC	BRIGGS
N197 1946	79751	RAFVR		BRIGGS
N456 1948	812321	RAuxAF		BRIGHT
N329 1954	756762	RAFVR		BRIGHT
N673 1949	860338	RAuxAF		BRIGHT
N759 1945	90073	AAF		BRIGHTMORE
N573 1948	755842	RAFVR		BRIGINSHAW
N289 1947	1487	WAAF		BRIGSTOCKE
N93 1951	73099	RAFVR		BRILL
N805 1948	818014	RAuxAF		BRIND
N953 1949	799732	RAFVR		BRINKHURST
N1758 1942	85935	RAFVR		BRINTON
N805 1948	812035	RAuxAF		BRISCALL
N196 1959	2606533	RAFVR	wef 16/12/58	BRISLEY
N1355 1945	101502	RAFVR	DFC	BRISTOW
N289 1947	2433	WAAF		BRISTOW
N1638 1942	81673	RAFVR		BRISTOW
N381 1945	126138	RAFVR		BRISTOW
N716 1958	48749	RAFVR	wef 21/10/52	BRITCHFORD
N759 1945	111488	RAFVR	DFC,AFC	BRITTAIN
N934 1948	46684	RAFVR		BRITTON
N1225 1944	88425	RAFVR		BROAD
N358 1947	811080	AAF		BROADBENT
N43 1956	2676605	RAuxAF		BROADBENT
N358 1947	90707	AAF		BROADBENT
N258 1958	78097	RAFVR	wef 21/3/44	BROADHEAD
YB 1998	n/a	RAuxAF		BROADRIBB
N1680 1942	70085	RAFVR		BROCK
N799 1954	70085	RAFVR	DFC; 1st Clasp	BROCK
N675 1950	586	WAAF		BRODIE
N895 1955	2601289	RAFVR		BRODIE
N456 1948	752675	RAuxAF		BROMAGE
N232 1948	869394	RAuxAF		BRONKS
N145 1951	858910	RAuxAF		BROOK
N131 1943	89979	RAFVR		BROOK
N358 1947	749688	RAFVR		BROOK
YB 2000	n/a	RAuxAF		BROOKE
N35 1952	2688501	RAuxAF		BROOKES
N1219 1951	849023	RAuxAF		BROOKES
N232 1948	868600	RAuxAF		BROOKES
N780 1950	759339	RAFVR		BROOKIN
N1 1955	90881	RAuxAF		BROOKING
N1004 1946	159859	RAFVR		BROOKS
N179 1950	801436	RAuxAF		BROOKS
N155 1963	2651089	RAuxAF	wef 19/11/62	BROOKS
YB 1999	n/a	RAuxAF		BROOKS
N87 1948	856094	RAuxAF		BROOME
N671 1945	126546	RAFVR	DFM	BROOMFIELD
N1 1950	855186	RAuxAF		BROUGH
N724 1951	867801	RAuxAF		BROUGH

Surname	Initials			Rank	Status	Date
BROUGHTON	C	W		LAC		16/03/1950
BROUGHTON	L			CPL		01/07/1948
BROUSSON	R	H	C	GRP CAP		31/10/1956
BROWN	A	E		FL		04/10/1945
BROWN	A	E		FL		28/02/1946
BROWN	A	E		SQD LDR	Acting	01/10/1953
BROWN	A			FLT SGT		30/03/1943
BROWN	A			WING COMM	Acting	28/02/1946
BROWN	A			CPL		01/07/1948
BROWN	B	E		FL		15/09/1949
BROWN	C	B		SQD LDR	Acting	24/01/1946
BROWN	C	D		FL		01/05/1963
BROWN	C	J		SQD LDR	Acting	10/06/1948
BROWN	C	L		CPL		30/09/1948
BROWN	D	B		FL		22/10/1958
BROWN	D	H		FO		08/05/1947
BROWN	D	R		FLT SGT		21/03/1946
BROWN	D	S		FL		28/02/1946
BROWN	D	W		FL		05/10/1944
BROWN	D			CPL		23/02/1950
BROWN	D			WING COMM	Acting	16/04/1958
BROWN	D			CPL		03/09/1989
BROWN	D			CPL		03/09/1990
BROWN	E	A		WO		26/08/1948
BROWN	E	C		SQD LDR		04/11/1943
BROWN	E	R		FLT SGT		23/08/1951
BROWN	E	W		LAC		24/11/1949
BROWN	E			FL		10/06/1948
BROWN	E			LAC		23/02/1950
BROWN	G	E		FO		10/05/1945
BROWN	G	W		CPL		05/01/1950
BROWN	H	H		FL		15/07/1948
BROWN	H	J	R	FO		10/05/1945
BROWN	H	O		SGT		25/01/1951
BROWN	J	B		CPL		17/01/1952
BROWN	J	G	L	WING COMM		26/01/1950
BROWN	J	H		FL		25/06/1953
BROWN	J	R		REV		17/06/1943
BROWN	J	S		FLT SGT		15/07/1948
BROWN	J	S		SGT		16/03/1950
BROWN	J	W		SGT		03/06/1943
BROWN	J			SQD LDR	Acting	08/11/1945
BROWN	J			FL		26/11/1958
BROWN	K	L		FLT SGT		28/02/1946
BROWN	K			FL		22/09/1955
BROWN	P	G		FL		19/10/1960
BROWN	P	J		CPL		05/05/1997
BROWN	P			SAC		16/04/1958
BROWN	R	A		FL		11/01/1945
BROWN	R	C		SGT		07/07/1949
BROWN	R	E		FL		20/09/1945
BROWN	R	E		WO		28/02/1946
BROWN	R	F		CPL		07/07/1949
BROWN	R	G		FO		14/10/1943
BROWN	R	I		FO		27/03/1947
BROWN	R			SQD LDR	Acting	27/01/1944
BROWN	S	W		LAC		28/07/1955
BROWN	S			WO		30/09/1948
BROWN	T	E	W	WING COMM		28/10/1943
BROWN	T			FL		30/09/1948

Source	Service No	Force	Notes	Surname
N257 1950	750249	RAFVR		BROUGHTON
N527 1948	853558	RAuxAF		BROUGHTON
N761 1956	n/a	RAF	OBE,MA,AFRAeS	BROUSSON
N1089 1945	118424	RAFVR	DFC	BROWN
N197 1946	122389	RAFVR		BROWN
N796 1953	81550	RAFVR		BROWN
N1014 1943	808013	AAF		BROWN
N197 1946	80843	RAFVR	DFC	BROWN
N527 1948	852160	RAuxAF		BROWN
N953 1949	185329	RAFVR	DFC	BROWN
N82 1946	109525	RAFVR		BROWN
N333 1963	199616	RAFVR	wef 20/3/55	BROWN
N456 1948	68295	RAFVR		BROWN
N805 1948	846575	RAuxAF		BROWN
N716 1958	174776	RAFVR	DFC; wef 18/7/52	BROWN
N358 1947	137810	RAFVR		BROWN
N263 1946	807109	AAF		BROWN
N197 1946	77917	RAFVR		BROWN
N1007 1944	86662	RAFVR		BROWN
N179 1950	816235	RAuxAF		BROWN
N258 1958	130225	RAuxAF	DSO; wef 1/7/54	BROWN
YB 1996	n/a	RAuxAF		BROWN
YB 2000	n/a	RAuxAF	1st Clasp	BROWN
N694 1948	751798	RAFVR		BROWN
N1144 1943	70088	RAFVR		BROWN
N866 1951	808408	RAuxAF		BROWN
N1193 1949	800644	RAuxAF		BROWN
N456 1948	142378	RAFVR		BROWN
N179 1950	870020	RAuxAF		BROWN
N485 1945	148173	RAFVR	AFC	BROWN
N1 1950	844330	RAuxAF		BROWSN
N573 1948	121477	RAFVR	DFM	BROWN
N485 1945	139146	RAFVR		BROWN
N93 1951	803491	RAuxAF		BROWN
N35 1952	2688500	RAuxAF		BROWN
N94 1950	72572	RAFVR	MB,ChB	BROWN
N527 1953	86262	RAFVR		BROWN
N653 1943	90202	AAF	MA	BROWN
N573 1948	846231	RAuxAF		BROWN
N257 1950	803604	RAuxAF		BROWN
N609 1943	800416	AAF		BROWN
N1232 1945	84952	RAFVR	DFC	BROWN
N807 1958	154697	RAFVR	wef 22/10/54	BROWN
N197 1946	807110	AAF		BROWN
N708 1955	163363	RAuxAF		BROWN
N775 1960	199711	RAFVR	wef 29/3/60	BROWN
YB 2002	n/a	RAuxAF		BROWN
N258 1958	2686520	RAuxAF	wef 16/5/54	BROWN
N14 1945	114156	RAFVR		BROWN
N673 1949	812314	RAuxAF		BROWN
N1036 1945	133875	RAFVR		BROWN
N197 1946	748253	RAFVR		BROWN
N673 1949	860362	RAuxAF		BROWN
N1065 1943	128366	RAFVR		BROWN
N250 1947	130441	RAFVR		BROWN
N59 1944	78250	RAFVR		BROWN
N565 1955	869393	RAuxAF		BROWN
N805 1948	745380	RAFVR		BROWN
N1118 1943	70091	RAFVR		BROWN
N805 1948	82226	RAFVR		BROWN

Surname	Initials			Rank	Status	Date
BROWN	W	A		WING COMM	Acting	09/11/1944
BROWN	W	E	H	FL		16/08/1961
BROWN	W	F		LAC		24/11/1949
BROWN	W	G		FO		06/03/1947
BROWN	W	McK		CPL		01/07/1948
BROWN	W	S		FL		24/07/1952
BROWNE	A			FL		06/03/1947
BROWNE	E	H		FL		25/03/1959
BROWNE	J	J		FL		21/06/1945
BROWNE	J	J		FL		25/06/1953
BROWNE	J	P	J	FL		20/11/1957
BROWNE	T	D		FL		21/10/1959
BROWNELL	C	R		FL	Acting	05/06/1947
BROWNILL	R	W		CPL		14/10/1948
BROWNING	D	W	P	WO		24/12/1942
BROWNING	K	S		FO		24/01/1946
BROWNING	N	M		SQD LDR		04/02/1943
BROWNING	R	A		SQD LDR	Acting	08/05/1947
BROWNLIE	R			WO		01/05/1993
BROWNRIGG	D	W	P	SQD LDR	Acting	04/09/1957
BROWNSILL	A	G		FL		25/01/1945
BROXUP	H	J		FL		10/08/1944
BRUCE	R	D		WING COMM	Acting	20/05/1948
BRUCE	W	N		SAC		19/10/1960
BRUMBLEY	C	S		FLT SGT		01/07/1948
BRUMBY	D			SGT		23/09/1993
BRUMWELL	R	W		FL		24/01/1946
BRUNI	R	L		CPL		31/10/1956
BRUTON	J	E		FO		04/05/1944
BRUTON	W			SGT		18/03/1948
BRUTON	W			FLT SGT		29/04/1954
BRUZAUD	A	W	P	SQD OFF		20/07/1960
BRYAN	I	M		SQD OFF	Acting	08/05/1947
BRYAN	T	B		FLT SGT		02/03/1988
BRYAN	T	B		FLT SGT		02/03/1998
BRYAN	W	H		WO		03/10/1946
BRYANT	H	E		FO		17/01/1952
BRYANT	J	E		LAC		30/09/1948
BRYANT	J	H		FL		19/08/1954
BRYANT	J	K		CPL		30/03/1990
BRYANT	J	K		SGT		04/03/1990
BRYCESON	K	E		FL		08/11/1945
BRYDEN-BROWN	D			FL		16/10/1952
BUCHANAN	A			SAC		05/07/1995
BUCHANAN	D	C		SQD LDR		05/01/1950
BUCKHAM	A			FLT SGT		26/08/1948
BUCKINGHAM	A	H		FO		01/10/1953
BUCKINGHAM	M	P	E	WO		05/06/1947
BUCKINGHAM	P	T		FO		02/09/1943
BUCKLAND	L	A		FL		18/03/1948
BUCKLER	E	W		SQD LDR	Acting	16/10/1947
BUCKLEY	F			CPL	Temporary	18/02/1954
BUCKLEY	R	T		WO		18/04/1946
BUCKLEY	W	H		LAC		06/09/1951
BUDD	B	A	T	FL	Acting	13/01/1949
BUDD	G	O		WING COMM		03/12/1942
BUDD	P	W		FL		18/04/1946
BUDD	R	A		WING COMM		10/12/1942
BUDD	R	A		SGT		05/02/1948

Source	Service No	Force	Notes	Surname
N1150 1944	90309	AAF		BROWN
N624 1961	2607635	RAFVR	wef 4/7/61	BROWN
N1193 1949	869264	RAuxAF		BROWN
N187 1947	170704	RAFVR		BROWN
N527 1948	844881	RAuxAF		BROWN
N569 1952	78502	RAFVR		BROWN
N187 1947	126811	RAFVR		BROWNE
N196 1959	144420	RAFVR	wef 19/6/54	BROWNE
N671 1945	146142	RAFVR		BROWNE
N527 1953	146142	RAuxAF	1st Clasp	BROWNE
N835 1957	133663	RAFVR	DFC; wef 26/8/53	BROWNE
N708 1959	2600929	RAF	wef 6/9/55	BROWNE
N460 1947	187386	RAFVR		BROWNELL
N856 1948	868723	RAuxAF		BROWNILL
N1758 1942	801171	AAF		BROWNING
N82 1946	186329	RAFVR		BROWNING
N131 1943	70092	RAFVR		BROWNING
N358 1947	69352	RAFVR	MBE	BROWNING
YB 1996	n/a	RAuxAF		BROWNLIE
N611 1957	203090	RAFVR(T)	1st Clasp; wef 19/9/44; no record of first award	BROWNRIGG
N70 1945	88615	RAFVR		BROWNSILL
N812 1944	80557	RAFVR		BROXUP
N395 1948	70094	RAFVR		BRUCE
N775 1960	2684154	RAuxAF	wef 4/5/60	BRUCE
N527 1948	771116	RAFVR		BRUMBLEY
YB 1996	n/a	RAuxAF		BRUMBY
N82 1946	127140	RAFVR	DFC	BRUMWELL
N761 1956	770720	RAFVR		BRUNI
N425 1944	133996	RAFVR		BRUTON
N232 1948	816081	RAuxAF		BRUTON
N329 1954	2681009	RAuxAF	1st Clasp	BRUTON
N531 1960	415	WAAF	OBE; wef 16/4/44	BRUZAUD
N358 1947	1488	WAAF		BRYAN
YB 1996	n/a	RAuxAF		BRYAN
YB 1999	n/a	RAuxAF	1st Clasp	BRYAN
N837 1946	749513	RAFVR		BRYAN
N35 1952	136467	RAFVR		BRYANT
N805 1948	861182	RAuxAF		BRYANT
N672 1954	143702	RAFVR	DFC	BRYANT
YB 1996	n/a	RAuxAF		BRYANT
YB 1999	n/a	RAuxAF		BRYANT
N1232 1945	122991	RAFVR	AFC	BRYCESON
N798 1952	111432	RAuxAF		BRYDEN-BROWN
YB 1997	n/a	RAuxAF		BUCHANAN
N1 1950	90948	RAuxAF	MA,MB,ChB	BUCHANAN
N694 1948	800441	RAuxAF		BUCKHAM
N796 1953	159179	RAFVR		BUCKINGHAM
N460 1947	813194	AAF		BUCKINGHAM
N918 1943	116143	RAFVR		BUCKINGHAM
N232 1948	162686	RAFVR		BUCKLAND
N846 1947	106780	RAFVR		BUCKLER
N137 1954	799932	RAFVR		BUCKLEY
N358 1946	751302	RAFVR		BUCKLEY
N913 1951	809032	RAuxAF		BUCKLEY
N30 1949	106243	RAFVR		BUDD
N1638 1942	90209	AAF	DFC	BUDD
N358 1946	157007	RAFVR		BUDD
N1680 1942	90208	AAF		BUDD
N87 1948	812123	RAuxAF		BUDD

Surname	Initials			Rank	Status	Date
BUDGE	R	K		WING COMM	Acting	20/05/1948
BUFTON	G	B		SAC		02/12/1992
BUGGS	A	G		CPL		30/09/1948
BULL	D	H		FO		10/02/1949
BULL	G			WO		27/01/1944
BULLEN	F	O		FL		05/09/1946
BULLEN	G	F		FL		21/06/2005
BULLEN	J			LAC		22/02/1961
BULLEYMENT	J	C		SQD LDR	Acting	16/03/1950
BULLEYMENT	J	C		SQD LDR		08/03/1956
BULLOCK	J	D		CPL		26/01/1950
BULLOCK	J	S		CPL		20/08/1958
BULMAN	C	T		FLT SGT		22/04/1943
BULMAN	G	C		WING COMM	Acting	25/01/1951
BULMER	R			LAC		16/03/1950
BUNCE-PHILLIPS	E	G		WING COMM		05/06/1947
BUNCH	D	C		FL		30/11/1944
BUNGEY	D	K		FL		31/10/1956
BUNKELL	D	W		SGT		26/02/1958
BUNKELL	G	W		FL		01/05/1994
BUNKER	C	A		CPL		04/11/1948
BUNN	W	A		FL		24/01/1946
BUNTING	L	E	R	FO		10/12/1942
BUNTING	L	E	R	FL		10/04/1952
BUNTING	M	C		SQD LDR	Acting	13/12/1945
BUNTON	T	H		SGT		07/07/1949
BURBIDGE	C	H		FL		05/01/1950
BURDEKIN	A	G		FO		01/10/1953
BURDETT	G	A	T	FL	Acting	26/01/1950
BURDON	G	R		FO		30/11/1944
BURGE	A	T		FO		01/07/1948
BURGE	E			FO		10/04/1952
BURGE	E			FL		26/04/1961
BURGESS	C	G		SQD LDR		05/06/1947
BURGESS	F	T		CPL		17/04/1947
BURGESS	G			CPL		03/07/1952
BURGESS	J	A		FL		11/07/1946
BURGESS	J	H	B	FL		09/08/1945
BURGESS	R	F		FO		23/02/1950
BURGESS	T	E		WING COMM		17/04/1947
BURKE	D			CPL		30/09/1954
BURKETT	P	E		CPL		01/10/1953
BURKITT	R	W		FL	Acting	10/12/1942
BURLAND	W			CPL		13/01/1949
BURLEIGH	V			FLT OFF		17/04/1947
BURLEY	D			CPL		14/09/1950
BURLEY	S	F		FL		19/08/1954
BURMAN	J	W		SQD LDR	Acting	12/08/1943
BURMAN	J	W		FL		25/07/1962
BURMAN	W	A	W	FL		04/10/1945
BURNARD	R			SAC		22/03/1997
BURNETT	G			CPL		27/04/1950
BURNETT	G			FL	Temporary	01/12/1955
BURNETT	I			WO		07/08/1947
BURNEY	H	F		PO		17/03/1949
BURNHAM	C	J		CPL		26/08/1948
BURNHAM	K	N		FO		07/01/1943
BURNHAM	M	E		FL		06/05/1943
BURNHAM	R			SAC		22/02/1961

Source	Service No	Force	Notes	Surname
N395 1948	73111	RAFVR	OBE	BUDGE
YB 1996	n/a	RAuxAF		BUFTON
N805 1948	812333	RAuxAF		BUGGS
N123 1949	150371	RAFVR	Deceased	BULL
N59 1944	748719	RAFVR		BULL
N751 1946	130206	RAFVR		BULLEN
LG 2005	127906	RAFVR		BULLEN
N153 1961	853657	AAF	wef 27/5/44	BULLEN
N257 1950	104919	RAFVR		BULLEYMENT
N187 1956	104919	RAuxAF	1st Clasp	BULLEYMENT
N94 1950	815022	RAuxAF		BULLOCK
N559 1958	2694017	RAuxAF	wef 6/5/58	BULLOCK
N475 1943	801271	AAF		BULMAN
N93 1951	72255	RAFVR		BULMAN
N257 1950	808109	RAuxAF		BULMER
N460 1947	90434	AAF		BUNCE-PHILLIPS
N1225 1944	115674	RAFVR	DFC	BUNCH
N761 1956	164165	RAuxAF		BUNGEY
N149 1958	2694010	RAuxAF	wef 22/1/58	BUNKELL
YB 1997	2631309	RAuxAF		BUNKELL
N934 1948	843988	RAuxAF		BUNKER
N82 1946	129114	RAFVR	DFM	BUNN
N1680 1942	63780	RAFVR		BUNTING
N278 1952	63780	RAFVR	1st Clasp	BUNTING
N1355 1945	117862	RAFVR	MBE	BUNTING
N673 1949	800465	RAuxAF		BUNTON
N1 1950	103001	RAFVR	DFC	BURBIDGE
N796 1953	143405	RAFVR		BURDEKIN
N94 1950	158447	RAFVR		BURDETT
N1225 1944	137589	RAFVR	AFM	BURDON
N527 1948	183351	RAFVR		BURGE
N278 1952	190115	RAFVR		BURGE
N339 1961	190115	RAFVR	1st Clasp; wef 13/2/61	BURGE
N460 1947	72092	RAFVR		BURGESS
N289 1947	770585	RAFVR		BURGESS
N513 1952	799890	RAFVR		BURGESS
N594 1946	157279	RAFVR	DFC, DFM	BURGESS
N874 1945	67601	RAFVR		BURGESS
N179 1950	158157	RAFVR		BURGESS
N289 1947	72124	RAFVR		BURGESS
N799 1954	2688621	RAuxAF		BURKE
N796 1953	843173	RAuxAF		BURKETT
N1680 1942	64464	RAFVR		BURKITT
N30 1949	870140	RAuxAF		BURLAND
N289 1947	677	WAAF		BURLEIGH
N926 1950	872252	RAuxAF		BURLEY
N672 1954	108656	RAuxAF		BURLEY
N850 1943	77459	RAFVR	AFC	BURMAN
N549 1962	77459	RAFVR	AFC; 1st Clasp; wef 15/5/55	BURMAN
N1089 1945	81659	RAFVR		BURMAN
YB 1998	n/a	RAuxAF		BURNARD
N406 1950	750372	RAFVR		BURNETT
N895 1955	110190	RAFVR		BURNETT
N662 1947	882051	WAAF		BURNETT
N231 1949	145808	RAFVR	Deceased	BURNEY
N694 1948	861063	RAuxAF		BURNHAM
N3 1943	61466	RAFVR		BURNHAM
N526 1943	72982	RAFVR		BURNHAM
N153 1961	2687040	RAFVR	wef 23/10/58	BURNHAM

Surname	Initials			Rank	Status	Date
BURNINGHAM	J			FL		31/05/1945
BURNS	D	A		SGT	Acting	12/06/1947
BURNS	D	A		FL	Acting	01/05/1957
BURNS	W	A		FLT SGT		29/04/1954
BURR	C	F		CPL		06/09/1951
BURR	H	W		FL		20/09/1945
BURRELL	R			SQD LDR		20/09/1945
BURRIDGE	T			LAC		15/09/1949
BURROWES	J			SQD LDR		20/09/1961
BURROWS	A	E	J	FL		11/07/1946
BURROWS	J	C		SGT		06/01/1955
BURROWS	L			WO	Acting	23/02/1950
BURRY	D	W		FL		22/10/1958
BURSTON	H	J		FL		21/03/1946
BURTON	A	H		SQD LDR	Acting	06/03/1947
BURTON	C			SGT		01/07/1948
BURTON	D	Y		SEC OFF		07/08/1947
BURTON	J			FL		04/10/1945
BURTON	K	D		LAC		08/03/1951
BURTON	R	G		FL		16/10/1952
BURTON	R	J		FO		24/01/1946
BURTON	W	T		LAC		04/11/1948
BURVILL	K	I		FO		16/01/1957
BURY	N			WING COMM		06/03/1947
BUSBRIDGE	D	C		FL		30/10/1957
BUSBY	T	S	C	FL		28/11/1946
BUSH	C	E		LAC		13/01/1949
BUSH	W	B		FL		26/01/1950
BUSH	W	F		WO	Acting	15/07/1948
BUSHBY	J	R		FL		18/03/1948
BUSHE-CARYESFORD	B	J		WING COMM	Acting	06/03/1947
BUSKELL	R	O		WING COMM	Acting	08/11/1945
BUTCHER	M	C		FL		13/10/1955
BUTCHER	R	H	F	FL		26/01/1950
BUTLAND	D			FLT OFF	Acting	08/05/1947
BUTLER	E	J		SGT		15/11/1951
BUTLER	F			FO		25/06/1953
BUTLER	I	B		FL		15/04/1943
BUTLER	L	B		WO		30/03/1943
BUTLER	R	E		CPL		12/06/1947
BUTLER	S	M		CPL		16/10/1947
BUTT	B			LAC		18/11/1948
BUTT	K	J		FL		30/08/1945
BUTTERELL	J	M		FLT OFF	Acting	08/05/1947
BUTTERFIELD	W	A		WO		08/05/1947
BUTTERICK	A	F		FO		28/11/1946
BUTTERIES	J			FL		18/03/1948
BUTTERWORTH	H			AC1		01/05/1952
BUTTOLPH	N	R		SAC		25/09/1990
BUTTON	C	H		FL		15/07/1948
BUTTON	P	G		CPL		18/11/1948
BUTT-REED	J	G		SQD LDR		30/03/1943
BUXTON	J	T		LAC		21/10/1959
BUXTON	R	S		FL		19/06/1998
BYATT	K	J		FL		22/10/1958
BYRD	S	A		WING COMM	Acting	27/03/1947
BYRNE	E	A		CPL		30/09/1948
BYRNE	F			FL		07/08/1947
BYRNE	J	D		FL		03/07/1993
BYRNE	O	T		LAC		01/05/1957

Source	Service No	Force	Notes	Surname
N573 1945	87405	RAFVR	AFC	BURNINGHAM
N478 1947	811140	AAF		BURNS
N277 1957	2604508	RAFVR	1st Clasp; wef 21/12/54	BURNS
N329 1954	2681064	RAuxAF		BURNS
N913 1951	747241	RAFVR		BURR
N1036 1945	120978	RAFVR	AFC	BURR
N1036 1945	90603	AAF	DFC	BURRELL
N953 1949	865654	RAuxAF		BURRIDGE
N720 1961	184687	RAuxAF	wef 19/7/61	BURROWES
N594 1946	152485	RAFVR		BURROWS
N1 1955	744297	RAFVR		BURROWS
N179 1950	855245	RAuxAF		BURROWS
N716 1958	189555	RAFVR	wef 20/9/52	BURRY
N263 1946	149927	RAFVR		BURSTON
N187 1947	84007	RAFVR	DSO, DFC	BURTON
N527 1948	743431	RAFVR		BURTON
N662 1947	8347	WAAF		BURTON
N1089 1945	100620	RAFVR		BURTON
N244 1951	819187	RAuxAF		BURTON
N798 1952	91260	RAuxAF		BURTON
N82 1946	185882	RAFVR		BURTON
N934 1948	812334	RAuxAF		BURTON
N24 1957	2682054	RAuxAF		BURVILL
N187 1947	73115	RAFVR		BURY
N775 1957	120169	RAFVR	wef 17/7/44	BUSBRIDGE
N1004 1946	122974	RAFVR		BUSBY
N30 1949	846301	RAuxAF		BUSH
N94 1950	90708	RAuxAF		BUSH
N573 1948	840140	RAuxAF		BUSH
N232 1948	141760	RAFVR		BUSHBY
N187 1947	72169	RAFVR		BUSHE-CARYESFORD
N1232 1945	77460	RAFVR	DFC	BUSKELL
N762 1955	193374	RAuxAF		BUTCHER
N94 1950	61068	RAFVR		BUTCHER
N358 1947	3085	WAAF		BUTLAND
N1185 1951	864113	RAuxAF		BUTLER
N527 1953	109163	RAFVR		BUTLER
N454 1943	84700	RAFVR	AFC	BUTLER
N1014 1943	808102	AAF		BUTLER
N478 1947	811136	AAF	BEM	BUTLER
N846 1947	747068	RAFVR		BUTLER
N984 1948	845351	RAuxAF		BUTT
N961 1945	130797	RAFVR		BUTT
N358 1947	2705	WAAF		BUTTERELL
N358 1947	811054	AAF		BUTTERFIELD
N1004 1946	202121	RAFVR		BUTTERICK
N232 1948	168514	RAFVR		BUTTERIES
N336 1952	858148	RAuxAF		BUTTERWORTH
YB 1996	n/a	RAuxAF		BUTTOLPH
N573 1948	139732	RAFVR		BUTTON
N984 1948	815185	RAuxAF		BUTTON
N1014 1943	70105	RAFVR	AFC	BUTT-REED
N708 1959	752084	RAFVR	wef 16/7/44	BUXTON
YB 1999	8032125	RAuxAF		BUXTON
N716 1958	201097	RAFVR	wef 30/6/52	BYATT
N250 1947	72211	RAFVR		BYRD
N805 1948	813211	RAuxAF		BYRNE
N662 1947	127565	RAFVR		BYRNE
YB 1996	2632617	RAuxAF		BYRNE
N277 1957	2689504	RAuxAF	wef 23/11/56	BYRNE

Surname	Initials				Rank	Status	Date
BYRON	C				WING COMM	Acting	12/08/1943
BYRON	D	H			FL		28/02/1946
CACKETT	R	T			WING COMM		06/03/1947
CADDE	E	H			LAC		18/11/1948
CADDEN	J	H	G	C	CPL		16/01/1957
CADMAN	A	J			LAC		14/10/1948
CADMAN	F	S			SQD LDR	Acting	28/02/1946
CADWALLADER	I	F			FO		01/03/1945
CAEN	L	G			FL		11/07/1946
CAFFREY	M	L			FLT SGT		28/03/1997
CAGIENARD	C	E			FL	Acting	05/06/1947
CAINE	D				CPL		12/02/1953
CAIRNS	J	A			LAC		08/07/1959
CAIRNS	J	P	W		FL		05/01/1950
CAIRNS	M	J			SQD LDR		01/01/2002
CAIRNS	R	M	B		FO		05/10/1944
CALDER	A	S			FL		12/06/1947
CALDERWOOD	T	M			FL		09/05/1946
CALDWELL	G	E			LAC		15/09/1949
CALE	R	H			SQD LDR	Acting	26/01/1950
CALLANDER	J	J			FL		01/05/1957
CALLINGHAM	J				FL		23/05/1946
CALLON	E	C			FL		22/10/1958
CALLUM	P	S			LAC		13/11/1947
CALVER	A	E			SGT		28/11/1946
CALVER	F	W			FLT SGT	Acting	15/07/1948
CALVER	N	A			SGT		11/07/1946
CALWELL	G				FL		07/12/1950
CAMERON	D	T			CPL	Temporary	30/10/1957
CAMERON	D				CPL		18/10/1951
CAMERON	N				WO		04/05/1944
CAMERON	N				SQD LDR		06/03/1947
CAMERON	W				WING COMM	Acting	17/02/1944
CAMERON	W				CPL		17/04/1947
CAMISH	B	G			SGT		01/12/1955
CAMPBELL	C	F			FL		16/01/1957
CAMPBELL	C	V	T		FL		03/10/1962
CAMPBELL	C				SAC		05/02/1993
CAMPBELL	C				SAC		09/10/1998
CAMPBELL	D	C	O		WO		09/08/1945
CAMPBELL	D				CPL		20/05/1948
CAMPBELL	D				CPL		16/04/1958
CAMPBELL	D				CPL		27/03/1963
CAMPBELL	G				FL		20/03/1957
CAMPBELL	I	G			FO		06/05/1943
CAMPBELL	I	M			GRP OFF		22/02/1951
CAMPBELL	I				CPL		20/08/1958
CAMPBELL	N	G			SGT		15/11/1961
CAMPBELL	P	C			FL		08/11/1945
CAMPBELL	P	C			SQD LDR		25/01/1951
CAMPBELL	S	A			SQD LDR		16/10/1947
CAMPBELL	S	D			CPL		09/05/1997
CAMPBELL	T	W			WING COMM		14/10/1943
CAMPBELL	W	S			FL		06/01/1955
CAMPBELL-MacMILLAN-COLLINS	B	W			SGT	Acting	19/08/1959
CAMPBELL-ORDE	I	R			WING COMM		03/12/1942
CAMPBELL-ORDE	I	R			WING COMM		16/04/1958
CAMPION	H	W			SGT		08/07/1959
CAMPLIN	G	E			FL		28/02/1946
CANBY	W	T	E	C	WO	Acting	01/07/1948

Source	Service No	Force	Notes	Surname
N850 1943	70108	RAFVR		BYRON
N197 1946	112685	RAFVR		BYRON
N187 1947	72256	RAFVR		CACKETT
N984 1948	841529	RAuxAF		CADDE
N24 1957	2650777	RAuxAF		CADDEN
N856 1948	856546	RAuxAF		CADMAN
N197 1946	104425	RAFVR		CADMAN
N225 1945	142524	RAFVR		CADWALLADER
N594 1946	100043	RAFVR		CAEN
YB 1998	n/a	RAuxAF		CAFFREY
N460 1947	110981	RAFVR		CAGIENARD
N115 1953	750817	RAFVR		CAINE
N454 1959	2684608	RAuxAF	wef 24/2/59	CAIRNS
N1 1950	124540	RAFVR	DFC	CAIRNS
YB 2003	211142	RAuxAF		CAIRNS
N1007 1944	133623	RAFVR		CAIRNS
N478 1947	155568	RAFVR		CALDER
N415 1946	106757	RAFVR		CALDERWOOD
N953 1949	865599	RAuxAF		CALDWELL
N94 1950	90949	RAuxAF	MC	CALE
N277 1957	142239	RAuxAF	wef 1/9/54	CALLANDER
N468 1946	143216	RAFVR		CALLINGHAM
N716 1958	176027	RAFVR	wef 6/1/54	CALLON
N933 1947	868637	AAF		CALLUM
N1004 1946	805085	AAF		CALVER
N573 1948	805400	RAuxAF		CALVER
N594 1946	805468	AAF		CALVER
N1210 1950	157946	RAFVR		CALWELL
N775 1957	799976	RAFVR	wef 29/5/44	CAMERON
N1077 1951	816252	RAuxAF		CAMERON
N425 1944	755392	RAFVR		CAMERON
N187 1947	102585	RAFVR	DSO,DFC	CAMERON
N132 1944	70770	RAFVR		CAMERON
N289 1947	755406	RAFVR		CAMERON
N895 1955	749773	RAFVR		CAMISH
N24 1957	152312	RAuxAF		CAMPBELL
N722 1962	83333	RAFVR	wef 21/7/62	CAMPBELL
YB 1996	n/a	RAuxAF		CAMPBELL
YB 1999	n/a	RAuxAF		CAMPBELL
N874 1945	741676	RAFVR		CAMPBELL
N395 1948	803552	RAuxAF		CAMPBELL
N258 1958	2683523	RAuxAF	1st Clasp; wef 18/2/53	CAMPBELL
N239 1963	2650763	RAuxAF	wef 3/2/63	CAMPBELL
N198 1957	139404	RAF	wef 10/2/54	CAMPBELL
N526 1943	67022	RAFVR		CAMPBELL
N194 1951	79	WAAF	CBE	CAMPBELL
N559 1958	2677416	RAuxAF	wef 26/6/53	CAMPBELL
N902 1961	2662149	WRAAF	wef 18/10/61	CAMPBELL
N1232 1945	88433	RAFVR		CAMPBELL
N93 1951	90822	RAuxAF		CAMPBELL
N846 1947	73119	RAFVR		CAMPBELL
YB 1998	n/a	RAuxAF		CAMPBELL
N1065 1943	70111	RAFVR	AFC	CAMPBELL
N1 1955	1568523	RAuxAF		CAMPBELL
N552 1959	2682607	RAFVR	wef 14/7/59	CAMPBELL-MacMILLAN-COLLINS
N1638 1942	90079	AAF		CAMPBELL-ORDE
N258 1958	90079	AAF	1st Clasp; wef 20/8/41	CAMPBELL-ORDE
N454 1959	23508932	TA (RE)	wef 26/10/54	CAMPION
N197 1946	116438	RAFVR		CAMPLIN
N527 1948	749833	RAFVR		CANBY

Surname	Initials			Rank	Status	Date
CANN	E	E	C	SGT		26/08/1948
CANSDALE	A	C		FLT SGT		20/05/1948
CANSICK	A	C		FO		20/09/1945
CANTON	J			FL		12/04/1945
CANTRELL	E			SQD LDR	Acting	15/09/1949
CAPES	R	B		FO		15/04/1943
CAPUCCI	R			CPL		24/11/1949
CARBERY	M	C		SGT	Acting	07/12/1960
CARDEN	P	D		SQD LDR	Acting	23/05/1946
CARDER	S	H		FL		08/11/1945
CARDEW	W	A	T	FO	Acting	03/12/1942
CARDUS	M			FLT OFF	Acting	08/05/1947
CARDY	L	H	J	WO		26/08/1948
CAREY	A	E		AC1		07/08/1947
CAREY	D	H		SGT		23/02/1950
CAREY	J	B		FLT SGT		27/07/1944
CARLILE	T	A		SQD LDR		21/09/1944
CARLIN	J	M		FLT OFF		08/05/1947
CARMICHAEL	D	A		FL		08/11/1945
CARMICHAEL	J			FO		18/04/1946
CARNEY	R	C		CPL		13/01/1949
CARPENTER	F	G		FL		07/07/1949
CARPENTER	L			FL		22/10/1958
CARPENTER	N			FL		18/04/1946
CARPENTER	R	G		FL		05/01/1950
CARR	E	G		FO		10/06/1948
CARR	F	R		SQD LDR		01/02/1990
CARR	K	F		WING OFF	Acting	10/08/1950
CARR	W	A	K	FL		24/01/1946
CARR	W	G		FL	Acting	05/06/1947
CARRAN	J	H		CPL		10/02/1949
CARRESS	S			CPL		07/12/1960
CARRICK	E	D		SGT		26/08/1948
CARRICK	W			CPL		15/07/1948
CARRIER	S	T		SQD LDR	Acting	05/06/1947
CARRIGAN	J	E		LAC		22/02/1951
CARROLL	G	J		CPL		07/07/1949
CARROLL	M	B		FO		25/06/1953
CARRUTHERS	J	R		SQD LDR	Acting	20/09/1945
CARSLAW	R	H		CPL		17/03/1949
CARSON	F	S		SGT		16/10/1947
CARSON	F			LAC		09/01/1963
CARSON	R	R	L	K	FL	22/10/1958
CARSWELL	W	C		FL		30/08/1945
CARTER	A	B		SQD LDR	Acting	06/03/1947
CARTER	A	C		FL		30/03/1944
CARTER	A	W		FL		23/05/1946
CARTER	C	W		WO		25/05/1950
CARTER	F			FL		24/01/1946
CARTER	H	T		FL		01/05/1957
CARTER	J			FL		06/03/1947
CARTER	J			CPL	Acting	15/11/1951
CARTER	L	C		LAC		05/06/1947
CARTER	L	C		FL		13/11/1947
CARTER	M	H		FL	Acting	13/11/1947
CARTER	M	L	H	FL		08/02/1945
CARTER	P	J		FLT OFF	Acting	07/08/1947
CARTER	R	F		FL		20/09/1961
CARTER	S	J	H	FO		06/03/1947
CARTER	T	C		WING COMM		07/06/1945

Source	Service No	Force	Notes	Surname
N694 1948	813068	RAuxAF		CANN
N395 1948	770489	RAFVR		CANSDALE
N1036 1945	177246	RAFVR		CANSICK
N381 1945	116061	RAFVR		CANTON
N953 1949	81549	RAFVR		CANTRELL
N454 1943	114117	RAFVR		CAPES
N1193 1949	743872	RAFVR		CAPUCCI
N933 1960	2663732	WRAAF	wef 19/9/60	CARBERY
N468 1946	84316	RAFVR	DFC	CARDEN
N1232 1945	129237	RAFVR		CARDER
N1638 1942	112055	RAFVR		CARDEW
N358 1947	867	WAAF		CARDUS
N694 1948	747869	RAFVR		CARDY
N662 1947	844715	AAF		CAREY
N179 1950	752207	RAFVR		CAREY
N763 1944	802391	AAF		CAREY
N961 1944	90966	AAF		CARLILE
N358 1947	587	WAAF		CARLIN
N1232 1945	129718	RAFVR		CARMICHAEL
N358 1946	178498	RAFVR		CARMICHAEL
N30 1949	866340	RAuxAF		CARNEY
N673 1949	72624	RAFVR		CARPENTER
N716 1958	184677	RAFVR	wef 2/12/52	CARPENTER
N358 1946	127135	RAFVR	DFC	CARPENTER
N1 1950	144351	RAFVR		CARPENTER
N456 1948	130748	RAFVR		CARR
YB 1996	2629698	RAuxAF		CARR
N780 1950	439	WAAF		CARR
N82 1946	78269	RAFVR		CARR
N460 1947	199796	RAFVR		CARR
N123 1949	853757	RAuxAF		CARRAN
N933 1960	2691084	RAuxAF	wef 3/9/60	CARRESS
N694 1948	751953	RAFVR		CARRICK
N573 1948	843032	RAuxAF		CARRICK
N460 1947	90787	AAF		CARRIER
N194 1951	866455	RAuxAF		CARRIGAN
N673 1949	753237	RAFVR		CARROLL
N527 1953	204360	RAFVR		CARROLL
N1036 1945	122924	RAFVR		CARRUTHERS
N231 1949	873980	RAuxAF		CARSLAW
N846 1947	816239	AAF		CARSON
N22 1963	2676945	RAuxAF	wef 24/11/60	CARSON
N716 1958	187936	RAFVR	wef 10/2/54	CARSON
N961 1945	87638	RAFVR		CARSWELL
N187 1947	72626	RAFVR		CARTER
N281 1944	61006	RAFVR		CARTER
N468 1946	79745	RAFVR		CARTER
N517 1950	755251	RAFVR		CARTER
N82 1946	129973	RAFVR	DFC	CARTER
N277 1957	157620	RAFVR	wef 1/2/57	CARTER
N187 1947	128385	RAFVR		CARTER
N1185 1951	856056	RAuxAF		CARTER
N460 1947	811219	AAF		CARTER
N933 1947	79591	RAFVR		CARTER
N933 1947	67735	RAFVR		CARTER
N139 1945	89061	RAFVR		CARTER
N662 1947	6218	WAAF		CARTER
N720 1961	116499	RAFVR	wef 31/7/60	CARTER
N187 1947	182183	RAFVR		CARTER
N608 1945	72012	RAFVR		CARTER

Surname	Initials			Rank	Status	Date
CARTWRIGHT	H	K		FL		11/05/1944
CARTWRIGHT	J			FL		18/04/1946
CARWELL	A	R		WO		20/05/1948
CARWELL	A	R		FO		22/05/1952
CARVER	K	M		SQD LDR	Acting	07/06/1945
CASE	A	C		WING COMM		17/06/1954
CASEY	H	J		SGT		24/06/1943
CASEY	J	E		LAC		15/07/1948
CASEY	M			CPL		13/01/1949
CASS	T	L		SQD LDR	Acting	24/01/1946
CASSIDY	T	H		WO		09/05/1946
CASSIDY	T	H		CPL		01/11/1951
CASSIE	G	E		FL		30/03/1944
CASSINGHAM	B			LAC		26/01/1950
CASSON	L	H		FL		09/11/1950
CASSON	L	H		SQD LDR		27/11/1952
CAST	H			FL		12/08/1943
CASTELL	E	F		FL		09/05/1946
CASTLE	H	W		FL		30/03/1944
CASTLE	J			CPL		13/11/1947
CATER	J	H		SQD LDR	Acting	21/06/1945
CATER	R	S	C	FL		17/04/1947
CATFORD	J	M		FL		20/07/1960
CATHCART	H	T		WING COMM	Acting	22/02/1961
CATLEY	H	P		CPL		04/11/1948
CATLEY	W			CPL	Temporary	23/06/1955
CATO	E	T		WING COMM		15/11/1951
CATT	R	H		SGT		18/02/1954
CATTANACH	J	A		WO		24/11/1949
CATTERICK	M			CPL		10/06/1948
CAUNT	A	S		FL		08/11/1945
CAUNTER	J	W	T	FO		26/11/1958
CAVE	A	M		WING COMM	Acting	28/02/1946
CAVE	D	C	F	SQD LDR	Acting	05/06/1947
CAVENDISH	H	S		SQD LDR		10/12/1942
CAZENOVE	P	F		FL		08/05/1947
CHADWICK	D	F		FL		30/03/1944
CHADWICK	J			FL		05/08/1964
CHAFFER	H	C		SGT		17/01/1952
CHALK	B	W		SGT		07/08/1947
CHALK	D	F		FL		26/11/1958
CHALLENGER	A			CPL		26/08/1948
CHALLENGER	W			AC2		22/02/1951
CHALLINOR	B	J		SGT		19/08/1954
CHALLIS	A	W		AC1		19/07/1951
CHALLIS	P	L		SAC		26/07/1996
CHALMERS	H	M		FO		15/03/1944
CHALMERS	J	A		SQD LDR		16/10/1947
CHALMERS	J	S	H	SQD LDR		05/01/1950
CHALMERS	R	A	R	WING COMM	Acting	11/05/1944
CHALMERS	R	A	R	FL		31/10/1956
CHALMERS	W			FL		09/11/1963
CHAMBERLAIN	E	L		FO		21/09/1944
CHAMBERLAIN	R	W		FL		25/05/1950
CHAMBERLAIN	S			SAC		21/02/1997
CHAMBERS	E			FL		13/02/1957
CHAMBERS	E			FL		14/12/1964
CHAMBERS	R			SAC		08/07/1959
CHAMBERS	T	L		SGT		12/06/1947
CHANDLER	E	F		FL		01/08/1946

Source	Service No	Force	Notes	Surname
N444 1944	89775	RAFVR		CARTWRIGHT
N358 1946	113839	RAFVR	AFC	CARTWRIGHT
N395 1948	742302	RAFVR		CARWELL
N402 1952	2603009	RAFVR	1st Clasp	CARWELL
N608 1945	79730	RAFVR	DFC	CARVER
N471 1954	74074	RAuxAF	MBE	CASE
N677 1943	816037	AAF		CASEY
N573 1948	850502	RAuxAF		CASEY
N30 1949	840548	RAuxAF		CASEY
N82 1946	137343	RAFVR	DFM	CASS
N415 1946	815138	AAF		CASSIDY
N1126 1951	2681500	RAuxAF	1st Clasp	CASSIDY
N281 1944	108836	RAFVR		CASSIE
N94 1950	812344	RAuxAF		CASSINGHAM
N1113 1950	91000	RAuxAF	DFC	CASSON
N915 1952	91000	RAuxAF	DFC; 1st Clasp	CASSON
N850 1943	83720	RAFVR		CAST
N415 1946	117582	RAFVR	DFC	CASTELL
N281 1944	122360	RAFVR		CASTLE
N933 1947	850176	AAF		CASTLE
N671 1945	86675	RAFVR	AFC	CATER
N289 1947	73123	RAFVR		CATER
N531 1960	186322	RAFVR	DFC; wef 8/4/59	CATFORD
N153 1961	2650862	RAuxAF	wef 9/11/60	CATHCART
N934 1948	869304	RAuxAF		CATLEY
N473 1955	743287	RAuxAF		CATLEY
N1185 1951	77020	RAFVR	MB,BS,FRCS	CATO
N137 1954	2611718	RAFVR		CATT
N1193 1949	799959	RAFVR		CATTANACH
N456 1948	808335	RAuxAF		CATTERICK
N1232 1945	105165	RAFVR	DFC	CAUNT
N807 1958	2601015	RAFVR	wef 6/12/52	CAUNTER
N197 1946	90682	AAF		CAVE
N460 1947	110895	RAFVR		CAVE
N1680 1942	90126	AAF		CAVENDISH
N358 1947	73727	RAFVR		CAZENOVE
N281 1944	101519	RAFVR		CHADWICK
MOD 1964	123090	RAFVR		CHADWICK
N35 1952	747185	RAFVR		CHAFFER
N662 1947	749100	RAFVR		CHALK
N807 1958	185750	RAFVR	wef 18/11/52	CHALK
N694 1948	743119	RAFVR		CHALLENGER
N194 1951	868724	RAuxAF		CHALLENGER
N672 1954	2695011	RAuxAF		CHALLINOR
N724 1951	841175	RAuxAF		CHALLIS
YB 1997	n/a	RAuxAF		CHALLIS
MOD 1964	133261	RAFVR		CHALMERS
N846 1947	72744	RAFVR		CHALMERS
N1 1950	90761	RAuxAF		CHALMERS
N444 1944	72983	RAFVR		CHALMERS
N761 1956	72983	RAFVR	1st Clasp	CHALMERS
MOD 1964	2612510	RAFVR		CHALMERS
N961 1944	139937	RAFVR		CHAMBERLAIN
N517 1950	91116	RAuxAF		CHAMBERLAIN
YB 1998	n/a	RAuxAF		CHAMBERLAIN
N104 1957	132210	RAFVR	MBE; wef 14/12/54	CHAMBERS
MOD 1967	132210	RAFVR	MBE; 1st Clasp	CHAMBERS
N454 1959	2685561	RAFVR	wef 7/11/58	CHAMBERS
N478 1947	850545	AAF		CHAMBERS
N652 1946	115851	RAFVR	DFM	CHANDLER

The Air Efficiency Award 1942-2005

Surname	Initials			Rank	Status	Date
CHANDLER	E	S		FLT SGT		24/01/1946
CHANDLER	H	H		FL		17/02/1944
CHANDLER	I	E	J	GNR		17/04/1947
CHANDLER	R	F		FL		28/02/1946
CHANNING	F	A		FLT SGT		14/10/1948
CHANTRY	A			SGT		17/04/1947
CHAPE	W	F		SQD LDR	Acting	30/03/1944
CHAPMAN	A	A		SGT		07/08/1947
CHAPMAN	A	F		AC1		16/10/1947
CHAPMAN	A	S		CPL		27/11/1952
CHAPMAN	A	V		FL	Acting	27/04/1950
CHAPMAN	A			SGT		30/09/1948
CHAPMAN	B	D		CPL		12/09/1997
CHAPMAN	B			FL		23/05/1946
CHAPMAN	D	J		FL	Acting	01/07/1948
CHAPMAN	F	A		FL		13/10/1955
CHAPMAN	G	B		FL		18/11/1948
CHAPMAN	H	L	J	FO		04/10/1945
CHAPMAN	J	W		FL		15/01/1953
CHAPMAN	J			SGT		07/06/1951
CHAPMAN	M	A		FO		10/02/1949
CHAPMAN	R	J		FLT SGT	Acting	04/10/1945
CHAPMAN	W	E		CPL		07/12/1960
CHAPMAN	W	J		CPL	Temporary	13/10/1955
CHAPPELL	A	K		FL		21/09/1944
CHAPPELL	J	I		SQD LDR		01/01/2002
CHARLES	A	B		SQD LDR		09/12/1959
CHARLES	R	J		CPL		06/03/1947
CHARLESWORTH	T	M		FL	Acting	05/06/1947
CHARLESWORTH	W	H	M	FL	Acting	05/06/1947
CHARLTON	J	H		WO	Acting	25/03/1954
CHARTERIS	L	C		SGT		16/03/1950
CHARTERIS	M	M		FL		15/09/1949
CHARTERS	S	E		FL		10/02/2000
CHARTRES	H			FL	Acting	17/04/1947
CHASE	F	J	A	WING COMM	Acting	25/01/1945
CHATFIELD	R	M		SQD LDR	Acting	23/05/1946
CHATTEN	D			CPL		10/06/1948
CHEESEBOROUGH	J	W		FL		08/07/1959
CHEESEMAN	R	E		SQD LDR	Acting	05/01/1950
CHEGWIDDEN	P			FL		11/06/2001
CHENERY	J	A		FO		06/03/1947
CHERRY	J			WING COMM	Acting	03/12/1942
CHESSWORTH	H	C		SGT		12/06/1947
CHEYNE	E	S		FL		16/01/1957
CHICK	A	H		SGT	Acting	26/04/1956
CHICK	C	G		LAC		25/05/1950
CHIDELL	D	C	L	SQD LDR	Acting	08/11/1945
CHILARD	J	P		LAC		26/02/1958
CHILTON	S	N		FL		26/11/1958
CHIPPERFIELD	E	W	G	WO		07/07/1949
CHIPPING	D	J		FL		21/03/1946
CHISHOLM	R	A		WING COMM	Acting	17/12/1942
CHISHOLM	T	W		SGT		13/01/1949
CHIVERS	W	E	J	FL		26/11/1958
CHOPPING	R	C		SQD LDR	Acting	26/01/1950
CHORLEY	H	T	J K	FL		08/03/1956
CHORLTON	W	L		CPL		07/07/1949
CHOUFFOT	G	C		FL		21/03/1946
CHOWN	J	D		SQD LDR	Acting	04/10/1945

Source	Service No	Force	Notes	Surname
N82 1946	807045	AAF		CHANDLER
N132 1944	106245	RAFVR	AFC,DFM	CHANDLER
N289 1947	750665	RAFVR		CHANDLER
N197 1946	126720	RAFVR		CHANDLER
N856 1948	844885	RAuxAF		CHANNING
N289 1947	800055	AAF		CHANTRY
N281 1944	81028	RAFVR		CHAPE
N662 1947	845316	AAF		CHAPMAN
N846 1947	845105	AAF		CHAPMAN
N915 1952	859503	RAuxAF		CHAPMAN
N406 1950	139332	RAFVR		CHAPMAN
N805 1948	747504	RAFVR		CHAPMAN
YB 2001	n/a	RAuxAF		CHAPMAN
N468 1946	144697	RAFVR		CHAPMAN
N527 1948	138615	RAFVR		CHAPMAN
N762 1955	199499	RAFVR	AFM	CHAPMAN
N984 1948	139594	RAFVR		CHAPMAN
N1089 1945	162957	RAFVR		CHAPMAN
N28 1953	90632	RAuxAF		CHAPMAN
N560 1951	815059	RAuxAF		CHAPMAN
N123 1949	134746	RAFVR	Deceased	CHAPMAN
N1089 1945	813020	AAF		CHAPMAN
N933 1960	2691085	RAuxAF	wef 3/9/60	CHAPMAN
N762 1955	800591	RAuxAF		CHAPMAN
N961 1944	80808	RAFVR		CHAPPELL
YB 2003	5206775	RAuxAF		CHAPPELL
N843 1959	76005	RAF	wef 17/8/44	CHARLES
N187 1947	803459	AAF		CHARLES
N460 1947	187196	RAFVR		CHARLESWORTH
N460 1947	147722	RAFVR		CHARLESWORTH
N249 1954	2678001	RAuxAF		CHARLTON
N257 1950	749781	RAFVR		CHARTERIS
N953 1949	112063	RAFVR	MBE	CHARTERIS
YB 2001	2640499	RAuxAF		CHARTERS
N289 1947	144995	RAFVR	MBE	CHARTRES
N70 1945	72283	RAFVR	DFC	CHASE
N468 1946	86417	RAFVR	DFC	CHATFIELD
N456 1948	808139	RAuxAF		CHATTEN
N454 1959	91375	RAuxAF	wef 3/5/59	CHEESEBOROUGH
N1 1950	77910	RAFVR		CHEESEMAN
YB 2002	2641772	RAuxAF		CHEGWIDDEN
N187 1947	189586	RAFVR		CHENERY
N1638 1942	90231	AAF	OBE	CHERRY
N478 1947	811006	AAF		CHESSWORTH
N24 1957	184941	RAuxAF		CHEYNE
N318 1956	2676602	RAuxAF		CHICK
N517 1950	813240	RAuxAF		CHICK
N1232 1945	83262	RAFVR	MBE	CHIDELL
N149 1958	841210	AAF	wef 20/4/44	CHILARD
N807 1958	120927	RAFVR	DFC; wef 25/8/52	CHILTON
N673 1949	752109	RAFVR		CHIPPERFIELD
N263 1946	67603	RAFVR	AFC	CHIPPING
N1708 1942	90233	AAF	DFC	CHISHOLM
N30 1949	866325	RAuxAF		CHISHOLM
N807 1958	135145	RAFVR	wef 23/6/52	CHIVERS
N94 1950	70126	RAFVR	DFC; Deceased	CHOPPING
N187 1956	156165	RAFVR		CHORLEY
N673 1949	857448	RAuxAF		CHORLTON
N263 1946	126777	RAFVR		CHOUFFOT
N1089 1945	79531	RAFVR		CHOWN

Surname	Initials			Rank	Status	Date
CHRISTIAN	C	W		SGT		05/06/1947
CHRISTIE	G			CPL		09/12/1959
CHRISTIE	M	G		FLT OFF	Acting	16/10/1947
CHRISTMAS	G	J		SGT		15/07/1948
CHURCH	C	E		WO		13/11/1947
CHURCH	E			FL		26/08/1948
CHURCHER	E	E		CPL		01/07/1948
CHURCHILL-BALDWIN	M	L		FLT OFF	Acting	17/04/1947
CLACKSON	D	L		SQD LDR	Acting	03/12/1942
CLACY	A	E		SGT		26/04/1951
CLAESENS	A	P		FL		01/01/2003
CLAGUE	A			WO		11/07/1946
CLAMPETT	B	H		CPL		25/03/1954
CLANCEY	M	E		WING OFF	Acting	27/03/1947
CLAPHAM	E	M		FLT OFF		21/02/1952
CLAPHAM	E	M		FLT OFF		22/02/1961
CLAPHAM	S			FLT SGT		15/01/1958
CLAREY	J			CPL		19/07/1951
CLARK	A	G		CPL		21/12/1950
CLARK	C	A		FO		05/02/1948
CLARK	C	B		CPL	Acting	07/12/1960
CLARK	C	F		LAC		21/06/1951
CLARK	C	P		FL	Acting	13/01/1944
CLARK	C	R		SQD LDR		27/03/1947
CLARK	C	R		FL		18/02/1954
CLARK	D	de B		WING COMM		04/02/1943
CLARK	D	G		CPL		01/07/1948
CLARK	D	G		FLT SGT	Acting	25/03/1954
CLARK	E	G		FL		15/09/1949
CLARK	E	W	G	LAC		10/06/1948
CLARK	F	E		FL		28/02/1946
CLARK	G	W		CPL		26/08/1948
CLARK	J	E		FLT SGT		13/12/1945
CLARK	J	H		FO		20/04/1944
CLARK	J			CPL		07/08/1947
CLARK	L			FO		29/04/1954
CLARK	M	H		FL		26/11/1953
CLARK	R			CPL		28/06/1945
CLARK	R			FL		29/04/1954
CLARK	S			SGT		12/04/1945
CLARK	W	E	N	WING COMM	Acting	26/01/1950
CLARK	W	T	M	FL		28/06/1945
CLARKE	A	J		WO		13/01/1949
CLARKE	B	D		FLT OFF		05/02/1948
CLARKE	B	G		CPL		01/07/1948
CLARKE	C	S		FL		18/04/1946
CLARKE	D	W		FL		20/11/1957
CLARKE	F	A		JNR TECH		21/10/1959
CLARKE	G	G		FO		24/06/1943
CLARKE	G	T		PILOT 1		18/03/1948
CLARKE	H	H		SGT		18/03/1948
CLARKE	H	R		FL		18/04/1946
CLARKE	J	M		SQD LDR		12/06/1947
CLARKE	R	C		FL		09/08/1945
CLARKE	R	C		FO		04/09/1947
CLARKE	R	F		WO	Acting	05/06/1947
CLARKE	R	J		LAC		29/11/1951
CLARKE	R			FLT SGT		08/02/1945
CLARKE	R			CPL		10/06/1948
CLARKE	W	A		FL		11/05/1944

Source	Service No	Force	Notes	Surname
N460 1947	872480	AAF		CHRISTIAN
N843 1959	2676436	RAFVR	wef 15/10/58	CHRISTIE
N846 1947	3048	WAAF		CHRISTIE
N573 1948	748936	RAFVR		CHRISTMAS
N933 1947	812125	AAF		CHURCH
N694 1948	116514	RAFVR		CHURCH
N527 1948	749784	RAFVR		CHURCHER
N289 1947	3441	WAAF		CHURCHILL-BALDWIN
N1638 1942	90087	AAF		CLACKSON
N402 1951	800469	RAuxAF		CLACY
YB 2003	2632164	RAuxAF		CLAESENS
N594 1946	751339	RAFVR		CLAGUE
N249 1954	2603823	RAFVR		CLAMPETT
N250 1947	839	WAAF		CLANCEY
N134 1952	106	WAAF		CLAPHAM
N153 1961	106	WRAAF	1st Clasp; wef 12/3/57	CLAPHAM
N40 1958	2686531	RAuxAF	wef 27/11/57	CLAPHAM
N724 1951	854627	RAuxAF		CLAREY
N1270 1950	702080	RAFVR		CLARK
N87 1948	134265	RAFVR		CLARK
N933 1960	2693550	RAuxAF	wef 2/10/60	CLARK
N621 1951	867782	RAuxAF		CLARK
N11 1944	106151	RAFVR		CLARK
N250 1947	72156	RAFVR		CLARK
N137 1954	187889	RAFVR		CLARK
N131 1943	90086	AAF		CLARK
N527 1948	752096	RAFVR		CLARK
N249 1954	2693501	RAuxAF	1st Clasp	CLARK
N953 1949	145324	RAFVR		CLARK
N456 1948	840007	RAuxAF		CLARK
N197 1946	90821	AAF		CLARK
N694 1948	847292	RAuxAF		CLARK
N1355 1945	807117	AAF		CLARK
N363 1944	133480	RAFVR		CLARK
N662 1947	747174	RAFVR		CLARK
N329 1954	2603550	RAFVR		CLARK
N952 1953	91316	RAuxAF		CLARK
N706 1945	80248	AAF		CLARK
N329 1954	195418	RAFVR		CLARK
N381 1945	807059	AAF		CLARK
N94 1950	74012	RAFVR		CLARK
N706 1945	126026	RAFVR	DFM	CLARK
N30 1949	751600	RAFVR		CLARKE
N87 1948	1947	WAAF		CLARKE
N527 1948	756766	RAFVR		CLARKE
N358 1946	162995	RAFVR		CLARKE
N835 1957	124388	RAFVR	DFC; wef 29/5/57	CLARKE
N708 1959	801570	RAF	wef 9/7/44	CLARKE
N677 1943	106805	RAFVR		CLARKE
N232 1948	748034	RAFVR		CLARKE
N232 1948	845284	RAuxAF		CLARKE
N358 1946	102587	RAFVR		CLARKE
N478 1947	73127	RAFVR		CLARKE
N874 1945	134043	RAFVR		CLARKE
N740 1947	182786	RAFVR		CLARKE
N460 1947	813089	AAF		CLARKE
N1219 1951	855835	RAuxAF		CLARKE
N139 1945	816035	AAF	BEM	CLARKE
N456 1948	850747	RAuxAF		CLARKE
N444 1944	112537	RAFVR		CLARKE

Surname	Initials			Rank	Status	Date
CLARKE	W	J		CPL		18/11/1948
CLARKE	W	L		SQD LDR	Acting	27/03/1947
CLARKE	W	W		LAC		17/01/1952
CLARKE	W			SGT		13/07/1950
CLARKSON	G	T		FL		15/09/1949
CLARKSON	G	T		FL		02/04/1953
CLARKSON	S	R		SGT		01/07/1948
CLATER	D	A		LAC		20/05/1948
CLATER	D	A		SAC		20/11/1957
CLAXTON	D	B		FL		30/03/1960
CLAY	R	A		WING COMM		04/02/1943
CLAYDON	R	E	J	CPL		17/03/1949
CLAYDON	R	J		PO		25/03/1954
CLAYFIELD	S	E		SGT		26/08/1948
CLAYSON	A	E		FL		09/05/1946
CLAYTON	E			SQD LDR		26/01/1950
CLAYTON	H			WING COMM	Acting	07/05/1953
CLEAR	E	F		SQD LDR	Acting	07/06/1951
CLEARE	F	J		FL		26/06/1947
CLEGG	G			FO		01/10/1953
CLEGG	O	C		SQD OFF	Acting	18/03/1948
CLEGG	R	A		SQD LDR		07/04/1955
CLEGGETT	L	J		LAC		13/02/1957
CLELLAND	E	J		FL		30/09/1948
CLEMENTI	C	M		WING COMM	Acting	09/08/1945
CLEMENTS	C	H		SQD LDR	Acting	26/08/1948
CLEMES	J	E		FO		07/04/1955
CLIBERY	A	H		LAC		20/01/1960
CLIFF	L	H	T	SQD LDR	Acting	31/05/1945
CLIFF	W			LAC		26/04/1961
CLIFFORD	F	G		FO		04/05/1944
CLIFFORD	F	J		LAC		07/07/1949
CLIFT	M	R	B	FL		13/12/1945
CLINCH	B	W	V	SGT		12/06/1947
CLINCH	S	J		SQD LDR		05/01/1950
CLINT	R	B		SGT	Acting	01/10/1953
CLINTON	J	A	T	FO		23/08/1951
CLIVE	P	J		SQD LDR	Acting	30/12/1943
CLOHOSEY	J			FLT SGT		05/06/1947
CLOSE	E	E		WO	Temporary	20/01/1960
CLOSE	W	J		SGT		15/09/1949
CLOSE	W	S		FL		06/03/1947
CLOUDER	N	T		FL		19/08/1959
CLOVER	T			CPL		20/05/1948
CLOWER	A	E		CPL		01/07/1948
CLOWES	K	E		FL		03/10/1946
CLUBE	M	V	M	GRP CAP		24/01/1946
CLUCAS	A	B		FLT SGT		01/03/1945
CLUCAS	L	J		LAC		09/08/1945
CLUROE	J			CPL		18/03/1948
CLYDE	W	P		WING COMM	Acting	15/04/1943
COARD	J			CPL		26/04/1956
COATES	B	C		FL		06/03/1947
COATES	D	A	J	FO		10/06/1948
COATES	R	S		PO		03/12/1942
COATES	S			FL		05/06/1947
COATH	T	R	A	SQD OFF	Acting	09/02/1951
COBB	G	T	F	LAC		10/08/1950
COBB	H	L	A	CPL		16/03/1950
COBB	R			FL		10/08/1950

Source	Service No	Force	Notes	Surname
N984 1948	841061	RAuxAF		CLARKE
N250 1947	86965	RAFVR		CLARKE
N35 1952	858058	RAuxAF		CLARKE
N675 1950	816212	RAuxAF		CLARKE
N953 1949	151488	RAFVR		CLARKSON
N298 1953	151488	RAFVR	1st Clasp	CLARKSON
N527 1948	743410	RAFVR		CLARKSON
N395 1948	803627	RAuxAF		CLATER
N835 1957	2683505	RAuxAF	1st Clasp; wef 19/10/53	CLATER
N229 1960	145867	RAuxAF	DFM; wef 21/12/59	CLAXTON
N131 1943	90301	AAF		CLAY
N231 1949	756052	RAFVR		CLAYDON
N249 1954	2602028	RAFVR		CLAYDON
N694 1948	847620	RAuxAF		CLAYFIELD
N415 1946	127155	RAFVR		CLAYSON
N94 1950	90500	RAuxAF		CLAYTON
N395 1953	90984	RAuxAF		CLAYTON
N560 1951	72355	RAFVR		CLEAR
N519 1947	90515	AAF		CLEARE
N796 1953	163463	RAFVR		CLEGG
N232 1948	12111	WAAF		CLEGG
N277 1955	128379	RAFVR		CLEGG
N104 1957	2614678	RAFVR	wef 27/6/44	CLEGGETT
N805 1948	78085	RAFVR		CLELLAND
N874 1945	72465	RAFVR		CLEMENTI
N694 1948	90776	RAuxAF		CLEMENTS
N277 1955	2607057	RAFVR		CLEMES
N33 1960	855236	AAF	wef 13/7/44	CLIBERY
N573 1945	78720	RAFVR	AFC	CLIFF
N339 1961	852950	AAF	wef 20/6/44	CLIFF
N425 1944	131893	RAFVR		CLIFFORD
N673 1949	865563	RAuxAF		CLIFFORD
N1355 1945	141127	RAFVR		CLIFT
N478 1947	801571	AAF		CLINCH
N1 1950	72801	RAFVR		CLINCH
N796 1953	2677414	RAuxAF		CLINT
N866 1951	101757	RAFVR		CLINTON
N1350 1943	90152	AAFRO		CLIVE
N460 1947	811031	AAF		CLOHOSEY
N33 1960	745114	RAFVR	wef 30/5/44	CLOSE
N953 1949	816135	RAuxAF		CLOSE
N187 1947	130721	RAFVR	DFC	CLOSE
N552 1959	2601280	RAFVR	wef 11/9/53	CLOUDER
N395 1948	868656	RAuxAF		CLOVER
N527 1948	849238	RAuxAF		CLOWER
N837 1946	147693	RAFVR	DFM	CLOWES
N82 1946	90015	AAF		CLUBE
N225 1945	800376	AAF		CLUCAS
N874 1945	800383	AAF		CLUCAS
N232 1948	746603	RAFVR		CLUROE
N454 1943	90154	AAF	DFC	CLYDE
N318 1956	816078	RAuxAF		COARD
N187 1947	141824	RAFVR	DFC,DFM	COATES
N456 1948	142066	RAFVR		COATES
N1638 1942	107972	RAFVR		COATES
N460 1947	131497	RAFVR		COATES
N145 1951	1613	WAAF		COATH
N780 1950	864280	RAuxAF		COBB
N257 1950	864104	RAuxAF		COBB
N780 1950	62433	RAFVR		COBB

Surname	Initials			Rank	Status	Date
COCHLAN	A	J		SGT		16/10/1947
COCHRANE	E	M		SGT	Acting	26/11/1958
COCKCROFT	J	N		FL		23/06/1955
COCKELL	A	R		SGT		24/01/1946
COCKELL	R	D		CPL		23/02/1950
COCKERILL	H	E		FL		07/02/1962
COCKERLINE	P			SGT		01/11/1951
COCKS	H	M		FO		06/03/1947
CODD	N	W	E	FL		18/03/1948
CODRAI	A	W		FL		25/07/1962
COE	L	H		SGT		17/04/1947
COGHLAN	E	F	V	SGT		24/06/1943
COGHLAN	H	St.	J	WING COMM	Acting	25/05/1944
COLBORNE	D	G		LAC		28/08/1963
COLCLOUGH	H	R	G	FL		06/03/1947
COLE	A	L		FL		08/11/1945
COLE	C	F	J	WO		26/08/1948
COLE	C	W		FL		19/10/1960
COLE	E	J		SGT		17/04/1947
COLE	E			WO		21/09/1944
COLE	G	A	J	CPL		06/01/1955
COLE	G	W		SGT		15/09/1949
COLE	L	A		SGT		16/03/1950
COLE	L	W		CPL		26/08/1948
COLE	M	J		FL		19/10/1960
COLE	N			FL		27/11/1952
COLE	R	A	W	CPL		01/07/1948
COLE	R	V	W	SGT		15/01/1958
COLE	W	G		FL		04/10/1945
COLEMAN	C	G		SQD LDR	Acting	12/06/1947
COLEMAN	F	A		FL		16/10/1947
COLEMAN	G	L		SQD LDR	Acting	20/05/1948
COLEMAN	M	E		SGT		06/04/1994
COLEMAN	P	T		FL		28/03/1962
COLEMAN	R			FL		02/07/1958
COLEMAN	S	J		FL		09/08/1945
COLES	G	F		FL		25/03/1959
COLES	G	F		FL		10/12/1963
COLES	G	H		FL		18/04/1946
COLES	K	R		WO		17/06/1954
COLES	R	M	P	WO	Temporary	19/09/1956
COLES	T	R	J	FLT SGT		26/01/1950
COLHART	W	D		WING COMM		05/02/1948
COLHOUN	D	N	T	FL		01/05/1997
COLINESE	P	E		FL		29/04/1954
COLLEDGE	G			FL		18/11/1948
COLLETT	H	E		SQD OFF		05/02/1948
COLLEY	R	T		SQD LDR		20/05/1948
COLLICUTT	G			WO		09/07/1993
COLLIER	A			SGT	Temporary	28/11/1956
COLLIER	E	H		SGT		15/01/1953
COLLIER	I	M	De M G	SQD OFF	Acting	27/03/1947
COLLIER	J	A		SGT		07/07/1949
COLLIER	K			FLT SGT		26/11/1958
COLLIER	R			SQD LDR	Acting	07/07/1949
COLLING	W	E		LAC		10/06/1948
COLLINGBOURNE	S	E		FL		25/03/1959
COLLINGS	C	F	H	SAC		09/07/1993
COLLINGS	M	C		SQD LDR	Acting	26/04/1956
COLLINGWOOD	C	H		SAC		15/01/1958

Source	Service No	Force	Notes	Surname
N846 1947	813152	AAF		COCHLAN
N807 1958	2658612	WRAAF	wef 12/1/55	COCHRANE
N473 1955	186128	RAFVR	DFM	COCKCROFT
N82 1946	803342	AAF		COCKELL
N179 1950	803524	RAuxAF		COCKELL
N102 1962	187275	RAFVR	wef 14/11/60	COCKERILL
N1126 1951	743433	RAFVR		COCKERLINE
N187 1947	81437	RAFVR		COCKS
N232 1948	79586	RAFVR		CODD
N549 1962	189644	RAFVR	wef 21/7/55	CODRAI
N289 1947	746279	RAFVR		COE
N677 1943	813002	AAF		COGHLAN
N496 1944	90117	AAF	DFC	COGHLAN
N621 1963	5005304	RAuxAF	wef 2/8/63	COLBORNE
N187 1947	90910	AAF		COLCLOUGH
N1232 1945	141742	RAFVR		COLE
N694 1948	745971	RAFVR		COLE
N775 1960	108152	RAFVR	DFC; wef 3/8/59	COLE
N289 1947	804262	AAF		COLE
N961 1944	746858	RAFVR		COLE
N1 1955	749209	RAFVR		COLE
N953 1949	813244	RAuxAF		COLE
N257 1950	752778	RAFVR		COLE
N694 1948	843857	RAuxAF		COLE
N775 1960	163777	RAFVR	wef 13/12/58	COLE
N915 1952	193149	RAFVR		COLE
N527 1948	756766	RAFVR		COLE
N40 1958	2684522	RAuxAF	wef 9/3/57	COLE
N1089 1945	135480	RAFVR	DFC	COLE
N478 1947	69612	RAFVR		COLEMAN
N846 1947	73131	RAFVR		COLEMAN
N395 1948	72554	RAFVR		COLEMAN
YB 1996	n/a	RAuxAF		COLEMAN
N242 1962	190247	RAFVR	DFC; wef 16/1/62	COLEMAN
N452 1958	154531	RAFVR	wef 8/12/55	COLEMAN
N874 1945	129714	RAFVR	DFC,DFM	COLEMAN
N196 1959	147490	RAuxAF	wef 11/10/54	COLES
MOD 1964	147490	RAuxAF	1st Clasp	COLES
N358 1946	130319	RAFVR	DFM	COLES
N471 1954	751124	RAFVR	MBE	COLES
N666 1956	846511	RAuxAF		COLES
N94 1950	818044	RAuxAF		COLES
N87 1948	90229	RAuxAF		COLHART
YB 1999	4233559	RAuxAF		COLHOUN
N329 1954	132922	RAFVR		COLINESE
N984 1948	73632	RAFVR		COLLEDGE
N87 1948	537	WAAF		COLLETT
N395 1948	72803	RAFVR		COLLEY
YB 1997	n/a	RAuxAF		COLLICUTT
N833 1956	810042	RAuxAF		COLLIER
N28 1953	752113	RAFVR		COLLIER
N250 1947	2959	WAAF		COLLIER
N673 1949	863468	RAuxAF		COLLIER
N807 1958	2694513	RAuxAF	wef 13/11/52	COLLIER
N673 1949	145133	RAFVR		COLLIER
N456 1948	808335	RAuxAF		COLLING
N196 1959	142294	RAFVR	wef 22/12/53	COLLINGBOURNE
YB 1996	n/a	RAuxAF		COLLINGS
N318 1956	58756	RAuxAF		COLLINGS
N40 1958	2691001	RAuxAF	wef 21/11/57	COLLINGWOOD

Surname	Initials			Rank	Status	Date
COLLINGWOOD	J	W		FLT SGT		17/03/1949
COLLINGWOOD	W	E		FL		28/02/1946
COLLINS	C	W	S	SQD LDR		01/07/1948
COLLINS	C	W	S	FL		31/10/1956
COLLINS	C			CPL		01/10/1953
COLLINS	E	W		SQD LDR		21/08/1944
COLLINS	F	C		SGT		20/07/1960
COLLINS	F	T		SQD LDR	Acting	24/01/1946
COLLINS	K	M		WING OFF	Acting	07/12/1950
COLLINS	K	P		WO		26/06/1947
COLLINS	R	A	F	SGT		01/07/1948
COLLINS	V			LAC		10/06/1948
COLLINSON	G			FL		18/05/1944
COLLINSON	J	B		FL		27/07/1944
COLLIS	B	A		CPL		13/01/1949
COLLIS	L	J		SGT		12/10/1950
COLLISS	A	E		SGT		19/10/1960
COLSTON	A	A	L	FL	Acting	07/08/1947
COLSTON	F			LAC		13/01/1949
COLVILLE	W			CPL		12/06/1947
COLVIN	V	H		CPL		30/09/1948
COLVIN	V	H		CPL		22/02/1961
COMPTOM	H	J		FLT OFF		30/09/1948
COMPTON	G	H		SQD LDR	Acting	10/12/1942
CONAN-DOYLE	J	L	A	WING OFF	Acting	27/03/1947
CONNAL	K	I		SQD OFF		06/03/1947
CONNAL	M	J	W	FLT OFF		26/06/1947
CONNELL	F	L		SGT		16/03/1950
CONNELL	J	G		SQD LDR	Acting	01/07/1948
CONROY	J			FL		16/01/1957
CONSODINE	B	B		FL		20/09/1945
CONSTANTINE	T			SGT		23/02/1950
CONWAY	G			FO		17/12/1942
CONWAY	W	R		SGT	Acting	20/11/1957
COOK	A	E		FO		21/09/1944
COOK	A	F		LAC		18/03/1948
COOK	B	G		SQD LDR	Acting	09/05/1946
COOK	B			FL		19/01/1956
COOK	B			FL		23/12/1964
COOK	E	E		FL		17/03/1949
COOK	E	L	J	FL		12/04/1945
COOK	F	J	M	FL		28/10/1966
COOK	F			FL	Acting	07/07/1949
COOK	G	J		FO		27/11/1952
COOK	H			FL		21/03/1946
COOK	J	G	H	CPL		20/09/1951
COOK	J	S		FO		06/03/1947
COOK	L	C	F	SGT		23/02/1950
COOK	L	R		FL		05/06/1947
COOK	P	A		FO		19/06/1963
COOK	R			SQD LDR		26/01/1950
COOK	S	A		FLT SGT	Temporary	26/04/1961
COOK	W	D	G	FLT SGT	Acting	27/03/1952
COOKE	D	F		SGT		05/06/1947
COOKE	F	E		FL		07/12/1960
COOKE	F	J		FLT SGT		20/05/1948
COOKE	G	H		WO		24/06/1943
COOKE	H	R		FL		08/11/1945
COOKE	H			FL		03/10/1946
COOKE	H			LAC		20/09/1951

Source	Service No	Force	Notes	Surname
N231 1949	810000	RAuxAF		COLLINGWOOD
N197 1946	115636	RAFVR	DFM	COLLINGWOOD
N527 1948	72402	RAFVR		COLLINS
N761 1956	72402	RAFVR	1st Clasp	COLLINS
N796 1953	864137	RAuxAF		COLLINS
MOD 1964	91085	AAF		COLLINS
N531 1960	2684530	RAFVR	wef 24/4/57	COLLINS
N82 1946	77911	RAFVR	DSO	COLLINS
N1210 1950	378	WAAF	MBE	COLLINS
N519 1947	745164	RAFVR		COLLINS
N527 1948	752624	RAFVR		COLLINS
N456 1948	850748	RAuxAF		COLLINS
N473 1944	86627	RAFVR		COLLINSON
N763 1944	87015	RAFVR		COLLINSON
N30 1949	846122	RAuxAF		COLLIS
N1018 1950	752669	RAFVR		COLLIS
N775 1960	756613	RAF	wef 25/8/44	COLLISS
N662 1947	197784	RAFVR		COLSTON
N30 1949	818037	RAuxAF		COLSTON
N478 1947	811178	AAF		COLVILLE
N805 1948	744605	RAFVR		COLVIN
N153 1961	2682254	RAuxAF	1st Clasp; wef 28/7/60	COLVIN
N805 1948	1728	WAAF		COMPTOM
N1680 1942	90112	AAF		COMPTON
N250 1947	550	WAAF		CONAN-DOYLE
N187 1947	19	WAAF	OBE	CONNAL
N519 1947	528	WAAF		CONNAL
N257 1950	740600	RAFVR		CONNELL
N527 1948	91172	RAuxAF		CONNELL
N24 1957	179284	RAuxAF		CONROY
N1036 1945	79728	RAFVR		CONSODINE
N179 1950	844217	RAuxAF		CONSTANTINE
N1708 1942	115720	RAFVR		CONWAY
N835 1957	2650768	RAuxAF	wef 27/9/54	CONWAY
N961 1944	169571	RAFVR		COOK
N232 1948	840205	RAuxAF		COOK
N415 1946	102974	RAFVR	DFM	COOK
N43 1956	85099	RAuxAF		COOK
MOD 1966	2603539	RAuxAF	MBE; 1st Clasp	COOK
N231 1949	138497	RAFVR		COOK
N381 1945	104433	RAFVR		COOK
MOD 1967	148340	RAFVR		COOK
N673 1949	201116	RAFVR		COOK
N915 1952	159856	RAFVR		COOK
N263 1946	126096	RAFVR		COOK
N963 1951	752115	RAFVR		COOK
N187 1947	135211	RAFVR		COOK
N179 1950	743986	RAFVR		COOK
N460 1947	175996	RAFVR	DFM	COOK
N434 1963	2654891	WRAFVR	wef 1/10/59	COOK
N94 1950	72806	RAFVR		COOK
N339 1961	805176	AAF	wef 31/7/44	COOK
N224 1952	864163	RAuxAF		COOK
N460 1947	751172	RAFVR		COOKE
N933 1960	200980	RAuxAF	wef 21/9/60	COOKE
N395 1948	800392	RAuxAF		COOKE
N677 1943	815004	AAF		COOKE
N1232 1945	161352	RAFVR	DFC	COOKE
N837 1946	115312	RAFVR	DFC	COOKE
N963 1951	870298	RAuxAF		COOKE

Surname	Initials			Rank	Status	Date
COOKE	L	H		FL		05/02/1948
COOKE	R	G	T	FL		10/12/1942
COOKE	R	G	H	CPL		12/06/1947
COOKE	T	C		SQD LDR	Acting	08/05/1947
COOKE-SHILLITO	A			LAC		13/01/1949
COOKSEY	F	T	H	LAC		04/11/1948
COOKSEY	R	A		SQD LDR		13/12/1945
COOKSEY	W	G	H	CPL		07/07/1949
COOLING	R	D		FO		11/01/1945
COOLING	R	D		FL		26/11/1958
COOMBES	E	A		PO		22/04/1943
COOMBES	E	A		FO		23/08/1951
COOMBES	K	H	M	FLT SGT		20/05/1948
COOMBS	R	J		FO		17/12/1942
COONEY	E	E		FL		28/11/1956
COOPER	A	C		WING COMM		19/08/1954
COOPER	A	E	L	SQD LDR		05/06/1947
COOPER	A	E	L	FL		19/09/1956
COOPER	A	G		FL		10/05/1945
COOPER	D	C		FL		11/07/1946
COOPER	G	A	B	SQD LDR	Acting	30/03/1944
COOPER	G	B		FL		22/05/1952
COOPER	G	W		SQD LDR	Acting	10/06/1948
COOPER	H			LAC		15/07/1948
COOPER	J	A		FL		26/01/1950
COOPER	J	O		SGT	Temporary	26/04/1956
COOPER	J			WO		11/07/1946
COOPER	L	H	G	SGT		23/02/1950
COOPER	N	T		FL		04/10/1945
COOPER	R	S		FL		20/04/1944
COOPER	S	F		FO		10/05/1945
COOPER	V	R	V	WING COMM	Acting	05/06/1947
COOPER	W	J		SGT		20/09/1951
COPE	C			LAC		10/06/1948
COPE	C			SAC		07/02/1952
COPE	J			FL		28/02/1946
COPE	L	G		FO		10/06/1948
COPELAND	C	J		FLT SGT		22/10/1993
COPELAND	S			SGT		04/09/1947
COPLAND	A	S		CPL		01/07/1948
COPLESTON	F	W		FL		13/12/1961
COPLEY	J	B		SQD LDR	Acting	07/08/1947
COPLEY	R	A		WO		13/01/1949
COPLEY-SMITH	N	A		SQD LDR		26/01/1950
COPPERWAITE	V	J		FLT SGT		01/07/1948
COPPICK	H	A		LAC		16/03/1950
CORAM	F			CPL		25/03/1959
CORBETT	M			SAC		15/01/1953
CORBIN	W	J		FO		26/08/1948
CORDELL	H	A		FL		11/07/1946
CORDERY	A	W	H	AC2		13/07/1950
CORDES	J			FO		03/07/1952
CORIGAN	W	L	J	FL		21/10/1959
CORKEN	E	J		SQD LDR		16/10/1947
CORLETT	H	D		FL		23/02/1950
CORMACK	D			FLT SGT	Acting	10/06/1948
CORMACK	J	McR		SQD LDR	Acting	31/10/1956
CORNER	L			LAC		06/09/1951
CORNES	G	E		SQD LDR	Acting	24/01/1946
CORNEWALL-WALKER	T	J	R	SQD LDR		30/09/1954

Source	Service No	Force	Notes	Surname
N87 1948	86447	RAFVR		COOKE
N1680 1942	70137	RAFVR		COOKE
N478 1947	756160	RAFVR		COOKE
N358 1947	103506	RAFVR	DFC,AFC,DFM	COOKE
N30 1949	869301	RAuxAF		COOKE-SHILLITO
N934 1948	813133	RAuxAF		COOKSEY
N1355 1945	90411	AAF	TD	COOKSEY
N673 1949	800619	RAuxAF		COOKSEY
N14 1945	147991	RAFVR		COOLING
N807 1958	147991	RAFVR	1st Clasp; wef 12/4/50	COOLING
N475 1943	124112	RAFVR		COOMBES
N866 1951	124112	RAFVR	1st Clasp	COOMBES
N395 1948	880117	WAAF		COOMBES
N1708 1942	90324	RAFVR		COOMBS
N833 1956	171530	RAFVR		COONEY
N672 1954	147285	RAuxAF		COOPER
N460 1947	72201	RAFVR		COOPER
N666 1956	72201	RAFVR	1st Clasp	COOPER
N485 1945	81665	RAFVR		COOPER
N594 1946	155877	RAFVR		COOPER
N281 1944	90396	AAF		COOPER
N402 1952	125352	RAFVR	DFC	COOPER
N456 1948	67015	RAFVR		COOPER
N573 1948	868601	RAuxAF		COOPER
N94 1950	72632	RAFVR		COOPER
N318 1956	770207	RAFVR		COOPER
N594 1946	759135	RAFVR		COOPER
N179 1950	756163	RAFVR		COOPER
N1089 1945	89306	RAFVR		COOPER
N363 1944	82949	RAFVR		COOPER
N485 1945	174121	RAFVR		COOPER
N460 1947	90367	AAF		COOPER
N963 1951	701509	RAFVR		COOPER
N456 1948	808226	RAuxAF		COPE
N103 1952	2686012	RAuxAF	1st Clasp	COPE
N197 1946	103518	RAFVR	DFC,DFM	COPE
N456 1948	144144	RAFVR		COPE
YB 1996	n/a	RAuxAF		COPELAND
N740 1947	855838	AAF		COPELAND
N527 1948	841393	RAuxAF		COPLAND
N987 1961	201711	RAFVR	wef 6/11/61	COPLESTON
N662 1947	90825	AAF		COPLEY
N30 1949	748217	RAFVR		COPLEY
N94 1950	91133	RAuxAF		COPLEY-SMITH
N527 1948	747566	RAFVR		COPPERWAITE
N257 1950	808321	RAuxAF		COPPICK
N196 1959	2688601	RAFVR	wef 19/8/58	CORAM
N28 1953	2683709	RAuxAF		CORBETT
N694 1948	126536	RAFVR	DFC	CORBIN
N594 1946	100598	RAFVR		CORDELL
N675 1950	844075	RAuxAF		CORDERY
N513 1952	169952	RAFVR		CORDES
N708 1959	146246	RAFVR	wef 11/1/59	CORIGAN
N846 1947	84300	RAFVR		CORKEN
N179 1950	90502	RAuxAF	TD	CORLETT
N456 1948	873175	RAuxAF		CORMACK
N761 1956	193473	RAuxAF	AFC	CORMACK
N913 1951	870855	RAuxAF		CORNER
N82 1946	136806	RAFVR	DFM	CORNES
N799 1954	73765	RAFVR		CORNEWALL-WALKER

Surname	Initials			Rank	Status	Date
CORNFORTH	J	L		SGT		10/04/1952
CORNISH	F	E		CPL		27/03/1952
CORNISH	R			CPL		31/10/1956
CORNISH	T	H		WO		28/02/1946
CORNWALL	A	M		LAC		15/09/1949
CORNWALL-LEGH	C	L	S	FL		10/08/1950
CORRY	B	G		WING COMM		02/09/1943
CORRY	N	H		SQD LDR	Acting	17/04/1947
CORRY	S	T		WING COMM		24/06/1943
CORSER	D	S		FO		15/09/1949
COSBY	E	T		FL		24/01/1946
COSSAR	G	D		FL		09/11/1944
COSSEY	A			FLT SGT		07/02/1952
COSSEY	A			FLT SGT		07/02/1952
COTTER	A	J		LAC		01/12/1955
COTTER	J	M		CPL		03/12/1993
COTTERELL	A	P		FL		13/11/1947
COTTERELL	J			LAC		09/11/1950
COTTERILL	D			SAC		01/01/1994
COTTINGHAM	T			SGT		31/01/1998
COTTON	H	A		CPL		30/09/1948
COTTON	J	H		FL		26/01/1950
COTTON	K	R		FL		21/01/1959
COTTRELL	C	G	V	FO		13/04/1944
COUCH	R	H		CPL		15/07/1948
COUCHMAN	R	J		FL		23/05/1946
COUCHMAN	R	T		CPL		06/09/1951
COULSON	J			CPL		01/07/1948
COULTHARD	E	D		FLT OFF		27/03/1947
COUNTER	C	F		FL		06/03/1947
COURT	A	J		SGT		13/01/1949
COURT	T	E	G	FL	Acting	13/11/1947
COURTNEY	O	H	B	FL		30/03/1944
COUSINS	A	E		SGT		07/08/1947
COUSINS	B	C		FLT SGT		07/04/1955
COUSINS	G	H	F	FL		09/12/1959
COUSINS	G	W		SGT		27/04/1950
COUSINS	S	P		SGT		26/08/1948
COUSINS	W	H		SGT		16/03/1950
COUSSENS	H	W		FL		24/01/1946
COUSSENS	M	W		FL		04/03/1944
COUTTS	R	W		FL		28/02/1946
COUTTS	W	F		SGT	Temporary	15/11/1961
COUZENS	D	H		SGT		07/07/1949
COVELL	H	C		WING COMM		10/06/1948
COWAN	R	J		WO		07/01/1943
COWAN	R	J		WO		18/10/1951
COWAN	W	H		FLT SGT		08/05/1947
COWEN	F	B		FO		01/10/1953
COWEY	R	L		FO		13/01/1944
COWIE	G	M		CPL	Temporary	07/12/1960
COWLAM	C	R		FLT SGT		07/08/1947
COWLAM	J	L		FLT SGT		17/04/1947
COWLEY	D	M		FL		08/11/1945
COWLING	D	J		LAC		22/02/1951
COWLING	J	T		FO		01/12/2000
COX	B	F		SGT	Acting	20/01/1960
COX	C	A		SGT		29/06/1944
COX	D	F		FL		30/09/1954
COX	D	G	S	R	WING COMM	Acting	18/04/1946

Source	Service No	Force	Notes	Surname
N278 1952	857505	RAuxAF		CORNFORTH
N224 1952	846482	RAuxAF		CORNISH
N761 1956	701511	RAF		CORNISH
N197 1946	805033	AAF		CORNISH
N953 1949	803612	RAuxAF		CORNWALL
N780 1950	90837	RAuxAF		CORNWALL-LEGH
N918 1943	90033	AAF	DFC	CORRY
N289 1947	80544	RAFVR	DFC	CORRY
N677 1943	90034	AAF		CORRY
N953 1949	126106	RAFVR		CORSER
N82 1946	157403	RAFVR		COSBY
N1150 1944	116981	RAFVR		COSSAR
N103 1952	2686008	RAuxAF		COSSEY
N103 1952	2686008	RAuxAF	1st Clasp	COSSEY
N895 1955	844501	RAuxAF		COTTER
YB 1996	n/a	RAuxAF		COTTER
N933 1947	91033	AAF		COTTERELL
N1113 1950	863456	RAuxAF		COTTERELL
YB 2003	n/a	RAuxAF		COTTERILL
YB 2001	n/a	RAuxAF		COTTINGHAM
N805 1948	864893	RAuxAF		COTTON
N94 1950	73528	RAFVR		COTTON
N28 1959	154841	RAFVR	wef 28/6/57	COTTON
MOD 1964	119358	RAFVR		COTTRELL
N573 1948	845092	RAuxAF		COUCH
N468 1946	133205	RAFVR	DFM	COUCHMAN
N913 1951	841077	RAuxAF		COUCHMAN
N527 1948	866294	RAuxAF		COULSON
N250 1947	1616	WAAF	MBE	COULTHARD
N187 1947	85689	RAFVR	DFC	COUNTER
N30 1949	874071	RAuxAF		COURT
N933 1947	159800	RAFVR		COURT
N281 1944	84707	RAFVR		COURTNEY
N662 1947	869525	AAF		COUSINS
N277 1955	2605012	RAFVR		COUSINS
N843 1959	187009	RAFVR	wef 31/3/54	COUSINS
N406 1950	756078	RAFVR		COUSINS
N694 1948	747383	RAFVR		COUSINS
N257 1950	744445	RAFVR		COUSINS
N82 1946	120161	RAFVR		COUSSENS
MOD 1964	105871	RAFVR		COUSSENS
N197 1946	60516	RAFVR	DFC	COUTTS
N902 1961	750871	RAFVR	wef 24/7/44	COUTTS
N673 1949	818123	RAuxAF		COUZENS
N456 1948	90653	RAuxAF		COVELL
N3 1943	800131	AAF		COWAN
N1077 1951	800131	RAuxAF	1st Clasp	COWAN
N358 1947	802433	AAF		COWAN
N796 1953	149992	RAuxAF	MC,TD	COWEN
N11 1944	126097	RAFVR	DFC	COWEY
N933 1960	817152	AAF	wef 13/6/44	COWIE
N662 1947	770350	RAFVR		COWLAM
N289 1947	770348	RAFVR		COWLAM
N1232 1945	87033	RAFVR		COWLEY
N194 1951	810125	RAuxAF		COWLING
YB 2001	2639453	RAuxAF		COWLING
N33 1960	2692546	RAuxAF	wef 14/11/59	COX
N652 1944	863427	AAF		COX
N799 1954	162207	RAFVR		COX
N358 1946	101041	RAFVR	DFC	COX

Surname	Initials			Rank	Status	Date
COX	E	C		FL		28/02/1946
COX	F	E	C	FL		20/07/1960
COX	G	A	R	LAC		20/05/1948
COX	G	A		CPL		01/10/1953
COX	G	J		CPL		21/09/1995
COX	I	W	E	NAV		05/02/1948
COX	J	F		SQD LDR	Acting	28/02/1946
COX	J	F		SGT		05/02/1948
COX	L	J		FL	Acting	27/03/1947
COX	M	C		WO		13/11/1947
COX	N	O		SGT		16/03/1950
COX	O			LAC		27/04/1950
COX	R	D		SGT		01/07/1948
COX	S	A		CPL		04/11/1948
COX	W	G		SGT		20/03/1957
COXHILL	H	F		LAC		13/01/1949
CRABBE	C	W	S	SQD LDR	Acting	01/07/1948
CRABTREE	D	B		FL		01/03/1945
CRABTREE	K			FL		15/07/1948
CRACKNELL	A			LAC		25/03/1959
CRACKNELL	D	A		SQD LDR	Acting	12/08/1943
CRAGG	F	T		SQD LDR	Acting	20/09/1945
CRAIG	F	T		FO		17/01/1952
CRAIG	G	A	E	FL	Acting	10/06/1948
CRAIG	G	D		FL		20/09/1945
CRAIG	I	M		FLT OFF		01/07/1948
CRAIG	R	H		SQD LDR		06/03/1947
CRAIG	T	H		LAC		13/01/1949
CRAIG	W	M		CPL		10/02/1949
CRAIG-CAMERON	G			FLT SGT		07/08/1947
CRANE	F	H		CPL		10/02/1949
CRANE	H	E		FO		14/10/1948
CRANE	W	E	S	FLT SGT		01/05/1952
CRANMER	A	H	A	C	SQD LDR	20/04/1944
CRANMORE	E			LAC		13/01/1949
CRANWELL	E	W		FL		28/06/1945
CRAPPER	E			CPL		06/09/1951
CRAWFORD	D			LAC		10/02/1949
CRAWFORD	J			AC1		02/08/1951
CRAWFORD	V	L		SGT		01/07/1948
CRAWFORD	W	W		CPL		16/10/1963
CRAWLEY	A	M		SGT		17/03/1949
CREAN	E	J		FL		18/02/1954
CREAN	J			CPL		14/10/1943
CREASE	W	F		SGT		26/08/1948
CREASLEY	T			CPL		04/11/1948
CREED	D	A		LAC		07/02/1952
CRERAR	F			GRP CAP	Acting	03/12/1942
CRESSFORD	C	S		SGT		06/03/1947
CRICHTON	T	B		FL		26/04/1956
CRICKMORE	W	G		FL	Acting	05/02/1948
CRIDDLE	H			CPL		26/08/1948
CRIMMINGS	M			CPL		20/07/1960
CRIPPS	K	M		FL		24/11/1949
CRISP	R	J		FO		25/06/1990
CRITCHLEY	J	A	C	WING COMM	Acting	17/04/1947
CRITTEN	W	H		CPL		16/10/1947
CROCKER	F	C		SGT		05/06/1947
CROCKER	F	H		FL		26/11/1958
CROFT	A			SQD LDR	Acting	05/01/1950

Source	Service No	Force	Notes	Surname
N197 1946	77201	RAFVR		COX
N531 1960	03/02/2123	RAuxAF	wef 2/2/60	COX
N395 1948	840853	RAuxAF		COX
N796 1953	840224	RAuxAF		COX
YB 1996	n/a	RAuxAF		COX
N87 1948	747745	RAFVR		COX
N197 1946	81067	RAFVR		COX
N87 1948	852242	RAuxAF		COX
N250 1947	157819	RAFVR		COX
N933 1947	881189	WAAF		COX
N257 1950	841871	RAuxAF		COX
N406 1950	756147	RAFVR		COX
N527 1948	741248	RAFVR		COX
N934 1948	840904	RAuxAF		COX
N198 1957	2681557	RAuxAF	wef 5/7/56	COX
N30 1949	756416	RAFVR		COXHILL
N527 1948	91027	RAuxAF		CRABBE
N225 1945	125730	RAFVR		CRABTREE
N573 1948	172087	RAFVR		CRABTREE
N196 1959	870836	AAF	wef 3/6/44	CRACKNELL
N850 1943	72985	RAFVR	DFC	CRACKNELL
N1036 1945	72087	RAFVR		CRAGG
N35 1952	169099	RAFVR		CRAIG
N456 1948	110904	RAFVR		CRAIG
N1036 1945	92085	AAF		CRAIG
N527 1948	1826	WAAF		CRAIG
N187 1947	72068	RAFVR		CRAIG
N30 1949	873963	RAuxAF		CRAIG
N123 1949	873032	RAuxAF		CRAIG
N662 1947	820038	AAF		CRAIG-CAMERON
N123 1949	861143	RAuxAF		CRANE
N856 1948	176109	RAFVR		CRANE
N336 1952	816042	RAuxAF		CRANE
N363 1944	70147	RAFVR		CRANMER
N30 1949	873776	RAuxAF		CRANMORE
N706 1945	141532	RAFVR	DFC	CRANWELL
N913 1951	869321	RAuxAF		CRAPPER
N123 1949	874056	RAuxAF		CRAWFORD
N777 1951	873027	RAuxAF		CRAWFORD
N527 1948	752745	RAFVR		CRAWFORD
N756 1963	2685674	RAFVR	wef 11/7/63	CRAWFORD
N231 1949	800592	RAuxAF		CRAWLEY
N137 1954	202706	RAuxAF		CREAN
N1065 1943	803498	AAF		CREAN
N694 1948	846155	RAuxAF		CREASE
N934 1948	752707	RAFVR		CREASLEY
N103 1952	813253	RAuxAF		CREED
N1638 1942	90368	AAF	CBE	CRERAR
N187 1947	803508	AAF		CRESSFORD
N318 1956	188818	RAFVR		CRICHTON
N87 1948	120838	RAFVR		CRICKMORE
N694 1948	770415	RAFVR		CRIDDLE
N531 1960	2661079	WRAAF	wef 18/5/60	CRIMMINGS
N1193 1949	74815	RAFVR		CRIPPS
YB 1996	2625179	RAuxAF		CRISP
N289 1947	73139	RAFVR		CRITCHLEY
N846 1947	811179	AAF		CRITTEN
N460 1947	813061	AAF		CROCKER
N807 1958	164153	RAFVR	wef 10/7/53	CROCKER
N1 1950	72636	RAFVR		CROFT

Surname	Initials			Rank	Status	Date
CROFTS	A	L		CPL		30/09/1948
CROFTS	A			CPL		05/06/1947
CROFTS	J	M	F	PO		06/01/1955
CROMPTON	I	D		CPL		25/03/1959
CROMPTON	P	R		WING COMM	Acting	04/10/1945
CRONE	R	A		FL		24/01/1946
CROOK	H	K		FL		10/05/1945
CROOKS	H	E		CPL		10/02/1949
CROPPER	E	S	G	FL		07/08/1947
CROSBY	J	C	R	LAC		27/03/1947
CROSDALE	F			FO		29/04/1954
CROSKELL	M	E		FL		12/04/1945
CROSS	J	R		CPL		01/07/1948
CROSS	R	J		FO		19/10/1960
CROSS	R	W	S	WING COMM		17/04/1947
CROSSBY	J	W		FL		02/07/1958
CROSSE	A	A		FL		19/10/1960
CROSSE	A	A		FL		13/07/1964
CROSSFIELD	J	E		FL		15/01/1958
CROTTY	E	P		FL		01/05/1957
CROUCH	F	W	J	SGT	Temporary	20/01/1960
CROUCHER	G	V		CPL		23/02/1950
CROW	R			FO		08/11/1945
CROWLEY	T			SQD LDR		05/01/1950
CROWLEY-MILLING	D			WING COMM	Acting	26/10/1944
CROWTHER	L	M		GRP OFF		12/06/1947
CROWTHER	N	C		FL		30/03/1960
CROXALL	W			FO		19/08/1954
CROXALL	W			FL		14/11/1962
CROYDEN	S	B		SQD LDR	Acting	28/10/1943
CROZIER	L	A		FL		21/09/1944
CRUDGINGTON	G	C	B	LAC		10/02/1949
CRUDINGTON	S			FL		12/06/1993
CRUICKSHANK	H	L		SQD LDR	Acting	28/10/1943
CRUICKSHANKS	I	J	A	FL		29/06/1944
CRUM	H	V		WO		08/11/1945
CRUMP	R	I		SQD LDR	Acting	28/02/1946
CRUMPTON	F	G		AC1		19/08/1954
CRUSE	E	W	G	CPL	Temporary	26/02/1958
CRUSH	T	G		SQD LDR	Acting	10/06/1948
CUBITT	K	H		FL		18/04/1946
CUBITT	M	S		SAC		25/11/1997
CUFF	H	J		WING COMM		18/03/1948
CUFF	W	J	J	FL		26/11/1958
CULLEN	H	J		SGT		23/08/1951
CULLEN	M	J		WO		18/10/1951
CULLEN	W	J		SGT		26/04/1961
CULLERENE	P	F		SGT		27/04/1950
CULLEY	S			CPL		27/03/1952
CULMER	J	D		FO		28/11/1946
CULPITT	J	V		SQD LDR		31/01/1998
CUMBER	J	E		FL		06/03/1947
CUMBERS	A	B		FL		01/07/1948
CUMMINGS	J			LAC		16/10/1947
CUMMINGS	J			CPL		21/10/1959
CUNINGHAM	T			AC1		25/06/1953
CUNNINGHAM	B	H		SQD LDR		05/06/1947
CUNNINGHAM	C	G		LAC		12/06/1947
CUNNINGHAM	G	C		SQD LDR		10/08/1998
CUNNINGHAM	H	V		AC1		13/10/1955

Source	Service No	Force	Notes	Surname
N805 1948	868591	RAuxAF		CROFTS
N460 1947	870163	AAF		CROFTS
N1 1955	2693000	RAuxAF		CROFTS
N196 1959	2657122	WRAAF	wef 27/6/58	CROMPTON
N1089 1945	72250	RAFVR	DFC,AFC	CROMPTON
N82 1946	130416	RAFVR	DFC	CRONE
N485 1945	63789	RAFVR		CROOK
N123 1949	813231	RAuxAF		CROOKS
N662 1947	73141	RAFVR		CROPPER
N250 1947	808176	AAF		CROSBY
N329 1954	2603550	RAuxAF		CROSDALE
N381 1945	124118	RAFVR		CROSKELL
N527 1948	815084	RAuxAF		CROSS
N775 1960	143555	RAFVR	wef 17/10/59	CROSS
N289 1947	72126	RAFVR		CROSS
N452 1958	105341	RAuxAF	wef 23/9/54	CROSSBY
N775 1960	125492	RAFVR	wef 13/7/54	CROSSE
MOD 1967	125492	RAFVR	1st Clasp	CROSSE
N40 1958	143292	RAFVR	DFC; wef 30/9/57	CROSSFIELD
N277 1957	144424	RAFVR	wef 9/12/53	CROTTY
N33 1960	862508	AAF	wef 24/5/44	CROUCH
N179 1950	813073	RAuxAF		CROUCHER
N1232 1945	183575	RAFVR		CROW
N1 1950	72535	RAFVR	MB,ChB	CROWLEY
N1094 1944	78274	RAFVR	DSO,DFC	CROWLEY-MILLING
N478 1947	139	WAAF	MBE	CROWTHER
N229 1960	104626	RAFVR	wef 21/12/59	CROWTHER
N672 1954	173713	RAFVR		CROXALL
N837 1962	173713	RAuxAF	1st Clasp; wef 12/3/62	CROXALL
N1118 1943	70151	RAFVR		CROYDEN
N961 1944	141472	RAFVR		CROZIER
N123 1949	847045	RAuxAF		CRUDGINGTON
YB 1996	2629291	RAuxAF		CRUDINGTON
N1118 1943	70152	RAFVR		CRUICKSHANK
N652 1944	80819	RAFVR		CRUICKSHANKS
N1232 1945	740587	RAFVR	DFM	CRUM
N197 1946	112375	RAFVR	AFC	CRUMP
N672 1954	850660	RAuxAF		CRUMPTON
N149 1958	752299	RAuxAF	wef 30/7/44	CRUSE
N456 1948	141610	RAFVR		CRUSH
N358 1946	84297	RAFVR		CUBITT
YB 1998	n/a	RAuxAF		CUBITT
N232 1948	90706	RAuxAF		CUFF
N807 1958	166652	RAFVR	wef 9/3/53	CUFF
N866 1951	740321	RAFVR		CULLEN
N1077 1951	751461	RAFVR	DFC	CULLEN
N339 1961	2695600	RAFVR	wef 14/1/61	CULLEN
N406 1950	749855	RAFVR		CULLERENE
N224 1952	870110	RAuxAF		CULLEY
N1004 1946	177211	RAFVR	DFM	CULMER
YB 1999	2834721	RAuxAF		CULPITT
N187 1947	118536	RAFVR		CUMBER
N527 1948	118713	RAFVR	DFM	CUMBERS
N846 1947	816227	AAF		CUMMINGS
N708 1959	2681007	RAuxAF	1st Clasp; wef 19/2/54	CUMMINGS
N527 1953	843994	RAuxAF		CUNINGHAM
N460 1947	72183	RAFVR		CUNNINGHAM
N478 1947	843896	AAF		CUNNINGHAM
YB 1999	212244	RAuxAF		CUNNINGHAM
N762 1955	847276	RAuxAF		CUNNINGHAM

Surname	Initials			Rank	Status	Date
CUNNINGHAM	J			WING COMM		04/02/1943
CUNNINGHAM	J			FL	Acting	14/09/1950
CUNNINGHAM	J			FL		16/04/1958
CUNNINGHAM	R	A		CPL	Temporary	06/06/1962
CUNNINGHAM	R	G	H	SQD LDR		10/08/1950
CUPIT	J	W		CPL		01/07/1948
CURBESON	F	L		FO		28/02/1946
CURD	A	H	A	FL	Acting	05/02/1948
CURNOW	J			WING COMM		01/05/1998
CURRALL	G	H		WING COMM		06/03/1947
CURRAN	S	J		SAC		09/01/1963
CURRIE	W	S		CPL		26/01/1950
CURRY	G	W		WING COMM	Acting	06/03/1947
CURTIS	E	W	W	WO		05/01/1950
CURTIS	W	J		CPL		21/10/1959
CUTHBERT	A	J		SQD LDR	Acting	21/09/1944
CUTHBERT	C	J		SGT		07/07/1949
CUTHBERTSON	J	T		FL	Acting	15/07/1948
CUTHERTSON	R	W	S	FLT SGT		17/03/1949
CUTLER	A	W	J	LAC		17/03/1949
CUTLER	W	E	G	FL		14/10/1943
CUTMORE	R	H		FL		06/03/1947
CUTRESS	G	R		FLT SGT		27/03/1947
Da COSTA	A			SQD LDR	Acting	13/12/1945
DACKHAM	J	W		PO	Acting	08/05/1947
DACRE	D			FL		12/06/1947
DAGGETT	G	W		FL		26/11/1958
DAILEY	W	S		WING COMM		28/09/1950
DALE	M	W	A	FO		18/03/1948
DALE	P	J		SQD LDR	Acting	08/11/1945
DALES	A	H		CPL		05/06/1947
DALEY	F			LAC		01/07/1948
DALLEY	K	S		SAC		18/04/1999
DALRYMPLE	I	K		FL		06/05/1965
DALRYMPLE	W	G		SGT		23/02/1950
DALTON	R	W		SQD LDR	Acting	06/03/1947
DALTON	W	J		FL		08/11/1945
D'ALTON	D	J		SQD LDR	Acting	01/08/1946
DALY	G	A		FL		09/08/1945
DALY	T			LAC		05/01/1950
DALZIEL	J	G		FL		22/04/1943
DAMMS	H			SGT		14/10/1948
DAND	F	J		LAC		20/05/1948
DANDO	P	F		FLT SGT		01/07/1948
DANDY	W	H		FL		16/01/1957
DANIEL	J	W		LAC		04/11/1948
DANIELS	S	P		WING COMM	Acting	18/04/1946
DANKS	G	H	C	CPL		01/07/1948
DANTON	W	F		SQD LDR		06/03/1947
DARBY	E	R	B	FL	Acting	15/09/1949
DARBY	J			FL		07/04/1955
DARBY	L	H		FL		06/03/1947
DARE	T			CPL		27/11/1952
DARGAN	S			SQD LDR		11/03/1998
DARKE	P	J		FL		24/01/1946
DARKENS	S	B		SGT		01/05/1995
DARLASTON	A	C		FLT SGT	Acting	02/09/1943
DARLING	G	G		FO		04/09/1947
DARLINGTON	F	W	H	SQD LDR		12/04/1945
DARLINGTON	H			SGT	Acting	06/01/1955

Source	Service No	Force	Notes	Surname
N131 1943	90216	AAF	DSO,DFC	CUNNINGHAM
N926 1950	135776	RAFVR		CUNNINGHAM
N258 1958	135776	RAFVR	1st Clasp; wef 22/3/56	CUNNINGHAM
N435 1962	819051	AAF	wef 31/12/43	CUNNINGHAM
N780 1950	72014	RAFVR	MB,ChB	CUNNINGHAM
N527 1948	854251	RAuxAF		CUPIT
N197 1946	175217	RAFVR		CURBESON
N87 1948	139791	RAFVR		CURD
YB 2000	4231501	RAuxAF		CURNOW
N187 1947	73635	RAFVR	MM	CURRALL
N22 1963	2681158	RAuxAF	wef 19/10/62	CURRAN
N94 1950	816166	RAuxAF		CURRIE
N187 1947	86389	RAFVR	DSO,DFC	CURRY
N1 1950	745916	RAFVR		CURTIS
N708 1959	2689604	RAFVR	wef 4/2/59	CURTIS
N961 1944	80806	RAFVR		CUTHBERT
N673 1949	857559	RAuxAF		CUTHBERT
N573 1948	171531	RAFVR		CUTHBERTSON
N231 1949	743033	RAFVR		CUTHERTSON
N231 1949	845582	RAuxAF		CUTLER
N1065 1943	70157	RAFVR		CUTLER
N187 1947	86266	RAFVR		CUTMORE
N250 1947	800521	AAF		CUTRESS
N1355 1945	86409	RAFVR	AFC	Da COSTA
N358 1947	162757	RAFVR		DACKHAM
N478 1947	138096	RAFVR		DACRE
N807 1958	161271	RAFVR	DFC; wef 14/2/55	DAGGETT
N968 1950	72811	RAFVR	FORFEITED UNDER N777 1951	DAILEY
N232 1948	147854	RAFVR		DALE
N1232 1945	87406	RAFVR		DALE
N460 1947	841689	AAF		DALES
N527 1948	854308	RAuxAF		DALEY
YB 1999	n/a	RAuxAF		DALLEY
MOD 1966	2447739	RAFVR		DALRYMPLE
N179 1950	743768	RAFVR		DALRYMPLE
N187 1947	115715	RAFVR	DFM	DALTON
N1232 1945	146334	RAFVR	DFM	DALTON
N652 1946	77206	RAFVR	DFC,AFC	D'ALTON
N874 1945	106357	RAFVR		DALY
N1 1950	857462	RAuxAF		DALY
N475 1943	104593	RAFVR		DALZIEL
N856 1948	869364	RAuxAF		DAMMS
N395 1948	849073	RAuxAF		DAND
N527 1948	881436	WAAF		DANDO
N24 1957	131853	RAuxAF		DANDY
N934 1948	847345	RAuxAF		DANIEL
N358 1946	81676	RAFVR	DSO,DFC	DANIELS
N527 1948	864809	RAuxAF		DANKS
N187 1947	90917	AAF		DANTON
N953 1949	121961	RAFVR		DARBY
N277 1955	150680	RAFVR		DARBY
N187 1947	87086	RAFVR		DARBY
N915 1952	846607	RAuxAF		DARE
YB 1999	213718	RAuxAF		DARGAN
N82 1946	82694	RAFVR	AFC	DARKE
YB 1996	n/a	RAuxAF		DARKENS
N918 1943	805234	AAF		DARLASTON
N740 1947	127583	RAFVR		DARLING
N381 1945	80823	RAFVR	AFC	DARLINGTON
N1 1955	2650026	RAuxAF		DARLINGTON

Surname	Initials			Rank	Status	Date
DARLOW	A			FL		30/09/1948
DART	E	W		SGT		25/06/1953
DARWIN	G	W	L	SQD LDR		10/12/1942
DAVEY	H	W		FO		10/06/1948
DAVEY	J	B		FL		16/01/1957
DAVEY	J	B		SQD LDR	Acting	04/07/1965
DAVID	E	J		FL		02/07/1958
DAVIDSON	T	S		CPL		30/03/1944
DAVIDSON	W			LAC		01/10/1953
DAVIE	E			LAC		09/01/1963
DAVIES	B	D		WO		26/02/1993
DAVIES	C	D		FO		07/07/1949
DAVIES	C	N		FLT SGT		06/03/1947
DAVIES	C	N		FLT SGT		20/09/1951
DAVIES	C			FL		08/04/2000
DAVIES	D	E		WING COMM	Acting	20/04/1944
DAVIES	D	G		WO		08/07/1993
DAVIES	E	J		LAC		05/06/1947
DAVIES	E	J		LAC		01/12/1955
DAVIES	E	St.	H	FL		06/03/1947
DAVIES	E			SQD OFF		06/03/1947
DAVIES	G	M		SGT		20/05/1948
DAVIES	H	G		FL		07/07/1949
DAVIES	H	J		WING COMM	Acting	08/11/1945
DAVIES	H	P		FL		20/09/1951
DAVIES	J	M		FLT OFF	Acting	24/11/1949
DAVIES	J			CPL		05/06/1947
DAVIES	M	P		FL		05/09/1946
DAVIES	P	B		SQD LDR		06/03/1947
DAVIES	R			SAC		25/11/1996
DAVIES	S	T	C	SGT		23/02/1950
DAVIES	S			FL		01/08/1946
DAVIES	T	A		CPL		10/02/1949
DAVIES	V			FL		09/01/1963
DAVIES	W	H		LAC		30/10/1957
DAVIES	W	L		CPL	Temporary	13/10/1955
DAVIS	A	F	E	FL		13/12/1945
DAVIS	A	F		SAC		20/01/1960
DAVIS	F	C		SQD LDR		27/03/1947
DAVIS	F	P		WING COMM	Acting	25/05/1944
DAVIS	H	A		SGT		13/11/1947
DAVIS	H	S		CPL		07/07/1949
DAVIS	J	G		FL		09/08/1945
DAVIS	J	G		WING COMM	Acting	07/07/1949
DAVIS	J	G		SQD LDR	Acting	15/01/1958
DAVIS	J			FL		18/03/1948
DAVIS	L	F		CPL		27/03/1947
DAVIS	N			FL		09/12/1959
DAVIS	P	B	N	WING COMM	Acting	30/12/1943
DAVIS	S	L		FL	Acting	07/07/1949
DAVIS	W	J		WO		30/03/1943
DAVIS	W	L		FL		20/09/1945
DAVISON	B	M		SQD LDR	Acting	18/03/1948
DAVISON	J			FL		05/09/1962
DAVISON	W	K		WING COMM	Acting	24/11/1949
DAVY	S			WO		12/04/1945
DAWES	J	S		FLT SGT		23/02/1950
DAWES	L	G		FO		28/02/1946
DAWES	L	H		FL		17/04/1947
DAWSON	C	L		SQD LDR		10/06/1948

Source	Service No	Force	Notes	Surname
N805 1948	84341	RAFVR		DARLOW
N527 1953	840196	RAuxAF		DART
N1680 1942	90065	AAF		DARWIN
N456 1948	191955	RAFVR		DAVEY
N24 1957	164526	RAFVR		DAVEY
MOD 1967	164526	RAFVR	1st Clasp	DAVEY
N452 1958	63885	RAuxAF	BSc; wef 20/9/54	DAVID
N281 1944	803351	AAF		DAVIDSON
N796 1953	874126	RAuxAF		DAVIDSON
N22 1963	802541	AAF	wef 29/11/43	DAVIE
YB 1997	C3523903	RAuxAF		DAVIES
N673 1949	110402	RAFVR		DAVIES
N187 1947	810054	AAF		DAVIES
N963 1951	810054	RAuxAF	1st Clasp	DAVIES
YB 2002	n/a	RAuxAF		DAVIES
N363 1944	72987	RAFVR	DFC,AFC	DAVIES
YB 1996	n/a	RAuxAF		DAVIES
N460 1947	845433	AAF		DAVIES
N895 1955	845433	RAuxAF		DAVIES
N187 1947	90505	AAF		DAVIES
N187 1947	299	WAAF		DAVIES
N395 1948	812146	RAuxAF		DAVIES
N673 1949	150046	RAFVR		DAVIES
N1232 1945	102577	RAFVR	DSO,DFC	DAVIES
N963 1951	91091	RAuxAF		DAVIES
N1193 1949	820	WAAF		DAVIES
N460 1947	853704	AAF		DAVIES
N751 1946	119872	RAFVR		DAVIES
N187 1947	73154	RAFVR		DAVIES
YB 1997	n/a	RAuxAF		DAVIES
N179 1950	863304	RAuxAF		DAVIES
N652 1946	144755	RAFVR	DFC	DAVIES
N123 1949	813123	RAuxAF		DAVIES
N22 1963	132228	RAFVR	wef 19/10/62	DAVIES
N775 1957	799667	RAFVR	wef 2/6/44	DAVIES
N762 1955	859650	RAuxAF		DAVIES
N1355 1945	129441	RAFVR		DAVIS
N33 1960	2683317	RAuxAF	wef 22/6/44	DAVIS
N250 1947	78878	RAFVR		DAVIS
N496 1944	78721	RAFVR		DAVIS
N933 1947	819001	AAF		DAVIS
N673 1949	865661	RAuxAF		DAVIS
N874 1945	148089	RAFVR	DFC	DAVIS
N673 1949	72313	RAFVR		DAVIS
N40 1958	72313	RAFVR	1st Clasp; wef 17/7/55	DAVIS
N232 1948	111700	RAFVR		DAVIS
N250 1947	861930	AAF		DAVIS
N843 1959	120676	RAFVR	wef 14/10/59	DAVIS
N1350 1943	70499	RAFVR		DAVIS
N673 1949	141617	RAFVR		DAVIS
N1014 1943	804060	AAF		DAVIS
N1036 1945	61922	RAFVR		DAVIS
N232 1948	78135	RAFVR		DAVISON
N655 1962	159135	RAFVR	wef 17/7/62	DAVISON
N1193 1949	73728	RAFVR	DSO,DFC; The Honorable	DAVISON
N381 1945	741590	RAFVR		DAVY
N179 1950	872299	RAuxAF		DAWES
N197 1946	145143	RAFVR		DAWES
N289 1947	112699	RAFVR	DFC	DAWES
N456 1948	73553	RAFVR		DAWSON

Surname	Initials				Rank	Status	Date
DAWSON	C				FL		21/01/1959
DAWSON	G	M			FLT SGT		20/05/1948
DAWSON	M	E	H		FL		01/08/1946
DAWSON	P	H			FL		30/11/1944
DAWSON	S	R			FL		28/02/1946
DAWSON	S				SQD LDR		01/05/1999
DAWSON	S				SQD LDR		01/04/1997
DAWSON	W	J			FL		09/11/1950
DAY	A	C			SGT		02/11/1995
DAY	A	H			SGT		26/08/1948
DAY	A	W			FL		19/08/1954
DAY	F	B			FL		08/11/1945
DAY	F	W	E		SGT		07/08/1947
DAY	F	W	E		FLT SGT		04/10/1951
DAY	R	H			FL		31/05/1945
DAY	R	S			SQD LDR	Acting	25/01/1945
DAYKIN	G	A			FL		20/04/1944
DAYKIN	J	W			FO		17/02/1944
DAYNES	F	W			CPL		30/09/1948
DAYUS	E	C			SQD LDR		26/11/1958
De ATH	R				FL		26/11/1958
De BEAUMONT	C	L	L	A	WING COMM	Acting	16/10/1947
De BROKE	W				GRP CAP	Acting	10/12/1942
De FAUQUE De JONQUIERES	M	V	J		SQD LDR		02/09/1943
De LARA	L	G	C		FL		05/10/1944
De LASZLO	J	A			WING COMM		06/03/1947
De LITTLE	J				WO		28/02/1946
De ROHAN	L	F			FL		16/10/1963
De VILLE	E				FL	Acting	30/09/1948
De VROME	F	W			SGT		18/11/1948
Del STROTHER	E	A			FO		07/07/1949
DEACON	G	A			WO		27/03/1947
DEAKIN	N	A			FL		20/11/1957
DEAN	C				LAC		24/11/1949
DEAN	H	P			SQD LDR		14/10/1943
DEAN	L	A			FL		26/11/1958
DEAN	R	G			FL		01/09/1960
DEAN	R	L			PO		28/02/1946
DEANE	D	H			FLT OFF		15/07/1948
DEANESLY	E	C			SQD LDR		03/12/1942
DEARDEN	J				CPL		21/06/1961
DEBENHAM	A	I	S		WING COMM	Acting	04/05/1944
DEBENHAM	C	H			FL		05/10/1944
DEHN	H	G			WING COMM	Acting	15/07/1948
DELLER	A	L	M		FL		30/08/1945
DEMEL	W	H			WING COMM		17/03/1949
DENBY	S				CPL		07/02/1952
DENHAM	C	H			FL		26/01/1950
DENHOLM	F	M			SQD OFF	Acting	27/03/1947
DENHOLM	G	L			WING COMM		03/12/1942
DENISON	A	A			WING COMM	Acting	26/08/1948
DENMAN	F	P			FL		05/06/1947
DENNETT	E	G			LAC		07/07/1949
DENNIS	A	L			FLT OFF		03/07/1952
DENNIS	B				SGT		14/10/1948
DENNIS	D	F			WING COMM	Acting	25/01/1945
DENNIS	J	N			FL		20/09/1945
DENNISON	A	G			WING COMM	Acting	04/02/1943
DENNISON	R	L			LAC		19/10/1960
DENT	A	W			FL		13/01/1949

Source	Service No	Force	Notes	Surname
N28 1959	164013	RAFVR	wef 29/4/53	DAWSON
N395 1948	815102	RAuxAF		DAWSON
N652 1946	112407	RAFVR	DFC,DFM	DAWSON
N1225 1944	84975	RAFVR		DAWSON
N197 1946	142531	RAFVR	DFC,DFM	DAWSON
YB 2000	4233276	RAuxAF		DAWSON
YB 1997	4233276	RAuxAF		DAWSON
N1113 1950	127161	RAFVR	DFC	DAWSON
YB 1997	n/a	RAuxAF		DAY
N694 1948	804275	RAuxAF		DAY
N672 1954	69574	RAFVR		DAY
N1232 1945	79368	RAFVR		DAY
N662 1947	803437	AAF		DAY
N1019 1951	2683508	RAuxAF	1st Clasp	DAY
N573 1945	132070	RAFVR		DAY
N70 1945	81922	RAFVR		DAY
N363 1944	89592	RAFVR		DAYKIN
N132 1944	149200	RAFVR		DAYKIN
N805 1948	840028	RAuxAF		DAYNES
N807 1958	91149	AAF	MRCS,LRCP; wef 29/8/44	DAYUS
N807 1958	202015	RAFVR	wef 21/5/52	De ATH
N846 1947	90520	AAF		De BEAUMONT
N1680 1942	90238	AAF	MC,AFC; Lord	De BROKE
N918 1943	72302	RAFVR		De FAUQUE De JONQUIERES
N1007 1944	132072	RAFVR	DFM	De LARA
N187 1947	72037	RAFVR		De LASZLO
N197 1946	745291	RAFVR	AFC	De LITTLE
N756 1963	2607065	RAFVR	wef 24/1/57	De ROHAN
N805 1948	72381	RAFVR		De VILLE
N984 1948	800490	RAuxAF	BEM	De VROME
N673 1949	155283	RAFVR		Del STROTHER
N250 1947	844744	AAF		DEACON
N835 1957	187142	RAFVR	wef 30/12/53	DEAKIN
N1193 1949	840303	RAuxAF		DEAN
N1065 1943	70166	RAFVR		DEAN
N807 1958	196980	RAFVR	wef 19/2/57	DEAN
MOD 1966	5000400	RAFVR		DEAN
N197 1946	196281	RAFVR		DEAN
N573 1948	379	WAAF		DEANE
N1638 1942	90251	AAF	DFC	DEANESLY
N478 1961	2688641	RAFVR	wef 24/6/59	DEARDEN
N425 1944	70167	RAFVR	DFC	DEBENHAM
N1007 1944	82669	RAFVR		DEBENHAM
N573 1948	90679	RAuxAF		DEHN
N961 1945	156643	RAFVR		DELLER
N231 1949	90773	RAuxAF	DFC	DEMEL
N103 1952	858163	RAuxAF		DENBY
N94 1950	91054	RAuxAF		DENHAM
N250 1947	414	WAAF		DENHOLM
N1638 1942	90190	AAF	DFC	DENHOLM
N694 1948	73612	RAFVR	MBE,MC	DENISON
N460 1947	149171	RAFVR		DENMAN
N673 1949	865609	RAuxAF		DENNETT
N513 1952	205	WAAF		DENNIS
N856 1948	841319	RAuxAF		DENNIS
N70 1945	64274	RAFVR	DSO,DFC	DENNIS
N1036 1945	134496	RAFVR		DENNIS
N131 1943	90264	AAF		DENNISON
N775 1960	2691053	RAuxAF	wef 24/2/60	DENNISON
N30 1949	91020	RAuxAF		DENT

Surname	Initials			Rank	Status	Date
DENT	W	E		CPL		26/08/1948
DENYER	E	W	G	FL		04/11/1948
DERBYSHIRE	A	H		SQD LDR		06/03/1947
DERBYSHIRE	J	N		SQD LDR		08/11/1945
DERRETT	A	E	C	SQD LDR	Acting	11/07/1946
DERRY	A	F		FL	Acting	13/11/1947
DESBOROUGH	E	R		SQD LDR		17/04/1947
DESMOND	P			FL		26/11/1958
DEUNTZER	D	C		FL		25/01/1945
DEVANEY	V	M		FLT SGT		10/08/1944
DEVENISH	W	J	J	L	SGT	01/07/1948
DEVINE	R			SGT	Temporary	09/01/1963
DEVITT	P	K		WING COMM	Acting	30/12/1943
DEVITT	P	K		WING COMM		19/08/1959
DEVLIN	J	A		SGT		07/02/1952
DEWAR	G	T		FO		01/12/1955
DEWAR	J			FL		25/01/1951
DEWEY	H	D		CAPT		21/10/1959
DEWSBURY	B	G		SQD LDR	Acting	10/02/1949
DEXTER	E	V		SGT		19/08/1954
DIAPER	E	T		FL		25/03/1959
DICCONSON	H	S		FL		26/11/1958
DICK	A			CPL		24/11/1949
DICK	D	C	M	FL		09/10/1987
DICK	I	A	G	L	WING COMM	12/07/1945
DICK	M	E		SAC		01/11/1985
DICK	M	E		SAC		01/11/1995
DICKEN	A			FL		19/09/1956
DICKEY	D	O'R		FL		15/01/1953
DICKIE	E	J		WING COMM	Acting	24/01/1946
DICKIN	E			FL		06/01/1955
DICKINSON	D	J		FL		20/03/1957
DICKINSON	W	F	R	FL		10/06/1948
DICKS	B	L		FO		24/07/1963
DICKS	T	A		FL		13/12/1945
DICKSON	A			SGT		30/03/1944
DICKSON	J	S		FL		28/02/1946
DIDWELL	R	R	J	SGT	Acting	30/03/1960
DIFFORD	H			FL		07/05/1991
DIGBY	D	M		SGT		05/02/1948
DIGBY-OVENS	G			FO		07/08/1947
DIGBY-WORSLEY	E	B		SQD LDR	Acting	26/01/1950
DIGGORY	R	F	R	FO		20/05/1948
DILLON	A			CPL		05/06/1947
DILLON	I	M		FLT SGT		06/05/1993
DILLON	J	E		CPL		28/09/1950
DIMSDALE	R	A		SQD LDR	Acting	12/08/1943
DING	D	P		FO		26/11/1958
DINGLE	G	H		FL		06/03/1947
DINNIE	N			WING OFF	Acting	08/05/1947
DINWOODIE	A			CPL	Acting	30/03/1960
DISHINGTON	J	L		FL		28/08/1963
DISNEY	H	A	S	WING COMM	Acting	30/11/1944
DISNEY-ROEBUCK	E	D		FLT OFF		15/09/1949
DIX	H	J		SGT		07/08/1947
DIX	P			FLT SGT		17/04/1947
DIXON	A	E		FLT SGT		06/03/1947
DIXON	B	W		FO		20/05/1948
DIXON	C	C		FLT SGT		01/07/1948
DIXON	D	E		FL		21/01/1959

Source	Service No	Force	Notes	Surname
N694 1948	859501	RAuxAF		DENT
N934 1948	138017	RAFVR		DENYER
N187 1947	84924	RAFVR		DERBYSHIRE
N1232 1945	77462	RAFVR	DFC	DERBYSHIRE
N594 1946	82961	RAFVR	DFC	DERRETT
N933 1947	172200	RAFVR		DERRY
N289 1947	72127	RAFVR		DESBOROUGH
N807 1958	168446	RAFVR	wef 25/7/52	DESMOND
N70 1945	111486	RAFVR		DEUNTZER
N812 1944	805236	AAF		DEVANEY
N527 1948	753248	RAFVR		DEVENISH
N22 1963	816070	AAF	wef 24/10/42	DEVINE
N1350 1943	90080	RAFRO		DEVITT
N552 1959	90080	RAuxAF	1st Clasp; wef 26/2/45	DEVITT
N103 1952	743714	RAFVR		DEVLIN
N895 1955	146639	RAuxAF		DEWAR
N93 1951	85257	RAFVR	MBE,GM	DEWAR
N708 1959	388123	AER	wef 17/11/57	DEWEY
N123 1949	91155	RAuxAF		DEWSBURY
N672 1954	801438	RAuxAF		DEXTER
N196 1959	147176	RAuxAF	wef 13/12/58	DIAPER
N807 1958	180183	RAFVR	wef 3/6/53	DICCONSON
N1193 1949	867388	RAuxAF		DICK
YB 1996	3519891	RAuxAF		DICK
N759 1945	90201	AAF		DICK
YB 1996	n/a	RAuxAF		DICK
YB 1997	n/a	RAuxAF	1st Clasp	DICK
N666 1956	114743	RAuxAF		DICKEN
N28 1953	150492	RAFVR		DICKEY
N82 1946	86423	RAFVR	MBE	DICKIE
N1 1955	134467	RAuxAF		DICKIN
N198 1957	180357	RAFVR	wef 3/1/56	DICKINSON
N456 1948	80570	RAFVR		DICKINSON
N525 1963	2680651	RAFVR(T)	wef 12/2/57	DICKS
N1355 1945	161027	RAFVR		DICKS
N281 1944	803241	AAF		DICKSON
N197 1946	90845	AAF		DICKSON
N229 1960	2694072	RAuxAF	wef 19/1/60	DIDWELL
YB 1997	210635	RAuxAF		DIFFORD
N87 1948	813157	RAuxAF		DIGBY
N662 1947	195675	RAFVR		DIGBY-OVENS
N94 1950	73636	RAFVR		DIGBY-WORSLEY
N395 1948	145616	RAFVR		DIGGORY
N460 1947	811156	AAF		DILLON
YB 1996	n/a	RAuxAF		DILLON
N968 1950	855979	RAuxAF		DILLON
N850 1943	73811	RAFVR		DIMSDALE
N807 1958	2601005	RAFVR	wef 28/6/52	DING
N187 1947	62448	RAFVR		DINGLE
N358 1947	240	WAAF		DINNIE
N229 1960	2659309	WRAAF	wef 19/1/60	DINWOODIE
N621 1963	184461	RAuxAF	DFC; wef 25/7/63	DISHINGTON
N1225 1944	72106	RAFVR		DISNEY
N953 1949	245	WAAF		DISNEY-ROEBUCK
N662 1947	750402	RAFVR		DIX
N289 1947	750367	RAFVR		DIX
N187 1947	868679	AAF		DIXON
N395 1948	183171	RAFVR		DIXON
N527 1948	770551	RAFVR		DIXON
N28 1959	103163	RAFVR	wef 6/11/54	DIXON

Surname	Initials			Rank	Status	Date
DIXON	F	A		FL	Acting	17/03/1949
DIXON	F	R		SQD LDR	Acting	05/06/1947
DIXON	F	S		FO		16/03/1950
DIXON	G			FLT SGT		10/08/1950
DIXON	J	C		FL		05/06/1947
DIXON	P	A		FO		26/11/1958
DIXON	P	T		FLT SGT		25/11/1989
DIXON	P			SQD LDR	Acting	28/02/1946
DIXON	S	R		FL		05/01/1950
DIXON	W	M		FL		06/03/1947
DOBBIN	J	N		SQD LDR	Acting	12/06/1947
DOBBS	H			FO		15/04/1943
DOBIE	R	D		AC1		23/02/1950
DOBSON	B	J		FL		28/02/1946
DOBSON	E	D	F	FL		13/02/1957
DOBSON	G	W		FO		15/06/1944
DOCHERTY	E			SAC		01/07/1989
DOCHERTY	E			CPL		01/07/1999
DODD	C	W	B	FL		09/12/1959
DODD	F	L		FL		10/05/1945
DODD	J	H		SQD LDR	Acting	22/09/1955
DODD	M			SQD LDR		16/03/1950
DODDS	J	C		SAC		22/02/1997
DODDS	L	V		GRP CAP	Acting	15/07/1948
DODKIN	W	H		SGT	Acting	22/02/1961
DODS	J	W		SGT		07/12/1950
DODWELL	S	L		FL		12/08/1943
DOELL	M	L		SAC		19/10/1994
DOLEY	A	S		FL		18/04/1946
DOLLIMORE	P			SGT		07/07/1949
DOMONEY	A	C		FL	Acting	25/01/1951
DON	R	S		FL		21/09/1944
DONALD	A			CPL		07/12/1960
DONALD	J	R		SQD LDR	Acting	24/01/1946
DONALDSON	A	S		WING COMM		07/01/1994
DONALDSON	D	W		SQD LDR		22/04/1943
DONALDSON	W	A		FLT SGT		26/08/1948
DONBAVAND	A	E		LAC		17/03/1949
DONBAVAND	E	G		AC1		12/06/1947
DONCASTER	B			LAC		28/07/1955
DONE	J	A		SIG 1		26/06/1947
DONIGER	N	A		FL		12/07/1945
DONKIN	H			CPL		07/07/1949
DONKIN	H			CPL		19/08/1959
DONLEY	W	L		FL		24/11/1949
DONNELLY	D	V		FO		06/03/1947
DONNELLY	J	C		FL		30/03/1944
DONNISON	W			FL	Acting	16/03/1950
DONOGHUE	J	P		SGT	Temporary	26/04/1956
DONOHUE	J			CPL		20/09/1945
DOPSON	G	W		FL		19/01/1956
DORE	A	S	W	GRP CAP	Acting	03/12/1942
DORE	J	B	W	FO		26/11/1953
DORLING	K			FL		21/01/1959
DORMON	T	S		FL		12/08/1943
DORMOR	R	B		FO		21/09/1944
DORRIAN	J			CPL		13/01/1949
DORRINGTON	E	P	M	FL		03/06/1959
DORRITY	D	J	A	FO		23/06/1955
DORWARD	C			FLT SGT		22/02/1951

Source	Service No	Force	Notes	Surname
N231 1949	135968	RAFVR		DIXON
N460 1947	81443	RAFVR		DIXON
N257 1950	131750	RAFVR		DIXON
N780 1950	867334	RAuxAF		DIXON
N460 1947	117289	RAFVR		DIXON
N807 1958	126764	RAFVR	wef 10/12/52	DIXON
YB 1996	n/a	RAuxAF		DIXON
N197 1946	68126	RAFVR		DIXON
N1 1950	113331	RAFVR		DIXON
N187 1947	86390	RAFVR	DSO,DFC	DIXON
N478 1947	67733	RAFVR	MC	DOBBIN
N454 1943	122817	RAFVR	DFM	DOBBS
N179 1950	853584	RAuxAF		DOBIE
N197 1946	109371	RAFVR	DFC	DOBSON
N104 1957	168874	RAFVR	wef 10/2/54	DOBSON
N596 1944	144573	RAFVR		DOBSON
YB 1996	n/a	RAuxAF		DOCHERTY
YB 2000	n/a	RAuxAF	1st Clasp	DOCHERTY
N843 1959	102343	RAFVR	wef 10/8/59	DODD
N485 1945	89766	RAFVR	DSO,AFC	DODD
N708 1955	145874	RAuxAF		DODD
N257 1950	72821	RAFVR	Deceased	DODD
YB 1998	n/a	RAuxAF		DODDS
N573 1948	73775	RAFVR		DODDS
N153 1961	2692583	RAuxAF	wef 13/11/60	DODKIN
N1210 1950	866293	RAuxAF		DODS
N850 1943	72147	RAFVR		DODWELL
YB 1996	n/a	RAuxAF		DOELL
N358 1946	159509	RAFVR		DOLEY
N673 1949	815049	RAuxAF		DOLLIMORE
N93 1951	110401	RAFVR		DOMONEY
N961 1944	81348	RAFVR		DON
N933 1960	2649730	RAuxAF	wef 1/10/60	DONALD
N82 1946	114253	RAFVR	AFC	DONALD
YB 1996	91448	RAuxAF		DONALDSON
N475 1943	70185	RAFVR	DFC	DONALDSON
N694 1948	761328	RAFVR		DONALDSON
N231 1949	857413	RAuxAF		DONBAVAND
N478 1947	811110	AAF		DONBAVAND
N565 1955	873147	RAuxAF		DONCASTER
N519 1947	746993	RAFVR		DONE
N759 1945	109048	RAFVR	DFC	DONIGER
N673 1949	807168	RAuxAF		DONKIN
N552 1959	2685528	RAuxAF	1st Clasp; wef 8/2/58	DONKIN
N1193 1949	162973	RAFVR	DFC,DFM	DONLEY
N187 1947	187670	RAFVR		DONNELLY
N281 1944	112425	RAFVR		DONNELLY
N257 1950	90636	RAuxAF		DONNISON
N318 1956	746112	RAFVR		DONOGHUE
N1036 1945	809030	AAF		DONOHUE
N43 1956	191606	RAFVR	DFC	DOPSON
N1638 1942	90230	AAF	DSO	DORE
N952 1953	193378	RAuxAF		DORE
N28 1959	193160	RAFVR	wef 5/8/53	DORLING
N850 1943	75283	RAFVR		DORMON
N961 1944	7059	RAFVR		DORMOR
N30 1949	854482	RAuxAF		DORRIAN
N368 1959	181682	RAFVR	wef 18/4/57	DORRINGTON
N473 1955	2606041	RAFVR		DORRITY
N194 1951	803428	RAuxAF		DORWARD

Surname	Initials			Rank	Status	Date
DOUCH	A	J		SQD LDR		09/08/1945
DOUGAN	F			CPL		07/12/1950
DOUGAN	R	M		FL		27/03/1947
DOUGHTY	G	O		FLT SGT		15/07/1948
DOUGHTY	H	A	S	FL		29/04/1954
DOUGLAS	A	G		SQD LDR		03/12/1942
DOUGLAS	J	F		WO		11/01/1945
DOUGLAS	K	J		FL		25/05/1944
DOUGLAS	R	J		SNR TECH		20/07/1960
DOUGLAS-HAMILTON	G	N		GRP CAP		22/04/1943
DOUGLAS-HAMILTON	M	A		WING COMM		14/10/1943
DOULTON	M	D		FO		10/02/1949
DOVE	G	G		SQD LDR	Acting	07/06/1945
DOVE	G	G		FO		26/11/1958
DOVER	M	R		FO		09/12/1992
DOVEY	F	L		SGT		01/07/1948
DOWARD	L	A		FL		07/04/1955
DOWDING	D	A		CPL		20/05/1948
DOWDING	H			CPL		08/05/1947
DOWDING	K	T		SQD LDR	Acting	26/01/1950
DOWELL	W			SQD LDR		06/01/1955
DOWLE	W	H	J	CPL		14/09/1950
DOWLER	G	W		FO		09/05/1946
DOWLING	J			CPL		20/08/1958
DOWLING	R	B		GRP CAP		26/01/1950
DOWNER	D	T		SQD LDR	Acting	21/09/1944
DOWNER	P	F		FO		20/04/1944
DOWNES	J			CPL		15/09/1949
DOWNIE	D			SGT		10/05/1945
DOWNING	C	H		SQD LDR		27/03/1947
DOWNING	W	V		FO		26/11/1958
DOWNTON	G	W		CPL		15/07/1948
DOWSE	S	H		FL		24/01/1946
DOWSETT	J			FL		24/01/1946
DOYLE	W			LAC		25/05/1960
DRACOTT	E	A		LAC		29/11/1951
DRAKE	F	V		GRP CAP		05/06/1947
DRAKES	D	M		CPL		14/10/1948
DRANE	J	W		FLT SGT		16/10/1947
DRANSFIELD	D			FO		01/12/1955
DRAPER	D	A	J	FL		27/11/1952
DRAPER	D	A	J	FL		26/04/1961
DRAPER	J	C		CPL		16/10/1947
DRAYTON	P			FLT SGT		15/11/1997
DREA	D	D		SGT		08/05/1947
DREDGE	A	S		FL		10/05/1945
DRESCH	D	T		FL		07/04/1955
DREW	B	K		FL		21/01/1959
DREW	J	L		FL		14/11/1998
DRINKWATER	H			FO		06/03/1947
DRUCE	A	C		CPL		15/07/1948
DRUCE	H	A		SQD LDR	Acting	25/01/1945
DRUMMOND	M	C		FO		08/07/1959
DRUMMOND	W	L		WING COMM	Acting	09/11/1944
DRY	S			CPL		01/07/1948
DRYDEN	D	M	M	FLT OFF		10/08/1950
DUCKER	G	H		WING COMM		20/07/1994
DUCLOS	V	S	H	SQD LDR		28/02/1946
DUDDY	S	J		FO		18/01/2001

Source	Service No	Force	Notes	Surname
N874 1945	72502	RAFVR		DOUCH
N1210 1950	873956	RAuxAF		DOUGAN
N250 1947	84588	RAFVR		DOUGAN
N573 1948	840777	RAuxAF		DOUGHTY
N329 1954	151001	RAFVR		DOUGHTY
N1638 1942	70188	RAFVR	DFC	DOUGLAS
N14 1945	741236	RAFVR		DOUGLAS
N496 1944	84928	RAFVR		DOUGLAS
N531 1960	746471	RAF	wef 13/7/44	DOUGLAS
N475 1943	90203	AAF	OBE,AFC; Lord	DOUGLAS-HAMILTON
N1065 1943	70189	RAFVR	OBE; Lord	DOUGLAS-HAMILTON
N123 1949	90235	RAuxAF	Deceased	DOULTON
N608 1945	62258	RAFVR	DFC	DOVE
N807 1958	62258	RAFVR	DFC; 1st Clasp; wef 1/3/50	DOVE
YB 1996	2631446	RAuxAF		DOVER
N527 1948	850588	RAuxAF		DOVEY
N277 1955	182242	RAFVR		DOWARD
N395 1948	749193	RAFVR		DOWDING
N358 1947	844551	AAF		DOWDING
N94 1950	72645	RAFVR		DOWDING
N1 1955	89222	RAuxAF		DOWELL
N926 1950	813107	RAuxAF		DOWLE
N415 1946	198172	RAFVR		DOWLER
N559 1958	2659561	WRAAF	wef 11/7/58	DOWLING
N94 1950	90605	RAuxAF		DOWLING
N961 1944	79156	RAFVR	DFC	DOWNER
N363 1944	141695	RAFVR		DOWNER
N953 1949	852211	RAuxAF		DOWNES
N485 1945	803370	AAF		DOWNIE
N250 1947	72405	RAFVR		DOWNING
N807 1958	2603313	RAFVR	wef 5/11/52	DOWNING
N573 1948	858814	RAuxAF		DOWNTON
N82 1946	86685	RAFVR		DOWSE
N82 1946	136705	RAFVR	DFC	DOWSETT
N375 1960	873925	RAFVR	wef 23/6/44	DOYLE
N1219 1951	844286	RAuxAF		DRACOTT
N460 1947	90591	AAF	MC	DRAKE
N856 1948	749079	RAFVR		DRAKES
N846 1947	845744	AAF		DRANE
N895 1955	2686600	RAuxAF		DRANSFIELD
N915 1952	112008	RAFVR		DRAPER
N339 1961	112008	RAFVR	DFC; 1st Clasp; wef 18/2/61	DRAPER
N846 1947	752559	RAFVR		DRAPER
YB 1998	n/a	RAuxAF		DRAYTON
N358 1947	801481	AAF		DREA
N485 1945	63785	RAFVR	DSO,AFC	DREDGE
N277 1955	197426	RAFVR		DRESCH
N28 1959	187521	RAFVR	wef 30/12/53	DREW
YB 2000	9946	RAuxAF		DREW
N187 1947	182219	RAFVR		DRINKWATER
N573 1948	846109	RAuxAF		DRUCE
N70 1945	81225	RAFVR		DRUCE
N454 1959	2659257	WRAAF	wef 12/5/59	DRUMMOND
N1150 1944	90390	AAF		DRUMMOND
N527 1948	870949	RAuxAF		DRY
N780 1950	966	WAAF		DRYDEN
YB 1997	209542	RAuxAF		DUCKER
N197 1946	79532	RAFVR	DFC	DUCLOS
YB 2002	2641348	RAuxAF		DUDDY

Surname	Initials			Rank	Status	Date
DUDLEY	S	C		FL		24/07/1963
DUDLISTON	G	A	S	SGT		29/06/1944
DUDLISTON	G	A	S	CPL	Acting	03/07/1952
DUDLISTON	G	A	S	SGT		26/02/1958
DUFF	F	G		FL		03/07/1952
DUFF	J	M		WO	Acting	24/06/1943
DUFFIELD	J			FL		16/03/1966
DUFFILL	A	V		SQD LDR	Acting	06/03/1947
DUFFILL	A	V		WING COMM	Acting	27/11/1952
DUFFIN	J			SQD LDR	Acting	13/12/1945
DUFF-MITCHELL	P	O	M	SQD LDR	Acting	27/04/1944
DUFFY	F			LAC		06/09/1951
DUKE	W	G		SQD LDR		26/06/1947
DUKE-WOOLLEY	H	B		FL		22/04/1943
DULSON	P	P		SQD LDR		22/10/1996
DUMBLE	G	S		FO		10/06/1948
DUMBRECK	W	V		WING COMM		17/03/1949
DUNBAR	R	J		SAC		08/07/1959
DUNCAN	A	L		FL		28/02/1946
DUNCAN	J	M		FL		28/02/1946
DUNCAN	J			SGT		05/02/1993
DUNCAN	M	V	A	FL		23/06/1955
DUNCAN	P	J		FL		10/05/1945
DUNCAN	R	H		FLT SGT		01/07/1993
DUNCAN	R	L		FL		23/05/1946
DUNFORD	A	B		SQD LDR		23/08/1951
DUNFORD	H	E		FL		23/05/1946
DUNHILL	W	E		FO		31/05/1945
DUNKLING	E	W		WO		15/09/1949
DUNLOP	I	D		FO		02/09/1943
DUNLOP	I	D	M	FL		21/01/1959
DUNLOP	J	H		SQD LDR	Acting	03/07/1952
DUNLOP	P	J		SGT		19/07/1951
DUNLOP	R	F		FO		13/02/1957
DUNLOP	R	P	G	SGT		12/06/1947
DUNMORE	R	E		SGT		26/11/1953
DUNN	A	C		FL		28/02/1946
DUNN	A	J		LAC		07/04/1955
DUNN	C	W		CPL		16/04/1958
DUNN	F	W		FL		13/02/1957
DUNN	M	J		FL		07/09/1994
DUNN	T	D	D	FL		28/02/1946
DUNNING	N	E		FL		22/09/1955
DUNNING	N	J		SAC		30/06/1995
DUNNING-WHITE	P	W		WING COMM	Acting	09/08/1945
DUNPHY	D	C		WO		18/04/1946
DUNPHY	P			FL		20/04/1944
DUNSIRE	R	B		FL		28/03/1962
DUNSTER	A	E		FO		26/04/1956
DUNSTER	E			FLT SGT	Acting	14/10/1948
DUNSTONE	G	L	P	FL		03/10/1946
DUNTHORNE	G	J		SQD LDR	Acting	10/06/1948
DUPEE	O	A		PO		24/12/1942
DUPREE	F	A		PO		21/01/1959
DURAND	R	W		WING COMM		20/05/1948
DURANDEAU	J	H		SGT		18/03/1948
DURHAM	J	R	O	FO		06/03/1947
DURNFORD	D	J		CPL		01/07/1948
DUROSE	G	J		WO		28/02/1946
DUTCH	G			CPL		15/07/1948

Source	Service No	Force	Notes	Surname
N525 1963	930760	RAFVR	wef 17/9/61	DUDLEY
N652 1944	812033	AAF		DUDLISTON
N513 1952	2680007	RAuxAF	1st Clasp	DUDLISTON
N149 1958	2680007	RAuxAF	2nd Clasp; wef 23/3/57	DUDLISTON
N513 1952	185480	RAFVR		DUFF
N677 1943	816043	AAF		DUFF
MOD 1967	312486	RAFVR		DUFFIELD
N187 1947	86398	RAFVR	DFC	DUFFILL
N915 1952	86398	RAFVR	1st Clasp	DUFFILL
N1355 1945	84028	RAFVR		DUFFIN
N396 1944	76310	RAFVR	AFC	DUFF-MITCHELL
N913 1951	750798	RAFVR		DUFFY
N519 1947	90027	AAF		DUKE
N475 1943	80809	RAFVR	DFC	DUKE-WOOLLEY
YB 1998	8025415	RAuxAF		DULSON
N456 1948	182556	RAFVR		DUMBLE
N231 1949	90413	RAuxAF		DUMBRECK
N454 1959	2684042	RAuxAF	wef 13/2/59	DUNBAR
N197 1946	79533	RAFVR		DUNCAN
N197 1946	127165	RAFVR	DFM	DUNCAN
YB 1996	n/a	RAuxAF		DUNCAN
N473 1955	198039	RAFVR		DUNCAN
N485 1945	83994	RAFVR	AFC	DUNCAN
YB 1996	n/a	RAuxAF		DUNCAN
N468 1946	138293	RAFVR		DUNCAN
N866 1951	91254	RAuxAF	DFC	DUNFORD
N468 1946	135481	RAFVR	DFC	DUNFORD
N573 1945	155828	RAFVR	DFM	DUNHILL
N953 1949	745304	RAFVR		DUNKLING
N918 1943	127024	RAFVR		DUNLOP
N28 1959	191450	RAFVR	wef 13/6/52	DUNLOP
N513 1952	104112	RAFVR		DUNLOP
N724 1951	812293	RAuxAF		DUNLOP
N104 1957	205353	RAuxAF	wef 30/12/56	DUNLOP
N478 1947	858034	AAF	BEM	DUNLOP
N952 1953	2685001	RAuxAF		DUNMORE
N197 1946	78455	RAFVR	MBE,DFC	DUNN
N277 1955	846850	RAuxAF		DUNN
N258 1958	819003	AAF	wef 6/12/43	DUNN
N104 1957	184324	RAFVR	wef 7/7/54	DUNN
YB 1998	5201602	RAuxAF		DUNN
N197 1946	115306	RAFVR	DFC	DUNN
N708 1955	163144	RAFVR		DUNNING
YB 1997	n/a	RAuxAF		DUNNING
N874 1945	90543	AAF	DFC	DUNNING-WHITE
N358 1946	741990	RAFVR		DUNPHY
N363 1944	70194	RAFVR		DUNPHY
N242 1962	176834	RAFVR	wef 26/9/59	DUNSIRE
N318 1956	2606056	RAFVR	DFM	DUNSTER
N856 1948	756103	RAFVR		DUNSTER
N837 1946	115718	RAFVR		DUNSTONE
N456 1948	80844	RAFVR		DUNTHORNE
N1758 1942	123298	RAFVR	DFM	DUPEE
N28 1959	2601299	RAFVR	wef 17/4/55	DUPREE
N395 1948	72455	RAFVR		DURAND
N232 1948	840207	RAuxAF		DURANDEAU
N187 1947	63863	RAFVR		DURHAM
N527 1948	756212	RAFVR		DURNFORD
N197 1946	747889	RAFVR		DUROSE
N573 1948	802533	RAuxAF		DUTCH

Surname	Initials			Rank	Status	Date
DUTHIE	G	H		SQD LDR	Acting	26/08/1948
DUTHOIT	R	J		FL		13/10/1955
DUTT	E	R		FL		15/06/1944
DUTTON	G	F	A	SNR TECH		17/06/1954
DUTTSON	G	F	C	FL		01/03/1945
DWELLY	E			FLT SGT		05/01/1950
DWYER	A			FLT SGT		01/05/1957
DYDE	J	A		FL	Acting	25/01/1951
DYE	A	P		FO		07/03/1997
DYKE	B	J	R	FLT OFF		19/01/1956
DYSON-SKINNER	P	de	L	WING COMM		04/12/1947
EACOTT	W	A		FL		21/01/1959
EADES-EACHUS	K	B		SQD OFF	Acting	04/11/1948
EAGLE	A	W		FLT SGT		21/10/1959
EAGLESTONE	C	W		SAC		18/05/1994
EAMES	E	A		FO		21/01/1959
EAMES	H	W	M	FO		15/09/1949
EAMES	J			FO		22/04/1943
EAMES	J			SQD LDR		17/01/1952
EARDLEY	D	A		SQD LDR	Acting	30/11/1944
EARL	G			FL	Acting	04/09/1947
EARL	K	C		FL		20/09/1945
EARLE	J	M		FLT OFF		13/02/1957
EARLE	J	V		CPL		09/11/1950
EARLEY	M	F		FL	Acting	06/03/1947
EARNSHAW	A	V		FL		23/02/1950
EARNSHAW	J	D		SGT		27/03/1947
EARTHROWL	C	H		FL		21/03/1946
EASBY	D	H		FL		21/01/1959
EASBY	R	S		SQD LDR		10/05/1945
EASSIE	W	J	D	FL		21/01/1959
EAST	A	G		CPL		24/06/1943
EAST	G	R		SGT		03/07/1952
EAST	G			FL		15/09/1949
EAST	P	F		SGT		07/07/1994
EASTICK	D	M		FL		26/06/1947
EASTWOOD	R	F		FL		27/07/1944
EATON	B			SGT		13/11/1947
EATON	L	E	W	FLT SGT		05/06/1947
ECCLES	H	H		SQD LDR		03/10/1946
ECCLES	N			FL		21/06/1961
ECCLES	W	T		SQD LDR	Acting	26/01/1950
EDDY	I			CPL		26/04/1951
EDDY	R			SGT		08/02/1995
EDELSTON	P	R		FL		19/01/1956
EDGAR	C	H		PO		06/03/1947
EDGAR	G	H	S	SQD LDR		01/07/1948
EDGAR	G	H		CPL		30/09/1948
EDGAR	I	F		FL		21/03/1946
EDGAR	J	A	D	SGT		18/03/1948
EDGAR	J			FLT SGT		26/11/1953
EDGE	A	R		WING COMM		30/03/1944
EDGE	F	D		FL		01/05/1957
EDGE	G	R		WING COMM	Acting	17/06/1943
EDGE	H			SQD LDR		26/01/1950
EDISON	C	J		CPL		20/05/1948
EDMISTON	G	A	F	FL		22/02/1951
EDMOND	R			FO		04/05/1944
EDMUNDS	A	G		FLT SGT		04/11/1948
EDNEY	H	S		FLT SGT		25/05/1950

Source	Service No	Force	Notes	Surname
N694 1948	91147	RAuxAF	OBE	DUTHIE
N762 1955	164939	RAFVR		DUTHOIT
N596 1944	78722	RAFVR	AFC	DUTT
N471 1954	751174	RAFVR		DUTTON
N225 1945	102954	RAFVR		DUTTSON
N1 1950	883449	WAAF		DWELLY
N277 1957	2688001	RAuxAF	BEM; wef 26/11/56	DWYER
N93 1951	110544	RAFVR		DYDE
YB 1997	2636656	RAuxAF		DYE
N43 1956	2655003	WRAAF		DYKE
N1007 1947	90445	AAF		DYSON-SKINNER
N28 1959	162637	RAFVR	wef 16/9/53	EACOTT
N934 1948	1534	WAAF		EADES-EACHUS
N708 1959	2689614	RAuxAF	wef 7/5/59	EAGLE
YB 1996	n/a	RAuxAF		EAGLESTONE
N28 1959	2603263	RAFVR	wef 28/6/52	EAMES
N953 1949	172504	RAFVR		EAMES
N475 1943	81548	RAFVR		EAMES
N35 1952	81548	RAFVR	DFC; 1st Clasp	EAMES
N1225 1944	108860	RAFVR		EARDLEY
N740 1947	170579	RAFVR		EARL
N1036 1945	129002	RAFVR	MBE	EARL
N104 1957	2329	WRAFVR	wef 23/9/54	EARLE
N1113 1950	753002	RAFVR		EARLE
N187 1947	146651	RAFVR		EARLEY
N179 1950	61599	RAFVR		EARNSHAW
N250 1947	809110	AAF		EARNSHAW
N263 1946	69450	RAFVR	DFC	EARTHROWL
N28 1959	193381	RAFVR	wef 21/6/53	EASBY
N485 1945	82670	RAFVR		EASBY
N28 1959	155415	RAFVR	wef 23/10/52	EASSIE
N677 1943	710025	AAF		EAST
N513 1952	799886	RAFVR		EAST
N953 1949	62788	RAFVR		EAST
YB 1996	n/a	RAuxAF		EAST
N519 1947	62032	RAFVR		EASTICK
N763 1944	87388	RAFVR	AFC	EASTWOOD
N933 1947	805447	AAF		EATON
N460 1947	743322	RAFVR		EATON
N837 1946	74330	RAFVR		ECCLES
N478 1961	58023	RAFVR(T)	wef 26/4/61	ECCLES
N94 1950	79767	RAFVR		ECCLES
N402 1951	881961	WAAF		EDDY
YB 1996	n/a	RAuxAF		EDDY
N43 1956	124575	RAuxAF	AFC	EDELSTON
N187 1947	203262	RAFVR		EDGAR
N527 1948	90671	RAuxAF		EDGAR
N805 1948	816092	RAuxAF	BEM	EDGAR
N263 1946	85642	RAFVR		EDGAR
N232 1948	816234	RAuxAF		EDGAR
N952 1953	2695500	RAuxAF		EDGAR
N281 1944	90325	AAF	AFC	EDGE
N277 1957	755148	RAF	MBE; wef 28/7/44	EDGE
N653 1943	90249	AAF	DFC	EDGE
N94 1950	73812	RAFVR		EDGE
N395 1948	841161	RAuxAF		EDISON
N194 1951	84955	RAFVR		EDMISTON
N425 1944	129960	RAFVR	DFC	EDMOND
N934 1948	804140	RAuxAF		EDMUNDS
N517 1950	840754	RAuxAF		EDNEY

Surname	Initials			Rank	Status	Date
EDWARD	G	S	M	FL		29/04/1954
EDWARD	G	S	M	FL		12/11/1963
EDWARD	T	E		FL		08/03/1956
EDWARDS	A	G		FL		07/07/1949
EDWARDS	A	G		LAC		13/12/1951
EDWARDS	A	H		WING COMM	Acting	06/03/1947
EDWARDS	C	F		SQD LDR		26/06/1947
EDWARDS	C	M		CPL		10/11/1995
EDWARDS	D	A		FL		11/07/1946
EDWARDS	G	W	F	CPL		27/04/1944
EDWARDS	H	W	F	FL		12/04/1945
EDWARDS	H			CPL		18/10/1951
EDWARDS	I	H		SQD LDR	Acting	30/12/1943
EDWARDS	J	G	S	SGT	Temporary	16/01/1957
EDWARDS	K	B	H	SQD LDR	Acting	26/06/1947
EDWARDS	K	C		FL		20/09/1945
EDWARDS	K	R		FO		24/01/1946
EDWARDS	M	J		FL		24/01/1946
EDWARDS	M	L		WING COMM	Acting	25/05/1944
EDWARDS	P	S		WING COMM	Acting	10/06/1948
EDWARDS	P	T		CPL		04/09/1947
EDWARDS	R			FL		08/11/1945
EDWARDS	R			WING COMM	Acting	20/05/1948
EDWARDS	S	J		FL		03/10/1946
EDWARDS	W	E		CPL		23/02/1950
EDWARDS	W	T	L	PO		30/09/1954
EDWARDS	W			WO		12/08/1943
EDYVEAN-WALKER	N			WING COMM		07/08/1947
EELES	R	A		SQD LDR	Acting	07/04/1955
EGAN	V	A		FLT SGT		24/11/1998
EGGLESTONE	W			SGT		21/09/1944
EGGLESTONE	W			CPL		15/01/1953
EGLIN	K			SGT		10/08/1950
EIDSFORTH	R			SQD LDR		20/05/1948
EILOART	G	M		FL		13/01/1949
ELAM	E	J	W	CPL		09/02/1951
ELDER	W			FO		23/05/1946
ELDERFIELD	C	F		SGT		18/03/1948
ELDON				FL		26/01/1950
ELDRED	J	C		FL		31/05/1945
ELDRIDGE	D	P		FO		21/01/1959
ELEY	N	J		FL		21/01/1959
ELFORD	P	M		SAC		17/05/1996
ELKES	R			CPL		24/11/1949
ELLAMS	W	A		CPL		07/08/1947
ELLERBECK	H	W		SGT		07/07/1949
ELLERTON	R	J		FL		29/04/1954
ELLES	F	W	T	CPL		04/11/1948
ELLINOR	H	S		SGT		06/05/1993
ELLIOT	R	L		SQD LDR	Acting	30/03/1943
ELLIOTT	C	S		CPL		26/08/1948
ELLIOTT	F	C		WO		09/05/1946
ELLIOTT	H	T		FL		15/01/1953
ELLIOTT	H	T		FL		09/01/1963
ELLIOTT	J	K		FL		04/09/1947
ELLIOTT	R	D		SQD LDR	Acting	25/05/1944
ELLIOTT	W	F	K	FL		10/06/1948
ELLIS	A	W		FLT SGT		17/03/1949
ELLIS	E	H		FO		30/09/1954
ELLIS	E	R		FO		23/02/1950

Source	Service No	Force	Notes	Surname
N329 1954	69547	RAFVR		EDWARD
MOD 1964	69547	RAFVR	1st Clasp	EDWARD
N187 1956	109076	RAFVR	DFC	EDWARD
N673 1949	118896	RAFVR		EDWARDS
N1271 1951	841302	RAuxAF		EDWARDS
N187 1947	72213	RAFVR		EDWARDS
N519 1947	72651	RAFVR		EDWARDS
YB 1996	n/a	RAuxAF		EDWARDS
N594 1946	110575	RAFVR		EDWARDS
N396 1944	813032	AAF		EDWARDS
N381 1945	85643	RAFVR	AFC	EDWARDS
N1077 1951	855982	RAuxAF		EDWARDS
N1350 1943	78979	RAFVR	DFC	EDWARDS
N24 1957	813223	RAuxAF		EDWARDS
N519 1947	90519	AAF		EDWARDS
N1036 1945	84680	RAFVR		EDWARDS
N82 1946	164527	RAFVR		EDWARDS
N82 1946	141139	RAFVR		EDWARDS
N496 1944	90380	AAF		EDWARDS
N456 1948	72129	RAFVR		EDWARDS
N740 1947	743163	RAFVR		EDWARDS
N1232 1945	115211	RAFVR	DSO	EDWARDS
N395 1948	73172	RAFVR		EDWARDS
N837 1946	84715	RAFVR	DFC	EDWARDS
N179 1950	752664	RAFVR		EDWARDS
N799 1954	2603660	RAFVR		EDWARDS
N850 1943	805162	AAF		EDWARDS
N662 1947	91018	AAF		EDYVEAN-WALKER
N277 1955	166751	RAuxAF		EELES
YB 2000	n/a	RAuxAF		EGAN
N961 1944	807060	AAF		EGGLESTONE
N28 1953	2685509	RAuxAF	1st Clasp	EGGLESTONE
N780 1950	810078	RAuxAF		EGLIN
N395 1948	79570	RAFVR		EIDSFORTH
N30 1949	72294	RAFVR	DFC	EILOART
N145 1951	840124	RAuxAF		ELAM
N468 1946	185069	RAFVR	DFM	ELDER
N232 1948	752226	RAFVR		ELDERFIELD
N94 1950	90777	RAuxAF	JP; The Earl of	ELDON
N573 1945	81029	RAFVR		ELDRED
N28 1959	2603758	RAFVR	wef 8/9/52	ELDRIDGE
N28 1959	163588	RAFVR	wef 14/7/54	ELEY
YB 1997	n/a	RAuxAF		ELFORD
N1193 1949	857564	RAuxAF		ELKES
N662 1947	811223	AAF		ELLAMS
N673 1949	842737	RAuxAF		ELLERBECK
N329 1954	185190	RAFVR		ELLERTON
N934 1948	861120	RAuxAF		ELLES
YB 1996	n/a	RAuxAF		ELLINOR
N1014 1943	88421	RAFVR	DFC	ELLIOT
N694 1948	857309	RAuxAF		ELLIOTT
N415 1946	749350	RAFVR		ELLIOTT
N28 1953	126867	RAFVR		ELLIOTT
N22 1963	126867	RAFVR	1st Clasp; wef 16/12/61	ELLIOTT
N740 1947	143527	RAFVR	DFM	ELLIOTT
N496 1944	76311	RAFVR	DFC	ELLIOTT
N456 1948	61145	RAFVR		ELLIOTT
N231 1949	846110	RAuxAF		ELLIS
N799 1954	203237	RAFVR		ELLIS
N179 1950	156495	RAFVR		ELLIS

Surname	Initials				Rank	Status	Date
ELLIS	H	E			SGT		07/04/1955
ELLIS	H	R			FL		30/03/1960
ELLIS	H	R			FL		30/03/1960
ELLIS	H	V			SQD LDR	Acting	09/05/1946
ELLIS	H				LAC		07/12/1960
ELLIS	J	L			CPL		20/05/1948
ELLIS	L	J			LAC		04/09/1947
ELLIS	P	J			SGT		01/05/1957
ELLIS	R	J			SGT		03/07/1957
ELLIS	W	T			FL		06/03/1947
ELLISON	A				FO		19/01/1956
ELLISON	H	M			SGT		12/04/1945
ELLISON	P	R	M		SQD LDR	Acting	06/03/1947
ELLOR	J	F			SQD LDR	Acting	14/10/1948
ELLYATT	R				FO		24/01/1946
ELMER	F	R			CPL		07/07/1949
ELSEY	C	W	C		FL		21/01/1959
ELSTOB	A	B			FLT SGT		17/04/1947
ELSTON	J	C			FL		17/04/1947
ELSTUBB	St.	J	C	de H	SQD LDR	Acting	20/04/1944
ELTON	H	S			SQD LDR		17/04/1947
ELVIDGE	K	H			FLT SGT		28/10/1943
ELWELL	E				AC1		13/12/1951
ELY	R				WO		05/01/1950
EMERSON	B	B			SQD OFF	Acting	17/03/1949
EMERY	M	S	C		FL		07/07/1949
EMMERSON	F	C			WING COMM	Acting	13/01/1949
EMMERSON	J	R			SQD LDR	Acting	22/02/1951
EMMERSON	J	R			FL		08/03/1956
EMMERSON	K	V			SGT		24/11/1949
EMSLIE	R				FLT SGT	Acting	19/10/1960
ENGLAND	J	L			FO		17/06/1954
ENNIS	A	S	R	E	SQD LDR	Acting	27/07/1944
ENNIS	W	H			AC2		12/04/1951
ENTICOTT	G	C			FO		05/01/1950
ERRINGTON	E				WING COMM	Acting	05/06/1947
ERSKINE	C	E			FLT SGT		30/03/1943
ERSKINE	C	E			FLT SGT		04/10/1951
ERSKINE	C	E			FLT SGT		04/10/1951
ERSKINE	C	E			FLT SGT		21/01/1959
ERSKINE	L	A	A		SGT		26/11/1953
ERSKINE	T	J			WING COMM		20/09/1945
ESHELBY	J	H			FL		24/01/1946
ESKRIGGE	R				CPL		18/03/1948
ESLER	R	A			SQD LDR	Acting	06/03/1947
ESLER	S	E			FL		09/11/1944
ESTHOP	H				FL		06/03/1947
ETCHES	C	J			CPL		05/02/1948
ETCHES	R	E			FL		01/05/1957
ETHERINGTON	H	J			FL		21/03/1946
ETHERINGTON	H				WING COMM	Acting	17/04/1947
ETTERIDGE	A	G			SQD LDR		06/03/1947
ETTWELL	E	J	D		WING COMM		19/08/1954
EVANS	A	L	L		SQD LDR	Acting	23/02/1950
EVANS	B				FL		01/07/1948
EVANS	C	E			CPL		30/09/1948
EVANS	C	F			FL		08/03/1956
EVANS	C	L			WO		10/05/1945
EVANS	C	W			FLT SGT	Acting	05/06/1947

Source	Service No	Force	Notes	Surname
N277 1955	819034	RAuxAF		ELLIS
N229 1960	62318	RAFVR	wef 11/7/52	ELLIS
N229 1960	62318	RAFVR	1st Clasp; wef 11/3/59	ELLIS
N415 1946	84968	RAFVR	DFC	ELLIS
N933 1960	2650015	RAuxAF	wef 7/8/54	ELLIS
N395 1948	746698	RAFVR		ELLIS
N740 1947	841669	AAF		ELLIS
N277 1957	2605297	RAuxAF	wef 16/1/54	ELLIS
N456 1957	2676403	RAuxAF	wef 30/6/53	ELLIS
N187 1947	110331	RAFVR		ELLIS
N43 1956	189818	RAuxAF		ELLISON
N381 1945	804152	AAF		ELLISON
N187 1947	89075	RAFVR		ELLISON
N856 1948	72654	RAFVR		ELLOR
N82 1946	173092	RAFVR		ELLYATT
N673 1949	812233	RAuxAF		ELMER
N28 1959	109115	RAFVR	wef 5/12/52	ELSEY
N289 1947	808165	AAF		ELSTOB
N289 1947	163746	RAFVR		ELSTON
N363 1944	62659	RAFVR		ELSTUBB
N289 1947	90667	AAF	MC	ELTON
N1118 1943	812060	AAF		ELVIDGE
N1271 1951	855884	RAuxAF		ELWELL
N1 1950	751858	RAFVR		ELY
N231 1949	1352	WAAF		EMERSON
N673 1949	143850	RAFVR		EMERY
N30 1949	72335	RAFVR		EMMERSON
N194 1951	128677	RAFVR	DFC,AFM	EMMERSON
N187 1956	128677	RAuxAF	DFC,AFM; 1st Clasp	EMMERSON
N1193 1949	756050	RAFVR		EMMERSON
N775 1960	872414	AAF	wef 9/7/44	EMSLIE
N471 1954	178345	RAFVR		ENGLAND
N763 1944	104249	RAFVR	DSO,DFC	ENNIS
N342 1951	861976	RAuxAF		ENNIS
N1 1950	104115	RAFVR		ENTICOTT
N460 1947	90760	AAF		ERRINGTON
N1014 1943	803177	AAF		ERSKINE
N1019 1951	2683515	RAuxAF	1st Clasp	ERSKINE
N1019 1951	2683515	RAuxAF	2nd Clasp	ERSKINE
N28 1959	2683515	RAuxAF	BEM; 3rd Clasp; wef 10/9/58	ERSKINE
N952 1953	803571	RAuxAF		ERSKINE
N1036 1945	72185	RAFVR		ERSKINE
N82 1946	124746	RAFVR	AFC	ESHELBY
N232 1948	811222	RAuxAF		ESKRIGGE
N187 1947	85012	RAFVR	DSO,DFC	ESLER
N1150 1944	121938	RAFVR	DFC	ESLER
N187 1947	159236	RAFVR		ESTHOP
N87 1948	812166	RAuxAF		ETCHES
N277 1957	111729	RAFVR	wef 28/7/44	ETCHES
N263 1946	160532	RAFVR		ETHERINGTON
N289 1947	91210	AAF	OBE	ETHERINGTON
N187 1947	72214	RAFVR		ETTERIDGE
N672 1954	90985	RAuxAF		ETTWELL
N179 1950	87722	RAFVR		EVANS
N527 1948	142015	RAFVR	DFM	EVANS
N805 1948	842505	RAuxAF	BEM	EVANS
N187 1956	73175	RAFVR		EVANS
N485 1945	816016	AAF		EVANS
N460 1947	743335	RAFVR		EVANS

Surname	Initials			Rank	Status	Date
EVANS	D	W	J	FLT SGT		31/10/1956
EVANS	E	A	J	FO		30/10/1957
EVANS	E	H		GRP CAP	Acting	03/12/1942
EVANS	F	A		FO		20/09/1945
EVANS	F	G		FL		02/04/1953
EVANS	G	J		FL		12/04/1945
EVANS	H	I		FO		23/06/1955
EVANS	H			FL	Acting	02/08/1951
EVANS	I	I		FLT OFF	Acting	08/05/1947
EVANS	K	W		FL		24/01/1946
EVANS	L	F		FL		21/01/1959
EVANS	M	B		SACW		21/01/1959
EVANS	M	J		FL		09/01/1963
EVANS	R	W		FLT SGT		05/06/1947
EVANS	R	W		FO		13/10/1955
EVANS	T	B		LAC		27/03/1947
EVANS	T	W		SQD LDR	Acting	28/02/1946
EVANS	W	R		FL		08/11/1945
EVANS	W	S	J	SGT		27/03/1947
EVELEIGH	R	E		CPL		13/01/1949
EVERARD	F	E		WING COMM		18/03/1948
EVERARD	M			FLT OFF		23/02/1950
EVERETT	A	D		WO		12/06/1947
EVERETT	D	J		SQD LDR	Acting	13/01/1949
EVERETT	E	V		FL		16/01/1957
EVERETT	H	R		FL	Acting	09/11/1950
EVERITT	G	H		WING COMM	Acting	01/03/1945
EVERITT	J	I		CPL		19/09/1956
EVERS	R	D	M	SQD LDR		06/12/1943
EWEN	J	F		FL		15/01/1958
EWEN	J	K		FLT OFF	Acting	13/07/1950
EWING	G	A		FO		09/11/1950
EWING	I	R	L	SGT		01/07/1948
EYRE	A			WING COMM	Acting	30/08/1945
EYRE	J	P		FL		01/08/1946
EYRE	S			FLT SGT		27/03/1952
EYRES	R	M		FO		31/10/1956
EYRES	T			CPL	Temporary	27/03/1963
FABIAN	R	G		WO		01/07/1948
FAHEY	J	J		SGT		26/02/1958
FAIERS	A	C	A	LAC		16/10/1952
FAIRBASS	H			CPL		16/10/1952
FAIRBROTHER	J			SGT		04/09/1947
FAIRBROTHER	L	H		WING COMM	Acting	07/07/1949
FAIRBROTHER	L	H		FL		17/06/1954
FAIRCLOTH	E	L		SGT		16/10/1947
FAIRCLOUGH	A	J		SQD LDR		20/05/1948
FAIRCLOUGH	A	T		CPL		20/05/1948
FAIRFIELD	G	S		LAC		14/10/1948
FAIRHURST	M			CPL		01/07/1948
FAIRLESS	A	R		SQD LDR		20/01/1960
FAIRS	G	R		FL		06/03/1947
FAIRWEATHER	W	P		SGT	Acting	03/07/1957
FALCON	C	H		SGT		05/06/1947
FALCONER	J	G		LAC		08/07/1959
FALLA	W	A	S	SQD LDR		16/10/1947
FANNON	T	E		FL		20/01/1960
FARAGHER	H			FL		29/11/1951
FARLEY	P			SGT		17/06/1954
FARMAN	R	H	J	CPL	Acting	20/08/1958

Source	Service No	Force	Notes	Surname
N761 1956	2699001	RAuxAF		EVANS
N775 1957	2606132	RAFVR	wef 7/11/56	EVANS
N1638 1942	90182	AAF		EVANS
N1036 1945	184563	RAFVR		EVANS
N298 1953	146698	RAFVR		EVANS
N381 1945	123995	RAFVR	DFM	EVANS
N473 1955	2682545	RAuxAF		EVANS
N777 1951	115815	RAFVR		EVANS
N358 1947	2947	WAAF		EVANS
N82 1946	125320	RAFVR	DFC	EVANS
N28 1959	69474	RAFVR	DSO,AFC; wef 7/8/54	EVANS
N28 1959	2655737	WRAAF	wef 8/11/58	EVANS
N22 1963	2494036	RAuxAF	wef 26/10/62	EVANS
N460 1947	811079	AAF		EVANS
N762 1955	811079	RAFVR	1st Clasp	EVANS
N250 1947	864926	AAF		EVANS
N197 1946	79373	RAFVR	DFC	EVANS
N1232 1945	67607	RAFVR		EVANS
N250 1947	863391	AAF		EVANS
N30 1949	840153	RAuxAF		EVELEIGH
N232 1948	73573	RAFVR	OBE	EVERARD
N179 1950	479	WAAF		EVERARD
N478 1947	751702	RAFVR		EVERETT
N30 1949	87566	RAFVR		EVERETT
N24 1957	193129	RAFVR		EVERETT
N1113 1950	133316	RAFVR		EVERETT
N225 1945	81032	RAFVR	DSO,DFC	EVERITT
N666 1956	743654	RAFVR		EVERITT
MOD 1964	110828	RAFVR		EVERS
N40 1958	180334	RAuxAF	wef 21/9/57	EWEN
N675 1950	2365	WAAF		EWEN
N1113 1950	168876	RAFVR		EWING
N527 1948	841174	RAuxAF		EWING
N961 1945	90408	AAF	DFC	EYRE
N652 1946	86335	RAFVR		EYRE
N224 1952	761062	RAFVR		EYRE
N761 1956	2649701	RAuxAF		EYRES
N239 1963	856522	AAF	wef 18/6/44	EYRES
N527 1948	745451	RAFVR		FABIAN
N149 1958	2688554	RAuxAF	wef 24/9/57	FAHEY
N798 1952	770481	RAFVR		FAIERS
N798 1952	812074	RAuxAF		FAIRBASS
N740 1947	811097	AAF		FAIRBROTHER
N673 1949	90909	RAuxAF		FAIRBROTHER
N471 1954	90909	RAFVR	1st Clasp	FAIRBROTHER
N846 1947	840217	AAF		FAIRCLOTH
N395 1948	72232	RAFVR		FAIRCLOUGH
N395 1948	840327	RAuxAF		FAIRCLOUGH
N856 1948	871686	RAuxAF		FAIRFIELD
N527 1948	867021	RAuxAF		FAIRHURST
N33 1960	173370	RAuxAF	wef 25/11/59	FAIRLESS
N187 1947	85718	RAFVR		FAIRS
N456 1957	2649524	RAuxAF	wef 11/12/54	FAIRWEATHER
N460 1947	811041	AAF		FALCON
N454 1959	867973	AAF	wef 7/7/44	FALCONER
N846 1947	72007	RAFVR		FALLA
N33 1960	201167	RAFVR	wef 11/1/55	FANNON
N1219 1951	62355	RAuxAF		FARAGHER
N471 1954	2649507	RAuxAF		FARLEY
N559 1958	2693508	RAuxAF	wef 3/5/58	FARMAN

Surname	Initials			Rank	Status	Date
FARMER	J	S	H	SGT		24/11/1949
FARMER	P	R		FL		06/03/1947
FARMER	R	B		FL		08/11/1945
FARMER	R			ENG IA		16/10/1947
FARNDALE	J	H		SGT		20/05/1948
FARNES	P	C	P	WING COMM	Acting	08/02/1945
FARNFIELD	H	B		FO		05/02/1948
FARQUHAR	A	D		GRP CAP		11/11/1943
FARQUHAR	A			SGT		15/03/1995
FARR	D	J		FO		27/01/1944
FARR	F	W		SGT		27/11/1952
FARRAND	E	S		SQD LDR	Acting	15/07/1948
FARRANDS	E	L		WO		05/02/1948
FARRANT	W	F		PO		24/08/1993
FARRAR	J			FL		21/01/1959
FARRER	G			LAC		04/09/1947
FARROW	C	W	C	FL		17/02/1944
FARROW	L	W		FL		01/03/1945
FARTHING	A	G		FL		28/02/1946
FARTHING	F	B		SQD LDR		17/04/1947
FARTHING	F	H		SQD LDR		17/04/1947
FAULDS	L	G		LAC		01/07/1948
FAULKNER	G	D		FL	Acting	07/08/1947
FAULKNER	W	R		WO		13/11/1947
FAWCETT	N	E		SQD LDR	Acting	07/08/1947
FAWSETT	K			SQD LDR		10/06/1948
FAZAKERLEY	J	A		FO		27/03/1947
FEARN	F			CPL		08/05/1947
FEARN	R	E		SQD LDR		10/06/1948
FEARNSIDES	E	R		SGT	Acting	21/06/1951
FEARS	L	R		FO		20/04/1944
FEATHER	J	S		FO		10/12/1942
FEATHERBY	H	F		FL		09/08/1945
FEATHERSTONE	S	G		LAC		13/07/1950
FEENEY	F			SGT		20/09/1945
FELKIN	S	D		GRP CAP	Acting	27/03/1947
FELLICK	D	S		FLT SGT		01/05/1952
FELLOWS	L			CPL		01/07/1948
FELLOWS	W	N		SGT	Temporary	30/10/1957
FELTHAM	E	M		SGT		04/11/1948
FENDICK	C	P		FL		06/03/1947
FENN	A	A		FL		08/11/1945
FENN	C	F		FL		12/04/1951
FENN	L	M		SEC OFF		26/06/1947
FENNEL	A	J		FO		02/03/2001
FENNELL	L	K		FL		28/02/1946
FENNELL	L	R		FO		30/09/1948
FENNING	E	E		FL		27/04/1944
FENTON	J	H		FO		08/03/1956
FENWICK	R			WING COMM	Acting	08/05/1947
FENWICK	S	G		SQD LDR	Acting	06/03/1947
FENWICK	W			CPL		04/09/1947
FEREDAY	I	A		CPL		07/12/1995
FERGUSON	D	I		FL		07/08/1947
FERGUSON	J	C		SQD LDR		27/03/1947
FERGUSON	P	J		SQD LDR		12/08/1943
FERGUSON	R	J		CPL	Acting	18/02/1954
FERGUSON	W			FL		21/01/1959
FERGUSSON	W	G		FL		13/12/1945
FERMOR	B	F		FL		28/11/1946

Source	Service No	Force	Notes	Surname
N1193 1949	743210	RAFVR		FARMER
N187 1947	143487	RAFVR		FARMER
N1232 1945	85253	RAFVR		FARMER
N846 1947	701225	RAFVR		FARMER
N395 1948	819195	RAuxAF		FARNDALE
N139 1945	88437	RAFVR	DFM	FARNES
N87 1948	86745	RAFVR		FARNFIELD
N1177 1943	90158	AAF	DFC	FARQUHAR
YB 1997	n/a	RAuxAF		FARQUHAR
N59 1944	130786	RAFVR		FARR
N915 1952	864007	RAuxAF		FARR
N573 1948	72830	RAFVR		FARRAND
N87 1948	740864	RAFVR		FARRANDS
YB 1997	n/a	RAuxAF		FARRANT
N28 1959	202230	RAFVR	wef 5/12/54	FARRAR
N740 1947	811175	AAF		FARRER
N132 1944	85685	RAFVR		FARROW
N225 1945	78980	RAFVR		FARROW
N197 1946	132755	RAFVR	DFC	FARTHING
N289 1947	73181	RAFVR		FARTHING
N289 1947	72336	RAFVR		FARTHING
N527 1948	874105	RAuxAF		FAULDS
N662 1947	67524	RAFVR		FAULKNER
N933 1947	754160	RAFVR		FAULKNER
N662 1947	91218	AAF		FAWCETT
N456 1948	72047	RAFVR		FAWSETT
N250 1947	181351	RAFVR		FAZAKERLEY
N358 1947	746058	RAFVR		FEARN
N456 1948	72512	RAFVR		FEARN
N621 1951	866499	RAuxAF		FEARNSIDES
N363 1944	129235	RAFVR		FEARS
N1680 1942	90159	RAFVR		FEATHER
N874 1945	101469	RAFVR		FEATHERBY
N675 1950	840982	RAuxAF		FEATHERSTONE
N1036 1945	803435	AAF		FEENEY
N250 1947	74650	RAFVR	OBE	FELKIN
N336 1952	746714	RAFVR		FELLICK
N527 1948	808214	RAuxAF		FELLOWS
N775 1957	800466	AAF	wef 28/3/43	FELLOWS
N934 1948	881797	WAAF		FELTHAM
N187 1947	78139	RAFVR		FENDICK
N1232 1945	115738	RAFVR		FENN
N342 1951	126029	RAFVR		FENN
N519 1947	4657	WAAF		FENN
YB 2001	2641340	RAuxAF		FENNEL
N197 1946	146696	RAFVR		FENNELL
N805 1948	110355	RAFVR		FENNELL
N396 1944	116804	RAFVR		FENNING
N187 1956	197890	RAFVR		FENTON
N358 1947	72151	RAFVR		FENWICK
N187 1947	91061	AAF		FENWICK
N740 1947	867124	AAF		FENWICK
YB 1997	n/a	RAuxAF		FEREDAY
N662 1947	84068	RAFVR		FERGUSON
N250 1947	72656	RAFVR		FERGUSON
N850 1943	90167	AAF		FERGUSON
N137 1954	2676845	RAuxAF		FERGUSON
N28 1959	188523	RAFVR	wef 31/10/52	FERGUSON
N1355 1945	117411	RAFVR		FERGUSSON
N1004 1946	130710	RAFVR	DFC	FERMOR

Surname	Initials		Rank	Status	Date
FERRABY	D	L	FL		28/02/1946
FERRIE	B		CPL		22/04/1986
FERRIE	B		CPL		22/04/1996
FERRIER	D	McL.	SQD LDR		05/06/1947
FETTES	R	S	WO		24/11/1949
FEWKES	F	L	FL		04/11/1948
FEWKES	J	E	FLT SGT		30/08/1945
FEWTRELL	L	M	SQD OFF		21/01/1959
FIELD	J	J	FL		25/05/1950
FIELD	J		SGT		10/02/1949
FIELD	W	F	WO		18/04/1946
FIELDEN	E	B	WING COMM	Acting	27/04/1944
FIELDHOUSE	L	A	FL		27/07/1944
FIELDHOUSE	L	A	FO		19/07/1951
FIELDING-JOHNSON	W	S	WING COMM		23/02/1950
FIELDSEND	E	C	SQD LDR	Acting	03/12/1942
FIFIELD	J	S	FL		10/05/1945
FIGG	R	H	LAC		24/11/1949
FILTNESS	A		CPL	Temporary	03/06/1959
FINCH	F	J F	SQD LDR	Acting	24/01/1946
FINCH	R	S	CPL		30/09/1948
FINDLATER	J		CPL		01/10/1993
FINDLAY	A	G	FO		05/06/1947
FINDLAY	J		SACW		14/11/1962
FINDLAY	R	W	LAC		07/08/1947
FINES	G	J	FO		06/11/1997
FINLAY	G	L	FL		09/05/1946
FINLAY	N	H	SGT		24/01/1946
FINLAYSON	R		LAC		06/06/1962
FINNEY	B	R	CPL		27/04/1950
FIREBRACE	F	W G	WING OFF	Acting	27/03/1947
FIRTH	H		SGT		07/07/1949
FISH	L	C	CPL		17/01/1952
FISHER	A	G A	SQD LDR		05/01/1950
FISHER	A	W	CPL		05/06/1947
FISHER	E	A	LAC		16/03/1950
FISHER	G	J	CPL		15/07/1948
FISHER	J	A	CPL		04/04/1997
FISHER	J		SAC		29/07/1999
FISHER	K	G	FL		01/07/1948
FISHER	L	J N	FL		01/07/1948
FISHER	S	H	FL		20/03/1957
FISHER	S	J	SGT		23/07/1998
FISHER	T	W	CPL		01/07/1948
FISHER	W		SGT		20/05/1948
FITZGERALD	A	F	SAC		20/11/1957
FITZGERALD	C	T	WING COMM		01/07/1948
FITZGERALD	E	W	LAC		12/10/1950
FITZGERALD	J	M	FLT SGT		07/12/1950
FITZGERALD	J		LAC		24/11/1949
FITZGERALD	T	A	FLT OFF		05/06/1947
FITZGIBBON	D		FLT SGT		05/01/1950
FITZGIBBON	P	J	FLT SGT		05/09/1946
FLACK	I	J	SQD LDR	Acting	27/01/1944
FLATMAN	E	H	CPL		27/11/1952
FLAVELL	A	C	FL		19/08/1954
FLAWITH	W	F	FO		04/09/1957
FLAXMAN	R	A M	LAC		01/07/1948
FLEMING	R		LAC		16/04/1958
FLEMING	W		CPL		07/07/1949

Source	Service No	Force	Notes	Surname
N197 1946	125759	RAFVR		FERRABY
YB 1996	n/a	RAuxAF		FERRIE
YB 1997	n/a	RAuxAF	1st Clasp	FERRIE
N460 1947	73183	RAFVR		FERRIER
N1193 1949	749437	RAFVR		FETTES
N934 1948	112405	RAFVR		FEWKES
N961 1945	815005	AAF		FEWKES
N28 1959	1487	WAAF	wef 6/8/44	FEWTRELL
N517 1950	144281	RAFVR		FIELD
N123 1949	863271	RAuxAF		FIELD
N358 1946	747778	RAFVR		FIELD
N396 1944	70214	RAFVR		FIELDEN
N763 1944	109081	RAFVR	DFC	FIELDHOUSE
N724 1951	109081	RAFVR	DFC; 1st Clasp	FIELDHOUSE
N179 1950	73613	RAFVR	MC,DFC	FIELDING-JOHNSON
N1638 1942	90397	AAF		FIELDSEND
N485 1945	83274	RAFVR	DFC	FIFIELD
N1193 1949	845461	RAuxAF		FIGG
N368 1959	846476	AAF	wef 2/6/44	FILTNESS
N82 1946	118041	RAFVR		FINCH
N805 1948	865201	RAuxAF		FINCH
YB 1996	n/a	RAuxAF		FINDLATER
N460 1947	134613	RAFVR		FINDLAY
N837 1962	2657866	WRAAF	wef 9/10/62	FINDLAY
N662 1947	811195	AAF		FINDLAY
YB 1999	2636265	RAuxAF		FINES
N415 1946	144776	RAFVR		FINLAY
N82 1946	816045	AAF		FINLAY
N435 1962	2609211	RAuxAF	wef 13/4/62	FINLAYSON
N406 1950	749528	RAFVR		FINNEY
N250 1947	843	WAAF		FIREBRACE
N673 1949	752830	RAFVR		FIRTH
N35 1952	770570	RAFVR		FISH
N1 1950	73708	RAFVR	AFC	FISHER
N460 1947	841939	AAF		FISHER
N257 1950	855154	RAuxAF		FISHER
N573 1948	840613	RAuxAF		FISHER
YB 1999	n/a	RAuxAF		FISHER
YB 2000	n/a	RAuxAF		FISHER
N527 1948	142004	RAFVR		FISHER
N527 1948	123433	RAFVR		FISHER
N198 1957	132290	RAFVR	wef 11/8/44	FISHER
YB 1999	n/a	RAuxAF		FISHER
N527 1948	805035	RAuxAF		FISHER
N395 1948	819064	RAuxAF		FISHER
N835 1957	2688549	RAuxAF	wef 4/10/57	FITZGERALD
N527 1948	90571	RAuxAF		FITZGERALD
N1018 1950	841209	RAuxAF		FITZGERALD
N1210 1950	809014	RAuxAF		FITZGERALD
N1193 1949	865582	RAuxAF		FITZGERALD
N460 1947	1827	WAAF		FITZGERALD
N1 1950	804249	RAuxAF		FITZGIBBON
N751 1946	804233	AAF		FITZGIBBON
N59 1944	84705	RAFVR	AFC	FLACK
N915 1952	799941	RAFVR		FLATMAN
N672 1954	76281	RAFVR		FLAVELL
N611 1957	205109	RAuxAF	wef 4/5/57	FLAWITH
N527 1948	843938	RAuxAF		FLAXMAN
N258 1958	816098	AAF	wef 23/12/43	FLEMING
N673 1949	874055	RAuxAF		FLEMING

Surname	Initials			Rank	Status	Date
FLEMMING	C	W		GRP CAP		19/08/1954
FLEMONS	H	C		FL		20/09/1945
FLETCHER	A	G		WO		04/10/1945
FLETCHER	C	W	B	LAC		19/07/1951
FLETCHER	G	E	C	CPL		01/07/1948
FLETCHER	J	M		SGT		25/03/1998
FLETCHER	J			FLT SGT		12/06/1947
FLETCHER	L	W		FL		20/05/1948
FLETCHER	R	A		FL		13/12/1945
FLETCHER	R	K		FL		26/01/1950
FLETCHER	R			WO		10/02/1949
FLETCHER	S	B		FL		09/08/1945
FLETT	J	H		FLT SGT	Acting	10/08/1950
FLIGHT	R	S		FO		20/09/1945
FLINDERS	D	C		FLT SGT		15/07/1948
FLINT	J			WING COMM	Acting	31/05/1945
FLOATE	W	C		FLT SGT		07/04/1955
FLOUNDERS	G	L		SGT		12/04/1945
FLOWER	R	E		FL		12/07/1945
FLOWER	V	L		FL		25/03/1954
FLOWERS	C	G		LAC		14/10/1948
FLOWERS	H	T		LAC		15/07/1948
FLOWERS	I	E		SGT		07/12/1960
FLOYD	F	G		FLT SGT		05/02/1948
FOGG	H			SGT		13/12/1945
FOGGIN	R	A		WING COMM		10/06/1948
FOKES	R	H		SQD LDR	Acting	13/01/1944
FOLEY	P	R		WING COMM	Acting	03/12/1942
FONTANNAZ	V	J		SQD LDR	Acting	06/03/1947
FOOTIT	I	W		FL		01/07/1948
FOPP	D			FL		20/09/1945
FORBES	C	M		FL		10/06/1948
FORBES	I	D		FL		21/01/1959
FORBES	N			FL		13/12/1945
FORBES	T	P	R	FO		12/06/1947
FORBES-SEMPILL	M			WING OFF		16/10/1947
FORD	F	W		SQD LDR		30/09/1954
FORD	H			CPL		27/03/1947
FORD	M	W	W	WING OFF	Acting	27/03/1947
FORD	M	W		SAC		02/12/1992
FORD	R	C		FL		21/06/1945
FORD	R	C		FO		03/07/1952
FORD	R	F		CPL		01/10/1953
FORD	S	T		LAC		01/07/1948
FORD	W	H		CPL		07/07/1949
FORD	W	H		FL		05/01/1950
FORD	W			CPL		10/06/1948
FORDHAM	H	E		LAC		20/05/1948
FORMBY	M	L		WING COMM	Acting	17/06/1943
FORMBY	R			CPL		05/06/1947
FORREST	D	H		FL		27/07/1944
FORREST	G	V		FL		27/01/1944
FORREST	G	V		FL		13/12/1951
FORREST	J	A		FO		03/07/1952
FORREST	J	S		CPL		30/03/1943
FORREST	W	F		FL		07/12/1960
FORRESTER	J			AC2		25/05/1950
FORRESTER	J	A		CPL		15/07/1948
FORSDYKE	D	K		FL		13/12/1945
FORSTER	A	D		WING COMM	Acting	01/03/1945

Source	Service No	Force	Notes	Surname
N672 1954	70218	RAFVR		FLEMMING
N1036 1945	76312	RAFVR		FLEMONS
N1089 1945	801362	AAF		FLETCHER
N724 1951	748098	RAFVR		FLETCHER
N527 1948	812180	RAuxAF		FLETCHER
YB 1999	n/a	RAuxAF		FLETCHER
N478 1947	811062	AAF		FLETCHER
N395 1948	199291	RAFVR		FLETCHER
N1355 1945	125317	RAFVR	DFC	FLETCHER
N94 1950	73185	RAFVR		FLETCHER
N123 1949	749431	RAFVR		FLETCHER
N874 1945	158034	RAFVR		FLETCHER
N780 1950	771347	RAFVR		FLETT
N1036 1945	159876	RAFVR	DFC	FLIGHT
N573 1948	749339	RAFVR		FLINDERS
N573 1945	121331	RAFVR	DFC,GM,DFM	FLINT
N277 1955	2607063	RAFVR		FLOATE
N381 1945	808136	AAF		FLOUNDERS
N759 1945	85262	RAFVR		FLOWER
N249 1954	106317	RAFVR		FLOWER
N856 1948	863292	RAuxAF		FLOWERS
N573 1948	864065	RAuxAF		FLOWERS
N933 1960	2693556	RAuxAF	wef 15/10/60	FLOWERS
N87 1948	800428	RAuxAF		FLOYD
N1355 1945	815037	AAF		FOGG
N456 1948	72832	RAFVR	OBE	FOGGIN
N11 1944	88439	RAFVR	DFM	FOKES
N1638 1942	90123	AAF		FOLEY
N187 1947	73614	RAFVR		FONTANNAZ
N527 1948	147141	RAFVR		FOOTIT
N1036 1945	112448	RAFVR	AFC	FOPP
N456 1948	121289	RAFVR	DFC	FORBES
N28 1959	172861	RAFVR	wef 15/6/54	FORBES
N1355 1945	90252	AAF		FORBES
N478 1947	127592	RAFVR		FORBES
N846 1947	1	WAAF	The Honorable	FORBES-SEMPILL
N799 1954	86985	RAFVR		FORD
N250 1947	869260	AAF		FORD
N250 1947	756	WAAF	OBE	FORD
YB 1996	n/a	RAuxAF		FORD
N671 1945	88214	RAFVR		FORD
N513 1952	88214	RAFVR	1st Clasp	FORD
N796 1953	743298	RAFVR		FORD
N527 1948	847730	RAuxAF		FORD
N673 1949	870113	RAuxAF		FORD
N1 1950	141013	RAFVR		FORD
N456 1948	857360	RAuxAF		FORD
N395 1948	841273	RAuxAF		FORDHAM
N653 1943	70221	RAFVR		FORMBY
N460 1947	811116	AAF		FORMBY
N763 1944	115218	RAFVR		FORREST
N59 1944	109482	RAFVR		FORREST
N1271 1951	109482	RAFVR	1st Clasp	FORREST
N513 1952	187680	RAuxAF		FORREST
N1014 1943	803363	AAF		FORREST
N933 1960	173912	RAuxAF	wef 2/5/56	FORREST
N517 1950	872457	RAuxAF		FORRESTER
N573 1948	841991	RAuxAF		FORRESTER
N1355 1945	77914	RAFVR		FORSDYKE
N225 1945	90290	AAF	DFC	FORSTER

Surname	Initials			Rank	Status	Date
FORSTER	A	T	F	FL		20/01/1960
FORSTER	G			CPL		17/03/1949
FORSTER	G			FLT SGT		08/02/1997
FORSTER	R	H	B	WING COMM		24/11/1949
FOSTER	A	R		FL		25/03/1959
FOSTER	C	H	L	FL		27/11/1952
FOSTER	D	S		FL		20/09/1945
FOSTER	D			SAC		18/01/1999
FOSTER	F	H		SGT		28/03/1962
FOSTER	H	R	V	REV		03/06/1943
FOSTER	J	M		SGT		28/09/1950
FOSTER	J			CPL		07/07/1949
FOSTER	L	F		FO		04/10/1945
FOSTER	M	G	L	WING COMM	Acting	13/01/1944
FOSTER	R	C		WO		15/01/1953
FOSTER	R	S	J	FL		30/09/1948
FOSTER	R	W		FL		06/03/1947
FOSTER	W	A		CPL		16/03/1950
FOSTER	W			CPL		10/02/1949
FOTHERINGHAM	W	B		CPL	Temporary	29/04/1954
FOWLE	T	G		CPL		01/07/1948
FOWLER	F			FL		19/08/1954
FOWLER	L	C		FL		06/03/1947
FOWLER	L	R		FLT SGT		15/07/1948
FOWLER	T	G		SGT		26/11/1953
FOWLIE	J			CPL	Temporary	09/12/1959
FOX	A	E		WO		13/01/1949
FOX	B	A		SGT		15/07/1948
FOX	D	E		FL		06/05/1943
FOX	F	A		FL		21/01/1959
FOX	G	E		FLT SGT		26/06/1947
FOX	J	R		FL		22/10/1958
FOX	J			SGT		09/12/1959
FOX	P	H		WO	Temporary	20/07/1960
FOX	S	A		SQD LDR	Acting	07/06/1945
FOX	W	M	S	SQD LDR		14/09/1950
FOXLEY	W	G		SQD LDR		07/07/1949
FOXLEY-NORRIS	C	N		SQD LDR		30/03/1943
FOY	J	T		FL		04/11/1948
FOY	J			FL		18/07/1956
FRAIN	G			SQD LDR	Acting	28/06/1945
FRANCE-COHEN	H			FL		11/01/1945
FRANCIS	C	A		WING COMM		17/04/1947
FRANCIS	C	C		CPL		25/05/1950
FRANCIS	C	W		FL		28/02/1946
FRANCIS	G	S		FL		13/10/1955
FRANCIS	J	P		FO		21/03/1946
FRANCIS	R	E		FO		20/05/1948
FRANK	E	M		SGT	Temporary	15/01/1958
FRANKLAND	R	N		SQD LDR		03/12/1942
FRANKLAND	W	E		CPL		16/10/1947
FRANKLIN	E	G		SQD LDR	Acting	23/05/1946
FRANKLIN	H			WO		30/03/1943
FRANKLIN	J	W		FL		15/04/1943
FRANKLIN	W	H		FLT SGT		26/01/1950
FRANKS	W	J		CPL	Acting	31/10/1956
FRASER	A	C		WING COMM		30/09/1948
FRASER	C			FL		01/02/1998
FRASER	J	L		FL		12/04/1945
FRASER	J	T		CPL		07/12/1950

Source	Service No	Force	Notes	Surname
N33 1960	205363	RAFVR	wef 18/4/58	FORSTER
N231 1949	866538	RAuxAF		FORSTER
YB 1998	n/a	RAuxAF		FORSTER
N1193 1949	73187	RAFVR		FORSTER
N196 1959	166234	RAFVR	wef 28/11/58	FOSTER
N915 1952	143455	RAFVR		FOSTER
N1036 1945	130445	RAFVR		FOSTER
YB 2000	n/a	RAuxAF		FOSTER
N242 1962	747202	RAFVR	wef 23/6/44	FOSTER
N609 1943	90104	AAF	MA	FOSTER
N968 1950	858018	RAuxAF		FOSTER
N673 1949	868689	RAuxAF		FOSTER
N1089 1945	186155	RAFVR		FOSTER
N11 1944	90007	AAF		FOSTER
N28 1953	864041	RAuxAF		FOSTER
N805 1948	77272	RAFVR		FOSTER
N187 1947	80815	RAFVR	DFC	FOSTER
N257 1950	808263	RAuxAF		FOSTER
N123 1949	854559	RAuxAF		FOSTER
N329 1954	2601823	RAFVR		FOTHERINGHAM
N527 1948	812361	RAuxAF		FOWLE
N672 1954	88498	RAuxAF		FOWLER
N187 1947	142459	RAFVR		FOWLER
N573 1948	805016	RAuxAF		FOWLER
N952 1953	841078	RAuxAF		FOWLER
N843 1959	817299	AAF	wef 28/6/44	FOWLIE
N30 1949	860446	RAuxAF	MM	FOX
N573 1948	857456	RAuxAF		FOX
N526 1943	85243	RAFVR		FOX
N28 1959	170708	RAFVR	wef 11/7/52	FOX
N519 1947	870924	AAF		FOX
N716 1958	120500	RAFVR	DFC; wef 20/4/44	FOX
N843 1959	2602785	RAFVR	wef 28/11/53	FOX
N531 1960	754399	RAFVR	wef 29/6/44	FOX
N608 1945	70224	RAFVR		FOX
N926 1950	90561	RAuxAF		FOX
N673 1949	76299	RAFVR		FOXLEY
N1014 1943	70225	RAFVR		FOXLEY-NORRIS
N934 1948	172932	RAFVR	DFC	FOY
N513 1956	104939	RAFVR		FOY
N706 1945	90473	AAF	MBE	FRAIN
N14 1945	116605	RAFVR		FRANCE-COHEN
N289 1947	91040	AAF		FRANCIS
N517 1950	844782	RAuxAF		FRANCIS
N197 1946	115712	RAFVR		FRANCIS
N762 1955	116104	RAFVR		FRANCIS
N263 1946	185748	RAFVR		FRANCIS
N395 1948	170889	RAFVR		FRANCIS
N40 1958	880514	WAAF	wef 25/8/44	FRANK
N1638 1942	90115	AAF	The Honorable	FRANKLAND
N846 1947	815113	AAF		FRANKLAND
N468 1946	79534	RAFVR	DFC,AFC	FRANKLIN
N1014 1943	808002	AAF		FRANKLIN
N454 1943	70228	RAFVR		FRANKLIN
N94 1950	746399	RAFVR		FRANKLIN
N761 1956	2699003	RAuxAF		FRANKS
N805 1948	90056	RAuxAF		FRASER
YB 1999	2797125	RAuxAF		FRASER
N381 1945	139207	RAFVR		FRASER
N1210 1950	807232	RAuxAF		FRASER

Surname	Initials			Rank	Status	Date
FRASER	L	N		SGT		12/06/1947
FRASER	W	K		SGT		16/08/1961
FRECKER	A	D		FL	Acting	10/12/1942
FREEDMAN	G	F		FL		21/03/1946
FREEDMAN	S	I		FL		21/06/1945
FREEMAN	A	L		LAC		16/08/1961
FREEMAN	F	A		SQD LDR		06/05/1993
FREEMAN	J			LAC		16/10/1947
FREEMAN	N	B		FL		28/02/1946
FREEMAN	N	H		LAC		04/09/1947
FREEMAN-PANNETT	P	G		SGT		07/08/1947
FREER	J	C		FL	Acting	17/03/1949
FREESTON	J	F		WING COMM		30/09/1948
FREESTONE	J	B		FL		29/06/1944
FREKE	T	A		FL		05/10/1944
FRENCH	A	W	L	FO		15/07/1948
FRENCH	C	L		FO		02/04/1953
FRENCH	F	J		SQD LDR		11/01/1945
FRESHWATER	R	M		SQD LDR	Acting	10/06/1948
FREW	R	C		CPL		01/04/1981
FREW	R	C		CPL		01/04/1991
FRICKER	D	R		SGT		13/01/1949
FRIEND	K	F		JNR TECH		22/05/1952
FRIENDSHIP	A	H	B	SQD LDR	Acting	09/08/1945
FRIPP	A	G		CPL		01/07/1948
FRIPP	J	H		FL		15/06/1944
FRITH	R	G		WING COMM		08/05/1947
FROST	B	C		FL		01/08/1946
FROST	H	L		SAC		02/05/1962
FROST	M	J		FL		21/03/1946
FROST	M	J		FL		07/12/1960
FROST	S	J		SQD LDR		06/03/1947
FROUD	T	R	W	FL		31/05/1945
FRY	H	S		LAC		06/09/1951
FRY	R	H		FL		06/03/1947
FRYER	W	M		LAC		21/02/1952
FUDGE	S	W		SGT		25/05/1960
FULFORD	D			SGT		30/09/1954
FULLER	K	B		FL		24/01/1946
FULLER	M	W		WING COMM	Acting	14/10/1948
FULLER	P	J	F	SEC OFF		10/05/1951
FULTON	F	F		GRP CAP	Acting	10/08/1944
FURNIVAL	J	W		LAC		26/01/1950
FURSE	A	W		FL		18/07/1956
FUTCHER	G	W		FL		28/07/1955
FYFE	C	F		SQD LDR		15/06/1944
FYFE	D	C		FL		08/11/1945
GABRIEL	C	P		GRP CAP		17/12/1942
GADD	D			FL		17/06/1954
GADD	J	E		WO		08/02/1945
GADNEY	C	H		SQD LDR		15/07/1948
GAFFRON	W	M		CPL		20/09/1951
GAGE	D	A		FL		21/01/1959
GAGE	J	E		CPL		10/06/1948
GAIR	I	D		FLT SGT		17/06/1954
GALBRAITH	A	C		FLT SGT		27/07/1944
GALE	M	D	C	CPL		15/09/1998
GALLOWAY	A			FL		18/11/1948
GALLOWAY	B			FLT OFF		08/05/1947
GALLOWAY	M	de	V	FLT OFF		25/05/1950

Source	Service No	Force	Notes	Surname
N478 1947	811073	AAF		FRASER
N624 1961	5006652	RAuxAF	wef 10/6/61	FRASER
N1680 1942	62647	RAFVR		FRECKER
N263 1946	105173	RAFVR		FREEDMAN
N671 1945	141087	RAFVR		FREEDMAN
N624 1961	840345	AAF	wef 11/3/44	FREEMAN
YB 1996	2692235	RAuxAF		FREEMAN
N846 1947	842613	AAF		FREEMAN
N197 1946	82697	RAFVR		FREEMAN
N740 1947	852276	AAF		FREEMAN
N662 1947	805442	AAF		FREEMAN-PANNETT
N231 1949	110361	RAFVR		FREER
N805 1948	90697	RAuxAF		FREESTON
N652 1944	64887	RAFVR		FREESTONE
N1007 1944	76313	RAFVR		FREKE
N573 1948	139153	RAFVR		FRENCH
N298 1953	141831	RAFVR	DFC	FRENCH
N14 1945	72120	RAFVR	DFC,AFC	FRENCH
N456 1948	72835	RAFVR		FRESHWATER
YB 1996	n/a	RAuxAF		FREW
YB 1996	n/a	RAuxAF	1st Clasp	FREW
N30 1949	813172	RAuxAF		FRICKER
N402 1952	2689510	RAuxAF		FRIEND
N874 1945	81637	RAFVR	DFM	FRIENDSHIP
N527 1948	756108	RAFVR		FRIPP
N596 1944	120212	RAFVR		FRIPP
N358 1947	90420	AAF		FRITH
N652 1946	77028	RAFVR		FROST
N314 1962	2684651	RAuxAF	wef 22/9/61	FROST
N263 1946	144786	RAFVR	DFC	FROST
N933 1960	144786	RAFVR	DFC; 1st Clasp; wef 8/6/52	FROST
N187 1947	72233	RAFVR		FROST
N573 1945	121203	RAFVR		FROUD
N913 1951	868718	RAuxAF		FRY
N187 1947	114139	RAFVR	DFC	FRY
N134 1952	872523	RAuxAF		FRYER
N375 1960	2649721	RAuxAF	wef 8/1/60	FUDGE
N799 1954	2650009	RAuxAF		FULFORD
N82 1946	141272	RAFVR		FULLER
N856 1948	72505	RAFVR		FULLER
N466 1951	616	WAAF		FULLER
N812 1944	74014	RAFVR	OBE	FULTON
N94 1950	853038	RAuxAF		FURNIVAL
N513 1956	201761	RAuxAF		FURSE
N565 1955	76521	RAFVR		FUTCHER
N596 1944	70234	RAFVR		FYFE
N1232 1945	86363	RAFVR		FYFE
N1708 1942	90204	AAF	OBE	GABRIEL
N471 1954	138665	RAFVR		GADD
N139 1945	801453	AAF		GADD
N573 1948	90827	RAuxAF	MBE	GADNEY
N963 1951	817153	RAuxAF		GAFFRON
N28 1959	187700	RAFVR	wef 17/9/52	GAGE
N456 1948	840417	RAuxAF		GAGE
N471 1954	2649513	RAuxAF		GAIR
N763 1944	807003	AAF		GALBRAITH
YB 1999	n/a	RAuxAF		GALE
N984 1948	149068	RAFVR	DFC	GALLOWAY
N358 1947	2444	WAAF		GALLOWAY
N517 1950	801	WAAF		GALLOWAY

Surname	Initials			Rank	Status	Date
GALLOWAY	M	de	V	FLT OFF		20/03/1957
GALLOWAY	R	E	C	SGT		07/06/1998
GALT	J	C		FL		21/09/1944
GAMBLE	A	W	H	FO		11/07/1946
GAMBLE	K	C		FO		07/05/1953
GAMBLE	L	G		CPL		26/04/1961
GAMBLE	W			PO		10/12/1942
GAMMELIEN	A	E		FO		15/09/1949
GAMMELIEN	A	E		FL		20/03/1957
GAMMELIEN	G	M		CPL		04/10/1951
GAMMON	J	R		FL		11/07/1946
GARBETT	W	J		CPL		07/08/1947
GARDEN	T	C		SQD LDR		08/02/1945
GARDINER	E	G	W	CPL		14/10/1948
GARDINER	F	T		WING COMM	Acting	04/10/1945
GARDINER	J	D		SQD LDR		05/01/1950
GARDINER	J	V	C	CPL		15/09/1949
GARDINER	W	N		FL		11/07/1946
GARDNER	C	H		FL		14/10/1948
GARDNER	E	J		FO		13/01/1949
GARDNER	E			FLT SGT		01/10/1953
GARDNER	E			CPL		19/08/1959
GARDNER	F	E	S	FO		27/07/1944
GARDNER	G	M		SQD LDR		24/06/1943
GARDNER	H	R		WING COMM	Acting	18/11/1948
GARDNER	J	B		FL		21/10/1959
GARDNER	J	D		FL		28/02/1946
GARDNER	R	S		SGT		27/03/1947
GARDNER	V	G	E	WING COMM		19/08/1954
GARGET	F	W		FL		19/10/1960
GARLAND	A	E		FL		21/03/1946
GARLAND	D	J		FL		17/04/1947
GARLAND	E			SQD LDR	Acting	13/07/1950
GARLAND	G	A	W	WING COMM		05/01/1950
GARMESON	E	H		FL		30/10/1957
GARMONSWAY	J	V		WING COMM		20/04/1944
GARNER	G	A		CPL		16/10/1947
GARNHAM	W	J		FL		09/05/1946
GARNWELL	F	L		LAC		17/03/1949
GARRATT	G	R	M	SQD LDR		15/04/1943
GARRETT	K	H		FL		26/04/1963
GARRIOCH	W	R		WO		18/04/1946
GARSIDE	E			FL		18/04/1946
GARSIDE	J	W		FL		20/11/1957
GARSIDE	T	F		FL		14/10/1948
GARTHWAITE	H			FL		01/07/1948
GARTON	C	D		SQD LDR		05/06/1947
GARTON	G	W		SQD LDR	Acting	20/04/1944
GARTON	J	C		FL	Acting	30/09/1948
GARTON	J	F		FL		08/11/1945
GARTON-HORSLEY	C	I	M	FLT OFF	Acting	27/03/1947
GARVIN	E			LAC		19/10/1960
GASCOIGNE	J	E		SGT		05/01/1950
GASKELL	R	C		FL		01/05/1957
GASKELL	S	W		LAC		13/07/1950
GASKIN	F	G		SQD LDR		18/11/1948
GASKIN	J			AC1		04/11/1948
GASKIN	W	J		CPL		13/07/1950
GASTON	J	L	C	FO		26/11/1958
GATE	J	L	E	P	CPL	27/03/1947

Source	Service No	Force	Notes	Surname
N198 1957	801	WRAAF	1st Clasp; wef 28/1/57	GALLOWAY
YB 2000	n/a	RAuxAF		GALLOWAY
N961 1944	118689	RAFVR		GALT
N594 1946	189464	RAFVR		GAMBLE
N395 1953	147169	RAFVR		GAMBLE
N339 1961	2685109	RAFVR	wef 21/12/60	GAMBLE
N1680 1942	118738	RAFVR		GAMBLE
N953 1949	148948	RAFVR		GAMMELIEN
N198 1957	148948	RAuxAF	1st Clasp; wef 13/1/56	GAMMELIEN
N1019 1951	800562	RAuxAF		GAMMELIEN
N594 1946	126689	RAFVR		GAMMON
N662 1947	840753	AAF		GARBETT
N139 1945	90200	AAF		GARDEN
N856 1948	749829	RAFVR		GARDINER
N1089 1945	72100	RAFVR	DFC	GARDINER
N1 1950	70236	RAFVR		GARDINER
N953 1949	812265	RAuxAF		GARDINER
N594 1946	121234	RAFVR		GARDINER
N856 1948	72836	RAFVR		GARDNER
N30 1949	122781	RAFVR		GARDNER
N796 1953	2601604	RAFVR		GARDNER
N552 1959	2601604	RAFVR	1st Clasp; wef 4/4/53	GARDNER
N763 1944	136321	RAFVR		GARDNER
N677 1943	90036	AAF		GARDNER
N984 1948	74015	RAFVR		GARDNER
N708 1959	164971	RAFVR	wef 26/5/57	GARDNER
N197 1946	82181	RAFVR		GARDNER
N250 1947	747396	RAFVR		GARDNER
N672 1954	90514	RAuxAF		GARDNER
N775 1960	175489	RAFVR	DFC; wef 14/1/54	GARGET
N263 1946	169714	RAFVR		GARLAND
N289 1947	73749	RAFVR		GARLAND
N675 1950	90906	RAuxAF		GARLAND
N1 1950	90476	RAuxAF	AFC	GARLAND
N775 1957	179352	RAFVR	wef 24/7/56	GARMESON
N363 1944	90681	AAF		GARMONSWAY
N846 1947	810105	AAF		GARNER
N415 1946	127027	RAFVR		GARNHAM
N231 1949	762160	RAFVR		GARNWELL
N454 1943	72215	RAFVR		GARRATT
MOD 1964	3132114	RAFVR		GARRETT
N358 1946	742039	RAFVR		GARRIOCH
N358 1946	89590	RAFVR		GARSIDE
N835 1957	196463	RAFVR	wef 21/9/56	GARSIDE
N856 1948	90597	RAuxAF		GARSIDE
N527 1948	144633	RAFVR		GARTHWAITE
N460 1947	72216	RAFVR		GARTON
N363 1944	67034	RAFVR	DFC	GARTON
N805 1948	133341	RAFVR		GARTON
N1232 1945	85230	RAFVR		GARTON
N250 1947	3928	WAAF		GARTON-HORSLEY
N775 1960	856019	AAF	wef 13/7/44	GARVIN
N1 1950	750146	RAFVR		GASCOIGNE
N277 1957	197599	RAuxAF	wef 21/8/56	GASKELL
N675 1950	810201	RAuxAF		GASKELL
N984 1948	90560	RAuxAF		GASKIN
N934 1948	750221	RAFVR		GASKIN
N675 1950	865551	RAuxAF		GASKIN
N807 1958	2650772	RAuxAF	wef 11/10/58	GASTON
N250 1947	814184	AAF		GATE

Surname	Initials			Rank	Status	Date
GATES	G			FL	Acting	17/04/1947
GATES	S	S		WING COMM		20/05/1948
GATHERCOLE	C	A		FO		26/08/1948
GATLISH	A	A		FO		26/11/1958
GATWARD	A	K		WING COMM	Acting	27/07/1944
GAUDIE	R	W		SQD LDR		05/01/1950
GAUL	P	G		LAC		20/05/1948
GAUNT	E	E		FLT OFF		27/03/1947
GAUNT	J	I	H	FLT OFF	Acting	08/05/1947
GAUNT	P	H		WO		15/07/1948
GAUNT	W	E		FL		01/05/1952
GAUNTLETT	D			FL		24/01/1946
GAUVAIN	S	de	P	FO		27/03/1952
GAVARD	A	S		SAC		19/10/1960
GAWN	J	D		PO		22/09/1955
GAY	K	D		FL		28/02/1946
GAYNER	J	R	H	WING COMM	Acting	27/07/1944
GEARD	L	H		FO		15/01/1953
GEARD	L	H		FL		01/05/1963
GEARY	J			WO		16/03/1950
GEARY	L	W		FL	Acting	27/03/1947
GEARY	P	C		FO		16/03/1950
GEE	J	W		SQD LDR	Acting	24/01/1946
GEE	N	H		LAC		28/09/1950
GEE	R	R		FL		18/04/1946
GEE	R	R		FO		01/12/1955
GELL	W	C	C	AVM	Acting	20/04/1944
GENDERS	G	E	C	FL		06/03/1947
GENT	J	F		CPL		26/08/1948
GENTLE	L	G		FLT SGT		26/08/1948
GEOGHEGAN	J	P		SQD LDR		08/11/1945
GEOGHEGAN	J	P		SGT		07/08/1947
GEOGHEGAN	M	G		FL		13/12/1945
GEORGE	A	P		WO		07/02/1952
GEORGE	D	C		SGT		27/03/1947
GEORGE	L	I		FL		18/04/1946
GEORGE	M	F	A	FL		28/02/1962
GEORGE	P	H		FL		30/03/1943
GEORGE	P	H		FL		30/03/1944
GERMAINE	S	G		FL	Acting	28/11/1946
GERRAGHTY	A	J	V	SQD LDR		24/05/2005
GERRISH	F	J		FL		02/04/1953
GERRY	C	G		FL		08/05/1947
GETHIN	R	C	J	FL	Acting	10/08/1950
GETHIN	R	T		FL		03/06/1959
GETHING	P	A		FL		16/10/1947
GHENT	J	H		LAC		01/07/1948
GHENT	W	A		SGT	Acting	21/02/1952
GIBB	W	C		CPL		05/06/1947
GIBBINS	D	G		FL		26/08/1948
GIBBINS	D	G		FL		26/11/1958
GIBBINS	I	D	H	SQD LDR	Acting	27/03/1947
GIBBONS	C	F		FL	Acting	30/03/1943
GIBBONS	C	M		SQD LDR	Acting	18/04/1946
GIBBS	F	H		SQD LDR		06/01/1955
GIBBS	W	J		FO		06/03/1947
GIBSON	A	C		FL		04/09/1957
GIBSON	A			FLT SGT		24/06/1943
GIBSON	A			SGT		27/03/1952
GIBSON	B	S		SQD LDR	Acting	24/01/1946

Source	Service No	Force	Notes	Surname
N289 1947	123533	RAFVR		GATES
N395 1948	90494	RAuxAF		GATES
N694 1948	188364	RAFVR		GATHERCOLE
N807 1958	164370	RAFVR	wef 14/6/58	GATLISH
N763 1944	83251	RAFVR	DSO,DFC	GATWARD
N1 1950	72390	RAuxAF		GAUDIE
N395 1948	840922	RAuxAF		GAUL
N250 1947	2731	WAAF		GAUNT
N358 1947	3308	WAAF		GAUNT
N573 1948	755625	RAFVR		GAUNT
N336 1952	115719	RAFVR		GAUNT
N82 1946	117156	RAFVR	DFC	GAUNTLETT
N224 1952	87724	RAFVR		GAUVAIN
N775 1960	2676442	RAuxAF	wef 18/10/58	GAVARD
N708 1955	2650780	RAuxAF		GAWN
N197 1946	119511	RAFVR		GAY
N763 1944	90399	AAF	DFC	GAYNER
N28 1953	106085	RAFVR		GEARD
N333 1963	106085	RAFVR	1st Clasp; wef 14/8/62	GEARD
N257 1950	759168	RAFVR		GEARY
N250 1947	86743	RAFVR		GEARY
N257 1950	153972	RAFVR		GEARY
N82 1946	60763	RAFVR	DFC	GEE
N968 1950	850712	RAuxAF		GEE
N358 1946	112388	RAFVR	DFC	GEE
N895 1955	112388	RAFVR	DFC; 1st Clasp	GEE
N363 1944	90580	AAF	DSO,MC,TD,DL	GELL
N187 1947	120165	RAFVR		GENDERS
N694 1948	753080	RAFVR		GENT
N694 1948	770859	RAFVR		GENTLE
N1232 1945	62263	RAFVR	DFC	GEOGHEGAN
N662 1947	853556	AAF		GEOGHEGAN
N1355 1945	83275	RAFVR		GEOGHEGAN
N103 1952	799891	RAFVR		GEORGE
N250 1947	751142	RAFVR		GEORGE
N358 1946	133522	RAFVR		GEORGE
N157 1962	2614414	RAFVR	wef 17/8/61	GEORGE
N1014 1943	81672	RAFVR		GEORGE
N281 1944	81672	RAFVR		GEORGE
N1004 1946	186324	RAFVR		GERMAINE
LG 2005	0210788D	RAFVR		GERRAGHTY
N298 1953	159671	RAFVR		GERRISH
N358 1947	80569	RAFVR		GERRY
N780 1950	113545	RAFVR		GETHIN
N368 1959	197492	RAFVR	wef 8/6/54	GETHIN
N846 1947	70239	RAFVR		GETHING
N527 1948	852957	RAuxAF		GHENT
N134 1952	744018	RAFVR		GHENT
N460 1947	811084	AAF		GIBB
N694 1948	119136	RAFVR		GIBBINS
N807 1958	119136	RAFVR	1st Clasp; wef 7/7/58	GIBBINS
N250 1947	73054	RAFVR	MBE	GIBBINS
N1014 1943	104397	RAFVR		GIBBONS
N358 1946	61012	RAFVR		GIBBONS
N1 1955	112606	RAuxAF		GIBBS
N187 1947	179658	RAFVR		GIBBS
N611 1957	159789	RAFVR	wef 4/7/55	GIBSON
N677 1943	816009	AAF		GIBSON
N224 1952	2681002	RAuxAF	1st Clasp	GIBSON
N82 1946	106684	RAFVR		GIBSON

Surname	Initials			Rank	Status	Date
GIBSON	C			FL		01/03/1945
GIBSON	D	A		WO		08/11/1945
GIBSON	E	J		SQD LDR	Acting	13/12/1945
GIBSON	E			WO		15/06/1944
GIBSON	G	M		WING COMM	Acting	05/01/1950
GIBSON	G	R		FLT OFF	Temporary	15/01/1958
GIBSON	H			FLT SGT		04/02/1943
GIBSON	H			WO	Temporary	21/10/1959
GIBSON	J	E		SQD LDR	Acting	05/10/1944
GIBSON	R	E		FLT SGT		13/12/1945
GIBSON	R	E		FLT SGT		31/10/1956
GIBSON	T	F		AC1		04/11/1948
GIBSON	W			FLT SGT	Acting	03/06/1959
GIDDINGS	M			SQD LDR		26/11/1953
GIDDINS	J	G		FL		10/12/1942
GIDMAN	M	J		FL		24/01/1946
GIFFORD	E			FL		09/05/1946
GIFFORD	R	A		FL		29/04/1954
GIFKIN	P	A		FL		27/11/1952
GILBERT	C	G		SQD LDR		27/04/1950
GILBERT	F			FL		28/11/1946
GILBERT	G	B		SQD LDR	Acting	17/02/1944
GILBERT	G	E		LAC		14/10/1948
GILBERT	G	E		CPL		03/06/1959
GILBERT	L	H		FL		04/10/1945
GILBERT	R	H		FL		04/10/1945
GILBODY	G	J		SQD LDR		30/09/1948
GILDER	F	E		SQD LDR		16/10/1952
GILES	E	E		SEC OFF		23/02/1950
GILES	J	F	L	WO		05/06/1947
GILKERSON	J	C		FL	Acting	21/12/1950
GILL	A	M		SQD LDR	Acting	27/01/1944
GILL	A	R		FLT SGT		13/12/1945
GILL	F	W	K	SGT		25/05/1950
GILL	W	H	G	FL		27/03/1947
GILLAM	F	C	C	FL		09/12/1959
GILLENDER	G	F	H	AC2		17/06/1954
GILLESPIE	A			WING COMM	Acting	06/03/1947
GILLESPIE	J	A		SGT		10/12/1942
GILLESPIE	J	P		PO	Acting	29/07/1965
GILLESPIE	J	W		FL		26/11/1953
GILLESPIE	J			CPL		20/05/1948
GILLETT	F	J		LAC		28/02/1946
GILLIES	A	C		LAC		01/07/1948
GILLIES	A			FLT SGT	Acting	11/01/1945
GILLIES	D	A		FL		23/02/1950
GILLIES	J	A		FL		20/09/1945
GILLIES	L	E		FLT SGT		12/08/1943
GILLINGHAM	F	E		SGT		21/06/1951
GILMER	T	R		FL		13/12/1945
GILMORE	J			CPL		01/07/1948
GILMOUR	D	S		JNR TECH		25/03/1954
GILMOUR	J			FLT SGT		07/06/1951
GILMOUR	W	MoM.		SQD LDR	Acting	13/12/1945
GILROY	G	K		GRP CAP	Acting	04/10/1945
GILSON	G	K		FL		18/04/1946
GIRDLER	W	T	C	FO		31/05/1945
GIRL	G	E		CPL	Temporary	19/08/1954
GLADSTONE	D	S		SQD LDR		30/03/1943
GLASBY	G	R		CPL		04/09/1957

Source	Service No	Force	Notes	Surname
N225 1945	104398	RAFVR	DFM	GIBSON
N1232 1945	742420	RAFVR		GIBSON
N1355 1945	77916	RAFVR		GIBSON
N596 1944	740626	RAFVR		GIBSON
N1 1950	72085	RAFVR	MB,ChB	GIBSON
N40 1958	129	WAAF	wef 13/4/44	GIBSON
N131 1943	808022	AAF		GIBSON
N708 1959	808022	AAF	1st Clasp; wef 12/4/45	GIBSON
N1007 1944	84030	RAFVR		GIBSON
N1355 1945	801119	AAF		GIBSON
N761 1956	2685502	RAuxAF	BEM; 1st Clasp	GIBSON
N934 1948	843874	RAuxAF		GIBSON
N368 1959	2696011	RAuxAF	wef 9/12/56	GIBSON
N952 1953	103753	RAuxAF		GIDDINGS
N1680 1942	70241	RAFVR		GIDDINS
N82 1946	85939	RAFVR		GIDMAN
N415 1946	126472	RAFVR		GIFFORD
N329 1954	102069	RAFVR		GIFFORD
N915 1952	64311	RAFVR		GIFKIN
N406 1950	72663	RAFVR		GILBERT
N1004 1946	109534	RAFVR		GILBERT
N132 1944	91175	AAF		GILBERT
N856 1948	842434	RAuxAF		GILBERT
N368 1959	2692018	RAuxAF	1st Clasp; wef 21/3/59	GILBERT
N1089 1945	141412	RAFVR	DFC	GILBERT
N1089 1945	101508	RAFVR	DFC	GILBERT
N805 1948	72386	RAFVR		GILBODY
N798 1952	83353	RAuxAF		GILDER
N179 1950	4660	WAAF		GILES
N460 1947	845472	AAF		GILES
N1270 1950	109156	RAFVR		GILKERSON
N59 1944	84709	RAFVR		GILL
N1355 1945	804176	AAF		GILL
N517 1950	760998	RAFVR		GILL
N250 1947	177170	RAFVR		GILL
N843 1959	200890	RAFVR	wef 7/11/52	GILLAM
N471 1954	846856	RAuxAF		GILLENDER
N187 1947	73198	RAFVR		GILLESPIE
N1680 1942	808069	AAF		GILLESPIE
MOD 1966	2617531	RAF		GILLESPIE
N952 1953	183121	RAuxAF		GILLESPIE
N395 1948	816063	RAuxAF		GILLESPIE
N197 1946	853718	AAF		GILLETT
N527 1948	847003	RAuxAF		GILLIES
N14 1945	803287	AAF		GILLIES
N179 1950	152031	RAFVR		GILLIES
N1036 1945	90900	AAF		GILLIES
N850 1943	741051	RAFVR		GILLIES
N621 1951	853686	RAuxAF		GILLINGHAM
N1355 1945	155591	RAFVR		GILMER
N527 1948	873307	RAuxAF		GILMORE
N249 1954	2601067	RAFVR		GILMOUR
N560 1951	803411	RAuxAF		GILMOUR
N1355 1945	107773	RAFVR	DFC,DFM	GILMOUR
N1089 1945	90481	AAF	DSO,DFC	GILROY
N358 1946	84722	RAFVR		GILSON
N573 1945	183403	RAFVR		GIRDLER
N672 1954	84355	RAuxAF		GIRL
N1014 1943	70250	RAFVR		GLADSTONE
N611 1957	747079	RAFVR	wef 3/7/44	GLASBY

Surname	Initials			Rank	Status	Date
GLASER	E	D		FL		28/02/1946
GLASS	J	N		FL		12/08/1943
GLASS	R	R		FL		09/05/1946
GLASSER	J	J		SGT		19/10/1960
GLEDALL	D	E	W	WO	Temporary	30/03/1960
GLEGG	A	J		SQD LDR	Acting	06/03/1947
GLEN	J	B		FL		20/09/1945
GLEN	U	M		FLT OFF		16/03/1950
GLEW	T	A		SAC		21/10/1993
GLINN	R	E		LAC		13/07/1950
GLOVER	A	H		SGT		04/09/1947
GLOVER	A	M		WING COMM		14/10/1943
GLOVER	N	D		FL		16/08/1961
GLOVER	W			SGT		15/09/1949
GOADBY	P	G		WO		09/05/1946
GOATHAM	R			SGT		15/09/1949
GODDARD	F	R		FL		13/01/1949
GODDARD	G	I		SAC		07/06/1999
GODDARD	G			FL		19/08/1959
GODDARD	H	G		SQD LDR		17/12/1942
GODDARD	L	M		FL		26/04/1961
GODDARD	R	H	T	FL		16/08/1961
GODDEN	C	J		SGT		03/07/1952
GODDEN	H	E	J	CPL		23/08/1951
GODFREY	D	F		CPL		24/11/1949
GODFREY	R			FL		05/01/1950
GODLEY	L	F		FLT SGT		02/04/1953
GOFF	R	W		CPL		07/07/1949
GOFF	S			CPL		07/08/1947
GOLD	W	J	H	FL	Acting	13/01/1949
GOLDIE	M			WING OFF	Acting	27/03/1947
GOLDING	D			FL		02/07/1958
GOLDING	G	W		FO		18/04/1946
GOLDMAN	A			PO		26/11/1953
GOLDUP	R	N		WO		28/02/1946
GOLLAN	R	H		SQD LDR		17/04/1947
GOMM	G	W		SQD LDR	Acting	10/06/1948
GONSE	R	L		FL		30/10/1957
GOOCH	L	J		FL		27/11/1952
GOOD	J	G		SQD LDR	Acting	13/12/1945
GOOD	S			FL		25/03/1959
GOODCHILD	J	P	F	FL		13/07/1950
GOODE	A	G		FL		12/06/1947
GOODE	C	S		SQD LDR		26/01/1950
GOODE	H	M		FL		23/02/1950
GOODE	H	W		CPL		06/03/1947
GOODE	T	J		FL		06/01/1955
GOODEY	P	W		CPL		07/07/1949
GOODFELLOW	C	A	J	WO		07/01/1943
GOODLIFFE	E	P		FL		25/03/1954
GOODMAN	G			FO		26/01/1950
GOODMAN	J	R		SQD LDR	Acting	13/11/1947
GOODMAN	J	S		FO		28/11/1956
GOODMAN	S	H		FLT SGT	Acting	26/11/1953
GOODRICH	F			FLT SGT		18/11/1948
GOODRICH	H	E		FO		24/01/1946
GOODRIDGE	F	R		SGT		01/10/1953
GOODRUN	R	W	H	FL		01/10/1953
GOODWIN	A			CPL		26/08/1948
GOODWIN	G	W		FL		12/07/1945

Source	Service No	Force	Notes	Surname
N197 1946	82178	RAFVR	DFC	GLASER
N850 1943	72158	RAFVR		GLASS
N415 1946	104445	RAFVR	DFC	GLASS
N775 1960	846280	RAF	wef 9/4/44	GLASSER
N229 1960	748677	RAFVR	wef 8/6/44	GLEDALL
N187 1947	84021	RAFVR	DFC	GLEGG
N1036 1945	120874	RAFVR		GLEN
N257 1950	509	WAAF		GLEN
YB 1996	n/a	RAuxAF		GLEW
N675 1950	746216	RAFVR		GLINN
N740 1947	845305	AAF		GLOVER
N1065 1943	15072	RAFVR		GLOVER
N624 1961	189048	RAF	wef 1/11/52	GLOVER
N953 1949	816059	RAuxAF		GLOVER
N415 1946	751243	RAFVR		GOADBY
N953 1949	812122	RAuxAF		GOATHAM
N30 1949	78990	RAFVR		GODDARD
YB 2001	n/a	RAuxAF		GODDARD
N552 1959	158895	RAFVR	wef 25/1/59	GODDARD
N1708 1942	70252	RAFVR	DFC,AFC	GODDARD
N339 1961	91368	RAuxAF	wef 6/3/61	GODDARD
N624 1961	172161	RAFVR	wef 6/12/60	GODDARD
N513 1952	750686	RAFVR		GODDEN
N866 1951	812187	RAuxAF		GODDEN
N1193 1949	750136	RAFVR		GODFREY
N1 1950	90880	RAuxAF		GODFREY
N298 1953	2602252	RAFVR		GODLEY
N673 1949	819036	RAuxAF		GOFF
N662 1947	840980	AAF		GOFF
N30 1949	113220	RAFVR		GOLD
N250 1947	692	WAAF		GOLDIE
N452 1958	186423	RAFVR	wef 2/6/53	GOLDING
N358 1946	181706	RAFVR		GOLDING
N952 1953	2603259	RAFVR		GOLDMAN
N197 1946	819070	AAF		GOLDUP
N289 1947	72320	RAFVR		GOLLAN
N456 1948	84245	RAFVR		GOMM
N775 1957	129569	RAFVR	wef 28/2/54	GONSE
N915 1952	87721	RAuxAF		GOOCH
N1355 1945	85026	RAFVR		GOOD
N196 1959	65628	RAFVR	wef 21/7/54	GOOD
N675 1950	109900	RAFVR		GOODCHILD
N478 1947	87713	RAFVR		GOODE
N94 1950	72837	RAFVR		GOODE
N179 1950	73203	RAFVR		GOODE
N187 1947	842482	AAF		GOODE
N1 1955	105438	RAFVR		GOODE
N673 1949	846070	RAuxAF		GOODEY
N3 1943	801035	AAF		GOODFELLOW
N249 1954	106320	RAFVR		GOODLIFFE
N94 1950	85042	RAFVR		GOODMAN
N933 1947	126540	RAFVR	DFC,AFC	GOODMAN
N833 1956	2601342	RAFVR		GOODMAN
N952 1953	2698022	RAuxAF		GOODMAN
N984 1948	840022	RAuxAF		GOODRICH
N82 1946	170353	RAFVR		GOODRICH
N796 1953	847516	RAuxAF		GOODRIDGE
N796 1953	185228	RAuxAF		GOODRUN
N694 1948	857405	RAuxAF		GOODWIN
N759 1945	141131	RAFVR		GOODWIN

Surname	Initials			Rank	Status	Date
GOODWIN	H	A		FL		10/06/1948
GOODWIN	H	B	N	FL		01/07/1948
GOODWIN	P			SGT		17/10/1995
GOODWIN	R	A		SGT		13/01/1999
GOODWIN	R	D		FL		24/01/1946
GOODWRIGHT	R	E		WO		30/03/1943
GOOLDING	T	H		FLT SGT	Temporary	26/04/1961
GOONAN	P			FL		16/09/1995
GORDON	A	P		LAC		02/04/1953
GORDON	J	R		FL		26/11/1958
GORDON	J	S		FL		14/11/1962
GORDON	K	J		FL		21/03/1946
GORDON	T			SGT		04/09/1947
GORE	R			FL		18/04/1946
GORE	W	E		FL		07/07/1949
GORMAN	J			FL		18/04/1946
GORTON	E	A		FL		27/03/1947
GOSLING	E	L		SQD LDR		28/06/1945
GOSLING	R	A		FL		28/02/1946
GOSLING	R	C		FL		24/11/1949
GOTHARD	E	A	S	SQD LDR		15/09/1949
GOTT	E	H		FL	Acting	26/06/1947
GOUGH	D	G	M	FL		25/01/1945
GOUGH	F	C	G	FL		28/07/1955
GOUGH	K	W		WING COMM		24/06/1943
GOUGH	R	E	J	FL		08/11/1945
GOULD	W			LAC		15/07/1948
GOULDING	J	A		CPL		13/07/1950
GOURLAY	H	W		FO		01/11/1951
GOVETT	R	T		FO		06/03/1947
GOWER	C			SQD LDR		09/05/1994
GOWING	R	P		FLT SGT		01/07/1948
GOYDER	D	G	R	FLT SGT	Acting	04/11/1948
GRACE	E	J		WO		27/02/1997
GRACEY	W	B	W	SQD LDR		17/06/1943
GRADLEY	R	W		SGT		25/05/1950
GRAESSER	R	F		WO		30/03/1943
GRAGE	W	R		FO		02/08/1954
GRAHAM	A	E		LAC		05/02/1948
GRAHAM	A	W		CPL		26/06/1947
GRAHAM	F	B		SEC OFF		12/06/1947
GRAHAM	G	G		FL		28/11/1946
GRAHAM	H	A		WING COMM		26/01/1950
GRAHAM	J	W	W	FL		12/07/1945
GRAHAM	J			CPL		06/03/1947
GRAHAM	R	D		LAC		23/02/1950
GRAHAM	R	K		SGT		10/06/1948
GRAHAM	W	J		LAC		21/01/1959
GRAHAM-GREEN	V	M	T	FL		12/12/1984
GRAHAM-GREEN	V	M	T	FL		12/12/1994
GRAIN	A	J		FL	Acting	17/03/1949
GRAINGER	D	N		SQD LDR	Acting	28/02/1946
GRAINGER	D			SAC		20/07/1960
GRANT	A	M		SQD LDR		00/00/1940
GRANT	A	M		WING COMM	Acting	26/11/1953
GRANT	A			CPL		19/10/1960
GRANT	I	H		FL		30/08/1945
GRANT	I	P		FL		28/02/1946
GRANT	I	P		FL	Acting	21/06/1951
GRANT	J	R		SGT		05/02/1993

Source	Service No	Force	Notes	Surname
N456 1948	79573	RAFVR	DFC	GOODWIN
N527 1948	90928	RAuxAF		GOODWIN
YB 1997	n/a	RAuxAF		GOODWIN
YB 2000	n/a	RAuxAF		GOODWIN
N82 1946	120495	RAFVR		GOODWIN
N1014 1943	741145	RAFVR		GOODWRIGHT
N339 1961	804216	AAF	wef 26/12/42	GOOLDING
YB 1997	211536	RAuxAF		GOONAN
N298 1953	871677	RAuxAF		GORDON
N807 1958	2691519	RAuxAF	wef 18/9/58	GORDON
N837 1962	1891163	RAFVR	wef 27/9/62	GORDON
N263 1946	158593	RAFVR	DFC	GORDON
N740 1947	811011	AAF		GORDON
N358 1946	100621	RAFVR		GORE
N673 1949	90279	RAuxAF	DFC; Deceased	GORE
N358 1946	156007	RAFVR		GORMAN
N250 1947	78507	RAFVR		GORTON
N706 1945	70256	RAFVR		GOSLING
N197 1946	161786	RAFVR	DFC	GOSLING
N1193 1949	85245	RAFVR		GOSLING
N953 1949	72130	RAFVR		GOTHARD
N519 1947	147431	RAFVR		GOTT
N70 1945	117790	RAFVR		GOUGH
N565 1955	155680	RAFVR		GOUGH
N677 1943	90067	AAF	AFC	GOUGH
N1232 1945	129272	RAFVR		GOUGH
N573 1948	864151	RAuxAF		GOULD
N675 1950	815195	RAuxAF		GOULDING
N1126 1951	60346	RAFVR		GOURLAY
N187 1947	162264	RAFVR		GOVETT
YB 1997	4078085	RAuxAF		GOWER
N527 1948	808396	RAuxAF		GOWING
N934 1948	804184	RAuxAF	BEM	GOYDER
YB 1997	n/a	RAuxAF		GRACE
N653 1943	90020	AAF		GRACEY
N517 1950	840596	RAuxAF		GRADLEY
N1014 1943	741109	RAFVR		GRAESSER
MOD 1964	2603322	RAFVR		GRAGE
N87 1948	842408	RAuxAF		GRAHAM
N519 1947	819126	AAF		GRAHAM
N478 1947	2767	WAAF		GRAHAM
N1004 1946	171166	RAFVR	DFM	GRAHAM
N94 1950	72574	RAFVR	BSc,MD,ChB	GRAHAM
N759 1945	84956	RAFVR		GRAHAM
N187 1947	816024	AAF		GRAHAM
N179 1950	749093	RAFVR		GRAHAM
N456 1948	808328	RAuxAF		GRAHAM
N28 1959	846183	RAFVR	wef 28/3/44	GRAHAM
YB 1996	9919	RAuxAF		GRAHAM-GREEN
YB 1996	9919	RAuxAF	1st Clasp	GRAHAM-GREEN
N231 1949	155255	RAFVR	DFM; Deceased	GRAIN
N197 1946	113860	RAFVR	DFC,DFM	GRAINGER
N531 1960	2683617	RAuxAF	wef 8/2/60	GRAINGER
N1014 1943	90181	AAFRO		GRANT
N952 1953	90181	RAuxAF	1st Clasp	GRANT
N775 1960	2692077	RAFVR	wef 20/8/60	GRANT
N961 1945	72515	RAFVR		GRANT
N197 1946	142573	RAFVR		GRANT
N621 1951	91245	RAuxAF	1st Clasp	GRANT
YB 1996	n/a	RAuxAF		GRANT

Surname	Initials			Rank	Status	Date
GRANT	P	N		SGT		07/12/1950
GRANT	R	D		FO		06/03/1947
GRANT	T	D		SQD LDR	Acting	08/11/1945
GRANT-FERRIS	R	G		WING COMM		03/12/1942
GRAVES	R	C		FL		28/11/1946
GRAVES	T	K		FL		17/04/1947
GRAVES	W	C		CPL		06/03/1947
GRAY	A	P		SQD LDR		12/08/1943
GRAY	C	K		FL		04/10/1945
GRAY	D	E		FL		23/05/1946
GRAY	D	E		SGT		15/11/1951
GRAY	G	J		SQD LDR	Acting	28/02/1946
GRAY	G			SGT		05/06/1947
GRAY	H	D		FL		30/12/1943
GRAY	J			FL		01/10/1995
GRAY	N	M		FL		03/10/1988
GRAY	N	M		FL		03/10/1998
GRAY	P			FL		26/04/1956
GRAY	R	W		WING COMM		12/06/1947
GRAY	R			SGT		25/01/1945
GRAY	S			FLT SGT	Acting	18/02/1954
GRAY	T			FL		08/11/1945
GRAY	W	G		FL		11/07/1946
GRAYLAND	H	T		FLT SGT	Acting	14/10/1948
GRAYLING	G	B		WING COMM	Acting	10/05/1951
GRAYSON	A	K		FL	Acting	02/09/1943
GRAYSON	P	L		SAC		07/05/1993
GRAYSON	W	S		FL		16/10/1947
GREATWICH	C	T	M	CPL		15/01/1953
GREAVES	D	E		FL		07/10/1964
GREAVES	G			WING COMM		22/04/1943
GREEHNAM	M	L		FL		26/08/1948
GREEN	A	J		FL		30/11/1944
GREEN	A			AC1		23/08/1951
GREEN	C	G		SGT		25/03/1959
GREEN	C	P		GRP CAP	Acting	18/05/1944
GREEN	D	A		SQD LDR	Acting	06/03/1947
GREEN	E			FL		07/08/1947
GREEN	F	L		FL		26/01/1950
GREEN	F	R		FL		29/06/1944
GREEN	F	W	W	WO		01/07/1948
GREEN	G	E		SGT		30/09/1948
GREEN	G	R		FL		17/06/1943
GREEN	J	A		FL		27/07/1944
GREEN	J	E		LAC		18/02/1954
GREEN	L	C		SGT		15/09/1949
GREEN	N	C		SQD LDR	Acting	09/08/1945
GREEN	N	W	F	FL		12/06/1947
GREEN	P			SQD LDR		24/06/1943
GREEN	R	A	W	FLT SGT	Acting	01/07/1948
GREEN	S	J		SAC		02/12/1992
GREEN	S			SGT	Temporary	19/10/1960
GREEN	V	F	L	LAC		17/01/1952
GREEN	W	J		FO		11/05/1944
GREEN	W	J		FL		07/02/1952
GREEN	W			SGT		18/03/1948
GREENE	K			FL		08/03/1951
GREENFIELD	B			SGT		23/02/1950
GREENFIELD	J	G		FL		20/09/1945
GREENHALGH	W	D		WO		20/05/1948

Source	Service No	Force	Notes	Surname
N1210 1950	749873	RAFVR		GRANT
N187 1947	174409	RAFVR		GRANT
N1232 1945	126524	RAFVR	DFM	GRANT
N1638 1942	90243	AAF	MP	GRANT-FERRIS
N1004 1946	83289	RAFVR		GRAVES
N289 1947	135413	RAFVR	DFC	GRAVES
N187 1947	750756	RAFVR		GRAVES
N850 1943	90155	AAF		GRAY
N1089 1945	81370	RAFVR	DFC	GRAY
N468 1946	161007	RAFVR		GRAY
N1185 1951	761103	RAFVR		GRAY
N197 1946	88209	RAFVR	DFC	GRAY
N460 1947	811022	AAF		GRAY
N1350 1943	86679	RAFVR		GRAY
YB 1996	2626709	RAuxAF		GRAY
YB 1997	2622305	RAuxAF		GRAY
YB 1999	2622305	RAuxAF	1st Clasp	GRAY
N318 1956	195018	RAFVR		GRAY
N478 1947	72234	RAFVR	OBE	GRAY
N70 1945	803405	AAF		GRAY
N137 1954	2678817	RAuxAF		GRAY
N1232 1945	85236	RAFVR		GRAY
N594 1946	143994	RAFVR		GRAY
N856 1948	812148	RAuxAF		GRAYLAND
N466 1951	72117	RAFVR	OBE,MRCS,LRCP	GRAYLING
N918 1943	113988	RAFVR		GRAYSON
YB 1996	n/a	RAuxAF		GRAYSON
N846 1947	100105	RAFVR		GRAYSON
N28 1953	849198	RAuxAF		GREATWICH
MOD 1964	3121545	RAFVR		GREAVES
N475 1943	90063	AAF	AFC	GREAVES
N694 1948	86738	RAFVR		GREEHNAM
N1225 1944	62255	RAFVR		GREEN
N866 1951	858111	RAuxAF		GREEN
N196 1959	2603416	RAuxAF	wef 20/5/54	GREEN
N473 1944	90134	AAF	DSO,DFC	GREEN
N187 1947	82179	RAFVR	DSO,DFC	GREEN
N662 1947	84042	RAFVR		GREEN
N94 1950	81944	RAFVR		GREEN
N652 1944	86639	RAFVR		GREEN
N527 1948	747797	RAFVR		GREEN
N805 1948	813118	RAuxAF		GREEN
N653 1943	76315	RAFVR		GREEN
N763 1944	117994	RAFVR		GREEN
N137 1954	810001	RAuxAF		GREEN
N953 1949	744663	RAFVR		GREEN
N874 1945	104438	RAFVR	DSO,DFC	GREEN
N478 1947	123199	RAFVR	DFC	GREEN
N677 1943	90005	AAF	AFC	GREEN
N527 1948	770409	RAFVR		GREEN
YB 1996	n/a	RAuxAF		GREEN
N775 1960	869443	AAF	wef 4/7/44	GREEN
N35 1952	752428	RAFVR		GREEN
N444 1944	135002	RAFVR		GREEN
N103 1952	135002	RAFVR	1st Clasp	GREEN
N232 1948	857339	RAuxAF		GREEN
N244 1951	149479	RAFVR	MBE,DFM	GREENE
N179 1950	807172	RAuxAF		GREENFIELD
N1036 1945	122973	RAFVR		GREENFIELD
N395 1948	748670	RAFVR		GREENHALGH

Surname	Initials			Rank	Status	Date
GREENLAND	B	G		FL	Acting	04/09/1947
GREENLAND	H	J		SQD LDR		03/12/1942
GREENLEY	G	H		CPL		15/01/1953
GREENMAN	L	H		WING COMM	Acting	08/11/1945
GREENOW	L	R		CPL		16/03/1950
GREENSHIELDS	J	W		SQD LDR		26/01/1950
GREENSLADE	P			SAC		29/07/1999
GREENWOOD	D	N		CPL		05/06/1947
GREENWOOD	H	A	J	SGT		24/11/1949
GREER	W			SGT		26/06/1947
GREGG	A	J		CPL		01/07/1948
GREGORCYZK	R	F		CPL		05/04/1997
GREGORY	A	E		FL		24/01/1946
GREGORY	C			SGT	Temporary	16/01/1957
GREGORY	D	S		FL		15/07/1948
GREGORY	D	S		FL		19/10/1960
GREGORY	E			CPL		26/01/1950
GREGORY	F			WING COMM		16/10/1947
GREGORY	G			CPL		17/01/1952
GREGORY	R			CPL	Temporary	01/12/1955
GREGORY	W	J		SQD LDR	Acting	21/03/1946
GREGORY	W			CPL	Temporary	25/07/1962
GREIG	P	C		GRP OFF	Acting	06/03/1947
GRESSWELL	C	H		SQD LDR		26/06/1947
GRESTY	W			SQD LDR		06/03/1947
GRETTON	R			SGT		20/05/1948
GREY	J	E	I	SQD LDR	Acting	13/12/1945
GRIBBEN	J	H		SGT		26/02/1958
GRICE	J			SGT		07/12/1950
GRIFFIN	C	J		FL	Acting	26/06/1947
GRIFFIN	H			FL		01/07/1948
GRIFFIN	J	C		SGT		14/09/1950
GRIFFIN	R	E		FL		06/03/1947
GRIFFIN	R	F	E	SGT		27/04/1950
GRIFFIN	R	T		FL	Acting	27/04/1950
GRIFFITH	C	C		FL		09/05/1946
GRIFFITHS	C	D		WING COMM		12/08/1943
GRIFFITHS	C	D		SQD LDR		01/10/1953
GRIFFITHS	C	R		SQD LDR		30/12/1943
GRIFFITHS	J	A		FLT OFF		17/04/1947
GRIFFITHS	L	W		FLT SGT	Acting	12/06/1947
GRIFFITHS	N			SAC		12/10/1995
GRIFFITHS	T	E		CPL		26/08/1948
GRIFFITHS	W			SQD LDR		17/04/1947
GRIMOLDBY	G	R		CPL		25/05/1950
GRINDROD	S	J		WO	Acting	25/03/1954
GROGAN	N			FL		25/05/1944
GRONER	B	G		SQD LDR	Acting	20/05/1948
GROOM	L	D		FL		18/04/1946
GROOMBRIDGE	A	D	J	FL		07/12/1960
GROSE	F			FL		07/07/1949
GROVE	A	W		FL		14/08/1968
GROVE	T	E	H	GRP CAP		26/06/1947
GROVER	J	B		SACW		15/01/1958
GROVES	R	J		FL		08/07/1959
GROVES	W	E		SGT		10/06/1948
GRUBB	E	G		FO		13/01/1944
GRUBB	H	F		FO		04/05/1944
GRUBB	W	J		FLT SGT		02/08/1951
GRUNDY	C	M	L	FO		30/09/1948

Source	Service No	Force	Notes	Surname
N740 1947	170774	RAFVR		GREENLAND
N1638 1942	70263	RAFVR		GREENLAND
N28 1953	871714	RAuxAF		GREENLEY
N1232 1945	79544	RAFVR		GREENMAN
N257 1950	747155	RAFVR		GREENOW
N94 1950	72838	RAFVR		GREENSHIELDS
YB 2000	n/a	RAuxAF		GREENSLADE
N460 1947	811180	AAF		GREENWOOD
N1193 1949	743606	RAFVR		GREENWOOD
N519 1947	816095	AAF		GREER
N527 1948	842289	RAuxAF		GREGG
YB 1998	n/a	RAuxAF		GREGORCYZK
N82 1946	133005	RAFVR	DFC	GREGORY
N24 1957	805320	RAuxAF		GREGORY
N573 1948	114335	RAFVR		GREGORY
N775 1960	114335	RAFVR	1st Clasp; wef 3/2/60	GREGORY
N94 1950	857507	RAuxAF		GREGORY
N846 1947	90731	AAF		GREGORY
N35 1952	858202	RAuxAF		GREGORY
N895 1955	756624	RAFVR		GREGORY
N263 1946	115577	RAFVR	DSO,DFC,DFM	GREGORY
N549 1962	805262	AAF	wef 9/4/41	GREGORY
N187 1947	80	WAAF		GREIG
N519 1947	72202	RAFVR		GRESSWELL
N187 1947	90614	AAF	MC	GRESTY
N395 1948	750103	RAFVR		GRETTON
N1355 1945	86368	RAFVR		GREY
N149 1958	2688058	RAuxAF	wef 11/1/58	GRIBBEN
N1210 1950	843862	RAuxAF		GRICE
N519 1947	60579	RAFVR		GRIFFIN
N527 1948	134524	RAFVR	AFC	GRIFFIN
N926 1950	801538	RAuxAF		GRIFFIN
N187 1947	130690	RAFVR	DFM	GRIFFIN
N406 1950	743194	RAFVR		GRIFFIN
N406 1950	72665	RAFVR		GRIFFIN
N415 1946	157079	RAFVR		GRIFFITH
N850 1943	90234	AAF	DFC	GRIFFITHS
N796 1953	72839	RAFVR		GRIFFITHS
N1350 1943	90228	AAFRO		GRIFFITHS
N289 1947	2287	WAAF		GRIFFITHS
N478 1947	811086	AAF		GRIFFITHS
YB 1996	n/a	RAuxAF		GRIFFITHS
N694 1948	849084	RAuxAF		GRIFFITHS
N289 1947	72525	RAFVR		GRIFFITHS
N517 1950	864172	RAuxAF		GRIMOLDBY
N249 1954	2678213	RAuxAF		GRINDROD
N496 1944	116151	RAFVR		GROGAN
N395 1948	72841	RAFVR		GRONER
N358 1946	156392	RAFVR	DFC	GROOM
N933 1960	174101	RAFVR	DFM; wef 26/10/60	GROOMBRIDGE
N673 1949	110306	RAFVR		GROSE
MOD 1968	135830	RAFVR		GROVE
N519 1947	90716	AAF	GM,DSC	GROVE
N40 1958	2655669	WRAAF	wef 24/11/57	GROVER
N454 1959	136112	RAFVR	wef 7/2/54	GROVES
N456 1948	752675	RAFVR		GROVES
N11 1944	123639	RAFVR		GRUBB
N425 1944	123640	RAFVR		GRUBB
N777 1951	852283	RAuxAF		GRUBB
N805 1948	67737	RAFVR		GRUNDY

Surname	Initials				Rank	Status	Date
GRUNDY	T				CPL		17/01/1952
GRUTCHFIELD	A	R			SQD LDR		05/06/1947
GUBBINS	J	J			FL		09/08/1945
GUEST	G	H	W		FO		25/05/1944
GUEST	R	G			FL		10/12/1942
GUEST	T	McF			SGT	Acting	18/02/1954
GUILDFORD	J	W			LAC		30/09/1948
GUINNESS	A	E			FLT OFF	Acting	17/04/1947
GUIVER	G	W			FL		19/08/1959
GUNN	G	R			WING COMM	Acting	27/03/1947
GUNN	J	G			SQD LDR	Acting	01/07/1948
GUNTON	T	F	W		FL		03/07/1957
GURNELL-SPARKS	F				LAC		12/06/1947
GUTHRIE	N	H			FL		21/06/1945
GUTTERIDGE	M				FLT OFF		19/08/1954
GUY	M	R			FL		03/07/1998
GWILLIAM	A	A			SGT		16/10/1952
GWYN-JONES	W	T			FL		20/05/1948
H. D'A. De FROBERVILLE	M	J	P	F	WING COMM	Acting	08/05/1947
HACK	C	F			LAC		23/08/1951
HACKER	H	W			FO		16/01/1957
HACKER	H	W			FL		02/08/1965
HACKER	M	E			CPL		15/03/1995
HACKETT	L	A			FLT SGT		24/06/1943
HACKETT	P	D	V		SQD LDR		10/06/1948
HACKETT	R				FO		14/10/1948
HADDOCK	F	J			FL		29/04/1954
HADLAND	G	H			FO		10/12/1942
HADLAND	J	V			SQD LDR	Acting	09/08/1945
HADLER	A	D			SQD LDR		05/01/1950
HADLEY	H				FLT SGT		04/10/1945
HADLEY	J				FL		31/10/1956
HADLEY	T	P			FLT SGT		03/03/1994
HADLINGTON	A	T	J		FLT SGT		21/06/1945
HADWEN	G	D			SQD LDR		06/01/1955
HAGART	T				FLT SGT	Acting	25/06/1953
HAGGARTY	J	M	W		FL		01/12/1955
HAGGER	R	N			PO	Acting	15/06/1944
HAGGER	S	P			SQD LDR		30/03/1944
HAGGO	S	J			SQD LDR		04/01/2005
HAIG	J	G	E		SQD LDR	Acting	27/04/1944
HAIGH	S	H			SGT		19/10/1960
HAIGHTON	W	R			CPL		26/08/1948
HAILEY	A	F			FL		06/03/1947
HAINES	F				CPL		12/06/1947
HAINES	H	G	M		FL		05/06/1947
HAINES	H	J			CPL		23/02/1950
HAINES	S				SGT		01/07/1948
HAINGE	E	D			SQD LDR		26/01/1950
HAINSWORTH	A	P			WO		04/10/1945
HAIR	E	M			FLT OFF		14/09/1950
HAIRS	P	R			FL		29/06/1944
HAISELL	F				FI		13/12/1945
HAKES	W	B			FLT SGT		14/10/1948
HALE	L	V			LAC		14/09/1950
HALE	N	B			LAC		28/03/1962
HALE	P	H	G		FO		25/05/1950
HALES	R				FL		09/12/1959
HALES	S	G	R		FL		07/12/1960
HALIFAX	J				FL		26/10/1944

Source	Service No	Force	Notes	Surname
N35 1952	858212	RAuxAF		GRUNDY
N460 1947	72219	RAFVR		GRUTCHFIELD
N874 1945	81645	RAFVR		GUBBINS
N496 1944	89186	EX-AAF		GUEST
N1680 1942	83737	RAFVR		GUEST
N137 1954	2677409	RAuxAF		GUEST
N805 1948	855052	RAuxAF		GUILDFORD
N289 1947	2673	WAAF		GUINNESS
N552 1959	112676	RAFVR	wef 26/5/55	GUIVER
N250 1947	73025	RAFVR	OBE	GUNN
N527 1948	68207	RAFVR	MBE	GUNN
N456 1957	104400	RAuxAF	wef 15/1/57	GUNTON
N478 1947	811085	AAF		GURNELL-SPARKS
N671 1945	133533	RAFVR	DFM	GUTHRIE
N672 1954	2657501	WRAAF		GUTTERIDGE
YB 1999	212232	RAuxAF		GUY
N798 1952	2611887	RAFVR		GWILLIAM
N395 1948	90423	RAuxAF		GWYN-JONES
N358 1947	72358	RAFVR		H. D'A. De FROBERVILLE
N866 1951	850615	RAuxAF		HACK
N24 1957	2602320	RAFVR		HACKER
MOD 1966	2602320	RAFVR	1st Clasp	HACKER
YB 1997	n/a	RAuxAF		HACKER
N677 1943	815012	AAF		HACKETT
N456 1948	72171	RAFVR		HACKETT
N856 1948	106017	RAFVR		HACKETT
N329 1954	199964	RAFVR		HADDOCK
N1680 1942	85940	RAFVR		HADLAND
N874 1945	106845	RAFVR	DFC	HADLAND
N1 1950	72298	RAFVR		HADLER
N1089 1945	805330	AAF		HADLEY
N761 1956	196562	RAFVR		HADLEY
YB 1996	n/a	RAuxAF		HADLEY
N671 1945	805314	AAF		HADLINGTON
N1 1955	83357	RAuxAF		HADWEN
N527 1953	743200	RAFVR		HAGART
N895 1955	199452	RAuxAF		HAGGARTY
N596 1944	170773	RAFVR		HAGGER
N281 1944	77041	RAFVR		HAGGER
LG 2005	0212611E	RAuxAF		HAGGO
N396 1944	90189	AAF		HAIG
N775 1960	2691054	RAuxAF	wef 1/2/60	HAIGH
N694 1948	840593	RAuxAF		HAIGHTON
N187 1947	172600	RAFVR		HAILEY
N478 1947	850233	AAF		HAINES
N460 1947	176105	RAFVR		HAINES
N179 1950	842567	RAuxAF		HAINES
N527 1948	858066	RAuxAF		HAINES
N94 1950	86969	RAFVR		HAINGE
N1089 1945	809028	AAF		HAINSWORTH
N926 1950	5131	WAAF		HAIR
N652 1944	76316	RAFVR		HAIRS
N1355 1945	145497	RAFVR		HAISELL
N856 1948	845647	RAuxAF		HAKES
N926 1950	857573	RAuxAF		HALE
N242 1962	2614217	RAFVR	wef 7/2/62	HALE
N517 1950	150020	RAFVR		HALE
N843 1959	137245	RAuxAF	wef 4/10/59	HALES
N933 1960	114280	RAuxAF	wef 24/10/60	HALES
N1094 1944	129467	RAFVR		HALIFAX

Surname	Initials				Rank	Status	Date
HALL	B	T	F		SQD LDR		10/07/1994
HALL	C	H	B		FL		18/04/1946
HALL	C	R			FLT SGT		22/05/1952
HALL	D	R			FL		11/07/1946
HALL	E	W	D		LAC		07/07/1949
HALL	E	W			LAC		10/08/1950
HALL	F				CPL		12/06/1947
HALL	G				FL		28/02/1946
HALL	H	G	A		CPL		26/08/1948
HALL	I	G	C		FL		01/08/1946
HALL	J	W	G		PILOT 1		24/11/1949
HALL	J				SGT		10/06/1948
HALL	K	G			FO		21/03/1946
HALL	M				FLT OFF		06/03/1947
HALL	P	F			FL		13/12/1945
HALL	R	A			WING COMM		27/04/1950
HALL	R	McD.			FL		28/02/1946
HALL	R	W			LAC		18/03/1948
HALL	R				SGT		24/01/1946
HALL	R				FL	Acting	26/01/1950
HALL	W	G			CPL		07/08/1947
HALLETT	A	J	N		SGT		18/11/1948
HALLEY	A	A			SQD LDR	Acting	30/08/1945
HALLOWS	B	R	W		WING COMM	Acting	12/04/1945
HALSE	G				FL		04/02/1943
HAM	A	F	B		SQD LDR	Acting	17/01/1952
HAMBLET	L				FO		15/07/1948
HAMBLIN	J	F			CPL		02/04/1953
HAMBLING	F	W			CPL		13/07/1950
HAMBLYN	H	T			CPL		13/12/1951
HAMILTON	A	B			SACW		19/08/1959
HAMILTON	A	J	C		FL		11/07/1946
HAMILTON	A	M	M		SGT		23/08/1951
HAMILTON	C	W	C		SQD LDR	Acting	09/11/1944
HAMILTON	C	W	C		FL		26/04/1956
HAMILTON	C				LAC		30/03/1960
HAMILTON	G				SGT		16/10/1947
HAMILTON	H				SGT		05/06/1947
HAMILTON	J				CPL		07/12/1960
HAMILTON	P	J	A	C	FL		24/11/1949
HAMILTON	T				LAC		16/08/1961
HAMILTON	W	J	V		FL	Acting	09/11/1944
HAMILTON	W	S			LAC		13/07/1950
HAMILTON & BRANDON					GRP CAP		22/04/1943
HAMILTON-GRACE	A	V			FLT OFF		07/08/1947
HAMLEY	W				SQD LDR		17/04/1947
HAMLIN	D				SGT		31/10/1956
HAMMERSLEY	G	H	A		LAC		12/06/1947
HAMMERTON	W	A			WING COMM	Acting	14/10/1943
HAMMOND	A	F			FO		19/01/1956
HAMMOND	F				LAC		16/03/1950
HAMPSON	C				LAC		16/10/1947
HAMPTON	F	P			FL		09/12/1959
HAMPTON	W	H			FO		01/05/1952
HANBURY	F	H			AIR COMMT	Acting	06/03/1947
HANCELL	J				SQD LDR		01/07/1948
HANCOCK	E	L			SQD LDR		27/01/1944
HANCOCK	F	W			SQD LDR	Acting	13/01/1944
HANCOCK	H	H			FL		16/01/1957
HANCOCK	H	H			SQD LDR	Acting	30/09/1966

Source	Service No	Force	Notes	Surname
YB 1996	2616111	RAuxAF		HALL
N358 1946	77207	RAFVR		HALL
N402 1952	801245	RAuxAF		HALL
N594 1946	84714	RAFVR		HALL
N673 1949	866637	RAuxAF		HALL
N780 1950	74331	RAFVR		HALL
N478 1947	811152	AAF		HALL
N197 1946	81638	RAFVR	DFC	HALL
N694 1948	840427	RAuxAF		HALL
N652 1946	129116	RAFVR		HALL
N1193 1949	754778	RAFVR		HALL
N456 1948	808137	RAuxAF		HALL
N263 1946	178873	RAFVR	DFC	HALL
N187 1947	1237	WAAF		HALL
N1355 1945	118042	RAFVR		HALL
N406 1950	90573	RAuxAF		HALL
N197 1946	147307	RAFVR		HALL
N232 1948	845296	RAuxAF		HALL
N82 1946	803452	AAF		HALL
N94 1950	86443	RAFVR		HALL
N662 1947	841268	AAF		HALL
N984 1948	846392	RAuxAF		HALLETT
N961 1945	79160	RAFVR		HALLEY
N381 1945	77787	RAFVR	DFC	HALLOWS
N131 1943	78724	RAFVR		HALSE
N35 1952	90711	RAuxAF		HAM
N573 1948	148273	RAFVR		HAMBLET
N298 1953	861117	RAuxAF		HAMBLIN
N675 1950	743577	RAFVR		HAMBLING
N1271 1951	864874	RAuxAF		HAMBLYN
N552 1959	2663757	WRAAF	wef 10/5/59	HAMILTON
N594 1946	60305	RAFVR		HAMILTON
N866 1951	802354	RAuxAF		HAMILTON
N1150 1944	88647	RAFVR		HAMILTON
N318 1956	88647	RAFVR	DFC; 1st Clasp	HAMILTON
N229 1960	858907	AAF	wef 25/6/44	HAMILTON
N846 1947	702852	RAFVR		HAMILTON
N460 1947	869551	AAF		HAMILTON
N933 1960	2696083	RAuxAF	wef 9/8/59	HAMILTON
N1193 1949	88253	RAFVR		HAMILTON
N624 1961	2650914	RAuxAF	wef 24/5/61	HAMILTON
N1150 1944	146031	RAFVR		HAMILTON
N675 1950	816222	RAuxAF		HAMILTON
N475 1943	90176	AAF	AFC; The Duke of	HAMILTON & BRANDON
N662 1947	513	WAAF		HAMILTON-GRACE
N289 1947	72847	RAFVR		HAMLEY
N761 1956	745885	RAF		HAMLIN
N478 1947	845372	AAF		HAMMERSLEY
N1065 1943	70276	RAFVR		HAMMERTON
N43 1956	106016	RAFVR		HAMMOND
N257 1950	870279	RAuxAF		HAMMOND
N846 1947	800667	AAF		HAMPSON
N843 1959	168140	RAFVR	wef 21/12/52	HAMPTON
N336 1952	105139	RAFVR		HAMPTON
N187 1947	216	WAAF	MBE	HANBURY
N527 1948	110917	RAFVR		HANCELL
N59 1944	70278	RAFVR		HANCOCK
N11 1944	90270	AAFRO		HANCOCK
N24 1957	151965	RAFVR	DFC	HANCOCK
MOD 1966	151965	RAFVR	DFC; 1st Clasp	HANCOCK

Surname	Initials			Rank	Status	Date
HANCOCK	J	E	G	SQD LDR	Acting	24/01/1946
HANCOCK	J	T	C	FO		05/01/1950
HANCOCK	J	W		FLT SGT		12/06/1947
HANCOCK	L	A		WO		28/02/1946
HANCOCK	P	G	M	WING COMM		19/08/1954
HANCOKCS	M	N		WING COMM		28/10/1943
HANDLEY	G			FO		25/03/1954
HANDLEY	G			FO		25/03/1954
HANDLEY	W	R		LAC		26/01/1950
HANDS	J	P		FO		10/12/1942
HANDY	A	H		SQD LDR	Acting	17/04/1947
HANHAM	B	F		SGT	Acting	20/01/1960
HANKIN	J			FL		02/07/1958
HANKINSON	G			CPL		15/09/1949
HANKS	H	G		SQD LDR		12/06/1947
HANNA	D			SGT		06/06/1962
HANNA	J	E		FL		24/01/1946
HANNAH	D	M		WING COMM		30/09/1954
HANNAH	K			SGT		28/11/1956
HANNON	W			SGT		08/11/1945
HANSON	L	G		FLT SGT		18/03/1948
HANSON	S	H		SGT		05/02/1948
HANSON	S	H		PO		16/10/1952
HARBORNE	C	V		CPL		10/02/1949
HARBOTT	C	H		SQD LDR		01/07/1948
HARCOURT	B	A	E	CPL		01/07/1948
HARCOURT	H	J		SQD LDR		02/07/1958
HARDCASTLE	D	A		FLT SGT		16/03/1950
HARDCASTLE	E	T		CPL		08/03/1951
HARDING	G	E		FL		30/09/1948
HARDING	N	D		FO		19/10/1960
HARDING	S	A		CPL		15/11/1951
HARDINGHAM	C	J		CPL		18/03/1948
HARDS	M	S		WING COMM	Acting	31/05/1945
HARDWICK	W	R	H	WO		07/07/1949
HARDY	G	A		FLT SGT		15/09/1949
HARDY	J			SQD LDR	Acting	04/11/1948
HARGREAVES	J	F	B	CPL		14/09/1950
HARKER	A	S		FL		27/07/1944
HARKNESS	H			WO		14/11/1962
HARLAND	R	E	W	SQD LDR	Acting	17/04/1947
HARLEY	G	H		SQD LDR		26/08/1948
HARLEY	H	S		SGT		26/06/1947
HARLEY-LOWE	P	R		FO		26/11/1996
HARMAN	A	H		SGT		01/07/1948
HARMAN	W			WING COMM	Acting	15/07/1948
HARMER	H	J	S	FL		20/01/1960
HARMON-MORGAN	J	V	V	FLT OFF		08/05/1947
HARNDEN	G	H		PO		12/08/1943
HARPER	A	W		SGT		24/11/1949
HARPER	E	M		CPL		08/07/1959
HARPER	F	M	H	WING COMM	Acting	06/03/1947
HARPER	G	T		LAC		12/06/1947
HARPER	H	S		SGT		26/04/1961
HARPER	R	H	L	FL	Acting	05/06/1947
HARRADENCE	C	W	J	FL		09/08/1945
HARRAP	F			SGT	Temporary	16/08/1961
HARRAWAY	M	T		FL		20/09/1945
HARRINGTON	E	J	F	WING COMM		30/11/1944
HARRINGTON	J	R		CPL		22/10/1958

Source	Service No	Force	Notes	Surname
N82 1946	79181	RAFVR	DFC	HANCOCK
N1 1950	101227	RAFVR		HANCOCK
N478 1947	770952	RAFVR		HANCOCK
N197 1946	751029	RAFVR		HANCOCK
N672 1954	90606	RAuxAF		HANCOCK
N1118 1943	70297	RAFVR		HANCOKCS
N249 1954	112134	RAFVR		HANDLEY
N249 1954	112134	RAFVR	1st Clasp	HANDLEY
N94 1950	856630	RAuxAF		HANDLEY
N1680 1942	83491	RAFVR		HANDS
N289 1947	63131	RAFVR		HANDY
N33 1960	2692551	RAuxAF	wef 12/12/59	HANHAM
N452 1958	189494	RAFVR	wef 15/4/53	HANKIN
N953 1949	852757	RAuxAF		HANKINSON
N478 1947	72299	RAFVR		HANKS
N435 1962	2601978	RAFVR	wef 5/4/62	HANNA
N82 1946	115854	RAFVR		HANNA
N799 1954	72318	RAFVR		HANNAH
N833 1956	2657107	WRAAF		HANNAH
N1232 1945	807084	AAF		HANNON
N232 1948	756175	RAFVR		HANSON
N87 1948	809117	RAuxAF		HANSON
N798 1952	2686508	RAuxAF	1st Clasp	HANSON
N123 1949	851367	RAuxAF		HARBORNE
N527 1948	130341	RAFVR	DFM	HARBOTT
N527 1948	843142	RAuxAF		HARCOURT
N452 1958	91168	AAF	wef 29/8/44	HARCOURT
N257 1950	749805	RAFVR		HARDCASTLE
N244 1951	868640	RAuxAF		HARDCASTLE
N805 1948	90882	RAuxAF		HARDING
N775 1960	116522	RAFVR	wef 10/3/44	HARDING
N1185 1951	841100	RAuxAF		HARDING
N232 1948	843429	RAuxAF		HARDINGHAM
N573 1945	122062	RAFVR	DFC,DFM	HARDS
N673 1949	801495	RAuxAF		HARDWICK
N953 1949	745036	RAFVR		HARDY
N934 1948	90778	RAuxAF		HARDY
N926 1950	748801	RAFVR		HARGREAVES
N763 1944	63791	RAFVR	DFM	HARKER
N837 1962	2677588	RAFVR	wef 13/3/62	HARKNESS
N289 1947	74676	RAFVR		HARLAND
N694 1948	80578	RAFVR	OBE	HARLEY
N519 1947	777151	RAFVR		HARLEY
YB 1997	2631202	RAuxAF		HARLEY-LOWE
N527 1948	749301	RAFVR		HARMAN
N573 1948	72321	RAFVR		HARMAN
N33 1960	125642	RAuxAF	DFC; wef 9/12/57	HARMER
N358 1947	313	WAAF		HARMON-MORGAN
N850 1943	131759	RAFVR	BEM	HARNDEN
N1193 1949	770510	RAFVR		HARPER
N454 1959	2659165	WRAAF	wef 9/9/58	HARPER
N187 1947	72131	RAFVR		HARPER
N478 1947	811168	AAF		HARPER
N339 1961	2650871	RAuxAF	wef 11/1/61	HARPER
N460 1947	159765	RAFVR		HARPER
N874 1945	117584	RAFVR		HARRADENCE
N624 1961	2614006	RAFVR	wef 31/3/61	HARRAP
N1036 1945	90236	AAF		HARRAWAY
N1225 1944	78725	RAFVR		HARRINGTON
N716 1958	2692009	RAuxAF	wef 24/8/58	HARRINGTON

Surname	Initials			Rank	Status	Date
HARRIS	C	E		CPL		10/02/1949
HARRIS	C	L		FL		31/05/1945
HARRIS	C	T		LAC		08/05/1947
HARRIS	E	N		SQD LDR		20/09/1945
HARRIS	F	N		FO		10/06/1948
HARRIS	G	P		CPL		05/05/1999
HARRIS	L	J		FL		28/02/1946
HARRIS	M	G		FL		25/01/1951
HARRIS	M	T		FL		28/02/1946
HARRIS	P	A		FLT SGT		02/11/1985
HARRIS	P	A		FLT SGT		16/06/1996
HARRIS	P			PO		16/10/1947
HARRISON	A	J		FL		15/08/1993
HARRISON	C	B		CPL		20/05/1948
HARRISON	D			SGT		14/10/1948
HARRISON	D			FLT OFF	Acting	02/04/1953
HARRISON	F	H		FL		09/08/1945
HARRISON	F	L		FL	Acting	26/08/1948
HARRISON	G	C		SGT		07/12/1960
HARRISON	G	W		FO		03/12/1942
HARRISON	G	W	H	FL	Acting	13/01/1949
HARRISON	H	T		FL	Acting	17/03/1949
HARRISON	J	H		SGT		24/07/1963
HARRISON	L	T		CPL	Temporary	15/11/1961
HARRISON	M			FO		07/07/1949
HARRISON	P	S		CPL		10/10/1995
HARRISON	P			FL		13/12/1945
HARRISON	R	N		FL		23/05/1946
HARRISON	S	E		FL		12/04/1945
HARRISON	S	J		WING COMM	Acting	17/02/1944
HARRISON-SLEEP	D	T		FL		15/09/1949
HARSTON	J			WING COMM	Acting	21/09/1944
HART	E	J		FL		28/11/1946
HART	G	W		SGT		15/09/1949
HART	H	W		CPL		01/11/1951
HART	J	F		WING COMM		26/08/1948
HART	L	R		SGT		07/07/1949
HART	W	D		SQD LDR		17/04/1947
HARTE	J	C		SGT		07/02/1962
HARTE-LOVELACE	F	E		FL		18/04/1946
HARTFORD	Y	H	B	FLT OFF		20/05/1948
HARTLEY	C	H		GRP CAP	Acting	31/05/1945
HARTLEY	D			CPL	Acting	13/12/1961
HARTLEY	G	H		SQD LDR	Acting	23/02/1950
HARTNELL-BEAVIS	F	J		FL		04/10/1945
HARTOP	W	C		FL		20/09/1945
HARTWELL	A	W	L	FL		18/04/1946
HARVEY	A	E		LAC		07/12/1950
HARVEY	A	G	E	SGT		01/07/1948
HARVEY	A	G	E	FL		15/11/1961
HARVEY	A	J		FL		05/06/1947
HARVEY	A	V		GRP CAP		30/03/1944
HARVEY	C	C		SQD LDR		25/03/1959
HARVEY	C	E		WING COMM		23/02/1950
HARVEY	E	J		FL		09/08/1945
HARVEY	I			SGT		27/03/1947
HARVEY	J	F		FL		18/04/1946
HARVEY	K			FO		18/07/1956
HARVEY	L	W		FO		28/06/1945
HARVEY	P	G	A	SQD LDR	Acting	27/03/1947

Source	Service No	Force	Notes	Surname
N123 1949	844208	RAuxAF		HARRIS
N573 1945	68131	RAFVR		HARRIS
N358 1947	818121	AAF		HARRIS
N1036 1945	90557	AAF	AFC	HARRIS
N456 1948	158138	RAFVR		HARRIS
YB 2001	n/a	RAuxAF		HARRIS
N197 1946	169894	RAFVR		HARRIS
N93 1951	102997	RAFVR	DFC,DFM	HARRIS
N197 1946	126094	RAFVR		HARRIS
YB 1996	n/a	RAuxAF		HARRIS
YB 1997	n/a	RAuxAF	1st Clasp	HARRIS
N846 1947	203540	RAFVR		HARRIS
YB 1997	4231793	RAuxAF		HARRISON
N395 1948	870071	RAuxAF		HARRISON
N856 1948	880012	WAAF		HARRISON
N298 1953	2657100	WRAAF	1st Clasp	HARRISON
N874 1945	148805	RAFVR	DFM	HARRISON
N694 1948	175029	RAFVR		HARRISON
N933 1960	2650808	RAuxAF	wef 9/3/60	HARRISON
N1638 1942	102093	RAFVR		HARRISON
N30 1949	89413	RAFVR	MBE	HARRISON
N231 1949	123538	RAFVR		HARRISON
N525 1963	2681189	RAuxAF	wef 16/4/63	HARRISON
N902 1961	818231	AAF	wef 4/7/44	HARRISON
N673 1949	146950	RAFVR		HARRISON
YB 1997	n/a	RAuxAF		HARRISON
N1355 1945	119171	RAFVR		HARRISON
N468 1946	84302	RAFVR	DFC	HARRISON
N381 1945	127143	RAFVR		HARRISON
N132 1944	90039	AAF	DFC	HARRISON
N953 1949	67262	RAFVR		HARRISON-SLEEP
N961 1944	17187	RAFVR		HARSTON
N1004 1946	147002	RAFVR		HART
N953 1949	809118	RAuxAF		HART
N1126 1951	844543	RAuxAF		HART
N694 1948	90493	RAuxAF	DSO	HART
N673 1949	748950	RAFVR		HART
N289 1947	72133	RAFVR		HART
N102 1962	2650933	RAuxAF	wef 29/11/61	HARTE
N358 1946	60291	RAFVR		HARTE-LOVELACE
N395 1948	827	WAAF		HARTFORD
N573 1945	72439	RAFVR	DFC,AFC	HARTLEY
N987 1961	5003008	RAFVR	wef 17/11/60	HARTLEY
N179 1950	122056	RAFVR	DFC	HARTLEY
N1089 1945	76456	RAFVR	DFC	HARTNELL-BEAVIS
N1036 1945	80551	RAFVR	DFC	HARTOP
N358 1946	131657	RAFVR		HARTWELL
N1210 1950	863382	RAuxAF		HARVEY
N527 1948	702283	RAFVR		HARVEY
N902 1961	702283	RAFVR	1st Clasp; wef 22/9/61	HARVEY
N460 1947	73534	RAFVR		HARVEY
N281 1944	90395	AAF	CBE	HARVEY
N196 1959	73598	RAFVR	MRCS,LRCP; wef 11/8/44	HARVEY
N179 1950	90495	RAuxAF		HARVEY
N874 1945	143693	RAFVR		HARVEY
N250 1947	740851	RAFVR		HARVEY
N358 1946	172365	RAFVR		HARVEY
N513 1956	1503638	RAFVR		HARVEY
N706 1945	176381	RAFVR		HARVEY
N250 1947	84315	RAFVR	DFC	HARVEY

Surname	Initials			Rank	Status	Date
HARVEY-BARNES	S	E		FL		26/01/1950
HARWOOD	D	H		SGT		10/02/1949
HARWOOD	W	F	J	FL		10/05/1945
HASLAM	G	M		WING COMM		06/03/1947
HASLER	E	J		SQD LDR		15/07/1948
HASSALL	C			FL		26/08/1948
HASSALL	C			FL		26/11/1958
HASSELL	K	A	H	WING COMM		23/02/1950
HASTIE	F			LAC		24/11/1949
HASTIE	R	M		FL		15/01/1953
HASTINGS	J			CPL		07/07/1949
HASTON	H	J		CPL		01/05/1952
HATCHER	F	W		FL	Acting	26/08/1948
HATCHER	R	S		CPL		25/05/1950
HATFIELD	K	G		FL		26/11/1953
HATHERSICH	G	K		FO		24/01/1946
HATTON	J	W		SQD LDR	Acting	21/03/1946
HATTON	R	F		LAC		25/05/1950
HAVERCROFT	H			SGT		21/06/1961
HAVERCROFT	R	E		FL		17/02/1944
HAW	C			SQD LDR	Acting	13/12/1945
HAWARD	C			FL	Acting	06/03/1947
HAWES	C	G		FL		13/02/1957
HAWES	F	H		SQD LDR		03/06/1943
HAWKE	J	S	T	SQD LDR		23/02/1950
HAWKE	P	S		FL		09/08/1945
HAWKER	C	G		SAC		18/04/1996
HAWKES	D	R		SQD LDR	Acting	21/03/1946
HAWKES	S			FLT SGT		30/09/1948
HAWKINS	A	A		SGT		23/02/1950
HAWKINS	L	A	W	SQD LDR		12/04/1951
HAWKINS	L	J	S	FO		08/05/1947
HAWKINS	L	J		LAC		23/02/1950
HAWKINS	V	M		FLT OFF		27/03/1947
HAWKLEY	G	E		CPL		04/11/1948
HAWKRIDGE	E	W		PO		28/11/1956
HAWKSLEY	J	R	B	FO		26/02/1958
HAWORTH-BOOTH	N	L		SQD LDR	Acting	25/01/1945
HAWTHORNE	W			CPL		20/05/1948
HAXTON	J	A		FLT SGT		01/08/1985
HAXTON	J	A		FLT SGT		01/08/1995
HAY	D			GRP CAP	Acting	04/12/1947
HAY	E	F	A	FO		06/03/1947
HAY	F	E		SACW		14/11/1962
HAY	G	R		SQD LDR	Acting	09/05/1946
HAY	J	G		LAC		20/09/1951
HAY	J	V		SQD LDR		01/03/1945
HAY	W	N		CPL		05/06/1947
HAYCROFT	M	C	C	FL		30/12/1943
HAYDEN	G			SGT		13/01/1949
HAYDEN	R	J		FLT SGT		13/07/1950
HAYES	C	R	S	SQD LDR		30/12/1943
HAYES	G	J		SQD LDR	Acting	14/09/1950
HAYES	J	F		SQD LDR		24/01/1946
HAYES	M	J		CPL	Temporary	19/09/1956
HAYES	R	H		SGT		21/01/1959
HAYES	R	J		CPL		04/01/1965
HAYES	T	N		WING COMM	Acting	06/05/1943
HAYHOE	G	W		SGT		05/02/1948

Source	Service No	Force	Notes	Surname
N94 1950	73640	RAFVR		HARVEY-BARNES
N123 1949	800578	RAuxAF		HARWOOD
N485 1945	61465	RAFVR		HARWOOD
N187 1947	90419	AAF	TD	HASLAM
N573 1948	118797	RAFVR		HASLER
N694 1948	134026	RAFVR	DSO,DFC	HASSALL
N807 1958	134026	RAFVR	DSO,DFC; 1st Clasp; wef 18/3/57	HASSALL
N179 1950	90437	RAuxAF		HASSELL
N1193 1949	867326	RAuxAF		HASTIE
N28 1953	158979	RAFVR	DFC	HASTIE
N673 1949	846028	RAuxAF		HASTINGS
N336 1952	803533	RAuxAF		HASTON
N694 1948	68846	RAFVR		HATCHER
N517 1950	846025	RAuxAF		HATCHER
N952 1953	181505	RAuxAF		HATFIELD
N82 1946	174771	RAFVR		HATHERSICH
N263 1946	142035	RAFVR	DFC	HATTON
N517 1950	842063	RAuxAF		HATTON
N478 1961	2691108	RAFVR	wef 22/4/61	HAVERCROFT
N132 1944	114000	RAFVR		HAVERCROFT
N1355 1945	117992	RAFVR	DFC,DFM	HAW
N187 1947	142304	RAFVR		HAWARD
N104 1957	136140	RAFVR	wef 16/8/55	HAWES
N609 1943	70285	RAFVR		HAWES
N179 1950	90885	RAuxAF	The Honorable	HAWKE
N874 1945	126862	RAFVR		HAWKE
YB 1997	n/a	RAuxAF		HAWKER
N263 1946	116451	RAFVR	DFC	HAWKES
N805 1948	711014	RAFVR		HAWKES
N179 1950	752138	RAFVR		HAWKINS
N342 1951	91201	RAuxAF		HAWKINS
N358 1947	182688	RAFVR		HAWKINS
N179 1950	861206	RAuxAF		HAWKINS
N250 1947	4188	WAAF		HAWKINS
N934 1948	869430	RAuxAF		HAWKLEY
N833 1956	2698014	RAuxAF		HAWKRIDGE
N149 1958	168269	RAuxAF	wef 20/2/55	HAWKSLEY
N70 1945	80552	RAFVR		HAWORTH-BOOTH
N395 1948	816089	RAuxAF		HAWTHORNE
YB 1996	n/a	RAuxAF		HAXTON
YB 1996	n/a	RAuxAF	1st Clasp	HAXTON
N1007 1947	73222	RAFVR		HAY
N187 1947	115249	RAFVR		HAY
N837 1962	2664316	WRAAF	wef 13/10/62	HAY
N415 1946	83258	RAFVR	DFC,AFC	HAY
N963 1951	803630	RAuxAF		HAY
N225 1945	70287	RAFVR		HAY
N460 1947	743566	RAFVR		HAY
N1350 1943	62660	RAFVR		HAYCROFT
N30 1949	702894	RAFVR		HAYDEN
N675 1950	800467	RAuxAF		HAYDEN
N1350 1943	70288	RAFVR		HAYES
N926 1950	78119	RAFVR		HAYES
N82 1946	87019	RAFVR		HAYES
N666 1956	840965	RAuxAF		HAYES
N28 1959	2603840	RAFVR	wef 4/12/58	HAYES
LG 2005	743371	RAuxAF		HAYES
N526 1943	90095	AAF	DFC	HAYES
N87 1948	804279	RAuxAF		HAYHOE

Surname	Initials			Rank	Status	Date
HAYHOE	H			SGT		08/05/1947
HAYHOE	N	J		CPL		14/11/1962
HAYLOCK	C			CPL		27/03/1947
HAYNE	W	W		FL		04/05/1944
HAYNES	F	J		CPL		26/01/1950
HAYNES	H	G		LAC		26/08/1948
HAYTER	G	R	E	SQD LDR		26/01/1950
HAYWARD	D	K		SGT		28/02/1946
HAYWARD	R	H		FL		13/01/1949
HAYWARD	S	C		FL		19/09/1956
HAYWOOD	C	E		CPL		26/06/1947
HAZEL	A	H		FLT SGT		10/02/1949
HAZELDEN	H	G		SQD LDR	Acting	01/08/1946
HEAD	P	J		FL		18/02/1954
HEADLEY	L	E		SQD LDR		11/11/1943
HEAL	P	W	D	WING COMM	Acting	10/08/1944
HEALEY	P	A		FO		06/03/1947
HEALY	J	B	C	FO		24/07/1952
HEALY	L	E	A	WING COMM		03/12/1942
HEAPE	J	S	H	FO		03/07/1957
HEAPE	J	S	H	FO		03/07/1957
HEAPS	J	R		SGT		01/07/1948
HEARD	J			SQD LDR		05/06/1947
HEARN	R	E		FL		13/02/1957
HEARN	R	E		FL		29/12/1965
HEATH	B			WING COMM	Acting	17/02/1944
HEATH	E	G	H	FL		02/08/1951
HEATH	E	G	H	FL		26/04/1961
HEATH	J	H		FL		08/07/1959
HEATH	J	W		FO		08/03/1956
HEATH	L	S		FL		28/11/1956
HEATH	R	W		LAC		14/01/1954
HEATH	W	J		FL		28/02/1946
HEATHER	L	H		FL		20/03/1957
HEATON	L	F		FL		28/03/1962
HEBRON	G	S		SQD LDR	Acting	13/12/1945
HEDGES	W	G	H	WING COMM	Acting	06/03/1947
HEELAS	E	T		FL		21/09/1944
HEENAN	M	E		FO		13/12/1961
HELLIER	R	K		FL		05/01/1950
HELLYER	R	J		FL		16/03/2004
HELLYER	R	O		FL		12/08/1943
HELMORE	D	E		FL	Acting	06/01/1955
HEMINGWAY	E			FL		30/09/1954
HEMMINGS	J	G		FL	Acting	26/06/1947
HEMMINGS	R			CPL		12/08/1998
HEMPKIN	J	K		SGT		04/11/1948
HEMPSTEAD	W	P		SGT		14/10/1948
HENCHMAN	J	S		FL		13/01/1949
HENDERSON	A	T		FL		06/01/1955
HENDERSON	F	E		LAC		13/01/1949
HENDERSON	F	J	A	FL		05/02/1948
HENDERSON	G	E		FL		09/08/1945
HENDERSON	G			WING COMM		11/01/1945
HENDERSON	H	J		WO		08/11/1945
HENDERSON	H	J		FLT SGT		04/10/1951
HENDERSON	H	J		WO		07/04/1955
HENDERSON	J	H		SQD LDR	Acting	01/07/1948
HENDERSON	J	L		FL		15/07/1948
HENDERSON	J			FL		28/02/1946

Source	Service No	Force	Notes	Surname
N358 1947	844199	AAF		HAYHOE
N837 1962	2607472	RAuxAF	wef 25/2/61	HAYHOE
N250 1947	743904	RAFVR		HAYLOCK
N425 1944	78726	RAFVR		HAYNE
N94 1950	752314	RAFVR		HAYNES
N694 1948	818203	RAuxAF		HAYNES
N94 1950	73223	RAFVR		HAYTER
N197 1946	801280	AAF		HAYWARD
N30 1949	91137	RAuxAF		HAYWARD
N666 1956	177514	RAFVR		HAYWARD
N519 1947	841269	AAF		HAYWOOD
N123 1949	484267	RAuxAF		HAZEL
N652 1946	60323	RAFVR	DFC	HAZELDEN
N137 1954	167864	RAFVR		HEAD
N1177 1943	70292	RAFVR		HEADLEY
N812 1944	90220	AAF		HEAL
N187 1947	137869	RAFVR		HEALEY
N569 1952	2684021	RAuxAF		HEALY
N1638 1942	90205	AAF	OBE	HEALY
N456 1957	2604767	RAuxAF	wef 16/6/43	HEAPE
N456 1957	2604767	RAuxAF	1st Clasp; wef 10/10/55	HEAPE
N527 1948	810106	RAuxAF		HEAPS
N460 1947	72425	RAFVR		HEARD
N104 1957	140283	RAFVR	wef 29/12/55	HEARN
MOD 1967	140283	RAFVR	1st Clasp	HEARN
N132 1944	90818	AAF	DFC	HEATH
N777 1951	79571	RAFVR		HEATH
N339 1961	79571	RAFVR	1st Clasp; wef 6/9/55	HEATH
N454 1959	101163	RAFVR	wef 18/10/54	HEATH
N187 1956	2686580	RAuxAF		HEATH
N833 1956	86411	RAF		HEATH
N18 1954	840891	RAuxAF		HEATH
N197 1946	133530	RAFVR		HEATH
N198 1957	85953	RAFVR	wef 25/3/44	HEATHER
N242 1962	104732	RAFVR	wef 30/12/58	HEATON
N1355 1945	78252	RAFVR		HEBRON
N187 1947	90878	AAF	OBE	HEDGES
N961 1944	60320	RAFVR		HEELAS
N987 1961	2663777	WRAAF	wef 1/11/61	HEENAN
N1 1950	156430	RAFVR		HELLIER
LG 2004	4220349D	RAuxAF		HELLYER
N850 1943	90054	AAF		HELLYER
N1 1955	2680027	RAuxAF		HELMORE
N799 1954	123909	RAFVR		HEMINGWAY
N519 1947	86747	RAFVR		HEMMINGS
YB 1999	n/a	RAuxAF		HEMMINGS
N934 1948	802503	RAuxAF		HEMPKIN
N856 1948	753160	RAFVR		HEMPSTEAD
N30 1949	72763	RAFVR		HENCHMAN
N1 1955	201399	RAFVR		HENDERSON
N30 1949	846077	RAuxAF		HENDERSON
N87 1948	161292	RAFVR		HENDERSON
N874 1945	130455	RAFVR		HENDERSON
N14 1945	90045	AAF		HENDERSON
N1232 1945	802291	AAF		HENDERSON
N1019 1951	802291	RAuxAF	1st Clasp	HENDERSON
N277 1955	2683303	RAuxAF	BEM; 2nd Clasp	HENDERSON
N527 1948	81554	RAFVR		HENDERSON
N573 1948	130216	RAFVR		HENDERSON
N197 1946	130798	RAFVR	DFM	HENDERSON

Surname	Initials			Rank	Status	Date
HENDERSON	J			FL	Acting	27/03/1952
HENDERSON	M	M		SQD LDR	Acting	30/03/1944
HENDERSON	T	D		FL		21/06/1945
HENDLEY	K	M		SGT	Temporary	27/03/1963
HENDRICK	R			SAC		01/10/1993
HENDRIE	H	A		FO		11/07/1946
HENDRIE	K	G	P	SQD LDR		26/01/1950
HENDRY	A	C		WING COMM		05/01/1950
HENDRY	P			FO		28/11/1946
HENDRY	W	E	R	FL		27/02/1963
HENNESSEY	M	P		FLT SGT		14/01/1954
HENNYS	W			FLT SGT		13/01/1949
HENRICO	G	F		CPL		05/01/1950
HENRY	A			FLT SGT		18/11/1948
HENRY	T	C		SQD LDR		13/07/1950
HENSHALL	G			CPL		05/06/1947
HENSON	F	G		FL	Acting	09/08/1945
HENTSCH	F	C		PO		30/03/1943
HEPBURN	D	G		FLT SGT		10/05/1945
HEPBURN	P			SAC		01/05/1995
HEPPELL	P	W	E	SQD LDR	Acting	18/04/1946
HEPTONSTALL	P			FL		01/12/1955
HERALD	G	L		CPL		20/03/1957
HERBERT	T			FL		28/10/1943
HERBERT	W	L		SGT		20/05/1948
HERD	J	L		FL		05/09/1946
HERITAGE	L	M		FL		19/08/1954
HERRIDGE	J	E		FO		07/04/1955
HESELTINE	G	C		SQD LDR		09/11/1950
HESTER	V	A		FL		07/06/1945
HESTER	V	C		FL		21/03/1946
HEWETT	B	A		WING COMM		03/06/1943
HEWETT	H			SGT		13/07/1950
HEWITT	A	E		SGT		20/08/1958
HEWITT	D	R		SGT		08/07/1959
HEWITT	E			SGT		26/01/1950
HEWITT	F	C		LAC		28/09/1950
HEWITT	H	F		LAC		16/04/1958
HEWITT	T	E	J	SQD LDR	Acting	17/04/1947
HEWLETT	R	E		FL		28/02/1946
HEWSON	E	P		FL		05/09/1946
HIBBERD	R	J		SQD LDR		14/10/1943
HIBBERD	W	H	A	FL		12/08/1943
HIBBERT	M	J		FO		07/04/1955
HIBBERT	M			SQD OFF	Acting	27/03/1947
HICK	D	T		WO		28/02/1946
HICKES	R	W	E	FL		21/10/1959
HICKEY	D	W		SAC		04/07/1998
HICKEY	W	J		CPL		01/07/1948
HICKIN	D	E		FL		28/02/1946
HICKING	G			SGT		17/04/1947
HICKMAN	C	A		FL		16/01/1957
HICKMOTT	J			SGT		07/12/1960
HICKS	H	H		CPL		24/07/1952
HICKS	H	J		FL		29/09/1965
HICKS	R			FL		13/12/1945
HICKS	W	H		SQD LDR	Acting	26/08/1948
HICKSON	B			SGT	Temporary	25/05/1960
HICKSON	J	C		WO		13/12/1945
HIGGENS	W	T		FO		22/10/1958

Source	Service No	Force	Notes	Surname
N224 1952	102186	RAFVR		HENDERSON
N281 1944	80825	RAFVR		HENDERSON
N671 1945	122975	RAFVR		HENDERSON
N239 1963	750453	RAFVR	wef 20/7/44	HENDLEY
YB 1996	n/a	RAuxAF		HENDRICK
N594 1946	189971	RAFVR		HENDRIE
N94 1950	72852	RAFVR		HENDRIE
N1 1950	90377	RAuxAF	MC,MD,ChB	HENDRY
N1004 1946	195852	RAFVR		HENDRY
N155 1963	82225	RAFVR	wef 24/6/44	HENDRY
N18 1954	2659158	WRAAF		HENNESSEY
N30 1949	859667	RAuxAF		HENNYS
N1 1950	748802	RAFVR		HENRICO
N984 1948	816001	RAuxAF		HENRY
N675 1950	73599	RAFVR	MRCS,LRCP	HENRY
N460 1947	811141	AAF		HENSHALL
N874 1945	119495	RAFVR		HENSON
N1014 1943	148014	RAFVR		HENTSCH
N485 1945	807069	AAF		HEPBURN
YB 1996	n/a	RAuxAF		HEPBURN
N358 1946	86370	RAFVR	DFC	HEPPELL
N895 1955	160699	RAFVR		HEPTONSTALL
N198 1957	874684	RAFVR	wef 28/6/44	HERALD
N1118 1943	70298	RAFVR		HERBERT
N395 1948	819028	RAuxAF		HERBERT
N751 1946	163036	RAFVR		HERD
N672 1954	80920	RAFVR		HERITAGE
N277 1955	88728	RAFVR		HERRIDGE
N1113 1950	72673	RAFVR		HESELTINE
N608 1945	112695	RAFVR	DFC	HESTER
N263 1946	137135	RAFVR		HESTER
N609 1943	90016	AAF		HEWETT
N675 1950	701542	RAFVR		HEWETT
N559 1958	2607416	RAFVR	wef 19/1/44	HEWITT
N454 1959	2655428	WRAAF	wef 19/3/59	HEWITT
N94 1950	857334	RAuxAF		HEWITT
N968 1950	801414	RAuxAF		HEWITT
N258 1958	743869	RAFVR	wef 3/7/44	HEWITT
N289 1947	73228	RAFVR		HEWITT
N197 1946	130728	RAFVR	DFC	HEWLETT
N751 1946	87400	RAFVR		HEWSON
N1065 1943	70300	RAFVR		HIBBERD
N850 1943	70301	RAFVR		HIBBERD
N277 1955	2681524	RAuxAF		HIBBERT
N250 1947	371	WAAF	OBE	HIBBERT
N197 1946	748104	RAFVR		HICK
N708 1959	182594	RAFVR	wef 11/7/52	HICKES
YB 2000	n/a	RAuxAF		HICKEY
N527 1948	844952	RAuxAF		HICKEY
N197 1946	106358	RAFVR		HICKIN
N289 1947	755240	RAFVR		HICKING
N24 1957	152280	RAuxAF		HICKMAN
N933 1960	2656738	WRAAF	wef 28/9/60	HICKMOTT
N569 1952	800560	RAuxAF		HICKS
MOD 1966	182061	RAFVR		HICKS
N1355 1945	82956	RAFVR		HICKS
N694 1948	79284	RAFVR		HICKS
N375 1960	869397	AAF	wef 1/7/44	HICKSON
N1355 1945	742873	RAFVR		HICKSON
N716 1958	122078	RAFVR	wef 10/10/57	HIGGENS

Surname	Initials			Rank	Status	Date
HIGGINS	C	G		SQD LDR		22/04/1943
HIGGINS	H	E		FLT SGT		05/06/1947
HIGGINS	J	F		FL		17/04/1947
HIGGS	W	G		WO		18/03/1948
HIGO	H			CPL		19/10/1960
HILBORNE	R	S		CPL		17/01/1997
HILDITCH	A	E		SQD LDR	Acting	21/03/1946
HILDRED	G	A		SQD LDR	Acting	09/08/1945
HILDRETH	F			FLT SGT		30/12/1943
HILDYARD	R	G	C	FO		26/08/1948
HILEY	H	E		CPL		15/09/1949
HILKEN	C	G		WO		13/01/1949
HILKEN	R	W		FL		15/04/1943
HILKEN	R	W		FL		12/02/1953
HILL	A	E	L	SQD LDR		07/07/1949
HILL	A	E		FL		08/03/1956
HILL	A	E		FL		14/01/1963
HILL	A	E		CPL		20/07/1960
HILL	A	M	M	FL		31/05/1945
HILL	A	M		FL		13/12/1945
HILL	A	McK		CPL	Temporary	23/06/1955
HILL	C	H		SGT		27/03/1947
HILL	C	L		FL		06/03/1947
HILL	C	R		FL		24/01/1946
HILL	C	S		LAC		17/03/1949
HILL	E	J		FL		01/07/1948
HILL	F	W		CPL		20/05/1948
HILL	H	B		SGT	Temporary	08/07/1959
HILL	H	W		FL		27/11/1952
HILL	J	H	W	FL		21/09/1944
HILL	J			SQD LDR		03/12/1942
HILL	L	C		FL		28/11/1956
HILL	R	J		FO		26/04/1956
HILL	R			WING COMM	Acting	07/07/1949
HILL	S	F		FLT SGT		01/07/1948
HILLHOUSE	S			FLT SGT		02/12/1992
HILLIER	F	C		CPL		04/11/1948
HILLMAN	G	E		FL		21/12/1950
HILLMON	L	B		SGT		17/10/1989
HILLS	E	F		CPL		05/06/1947
HILLS	H	E		GRP CAP	Acting	26/08/1948
HILLS	L	P	L	SQD OFF		12/06/1947
HILLWOOD	P			FL		31/05/1945
HILSON	R	K		CPL		17/05/1996
HILTON	H	A		CPL		01/07/1948
HILTON	H	M		FL		05/10/1944
HIND	S	N		WO		24/01/1946
HINDLE	D	L		FO		28/02/1946
HINDLE	F	K		FL		17/04/1947
HINDS	D	J		FL		13/02/1957
HINDS	K	W		FL		09/11/1944
HINDSON	E	B		FO		03/07/1952
HINE	C	J	W	FI		15/09/1949
HINGHAM	A			SQD LDR		30/12/1943
HINKLEY	V	W		FL		01/03/1945
HIPKIN	G			SQD LDR	Acting	24/01/1946
HIPPERSON	S	W		FO		08/11/1945
HIRD	J	C	M	SQD LDR		16/03/2004
HIRST	H			SGT		04/09/1957
HISCOX	R			WING COMM	Acting	03/12/1942

Source	Service No	Force	Notes	Surname
N475 1943	70302	RAFVR		HIGGINS
N460 1947	743957	RAFVR		HIGGINS
N289 1947	72675	RAFVR		HIGGINS
N232 1948	801501	RAuxAF		HIGGS
N775 1960	2691073	RAuxAF	wef 21/7/60	HIGO
YB 1998	n/a	RAuxAF		HILBORNE
N263 1946	87398	RAFVR	DFC	HILDITCH
N874 1945	130222	RAFVR	DFC	HILDRED
N1350 1943	805186	EX-AAF		HILDRETH
N694 1948	187100	RAFVR		HILDYARD
N953 1949	842740	RAuxAF		HILEY
N30 1949	745482	RAFVR		HILKEN
N454 1943	87028	RAFVR		HILKEN
N115 1953	87028	RAFVR	1st Clasp	HILKEN
N673 1949	90723	RAuxAF		HILL
N187 1956	183822	RAFVR		HILL
MOD 1966	183822	RAFVR	1st Clasp	HILL
N531 1960	2697522	RAFVR	wef 17/10/59	HILL
N573 1945	128686	RAFVR	DFC,DFM	HILL
N1355 1945	121333	RAFVR	DFC	HILL
N473 1955	874043	RAuxAF		HILL
N250 1947	842426	AAF		HILL
N187 1947	107813	RAFVR	MM	HILL
N82 1946	112518	RAFVR	DFC	HILL
N231 1949	846105	RAuxAF		HILL
N527 1948	84051	RAFVR		HILL
N395 1948	746196	RAFVR		HILL
N454 1959	744435	RAFVR	wef 16/7/44	HILL
N915 1952	161806	RAFVR		HILL
N961 1944	78727	RAFVR		HILL
N1638 1942	70307	RAFVR		HILL
N833 1956	174424	RAFVR	DFC	HILL
N318 1956	131273	RAFVR		HILL
N673 1949	72339	RAFVR	GM,MB,ChB	HILL
N527 1948	801430	RAuxAF		HILL
YB 1996	n/a	RAuxAF		HILLHOUSE
N934 1948	840151	RAuxAF		HILLIER
N1270 1950	73577	RAFVR		HILLMAN
YB 1998	n/a	RAuxAF		HILLMON
N460 1947	747257	RAFVR		HILLS
N694 1948	90463	RAuxAF	CBE	HILLS
N478 1947	693	WAAF		HILLS
N573 1945	120107	RAFVR	DFC	HILLWOOD
YB 1997	n/a	RAuxAF		HILSON
N527 1948	743368	RAFVR		HILTON
N1007 1944	108234	RAFVR		HILTON
N82 1946	745404	RAFVR		HIND
N197 1946	169654	RAFVR		HINDLE
N289 1947	119130	RAFVR	DFC	HINDLE
N104 1957	140083	RAFVR	wef 12/9/44	HINDS
N1150 1944	81921	RAFVR		HINDS
N513 1952	2685522	RAuxAF		HINDSON
N953 1949	120066	RAFVR		HINE
N1350 1943	70303	EX-RAFVR		HINGHAM
N225 1945	77463	RAFVR		HINKLEY
N82 1946	76010	RAFVR	DFC	HIPKIN
N1232 1945	182599	RAFVR		HIPPERSON
LG 2004	8064168L	RAuxAF		HIRD
N611 1957	2687010	RAFVR	wef 27/6/57	HIRST
N1638 1942	90083	AAF		HISCOX

Surname	Initials			Rank	Status	Date
HITCHCOCK	A	W		FL		19/09/1956
HITCHCOCK	G	D		FL		18/04/1946
HITCHCOCK	J	S		FL		30/11/1944
HITCHING	J	T		FL		23/05/1946
HITCHINGS	G	H		SQD LDR	Acting	26/08/1948
HITCHINGS	J	P		GRP CAP		13/01/1949
HITCHIN-KEMP	R	J	B	FLT SGT		01/07/1948
HOAD	R	J		FL		20/03/1957
HOAEN	H	S		FL		30/10/1957
HOAR	K	S		FLT SGT		18/02/1954
HOARE	H	J		FL		23/06/1955
HOBBS	D	R		SGT		21/03/1946
HOBBS	J	P		FL		28/07/1955
HOBBS	L	N		FL		19/08/1954
HOBBS	P	S		FL		16/10/1947
HOBBS	P	S		FL		30/10/1957
HOBDAY	K	C		FO		20/09/1945
HOBSON	J	V		CPL	Temporary	21/01/1959
HOCKEY	J			CPL		07/07/1949
HOCKEY	L	P	R	FL		24/01/1946
HOCKEY	R	C		WING COMM	Acting	17/12/1942
HOCKEY	R	C		SQD LDR		21/06/1951
HOCKEY	R	C		SQD LDR		20/08/1958
HODGE	A	E		CPL	Temporary	16/01/1957
HODGE	A	W		FLT SGT		27/03/1947
HODGE	D	F		FL		03/10/1946
HODGE	G			FO		14/10/1948
HODGE	J	H		WING COMM	Acting	03/12/1942
HODGES	A	E		FO		28/09/1950
HODGES	L	F		SQD LDR		06/03/1947
HODGES	T	C	P	FL		18/04/1946
HODGES	W	J		LAC		18/11/1948
HODGKINSON	J	A		SQD LDR	Acting	24/01/1946
HODGKINSON	J	W	G	FO		18/04/1946
HODGSON	E	E		CPL		18/03/1948
HODGSON	E			CPL		23/02/1950
HODGSON	E			LAC		10/08/1950
HODGSON	G	W		FL		20/05/1948
HODGSON	J	A	S	SQD LDR		29/06/1944
HODGSON	N	S		SQD LDR		05/06/1947
HODGSON	P			SQD LDR		08/03/1986
HODGSON	P			SQD LDR		08/03/1996
HODGSON	R	A		FO		21/03/1946
HODGSON	R	A		WING COMM		15/07/1948
HODGSON	T	P		FL		13/02/1957
HODKINSON	S	R		FL	Acting	05/06/1947
HODSON	C	V		WO		21/09/1944
HODSON	K	G		FL		18/03/1948
HODSON	R	N		FO		26/06/1947
HODSON	S	C		SGT		17/04/1947
HOFF	G	E		FO		11/01/1945
HOGG	B			CPL		28/08/1963
HOGG	H	G	F	FLT SGT		12/06/1947
HOGG	I	E		PO		02/04/1953
HOGG	J	M	S	SQD LDR		07/08/1947
HOGG	J			PO		11/05/1944
HOGG	W	S	M	CPL		07/07/1949
HOLDCROFT	G	J		SQD LDR		20/05/1948

Source	Service No	Force	Notes	Surname
N666 1956	126653	RAFVR		HITCHCOCK
N358 1946	122051	RAFVR		HITCHCOCK
N1225 1944	106813	RAFVR		HITCHCOCK
N468 1946	139031	RAFVR		HITCHING
N694 1948	90678	RAuxAF		HITCHINGS
N30 1949	90576	RAuxAF	CBE	HITCHINGS
N527 1948	804031	RAuxAF		HITCHIN-KEMP
N198 1957	2680080	RAuxAF	wef 6/10/56	HOAD
N775 1957	154406	RAuxAF	wef 10/12/53	HOAEN
N137 1954	749471	RAFVR		HOAR
N473 1955	113185	RAFVR		HOARE
N263 1946	740602	RAFVR		HOBBS
N565 1955	197013	RAFVR		HOBBS
N672 1954	136432	RAFVR		HOBBS
N846 1947	139887	RAFVR	DFC	HOBBS
N775 1957	139887	RAFVR	DFC; 1st Clasp; wef 2/1/57	HOBBS
N1036 1945	170475	RAFVR		HOBDAY
N28 1959	799895	RAFVR	wef 28/5/44	HOBSON
N673 1949	865506	RAuxAF		HOCKEY
N82 1946	76011	RAFVR		HOCKEY
N1708 1942	72993	RAFVR	DFC	HOCKEY
N621 1951	72993	RAFVR	DSO,DFC; 1st Clasp	HOCKEY
N559 1958	72993	RAFVR	DSO,DFC; 2nd Clasp; wef 31/12/55	HOCKEY
N24 1957	857495	RAuxAF		HODGE
N250 1947	865106	AAF		HODGE
N837 1946	179425	RAFVR	DFC	HODGE
N856 1948	121809	RAFVR		HODGE
N1638 1942	90179	AAF		HODGE
N968 1950	110517	RAFVR		HODGES
N187 1947	73461	RAFVR		HODGES
N358 1946	158290	RAFVR		HODGES
N984 1948	855024	RAuxAF		HODGES
N82 1946	112735	RAFVR	DFM	HODGKINSON
N358 1946	177912	RAFVR	DFC	HODGKINSON
N232 1948	868559	RAuxAF		HODGSON
N179 1950	872289	RAuxAF		HODGSON
N780 1950	751455	RAFVR		HODGSON
N395 1948	78143	RAFVR	MBE	HODGSON
N652 1944	70311	RAFVR		HODGSON
N460 1947	72203	RAFVR		HODGSON
YB 1997	209577	RAuxAF		HODGSON
YB 1997	209577	RAuxAF	1st Clasp	HODGSON
N263 1946	175647	RAFVR	DFM	HODGSON
N573 1948	90468	RAuxAF		HODGSON
N104 1957	58138	RAuxAF	wef 5/1/57	HODGSON
N460 1947	138970	RAFVR		HODKINSON
N961 1944	740843	RAFVR		HODSON
N232 1948	145106	RAFVR		HODSON
N519 1947	144367	RAFVR		HODSON
N289 1947	849172	AAF		HODSON
N14 1945	142976	RAFVR		HOFF
N621 1963	2664737	WRAFVR	wef 25/6/63	HOGG
N478 1947	812106	AAF		HOGG
N298 1953	2683013	RAuxAF		HOGG
N662 1947	91125	AAF		HOGG
N444 1944	170559	RAFVR		HOGG
N673 1949	844133	RAuxAF		HOGG
N395 1948	72677	RAFVR		HOLDCROFT

Surname	Initials			Rank	Status	Date
HOLDEN	G	W		SQD LDR		12/08/1943
HOLDEN	K			WING COMM	Acting	28/02/1946
HOLDER	J			FLT SGT		07/12/1950
HOLDING	A	R		SAC		01/05/1994
HOLDSTOCK	F	C		CPL	Temporary	28/07/1955
HOLE	K	W		FL		12/08/1943
HOLGATE	R	E		FL		28/02/1946
HOLLAND	J	E		SQD LDR		15/09/1949
HOLLAND	K	N		FL		09/05/1946
HOLLAND	M	L		SQD OFF		04/09/1947
HOLLAND	N	M	C	SQD OFF		12/10/1950
HOLLAND	R	J		FL		04/02/1943
HOLLAND	W	E		FL		29/06/1944
HOLLAND	W	H	H	FL		11/01/1945
HOLLAND	W	H		FL		24/01/1946
HOLLEY	A	T		CPL		03/06/1959
HOLLIDAY	H	L		FL		16/03/1950
HOLLIDAY	R	R		SGT		15/09/1949
HOLLIDGE	R	L		WO		28/02/1946
HOLLIDGE	R	L		FLT SGT		13/12/1951
HOLLIER	F	G		SQD LDR	Acting	27/03/1947
HOLLIMAN	S	T		FL		27/04/1950
HOLLIMAN	S	T		FL		06/01/1955
HOLLINGDALE	R	A	W	FO		18/07/1956
HOLLINGHURST	A			SQD LDR	Acting	24/11/1949
HOLLISTER	E	S		CPL		30/09/1948
HOLLOW	J	A		SAC		04/07/1994
HOLLOWAY	A	B		WING COMM		08/05/1947
HOLLOWAY	E			FL		12/04/1945
HOLLOWAY	J	F	A	FL		05/01/1950
HOLLOWAY	J	H		FL		24/11/1949
HOLLOWAY	R	T		PILOT 2		21/12/1950
HOLLOWAY	R	W		WING COMM		23/02/1950
HOLLOWAY	S	V		FL		11/01/1945
HOLLOWAY	W	S		LAC		27/03/1947
HOLLOWELL	A	W	G	CPL	Acting	07/08/1947
HOLLOWELL	K	B		FL		08/11/1945
HOLMAN	C	R		SGT	Acting	20/11/1957
HOLMES	A	B		SGT		23/02/1950
HOLMES	C	T		FL		27/01/1944
HOLMES	E	C		FL		10/12/1942
HOLMES	H	J	D	FO		06/01/1955
HOLMES	J	D	V	FL		20/10/1966
HOLMES	J	W		WO		01/07/1948
HOLMES	J			FO		06/03/1947
HOLMES	J			SGT		17/03/1949
HOLMES	K	J		SGT		01/07/1948
HOLMES	O	S		SQD LDR	Acting	26/08/1948
HOLMES	R	T		FL		28/10/1943
HOLNESS	E	J		LAC		14/11/1962
HOLT	A			CPL		10/08/1950
HOLT	F			FL		26/02/1958
HOLTBY	E			CPL	Acting	28/11/1956
HOLTOM	G	G		FL		08/11/1945
HOLTON	R	A		CPL		30/09/1948
HOLWAY	D	F		PO		01/03/1945
HOLWELL	G			FL		19/10/1960
HOLYWELL	J			FL		11/07/1946
HONE	D	H		FL		20/09/1945
HONEYCHURCH	J	L		FL		19/08/1959

Source	Service No	Force	Notes	Surname
N850 1943	103484	RAFVR	DSO,DFC	HOLDEN
N197 1946	90705	AAF	DFC	HOLDEN
N1210 1950	799887	RAFVR		HOLDER
YB 1996	n/a	RAuxAF		HOLDING
N565 1955	746879	RAFVR		HOLDSTOCK
N850 1943	70315	RAFVR		HOLE
N197 1946	130339	RAFVR	DFC	HOLGATE
N953 1949	81936	RAFVR		HOLLAND
N415 1946	84723	RAFVR		HOLLAND
N740 1947	702	WAAF		HOLLAND
N1018 1950	703	WAAF	OBE	HOLLAND
N131 1943	90151	AAF		HOLLAND
N652 1944	79152	RAFVR		HOLLAND
N14 1945	81648	RAFVR	DFC	HOLLAND
N82 1946	86664	RAFVR		HOLLAND
N368 1959	2692523	RAuxAF	wef 14/2/59	HOLLEY
N257 1950	66492	RAFVR	DFC	HOLLIDAY
N953 1949	851448	RAuxAF		HOLLIDAY
N197 1946	740910	RAFVR		HOLLIDGE
N1271 1951	740910	RAFVR	1st Clasp	HOLLIDGE
N250 1947	77905	RAFVR		HOLLIER
N406 1950	116539	RAFVR		HOLLIMAN
N1 1955	116539	RAFVR	1st Clasp	HOLLIMAN
N513 1956	171242	RAFVR		HOLLINGDALE
N1193 1949	91079	RAuxAF		HOLLINGHURST
N805 1948	861176	RAuxAF		HOLLISTER
YB 1996	n/a	RAuxAF		HOLLOW
N358 1947	72172	RAFVR	OBE	HOLLOWAY
N381 1945	115065	RAFVR		HOLLOWAY
N1 1950	78991	RAFVR		HOLLOWAY
N1193 1949	137492	RAFVR	MBE	HOLLOWAY
N1270 1950	701642	RAFVR		HOLLOWAY
N179 1950	90805	RAuxAF		HOLLOWAY
N14 1945	121329	RAFVR		HOLLOWAY
N250 1947	820059	AAF		HOLLOWAY
N662 1947	743503	RAFVR		HOLLOWELL
N1232 1945	113338	RAFVR	AFC	HOLLOWELL
N835 1957	2693618	RAuxAF	wef 28/7/57	HOLMAN
N179 1950	815176	RAuxAF		HOLMES
N59 1944	70319	RAFVR		HOLMES
N1680 1942	83304	RAFVR		HOLMES
N1 1955	2601282	RAFVR		HOLMES
MOD 1966	73816	RAFVR		HOLMES
N527 1948	809185	RAuxAF		HOLMES
N187 1947	180584	RAFVR		HOLMES
N231 1949	880199	WAAF		HOLMES
N527 1948	752143	RAFVR		HOLMES
N694 1948	90763	RAuxAF		HOLMES
N1118 1943	68730	RAFVR		HOLMES
N837 1962	840940	AAF	wef 7/3/44	HOLNESS
N780 1950	747434	RAFVR		HOLT
N149 1958	150130	RAuxAF	wef 11/10/56	HOLT
N833 1956	2691000	RAuxAF		HOLTBY
N1232 1945	145327	RAFVR	AFC	HOLTOM
N805 1948	863256	RAuxAF		HOLTON
N225 1945	182838	RAFVR		HOLWAY
N775 1960	154273	RAuxAF	wef 2/11/59	HOLWELL
N594 1946	156572	RAFVR		HOLYWELL
N1036 1945	80816	RAFVR		HONE
N552 1959	205191	RAFVR	wef 19/65/54	HONEYCHURCH

Surname	Initials			Rank	Status	Date
HOOD	A			SQD LDR	Acting	26/08/1948
HOOD	F	B		FL		26/08/1948
HOOD	T	W		PO		27/03/1947
HOOK	A	O		CPL		22/02/1961
HOOK	G	T		SGT		06/05/1943
HOOK	G	T		CPL		24/11/1949
HOOKWAY	D	N		SQD LDR	Acting	08/11/1945
HOOKWAY	S	G		SQD LDR	Acting	07/01/1943
HOOLEY	K	S		CPL		15/11/1951
HOOLEY	R	H		SIG 1		04/09/1947
HOOPER	R	F		CPL		30/09/1948
HOOPER	S			FL		26/01/1950
HOPE	A	P		WING COMM		03/12/1942
HOPE	J	J		CPL		05/01/1950
HOPE	J			SGT		27/03/1947
HOPEWELL	J	E	S	FO		04/02/1943
HOPKINS	F	E		CPL		07/08/1947
HOPKINS	G			CPL		30/12/1943
HOPWOOD	K			CPL		12/06/1947
HORLINGTON	I	K		FO		06/01/1955
HORN	J	A		FL		12/10/1988
HORN	J	A		FL		12/10/1998
HORNBY	K			LAC		12/06/1947
HORNBY	W	H		FL		30/12/1943
HORNE	I	D		FL		14/01/1954
HORNE	R	A	J	WING COMM	Acting	05/06/1947
HORNER	F	G		FL		23/05/1946
HORNER	R	H	J	FL		28/02/1946
HORNSBY	H	C		CPL		01/07/1948
HORNSBY	R	J		FLT SGT	Acting	25/01/1945
HORRELL	A	J		SQD LDR		20/05/1948
HORRELL	A	J		SQD LDR		15/01/1953
HORRELL	S	G		LAC		27/04/1950
HORROCKS	D	A		LAC		13/11/1947
HORSEY	F	B	S	WING COMM		01/07/1948
HORSLEY	A	I		SQD LDR	Acting	13/07/1950
HORSLEY	J	M		WING COMM	Acting	06/03/1947
HORSMAN	C	D		SQD LDR		19/08/1954
HORTON	A	E		LAC		14/10/1948
HORTON	R	J		FL		12/04/1951
HORTON	R	J		FL		20/07/1960
HOSBURN	J	C		SQD LDR		12/06/1947
HOSKING	T	F		WING COMM		26/01/1950
HOSKINS	J	H		WING COMM	Acting	31/05/1945
HOTSTON	P	S		FO		18/11/1960
HOUCHIN	J	F		SQD LDR		06/03/1947
HOUGHTON	C	L		FO		20/09/1945
HOUGHTON	C	L		FO		03/07/1952
HOUGHTON	J	T		FL		24/01/1946
HOUGHTON	R			SGT		05/06/1947
HOUGHTON	S	J		FL		24/01/1946
HOUGHTON	W			FO		03/06/1959
HOUSEMAN	A	E		WING COMM		19/09/1956
HOUSTON	W	R		SQD LDR	Acting	20/04/1944
HOW	N	A	W	FL		21/03/1946
HOWARD	A	E		WO	Acting	26/01/1950
HOWARD	E	S	E	SGT		01/07/1948
HOWARD	F	A		SAC		09/12/1959
HOWARD	G	C		SQD LDR	Acting	03/10/1946

Source	Service No	Force	Notes	Surname
N694 1948	131688	RAFVR		HOOD
N694 1948	90989	RAuxAF		HOOD
N250 1947	203720	RAFVR		HOOD
N153 1961	2684641	RAuxAF	wef 22/10/60	HOOK
N526 1943	804021	AAF		HOOK
N1193 1949	744966	RAFVR		HOOK
N1232 1945	82691	RAFVR		HOOKWAY
N3 1943	84339	RAFVR	DFC	HOOKWAY
N1185 1951	815187	RAuxAF		HOOLEY
N740 1947	751264	RAFVR		HOOLEY
N805 1948	864762	RAuxAF		HOOPER
N94 1950	73525	RAFVR	Deceased	HOOPER
N1638 1942	90127	AAF	DFC,Bart; Sir	HOPE
N1 1950	844739	RAuxAF		HOPE
N250 1947	857389	AAF		HOPE
N131 1943	115844	RAFVR		HOPEWELL
N662 1947	811102	AAF		HOPKINS
N1350 1943	807026	AAF		HOPKINS
N478 1947	743901	RAFVR		HOPWOOD
N1 1955	2685534	RAuxAF		HORLINGTON
YB 1997	2626146	RAuxAF		HORN
YB 1999	2626146	RAuxAF	1st Clasp	HORN
N478 1947	811218	AAF		HORNBY
N1350 1943	60513	RAFVR		HORNBY
N18 1954	802551	RAuxAF		HORNE
N460 1947	72267	RAFVR		HORNE
N468 1946	100034	RAFVR		HORNER
N197 1946	156421	RAFVR		HORNER
N527 1948	743370	RAFVR		HORNSBY
N70 1945	802368	AAF		HORNSBY
N395 1948	73616	RAFVR		HORRELL
N28 1953	73616	RAuxAF	1st Clasp	HORRELL
N406 1950	841879	RAuxAF		HORRELL
N933 1947	811190	AAF		HORROCKS
N527 1948	72134	RAFVR		HORSEY
N675 1950	76196	RAFVR		HORSLEY
N187 1947	91006	AAF		HORSLEY
N672 1954	132117	RAFVR		HORSMAN
N856 1948	818163	RAuxAF		HORTON
N342 1951	142018	RAFVR	DFC,DFM	HORTON
N531 1960	142018	RAFVR	DFC,DFM; 1st Clasp; wef 15/12/59	HORTON
N478 1947	72859	RAFVR		HOSBURN
N94 1950	73641	RAFVR	Deceased	HOSKING
N573 1945	100030	RAFVR	DSO,DFC	HOSKINS
MOD 1967	2611519	RAFVR		HOTSTON
N187 1947	72679	RAFVR	MBE	HOUCHIN
N1036 1945	162636	RAFVR		HOUGHTON
N513 1952	162636	RAFVR	1st Clasp	HOUGHTON
N82 1946	81685	RAFVR	DFC	HOUGHTON
N460 1947	811153	AAF		HOUGHTON
N82 1946	127136	RAFVR		HOUGHTON
N368 1959	122784	RAFVR	wef 5/8/44	HOUGHTON
N666 1956	73729	RAFVR	DFC	HOUSEMAN
N363 1944	75991	RAFVR		HOUSTON
N263 1946	78445	RAFVR		HOW
N94 1950	760075	RAFVR		HOWARD
N527 1948	847174	RAuxAF		HOWARD
N843 1959	2689562	RAFVR	wef 12/8/58	HOWARD
N837 1946	84944	RAFVR		HOWARD

Surname	Initials			Rank	Status	Date
HOWARD	G	L		SGT		13/12/1951
HOWARD	H	G		SQD LDR		01/07/1948
HOWARD	H	G		FL		21/01/1959
HOWARD	J	R		SQD LDR	Acting	28/02/1946
HOWARD	J	W		SQD LDR		30/09/1948
HOWARD	J	W		SGT		16/03/1950
HOWARD-RICE	G	B		SQD LDR	Acting	20/09/1945
HOWARTH	F	W		SGT		26/06/1947
HOWARTH	J	W		LAC		15/01/1953
HOWE	D	C		FL		24/01/1946
HOWE	H			LAC		15/09/1949
HOWE	R	B		WO		07/08/1947
HOWELL	D	R	S	FO		06/03/1947
HOWELL	J	E		WING COMM		03/12/1942
HOWELL	J			LAC		10/02/1949
HOWELL	S	W	R	SQD LDR	Acting	04/05/1944
HOWELL	T	H		FLT SGT		13/12/1945
HOWES	B	L	H	FL		03/06/1943
HOWES	D	A		FL		02/04/1953
HOWES	E	A		SGT		01/07/1948
HOWES	E	B		FL		25/03/1954
HOWES	L			WO	Acting	05/06/1947
HOWES	S	A	E	CPL		30/09/1954
HOWES	S	A	L	FL		31/10/1956
HOWITT	G	L		SQD LDR	Acting	11/11/1943
HOWITT	J	P		WO		21/03/1946
HOWLETT	E	H	T	FO		07/04/1955
HOWLETT	L	G		SGT		07/07/1949
HOWORTH	D	M		FL		05/01/1950
HOWSE	E	J		FL		18/11/1948
HOXLEY	T	A		LAC		01/07/1948
HUBBARD	G	C		FO		07/06/1945
HUBBARD	H	G		SQD LDR	Acting	13/12/1945
HUBBARD	L			FLT SGT		09/05/1946
HUBBARD	T	F		CPL		20/07/1960
HUCKLEBRIDGE	E	C		FL		06/03/1947
HUDSON	A			CPL		23/02/1950
HUDSON	C	O		SQD LDR		26/06/1947
HUDSON	C	P	M	SQD LDR		20/07/2004
HUDSON	J	G		WING COMM		12/06/1947
HUDSON	T	E		FLT SGT		15/09/1949
HUDSON	W	J		FL		16/03/1950
HUFFER	L	J		CPL		26/08/1948
HUGGETT	W	D	F	FL		08/11/1945
HUGGINS	J	W		FL		08/03/1956
HUGGINS	P	S		SQD LDR	Acting	09/08/1945
HUGGINS	W	W		FO		24/01/1946
HUGHES	E	E		FLT OFF		17/07/1944
HUGHES	E	G		WING COMM	Acting	23/05/1946
HUGHES	E	Y		WING COMM		06/03/1947
HUGHES	F	H		SGT		09/11/1950
HUGHES	G			LAC		26/06/1947
HUGHES	G			SGT		26/08/1948
HUGHES	H	K		FL		28/02/1946
HUGHES	J	E		SGT		07/08/1947
HUGHES	J	O		FL		03/07/1957
HUGHES	J	V		SGT	Acting	19/10/1960
HUGHES	K	A		SGT		01/07/1948
HUGHES	L	L		FLT SGT		15/07/1948
HUGHES	M	J		FO		20/11/1957

Source	Service No	Force	Notes	Surname
N1271 1951	771111	RAFVR		HOWARD
N527 1948	72456	RAFVR		HOWARD
N28 1959	72456	RAFVR	1st Clasp; wef 1/6/57	HOWARD
N197 1946	64273	RAFVR		HOWARD
N805 1948	91186	RAuxAF		HOWARD
N257 1950	801426	RAuxAF		HOWARD
N1036 1945	72495	RAFVR		HOWARD-RICE
N519 1947	870033	AAF		HOWARTH
N28 1953	858149	RAuxAF		HOWARTH
N82 1946	78253	RAFVR		HOWE
N953 1949	858799	RAuxAF		HOWE
N662 1947	749402	RAFVR		HOWE
N187 1947	198370	RAFVR		HOWELL
N1638 1942	90312	AAF	MB,ChB	HOWELL
N123 1949	850632	RAuxAF		HOWELL
N425 1944	72994	RAFVR	DFC	HOWELL
N1355 1945	804173	AAF		HOWELL
N609 1943	78981	RAFVR		HOWES
N298 1953	103795	RAuxAF		HOWES
N527 1948	756096	RAFVR		HOWES
N249 1954	193150	RAFVR		HOWES
N460 1947	804451	AAF		HOWES
N799 1954	2603860	RAFVR		HOWES
N761 1956	198722	RAFVR		HOWES
N1177 1943	81037	RAFVR	DFC	HOWITT
N263 1946	745757	RAFVR		HOWITT
N277 1955	2685596	RAuxAF		HOWLETT
N673 1949	859543	RAuxAF		HOWLETT
N1 1950	82678	RAFVR		HOWORTH
N984 1948	135010	RAFVR		HOWSE
N527 1948	840349	RAuxAF		HOXLEY
N608 1945	170891	RAFVR		HUBBARD
N1355 1945	81917	RAFVR		HUBBARD
N415 1946	805147	AAF		HUBBARD
N531 1960	2692083	RAuxAF	wef 11/5/60	HUBBARD
N187 1947	72861	RAFVR		HUCKLEBRIDGE
N179 1950	805455	RAuxAF		HUDSON
N519 1947	91208	AAF		HUDSON
LG 2004	0300868R	RAuxAF		HUDSON
N478 1947	90417	AAF		HUDSON
N953 1949	752049	RAFVR		HUDSON
N257 1950	148916	RAFVR		HUDSON
N694 1948	744383	RAFVR		HUFFER
N1232 1945	117988	RAFVR	DFC	HUGGETT
N187 1956	90884	RAuxAF		HUGGINS
N874 1945	66490	RAFVR		HUGGINS
N82 1946	186724	RAFVR		HUGGINS
MOD 1965	2731	WAAF		HUGHES
N468 1946	73047	RAFVR	DSO,DFC	HUGHES
N187 1947	74275	RAFVR		HUGHES
N1113 1950	854493	RAuxAF		HUGHES
N519 1947	853888	AAF		HUGHES
N694 1948	844008	RAuxAF		HUGHES
N197 1946	121439	RAFVR	DFC	HUGHES
N662 1947	811112	AAF		HUGHES
N456 1957	145244	RAuxAF	wef 27/4/57	HUGHES
N775 1960	2691570	RAuxAF	wef 17/8/60	HUGHES
N527 1948	804346	RAuxAF		HUGHES
N573 1948	843051	RAuxAF		HUGHES
N835 1957	2674950	WRAFVR	wef 5/11/55	HUGHES

Surname	Initials			Rank	Status	Date
HUGHES	N	M		AC1		17/04/1947
HUGHES	P	D		WO		17/03/1949
HUGHES	R			CPL		12/06/1947
HUGHES	R			LAC		09/01/1963
HUGHES	W	C		FL		05/09/1946
HUGHES	W	H		FL		22/02/1961
HUGHES	W	R	K	FL		01/07/1948
HUGILL	J			FLT SGT		06/05/1943
HUGILL	P	A		SQD OFF	Acting	15/07/1948
HUINS	J	P		WING COMM		03/12/1942
HUINS	R	L		WING COMM		27/03/1947
HULBERT	D	J		FO		08/11/1945
HULBERT	F	H	R	FL		04/10/1945
HULBERT	G	J		FL	Acting	06/03/1947
HULBERT	S	F	R	FL		23/02/1950
HULL	C	B		AC1		14/10/1948
HULL	C	G		FLT SGT		15/07/1948
HULL	F	T		SQD LDR		06/03/1947
HULL	J	C	G	SGT		20/09/1945
HULME	O			FO		25/05/1960
HUMAN	T	O		FO		04/11/1948
HUMBLE	R	L		FL		24/11/1953
HUMBY	K			SQD LDR	Acting	11/01/1945
HUMFREY	V	C		SQD LDR		15/07/1948
HUMMELL	L	J		FO		26/01/1950
HUMPHREY	A	J	R	FL		07/12/1960
HUMPHREY	A	J	R	FL		07/12/1960
HUMPHREYS	E			SAC		20/07/1960
HUMPHREYS	F	K		SQD LDR		01/07/1948
HUMPHREYS	J	E		LAC		13/01/1949
HUMPHREYS	K	H		FO		01/05/1957
HUMPHREYS	K	H		FL		01/07/1964
HUMPHREYS	P	C		FL		24/01/1946
HUMPHREYS	P	F		WING COMM	Acting	26/01/1950
HUMPHRIES	H			FL		28/11/1946
HUNGERFORD	A	R		SQD LDR		20/01/1960
HUNKIN	K			SAC		09/11/1995
HUNKIN	S	H		CPL		05/05/1994
HUNSLEY	K			SGT		25/05/1950
HUNT	A			LAC		10/06/1948
HUNT	A			FL		20/08/1958
HUNT	D	A	C	FL		18/05/1944
HUNT	E	C		FL		25/06/1953
HUNT	F	E	S	LAC		14/09/1950
HUNT	H	B		SQD LDR	Acting	04/10/1945
HUNT	J	L		FL		27/03/1947
HUNT	J			CPL		15/11/1951
HUNT	L	L		FL		30/08/1945
HUNT	M	D		FLT OFF		17/04/1947
HUNT	P	A		FL		28/02/1946
HUNT	P	R		SAC		30/09/1996
HUNT	T			SGT		07/04/1955
HUNTER	A	C		SGT		13/07/1960
HUNTER	C	H		FL		03/12/1942
HUNTER	D	J		FL		12/04/1945
HUNTER	D	S		CPL		30/03/1960
HUNTER	E	M		FL		28/02/1946
HUNTER	G	R		WO		24/01/1946
HUNTER	H			CPL		05/06/1947
HUNTER	J	H		CPL	Temporary	28/07/1955

Source	Service No	Force	Notes	Surname
N289 1947	814221	AAF		HUGHES
N231 1949	751868	RAFVR		HUGHES
N478 1947	811142	AAF		HUGHES
N22 1963	810200	AAF	wef 15/6/44	HUGHES
N751 1946	143714	RAFVR		HUGHES
N153 1961	159079	RAFVR	DFC,DFM; wef 13/5/53	HUGHES
N527 1948	137124	RAFVR	DFC	HUGHES
N526 1943	808039	AAF		HUGILL
N573 1948	619	WAAF		HUGILL
N1638 1942	90259	AAF	OBE,BCh,MRCS,LRCP	HUINS
N250 1947	90694	AAF		HUINS
N1232 1945	176565	RAFVR		HULBERT
N1089 1945	123641	RAFVR	AFC	HULBERT
N187 1947	111215	RAFVR		HULBERT
N179 1950	75288	RAFVR		HULBERT
N856 1948	844970	RAuxAF		HULL
N573 1948	749247	RAFVR		HULL
N187 1947	73241	RAFVR		HULL
N1036 1945	813034	AAF		HULL
N375 1960	2672058	WRAFVR	wef 17/4/60	HULME
N934 1948	116711	RAFVR		HUMAN
MOD 1964	109562	RAFVR		HUMBLE
N14 1945	81059	RAFVR	DSO,DFC	HUMBY
N573 1948	90908	RAuxAF		HUMFREY
N94 1950	177391	RAFVR	DFM	HUMMELL
N933 1960	158778	RAFVR	wef 14/6/44	HUMPHREY
N933 1960	158778	RAFVR	1st Clasp; wef 27/7/60	HUMPHREY
N531 1960	2606555	RAFVR	wef 1/2/55	HUMPHREYS
N527 1948	84673	RAFVR	AFC	HUMPHREYS
N30 1949	853800	RAuxAF		HUMPHREYS
N277 1957	189821	RAFVR	wef 1/7/54	HUMPHREYS
MOD 1967	189821	RAFVR	1st Clasp	HUMPHREYS
N82 1946	85272	RAFVR		HUMPHREYS
N94 1950	72293	RAFVR		HUMPHREYS
N1004 1946	138205	RAFVR		HUMPHRIES
N33 1960	73642	RAFVR	wef 3/8/44	HUNGERFORD
YB 1997	n/a	RAuxAF		HUNKIN
YB 1996	n/a	RAuxAF		HUNKIN
N517 1950	771036	RAFVR		HUNSLEY
N456 1948	850560	RAuxAF		HUNT
N559 1958	83363	RAFVR	wef 25/11/53	HUNT
N473 1944	111976	RAFVR		HUNT
N527 1953	1584516	RAFVR		HUNT
N926 1950	843043	RAuxAF		HUNT
N1089 1945	68721	RAFVR	DSO,DFC	HUNT
N250 1947	112668	RAFVR		HUNT
N1185 1951	856075	RAuxAF		HUNT
N961 1945	117349	RAFVR	DFC	HUNT
N289 1947	2924	WAAF		HUNT
N197 1946	79374	RAFVR		HUNT
YB 1997	n/a	RAuxAF		HUNT
N277 1955	2681509	RAuxAF		HUNT
N675 1950	809060	RAuxAF		HUNTER
N1638 1942	85249	RAFVR		HUNTER
N381 1945	101027	RAFVR		HUNTER
N229 1960	2687045	RAuxAF	wef 6/3/59	HUNTER
N197 1946	141141	RAFVR	DFC,DFM	HUNTER
N82 1946	744916	RAFVR		HUNTER
N460 1947	811010	AAF		HUNTER
N565 1955	803502	RAuxAF		HUNTER

Surname	Initials			Rank	Status	Date
HUNTER	M	A		SGT		07/12/1960
HUNTER	M	M		FLT OFF		24/11/1949
HUNTING	G	L		WING COMM		14/10/1943
HUNTLEY	E			WING COMM	Acting	06/03/1947
HURD	J			SGT		23/11/2004
HURD	L	S		FL		23/06/1955
HURST	A	J		FO		09/05/1946
HURST	D	J		FLT SGT		29/04/1954
HURST	F	C		FLT SGT		27/03/1947
HURST-BARNES	W			SQD LDR		26/01/1950
HUSBANDS	W			SGT	Acting	05/06/1947
HUSSEY	I	M		CPL		15/03/1995
HUSSEY	P	J		SQD LDR		16/03/2004
HUSSEY	W			LAC		27/03/1952
HUTCHEON	A	J		FO		02/07/1958
HUTCHINGS	E	L		FO		09/10/1942
HUTCHINGS	E			FLT SGT		26/08/1948
HUTCHINGS	F	G		AC2		15/11/1951
HUTCHINGS	F	G		SAC		21/01/1959
HUTCHINSON	A	R		FL		08/07/1959
HUTCHINSON	C	G		FL		21/10/1959
HUTCHINSON	H	L		SQD LDR		26/06/1947
HUTCHINSON	I			FO		04/09/1947
HUTCHINSON	J			LAC		23/02/1950
HUTCHINSON	M	M		SQD LDR		15/04/1943
HUTCHINSON	T	G	G	GRP CAP		17/04/1947
HUTCHINSON	W	J		SQD LDR	Acting	30/03/1943
HUTCHISON	J	R		FL		04/10/1945
HUTT	B	E	J	FL		06/03/1947
HUTT	S	L		FL		10/06/1948
HUTTON	A			CPL	Acting	16/04/1958
HUTTON	D	A		SAC		22/02/1961
HUTTON	H	D	M	SAC		20/07/1960
HUTTON	J	T		SGT		18/03/1948
HUTTON	L	M		SGT		28/03/1997
HUTTON	M			CPL		24/09/1998
HUTTON	T	C		FLT SGT		23/08/1951
HYATT	E	P		SGT		20/07/1960
HYDE	J	W		FO		12/08/1943
HYDE	R	J		SQD LDR		10/05/1945
HYETT	V	B		FO		19/08/1959
HYLAND	B	J		CPL		22/02/1961
I'ANSON	J	W		WO		30/03/1944
IGGULDEN	P	W		LAC		29/11/1951
IHEAGWARHM	E			SGT		17/01/1991
ILES	G	B		WING COMM		24/06/1943
IMLAH	K	C		FL		03/07/1952
IMPEY	K	B	G	FL		13/12/1945
IMRAY	H	S		WO		18/04/1946
INGALL	F	L		WING COMM		20/05/1948
INGALL	J	W		SGT		07/08/1947
INGHAM	J	A		FL		10/12/1942
INGLE	A			WING COMM	Acting	13/12/1945
INGLE-FINCH	M	R		WING COMM	Acting	21/03/1946
INGRAM	B	C	M	LAC		25/05/1950
INKSTER	J	F		SQD LDR		07/01/1943
INMAN	S	J		CPL		11/01/1991
INMAN	S	J		SGT		10/01/2001
INNES	R	A		FL		31/05/1945
INNES	W			FL		07/05/1953

Source	Service No	Force	Notes	Surname
N933 1960	2663727	WRAAF	wef 25/5/60	HUNTER
N1193 1949	1355	WAAF		HUNTER
N1065 1943	70336	RAFVR		HUNTING
N187 1947	72427	RAFVR		HUNTLEY
LG 2004	815021	RAFR		HURD
N473 1955	151299	RAFVR		HURD
N415 1946	195211	RAFVR		HURST
N329 1954	2602500	RAFVR		HURST
N250 1947	846547	AAF		HURST
N94 1950	72680	RAFVR		HURST-BARNES
N460 1947	750804	RAFVR		HUSBANDS
YB 1997	n/a	RAuxAF		HUSSEY
LG 2004	0300882N	RAuxAF		HUSSEY
N224 1952	868763	RAuxAF		HUSSEY
N452 1958	2603494	RAFVR	wef 3/3/58	HUTCHEON
MOD 1965	175650	RAFVR		HUTCHINGS
N694 1948	881718	WAAF		HUTCHINGS
N1185 1951	864162	RAuxAF		HUTCHINGS
N28 1959	2603821	RAFVR	1st Clasp; wef 30/3/56	HUTCHINGS
N454 1959	201755	RAFVR	wef 3/11/56	HUTCHINSON
N708 1959	68270	RAFVR	wef 7/7/59	HUTCHINSON
N519 1947	90809	AAF		HUTCHINSON
N740 1947	102960	RAFVR		HUTCHINSON
N179 1950	868694	RAuxAF		HUTCHINSON
N454 1943	90058	AAF	AFC	HUTCHINSON
N289 1947	73244	RAFVR		HUTCHINSON
N1014 1943	70337	RAFVR		HUTCHINSON
N1089 1945	106362	RAFVR	DFC	HUTCHISON
N187 1947	133809	RAFVR	DFC	HUTT
N456 1948	143680	RAFVR		HUTT
N258 1958	2688017	RAFVR	wef 26/3/52	HUTTON
N153 1961	2691049	RAuxAF	wef 11/12/59	HUTTON
N531 1960	2659328	WRAAF	wef 2/3/60	HUTTON
N232 1948	848472	RAuxAF		HUTTON
YB 1998	n/a	RAuxAF		HUTTON
YB 1999	n/a	RAuxAF		HUTTON
N866 1951	756180	RAFVR		HUTTON
N531 1960	2692081	RAuxAF	wef 11/5/60	HYATT
N850 1943	104766	RAFVR		HYDE
N485 1945	115301	RAFVR	AFC	HYDE
N552 1959	2601285	RAFVR	wef 25/9/53	HYETT
N153 1961	2649746	RAuxAF	wef 10/12/60	HYLAND
N281 1944	808023	AAF		I'ANSON
N1219 1951	841092	RAuxAF		IGGULDEN
YB 1998	n/a	RAuxAF		IHEAGWARHM
N677 1943	90001	AAF		ILES
N513 1952	150629	RAFVR		IMLAH
N1355 1945	86665	RAFVR		IMPEY
N358 1946	751235	RAFVR		IMRAY
N395 1948	90496	RAuxAF		INGALL
N662 1947	770339	RAFVR		INGALL
N1680 1942	81355	RAFVR		INGHAM
N1355 1945	83980	RAFVR	DFC,AFC	INGLE
N263 1946	84328	RAFVR	DFC,AFC	INGLE-FINCH
N517 1950	852080	RAuxAF		INGRAM
N3 1943	70339	RAFVR		INKSTER
YB 1996	n/a	RAuxAF		INMAN
YB 2002	n/a	RAuxAF	1st Clasp	INMAN
N573 1945	63784	RAFVR		INNES
N395 1953	91265	RAuxAF		INNES

Surname	Initials			Rank	Status	Date
INSTONE	W	J		SGT		22/04/1943
IONS	M	D		CPL		22/06/1996
IRELAND	E	A		FL		20/09/1945
IRELAND	W	E		SGT	Temporary	04/09/1957
IRELAND	W			CPL		12/06/1947
IRESON	W	H		SQD LDR	Acting	01/07/1948
IRVINE	A			SGT		17/03/1949
IRVINE	J	W		FL		11/07/1946
IRVING	E			CPL		02/08/1951
IRWIN	J			FL		24/01/1946
IRWIN	J			FL		15/07/1948
IRWIN	M	J		FO		02/09/1998
ISAAC	A	C	T	WING COMM	Acting	05/06/1947
ISAACS	L	L		FL		13/12/1945
ISABEL	J	E	E	FL		07/12/1960
ISMAY	T			SGT		01/07/1948
ISON	W	F	A	FL		25/05/1960
ISSAC	A			CPL		30/09/1948
IVES	L	J		LAC		15/11/1951
IVESON	D			WING COMM	Acting	18/04/1946
IVESON	T	C		FL		28/06/1945
IVEY	L	W		SQD LDR		27/04/1950
IVEY	R			FL		24/01/1946
JACK	D	M		SQD LDR		02/09/1943
JACK	J	L		WING COMM	Acting	03/12/1942
JACK	T	W		SGT		06/09/1951
JACKMAN	J			FO		30/03/1943
JACKS	H			LAC		05/06/1947
JACKSON	A	K		FO		05/06/1947
JACKSON	A	S		GRP CAP	Acting	27/03/1952
JACKSON	A	W	B	FL	Acting	15/07/1948
JACKSON	A			SQD LDR	Acting	18/04/1946
JACKSON	A			SGT		03/07/1952
JACKSON	B	S		SGT		11/07/1946
JACKSON	C	E		SQD LDR		05/06/1947
JACKSON	C	W		CPL	Temporary	01/05/1957
JACKSON	D	G		FL		03/10/1946
JACKSON	E	A		WING COMM	Acting	08/05/1947
JACKSON	G	T		CPL		07/04/1955
JACKSON	G	W		FL		19/08/1959
JACKSON	J	A		SGT		24/07/1952
JACKSON	J	E		FL		05/10/1944
JACKSON	J	E		FL		28/11/1956
JACKSON	J	K		FL	Acting	06/05/1943
JACKSON	J			SGT		26/01/1950
JACKSON	J			SAC		24/07/1963
JACKSON	K	V		FLT SGT		24/11/1949
JACKSON	R	V	K	SGT		26/06/1947
JACKSON	R	W		WING COMM		11/11/1943
JACKSON	S	P		SQD LDR		30/12/1943
JACKSON	S			SGT		29/06/1944
JACKSON	S			LAC		02/04/1953
JACKSON	T	C		FL		05/01/1950
JACKSON	W			CPL		09/02/1951
JACOB	A	N		SQD LDR	Acting	20/09/1945
JAMBLIN	C	A		SQD LDR		30/12/1943
JAMES	A	H		SGT	Acting	13/07/1950
JAMES	A	H		FLT SGT		21/01/1959
JAMES	A	I		FL		04/10/1945
JAMES	A	W	D	SQD LDR	Acting	12/04/1945

Source	Service No	Force	Notes	Surname
N475 1943	805220	AAF		INSTONE
YB 1997	n/a	RAuxAF		IONS
N1036 1945	106947	RAFVR		IRELAND
N611 1957	840836	AAF	wef 16/3/44	IRELAND
N478 1947	855239	AAF		IRELAND
N527 1948	85945	RAFVR	DFC	IRESON
N231 1949	816025	RAuxAF		IRVINE
N594 1946	176002	RAFVR		IRVINE
N777 1951	853826	RAuxAF		IRVING
N82 1946	142535	RAFVR		IRWIN
N573 1948	60289	RAFVR		IRWIN
YB 1999	2638061	RAuxAF		IRWIN
N460 1947	73866	RAFVR		ISAAC
N1355 1945	143215	RAFVR		ISAACS
N933 1960	196218	RAFVR	wef 21/9/52	ISABEL
N527 1948	749066	RAFVR		ISMAY
N375 1960	55588	RAFVR	wef 23/6/52	ISON
N805 1948	855015	RAuxAF		ISSAC
N1185 1951	840089	RAuxAF		IVES
N358 1946	86384	RAFVR	DSO,DFC	IVESON
N706 1945	128539	RAFVR	DFC	IVESON
N406 1950	70341	RAFVR		IVEY
N82 1946	122762	RAFVR	DFC	IVEY
N918 1943	90170	AAF		JACK
N1638 1942	90185	AAF	MBE,MC	JACK
N913 1951	872494	RAuxAF		JACK
N1014 1943	110812	RAFVR	DFC	JACKMAN
N460 1947	811131	AAF		JACKS
N460 1947	145206	RAFVR		JACKSON
N224 1952	90729	RAuxAF	OBE	JACKSON
N573 1948	62356	RAFVR		JACKSON
N358 1946	114299	RAFVR	DFC	JACKSON
N513 1952	812081	RAuxAF		JACKSON
N594 1946	804220	AAF		JACKSON
N460 1947	72506	RAFVR		JACKSON
N277 1957	750913	RAuxAF	wef 25/6/44	JACKSON
N837 1946	182190	RAFVR		JACKSON
N358 1947	72277	RAFVR		JACKSON
N277 1955	850686	RAuxAF		JACKSON
N552 1959	115846	RAFVR	wef 28/6/44	JACKSON
N569 1952	810202	RAuxAF		JACKSON
N1007 1944	72995	RAFVR		JACKSON
N833 1956	169777	RAFVR		JACKSON
N526 1943	102057	RAFVR		JACKSON
N94 1950	809214	RAuxAF	BEM	JACKSON
N525 1963	2681186	RAuxAF	wef 24/5/63	JACKSON
N1193 1949	884807	RAuxAF		JACKSON
N519 1947	756169	RAFVR		JACKSON
N1177 1943	70343	RAFVR		JACKSON
N1350 1943	70344	RAFVR	AFC	JACKSON
N652 1944	801292	AAF		JACKSON
N298 1953	870778	RAuxAF		JACKSON
N1 1950	90754	RAuxAF		JACKSON
N145 1951	858759	RAuxAF		JACKSON
N1036 1945	87394	RAFVR		JACOB
N1350 1943	70905	RAFVR	MBE	JAMBLIN
N675 1950	743345	RAFVR		JAMES
N28 1959	2692010	RAuxAF	1st Clasp; wef 5/7/58	JAMES
N1089 1945	60307	RAFVR		JAMES
N381 1945	72260	RAFVR		JAMES

Surname	Initials			Rank	Status	Date
JAMES	A	W		CPL		08/11/1945
JAMES	B	A		SQD LDR	Acting	26/08/1948
JAMES	B	E		FLT OFF		27/03/1947
JAMES	C	F		FL		11/07/1946
JAMES	E	A		SQD LDR	Acting	17/03/1949
JAMES	F	S		FL		19/08/1954
JAMES	G	C		SGT		05/06/1947
JAMES	G			SGT		20/03/1957
JAMES	H			SGT		13/11/1947
JAMES	J			FL		10/06/1948
JAMES	P	S		SQD LDR	Acting	11/01/1945
JAMES	R	W		CPL		04/09/1947
JAMES	R			FL	Acting	07/07/1949
JAMES	T	W		FLT SGT		15/06/1944
JAMES	T			WO	Acting	15/07/1948
JAMESON	T	S		FL		08/11/1945
JAMES-ROBERSTON	R	E		FL		21/10/1959
JAMIESON	J	A		SGT		26/04/1961
JAMIESON	J	B		FLT SGT		05/06/1947
JAMIESON	R	S		WO	Temporary	16/04/1958
JAMIESON	R	W		SQD LDR	Acting	27/07/1944
JANES	C	H	D	SGT		07/07/1949
JARMAN	H	L		FL		10/02/1949
JARMAN	H	T		FLT SGT		08/11/1945
JARRETT	J	H		CPL		15/07/1948
JARRETT	R	G		SGT		30/03/1943
JARVIS	A	N	H	LAC		27/03/1947
JARVIS	F	J		SGT		23/02/1950
JARVIS	G	J		FLT SGT		01/07/1948
JAY	A			FO		16/03/1950
JAY	E	A		FL	Acting	28/02/1946
JAY	R	E		WING COMM	Acting	27/07/1944
JEFFERISS	F	J	G	FL		07/04/1955
JEFFERSON	F	W		FL		28/07/1955
JEFFERSON	T	G		WING COMM	Acting	18/04/1946
JEFFERY	E	H		FL		20/08/1958
JEFFERY-CRIDGE	H	R		FL		21/03/1946
JEFFERYS	S	J		FL		12/02/1953
JEFFERYS	S	J		FL		28/08/1963
JEFFREY	A			FLT SGT		08/11/1945
JEFFREY	D	G		WO		01/11/1978
JEFFREY	D	G		WO		01/11/1988
JEFFREY	D	G		WO		01/11/1998
JEFFREY	J	M		SGT		01/07/1987
JEFFREY	J	M		SGT		01/07/1997
JEFFREY	N			CPL		17/03/1949
JEFFRIES	B	W		SQD LDR		06/03/1947
JELL	R	A		SQD LDR	Acting	24/01/1946
JELLEY	D			CPL		25/04/1998
JENKINS	C	R		WING COMM	Acting	12/06/1947
JENKINS	F	A	C	WO		11/01/1015
JENKINS	F	A	V	FL		06/03/1947
JENKINS	H	B		SQD LDR		07/08/1947
JENKINS	H	C		WO		27/03/1947
JENKINS	H	H		FL		11/07/1946
JENKINS	H	J		FO		01/10/1953
JENKINS	I			SGT		26/11/1958
JENKINS	J	G		FL		06/03/1947
JENKYNS	S	M		FL		31/05/1945

Source	Service No	Force	Notes	Surname
N1232 1945	803208	AAF		JAMES
N694 1948	104921	RAFVR		JAMES
N250 1947	913	WAAF		JAMES
N594 1946	172793	RAFVR		JAMES
N231 1949	81555	RAFVR		JAMES
N672 1954	80568	RAFVR		JAMES
N460 1947	811017	AAF		JAMES
N198 1957	747211	RAFVR	wef 24/7/44	JAMES
N933 1947	840369	AAF		JAMES
N456 1948	122795	RAFVR		JAMES
N14 1945	83276	RAFVR	DFC	JAMES
N740 1947	743732	RAFVR		JAMES
N673 1949	132113	RAFVR		JAMES
N596 1944	805181	AAF		JAMES
N573 1948	867914	RAuxAF	MBE	JAMES
N1232 1945	102083	RAFVR		JAMESON
N708 1959	201850	RAFVR	wef 19/8/52	JAMES-ROBERSTON
N339 1961	2685696	RAFVR	1st Clasp; wef 18/3/61 no record of first award	JAMIESON
N460 1947	811045	AAF		JAMIESON
N258 1958	755907	RAFVR	wef 17/8/44	JAMIESON
N763 1944	78728	RAFVR		JAMIESON
N673 1949	865605	RAuxAF		JANES
N123 1949	90631	RAuxAF		JARMAN
N1232 1945	801352	AAF		JARMAN
N573 1948	805464	RAuxAF		JARRETT
N1014 1943	808103	AAF		JARRETT
N250 1947	748531	RAFVR		JARVIS
N179 1950	850006	RAuxAF		JARVIS
N527 1948	812170	RAuxAF		JARVIS
N257 1950	158828	RAFVR		JAY
N197 1946	178281	RAFVR		JAY
N763 1944	90011	AAF		JAY
N277 1955	72460	RAFVR	MRCS,LRCP	JEFFERISS
N565 1955	165572	RAFVR		JEFFERSON
N358 1946	101521	RAFVR	DSO,AFC	JEFFERSON
N559 1958	127344	RAFVR	wef 18/9/52	JEFFERY
N263 1946	154490	RAFVR		JEFFERY-CRIDGE
N115 1953	170343	RAFVR		JEFFERYS
N621 1963	170343	RAuxAF	1st Clasp; wef 21/7/55	JEFFERYS
N1232 1945	802419	AAF		JEFFREY
YB 1996	n/a	RAuxAF		JEFFREY
YB 1996	n/a	RAuxAF	1st Clasp	JEFFREY
YB 1999	n/a	RAuxAF	2nd Clasp	JEFFREY
YB 1996	n/a	RAuxAF		JEFFREY
YB 1999	n/a	RAuxAF	1st Clasp	JEFFREY
N231 1949	866404	RAuxAF		JEFFREY
N187 1947	72868	RAFVR		JEFFRIES
N82 1946	64283	RAFVR	DFC	JELL
YB 1999	n/a	RAuxAF		JELLEY
N478 1947	72036	RAFVR		JENKINS
N14 1945	805184	AAF		JENKINS
N187 1947	163800	RAFVR		JENKINS
N662 1947	74019	RAFVR		JENKINS
N250 1947	755346	RAFVR		JENKINS
N594 1946	143651	RAFVR		JENKINS
N796 1953	159758	RAFVR		JENKINS
N807 1958	2659629	WRAAF	wef 6/4/57	JENKINS
N187 1947	130667	RAFVR	DFC	JENKINS
N573 1945	127144	RAFVR		JENKYNS

Surname	Initials			Rank	Status	Date
JENNENS	A	W		FLT SGT		07/04/1955
JENNINGS	J	H		FL		29/04/1954
JENNINGS	S	J		AC2		16/03/1950
JENNINGS	S	J		LAC		19/10/1960
JENSEN	F	W	M	FL	Acting	03/12/1942
JEPHSON	R			SGT	Acting	09/12/1959
JESS	A	S		FL		26/10/1944
JESSE	R	H		FL		05/09/1946
JESSOP	A	E		SGT		30/09/1948
JESSOP	R	A		SGT		01/07/1948
JESSOP	R	F		FL		13/12/1945
JEWELL	V	A		SGT		20/09/1961
JEWITT	R	W		SAC		19/10/1960
JINKINSON	A	E		FL		23/05/1946
JINKS	W	J		FL		03/06/1959
JOBSON	J	C		SGT		20/01/1960
JOBSON	S	G		FL		06/03/1947
JOCE	H	V	A	WO		30/12/1943
JOHNSON	A	H	E	FLT SGT		03/10/1962
JOHNSON	A	L		LAC		10/02/1949
JOHNSON	A	S		SQD LDR		06/05/1943
JOHNSON	B	T		SAC		12/09/1997
JOHNSON	C	S		FL		19/01/1956
JOHNSON	D	G		FL		26/04/1956
JOHNSON	D	K		SAC		19/09/1996
JOHNSON	E	A		FO		07/08/1947
JOHNSON	G	H		CPL		08/03/1951
JOHNSON	H	C		SGT		14/10/1948
JOHNSON	H	D		FL		05/10/1944
JOHNSON	J	E		GRP CAP	Acting	06/03/1947
JOHNSON	K	L		LAC		21/10/1959
JOHNSON	L	G		SQD LDR		27/01/1965
JOHNSON	L	R		FL		10/08/1950
JOHNSON	N	B		LAC		13/01/1949
JOHNSON	P	A	H	SGT		15/11/1951
JOHNSON	P	F		FO		19/08/1959
JOHNSON	R	A		FL		24/01/1946
JOHNSON	R	C		SGT		07/06/1951
JOHNSON	R	H		SQD LDR		27/03/1947
JOHNSON	R	R		FO		10/02/1949
JOHNSON	R	R		FLT OFF		02/07/1958
JOHNSON	R			CPL		23/02/1950
JOHNSON	S	C		FL		09/05/1946
JOHNSON	W	J		SQD LDR	Acting	06/03/1947
JOHNSON	W			SQD LDR		07/08/1947
JOHNSTON	A	B		SQD LDR		23/02/1950
JOHNSTON	C	C		FL		01/08/1946
JOHNSTON	E	J		LAC		12/06/1947
JOHNSTON	H	McP.		FL		28/02/1946
JOHNSTON	J	C		CPL		15/09/1949
JOHNSTON	K	W		FL		19/08/1959
JOHNSTON	R	J		FL		16/01/1957
JOHNSTON	S			WO		13/12/1945
JOHNSTON	T			LAC		13/12/1961
JOHNSTONE	A	V	R	WING COMM	Acting	10/12/1942
JOHNSTONE	G	R	A	McG. WING COMM	Acting	21/06/1945
JOHNSTONE	R	W		SGT		30/01/1994
JOLL	I	K	S	SQD LDR	Acting	07/08/1947
JOLLEY	K			FL		22/09/1955
JOLLY	S	C		SQD LDR	Acting	27/07/1944

Source	Service No	Force	Notes	Surname
N277 1955	850152	RAuxAF		JENNENS
N329 1954	81475	RAFVR		JENNINGS
N257 1950	872476	RAuxAF		JENNINGS
N775 1960	2691047	RAuxAF	1st Clasp; wef 8/1/60	JENNINGS
N1638 1942	102058	RAFVR		JENSEN
N843 1959	815066	AAF	wef 15/3/43	JEPHSON
N1094 1944	126141	RAFVR		JESS
N751 1946	104439	RAFVR		JESSE
N805 1948	819011	RAuxAF		JESSOP
N527 1948	747087	RAFVR		JESSOP
N1355 1945	121528	RAFVR		JESSOP
N720 1961	2692143	RAuxAF	wef 19/6/61	JEWELL
N775 1960	2691028	RAuxAF	wef 18/3/59	JEWITT
N468 1946	130440	RAFVR		JINKINSON
N368 1959	128085	RAFVR	wef 7/6/53	JINKS
N33 1960	2685578	RAFVR	wef 12/5/59	JOBSON
N187 1947	147923	RAFVR		JOBSON
N1350 1943	740981	RAFVR		JOCE
N722 1962	746408	RAFVR	wef 26/6/44	JOHNSON
N123 1949	857602	RAuxAF		JOHNSON
N526 1943	90305	AAF		JOHNSON
YB 2001	n/a	RAuxAF		JOHNSON
N43 1956	163925	RAFVR		JOHNSON
N318 1956	163188	RAFVR		JOHNSON
YB 1997	n/a	RAuxAF		JOHNSON
N662 1947	106794	RAFVR		JOHNSON
N244 1951	808203	RAuxAF		JOHNSON
N856 1948	853546	RAuxAF		JOHNSON
N1007 1944	119186	RAFVR		JOHNSON
N187 1947	83267	RAFVR	DSO,DFC	JOHNSON
N708 1959	2699503	RAuxAF	wef 13/4/57	JOHNSON
MOD 1966	76914	RAFVR	DFC	JOHNSON
N780 1950	120935	RAFVR	DFM	JOHNSON
N30 1949	846962	RAuxAF		JOHNSON
N1185 1951	742437	RAFVR		JOHNSON
N552 1959	233	WRAAF	wef 27/5/59	JOHNSON
N82 1946	116721	RAFVR	DFC	JOHNSON
N560 1951	753006	RAFVR		JOHNSON
N250 1947	72221	RAFVR	OBE	JOHNSON
N123 1949	136625	RAFVR		JOHNSON
N452 1958	246	WRAAF	wef 26/4/44	JOHNSON
N179 1950	770697	RAFVR		JOHNSON
N415 1946	62663	RAFVR		JOHNSON
N187 1947	115410	RAFVR	DFC	JOHNSON
N662 1947	113857	RAFVR		JOHNSON
N179 1950	90717	RAuxAF	MC	JOHNSTON
N652 1946	81926	RAFVR		JOHNSTON
N478 1947	756183	RAFVR		JOHNSTON
N197 1946	117127	RAFVR	DSO,DFC	JOHNSTON
N953 1949	802317	RAuxAF		JOHNSTON
N552 1959	185907	RAFVR	wef 5/6/59	JOHNSTON
N24 1957	55558	RAuxAF		JOHNSTON
N1355 1945	743048	RAFVR		JOHNSTON
N987 1961	2650927	RAuxAF	wef 23/10/61	JOHNSTON
N1680 1942	90163	AAF	DFC	JOHNSTONE
N671 1945	72082	RAFVR	DSO,DFC	JOHNSTONE
YB 1996	n/a	RAuxAF		JOHNSTONE
N662 1947	90951	AAF	DFC	JOLL
N708 1955	163448	RAuxAF		JOLLEY
N763 1944	88427	RAFVR		JOLLY

Surname	Initials			Rank	Status	Date
JONES	A	M		SGT		19/08/1954
JONES	A	P		FL		21/03/1946
JONES	A	R		SQD LDR	Acting	17/04/1947
JONES	A			SQD LDR		12/06/1947
JONES	A			CPL		12/06/1947
JONES	B			FL		26/11/1953
JONES	C	A		FL		19/08/1954
JONES	C	F	M	WING COMM	Acting	30/08/1945
JONES	C	R		CPL		05/09/1946
JONES	D	A		FL		25/03/1954
JONES	D	C		SQD LDR		16/10/1952
JONES	D	E		SQD LDR	Acting	24/01/1946
JONES	D	J		SGT		27/03/1947
JONES	D	L		SGT		13/11/1947
JONES	D	M	L	WING OFF		27/03/1947
JONES	D			LAC		03/07/1952
JONES	E	F		FLT SGT		01/10/1953
JONES	E	H		SQD LDR	Acting	21/09/1944
JONES	E	O		FL		09/08/1945
JONES	F	S		FO		01/10/1953
JONES	F			CPL		23/02/1950
JONES	G	D	K	WO		24/01/1946
JONES	G	M		CPL		07/07/1949
JONES	G	W		FO		10/02/1949
JONES	G			FL		04/10/1945
JONES	G			FLT SGT		18/12/1963
JONES	H	A		FL		05/09/1946
JONES	H	A		FL		20/09/1961
JONES	H	C		SGT		10/12/1942
JONES	H	C		FLT SGT		02/04/1953
JONES	H	E		FL		19/10/1960
JONES	H	J	L	FL		14/10/1943
JONES	H	S		SGT		16/01/1957
JONES	H	T		FL		04/10/1945
JONES	I	R		WING COMM	Acting	30/08/1945
JONES	I			CPL		12/04/1951
JONES	I			SGT		07/05/1953
JONES	J	A		FL	Acting	18/07/1956
JONES	J	F	R	FL		26/10/1944
JONES	J	K		FL		14/01/1954
JONES	J	S		WO		09/08/1945
JONES	J	T		CPL		07/07/1949
JONES	K			LAC		07/08/1947
JONES	K			SGT		06/01/1955
JONES	M	M		SGT		22/02/1961
JONES	N	J		CPL		05/01/1950
JONES	O	A	K	FL		03/07/1957
JONES	O	P		FO		04/11/1943
JONES	R	A		FL		01/10/1953
JONES	R	E		FL		25/01/1945
JONES	R	H		LAC		27/11/1952
JONES	R	L		FL		12/04/1945
JONES	R	M	W	FO		08/05/1947
JONES	R	S		FL		28/11/1946
JONES	R	W		WO		24/01/1946
JONES	R			LAC		21/06/1961
JONES	S	C		FO		21/03/1946
JONES	S	E		LAC		20/05/1948
JONES	S	G		FLT SGT		20/09/1945
JONES	S	J	L	AC1		21/12/1950

Source	Service No	Force	Notes	Surname
N672 1954	810110	RAuxAF		JONES
N263 1946	68732	RAFVR		JONES
N289 1947	84823	RAFVR		JONES
N478 1947	72340	RAFVR		JONES
N478 1947	811065	AAF		JONES
N952 1953	182962	RAuxAF		JONES
N672 1954	198467	RAuxAF		JONES
N961 1945	81916	RAFVR	DSO,DFC	JONES
N751 1946	805065	AAF		JONES
N249 1954	174372	RAFVR		JONES
N798 1952	90461	RAuxAF		JONES
N82 1946	125906	RAFVR		JONES
N250 1947	804195	AAF		JONES
N933 1947	815230	AAF		JONES
N250 1947	122	WAAF		JONES
N513 1952	853799	RAuxAF		JONES
N796 1953	853544	RAuxAF		JONES
N961 1944	82952	RAFVR		JONES
N874 1945	102534	RAFVR		JONES
N796 1953	174987	RAuxAF	DFC	JONES
N179 1950	853633	RAuxAF		JONES
N82 1946	741814	RAFVR		JONES
N673 1949	865632	RAuxAF		JONES
N123 1949	122785	RAFVR		JONES
N1089 1945	78854	RAFVR		JONES
N917 1963	2697600	RAFVR	wef 25/6/63	JONES
N751 1946	149293	RAFVR		JONES
N720 1961	149293	RAFVR	1st Clasp; wef 24/11/60	JONES
N1680 1942	805137	AAF		JONES
N298 1953	843097	RAuxAF		JONES
N775 1960	161087	RAFVR	DFC; wef 17/7/59	JONES
N1065 1943	70347	RAFVR		JONES
N24 1957	2682117	RAuxAF		JONES
N1089 1945	87643	RAFVR	DFC	JONES
N961 1945	90388	AAF		JONES
N342 1951	818059	RAuxAF		JONES
N395 1953	2602006	RAFVR		JONES
N513 1956	2689047	RAuxAF		JONES
N1094 1944	128559	RAFVR		JONES
N18 1954	56665	RAFVR		JONES
N874 1945	741779	RAFVR		JONES
N673 1949	855104	RAuxAF		JONES
N662 1947	858894	AAF		JONES
N1 1955	2688540	RAuxAF	1st Clasp	JONES
N153 1961	2656106	WRAAF	wef 16/7/60	JONES
N1 1950	812250	RAuxAF		JONES
N456 1957	188468	RAFVR	wef 12/2/54	JONES
N1144 1943	70349	RAFVR		JONES
N796 1953	106874	RAuxAF		JONES
N70 1945	83981	RAFVR		JONES
N915 1952	865531	RAuxAF		JONES
N381 1945	81362	RAFVR		JONES
N358 1947	161212	RAFVR		JONES
N1004 1946	159041	RAFVR		JONES
N82 1946	815068	AAF		JONES
N478 1961	802590	AAF	wef 22/6/44	JONES
N263 1946	199404	RAFVR		JONES
N395 1948	840905	RAuxAF		JONES
N1036 1945	748807	RAFVR		JONES
N1270 1950	770773	RAFVR		JONES

Surname	Initials			Rank	Status	Date
JONES	S	L		WO	Acting	07/07/1949
JONES	T	E		FL		12/02/1956
JONES	V	O		SQD LDR	Acting	17/04/1947
JONES	W	J		CPL		21/06/1961
JONES	W	M		FL		07/04/1955
JORDAN	H	A		FL		04/09/1957
JORDAN	H	A		FL		04/09/1957
JORDAN	F	R		SGT		04/09/1947
JORDAN	J	C		SQD LDR		02/09/1943
JORDAN	W	H		WO		28/02/1946
JORDAN	W			FL		06/03/1947
JOSEPH	E	W		SQD LDR	Acting	26/08/1948
JOUGHIN	E	W		CPL		30/09/1948
JOWITT	N	A	V	CPL	Temporary	20/01/1960
JOWLE	L			LAC		14/10/1948
JOYCE	E	J		SGT		07/07/1949
JOYCE	G	F		CPL		01/11/1951
JOYCE-CLARKE	D			FLT OFF	Acting	08/05/1947
JUDD	H	B	H	FL		04/05/1944
JUDD	M	T		GRP CAP	Acting	20/09/1945
JUDGE	J	W	C	FL		08/03/1956
JULEFF	J	R		FL	Acting	03/12/1942
JUPE	E	J	C	LAC		02/08/1951
JUPP	A	T		SQD LDR	Acting	24/01/1946
JUPP	H	R		LAC		14/10/1948
KAHN	W	A	H	FO		08/07/1959
KAISER	Y	M		PO		22/10/1958
KANE	D	R		CPL		05/01/1950
KANE	M			SGT		20/03/1957
KANE	M			SGT		25/07/1962
KATES	C	A	W	SGT		26/11/1953
KAVANAGH	M	A		LACW		07/12/1960
KAY	A	W		WING COMM	Acting	17/04/1947
KAY	D	C		FL		13/12/1945
KAY	H	B		LAC		16/10/1947
KAY	J	K		FL		06/03/1947
KAY	J			SGT		02/04/1953
KAY	S			CPL		30/09/1948
KAYE	J	J		FL		21/09/1944
KAYLL	C	H		CPL		05/06/1947
KAYLL	J	R		WING COMM	Acting	09/08/1945
KEANE	D	J		CPL		02/08/1951
KEAR	L	W		SGT	Temporary	28/07/1955
KEARNEY	M			SGT		24/12/1942
KEAST	H	A		SGT		07/08/1947
KEAT	G	H		WING COMM		19/08/1954
KEATES	J	K		FL	Acting	03/12/1942
KEATING	W			SAC		03/03/1994
KEEBLE	C	H		SQD LDR		12/06/1947
KEEGAN	B	V		SGT		08/05/1947
KEEGAN	J			FL		12/04/1945
KEEGAN	M	E		SAC		19/02/1999
KEELING	G	W		AC1		07/08/1947
KEELING	R	V		SQD LDR		05/01/1950
KEEN	R	S		CPL		06/09/1951
KEEN	V	B		CPL		03/07/1957
KEENE	S			CPL		12/09/1997
KEENSHAW	J			FL		06/03/1947
KEEP	G	R		FL		27/03/1947
KEEP	J	G		SQD LDR	Acting	30/12/1943

Source	Service No	Force	Notes	Surname
N673 1949	853574	RAuxAF		JONES
MOD 1965	200981	RAFVR		JONES
N289 1947	102758	RAFVR		JONES
N478 1961	2604368	RAFVR	wef 3/12/59	JONES
N277 1955	183943	RAFVR	DFC	JONES
N611 1957	72322	RAFVR	wef 6/2/44	JORDAN
N611 1957	72322	RAFVR	1st Clasp; wef 28/3/55	JORDAN
N740 1947	804234	AAF		JORDAN
N918 1943	72341	RAFVR		JORDAN
N197 1946	742138	RAFVR		JORDAN
N187 1947	146215	RAFVR		JORDAN
N694 1948	79595	RAFVR	OBE	JOSEPH
N805 1948	853569	RAuxAF		JOUGHIN
N33 1960	746682	RAFVR	wef 8/7/44	JOWITT
N856 1948	868731	RAuxAF		JOWLE
N673 1949	861047	RAuxAF		JOYCE
N1126 1951	753099	RAFVR		JOYCE
N358 1947	3998	WAAF		JOYCE-CLARKE
N425 1944	84979	RAFVR		JUDD
N1036 1945	72025	RAFVR	DSO,DFC,AFC	JUDD
N187 1956	142533	RAuxAF		JUDGE
N1638 1942	82717	RAFVR		JULEFF
N777 1951	846433	RAuxAF		JUPE
N82 1946	102654	RAFVR		JUPP
N856 1948	746353	RAFVR		JUPP
N454 1959	2606105	RAFVR	wef 26/6/56	KAHN
N716 1958	3814	WRAFVR	wef 9/6/54	KAISER
N1 1950	873833	RAuxAF		KANE
N198 1957	2683514	RAuxAF	wef 12/1/57	KANE
N549 1962	2683514	RAuxAF	1st Clasp; wef 14/4/62	KANE
N952 1953	2682523	RAuxAF		KATES
N933 1960	26637030	WRAAF	wef 28/3/60	KAVANAGH
N289 1947	90526	AAF		KAY
N1355 1945	114895	RAFVR	DFM	KAY
N846 1947	844184	AAF		KAY
N187 1947	83730	RAFVR		KAY
N298 1953	881539	WAAF		KAY
N805 1948	855911	RAuxAF		KAY
N961 1944	80813	RAFVR		KAYE
N460 1947	811187	AAF		KAYLL
N874 1945	90276	AAF	DSO,DFC	KAYLL
N777 1951	751761	RAFVR		KEANE
N565 1955	844728	RAuxAF		KEAR
N1758 1942	802202	AAF		KEARNEY
N662 1947	841551	AAF		KEAST
N672 1954	73716	RAFVR		KEAT
N1638 1942	113955	RAFVR	AFC	KEATES
YB 2002	n/a	RAuxAF		KEATING
N478 1947	72387	RAFVR		KEEBLE
N358 1947	750002	RAFVR		KEEGAN
N381 1945	86385	RAFVR		KEEGAN
YB 2000	n/a	RAuxAF		KEEGAN
N662 1947	840944	AAF		KEELING
N1 1950	82689	RAFVR	DFC	KEELING
N913 1951	759189	RAFVR		KEEN
N456 1957	2655671	WRAAF	wef 5/12/53	KEEN
YB 2001	n/a	RAuxAF		KEENE
N187 1947	137821	RAFVR		KEENSHAW
N250 1947	73258	RAFVR		KEEP
N1350 1943	78729	RAFVR		KEEP

Surname	Initials				Rank	Status	Date
KEEVIL	A	N			FL		20/05/1948
KEIR	C	D			FL		27/01/1944
KELLETT	M				WING COMM	Acting	13/12/1945
KELLETT	R	G			WING COMM		03/12/1942
KELLETT	R	G			SQD LDR		02/04/1953
KELLITT	W	H			FL		13/12/1945
KELLY	A	J			LAC		03/10/1962
KELLY	C	E	E		FL		28/02/1946
KELLY	D	W			FO		17/06/1943
KELLY	I	K			FLT OFF		01/10/1953
KELLY	J	F			FLT OFF		17/04/1947
KELLY	J	M			CPL		15/09/1949
KELLY	M	A			FL		16/01/1957
KELLY	P	A			FLT OFF	Acting	12/06/1947
KELLY	R	C			FL		25/06/1953
KELLY	R				CPL		08/05/1947
KELLY	S	G			FLT SGT		30/03/1943
KELLY	V	P			SQD LDR	Acting	06/03/1947
KELSEY	A	E			CPL		18/03/1948
KELWAY	A	C			SQD LDR		05/06/1947
KEMBLE	H	P			SGT		30/09/1948
KEMBLE	R	H			FLT SGT	Temporary	16/08/1961
KEMP	A	E	C		CPL		14/10/1948
KEMP	D	W			LAC		22/05/1952
KEMP	H	W			SGT	Temporary	29/04/1954
KEMP	N	L	D		FL		21/03/1946
KEMP	P	C			FLT SGT	Acting	04/05/1944
KEMP	P	R			FL		21/06/1951
KEMP	R	G			WING COMM		01/09/1996
KEMP	R	S			CPL		26/01/1950
KEMPE	J	W	R		WING COMM	Acting	25/01/1945
KEMP-LEWIS	R	H			FL		15/06/1944
KEMP-SCRIVEN	R	S			FL		07/07/1949
KENDALL	M				SGT		18/10/1997
KENDREW	G	H			CPL		27/11/1952
KENDRICK	J	E			CPL		26/01/1950
KENNAN	T	F			FL		21/09/1944
KENNAWAY	R				FL		17/06/1954
KENNEDY	D	E			FL		28/02/1946
KENNEDY	E	J			SGT		08/02/1997
KENNEDY	J	S			WING COMM	Acting	06/03/1947
KENNEDY	M	P			CPL		15/09/1949
KENNEDY-FINLAYSON	J				FL		08/11/1945
KENT	A	F			LAC		03/07/1957
KENT	H				LAC		25/05/1950
KENT	T				FL		28/03/1962
KENWORTHY	W				FL		06/03/1947
KEPPEL	W	B	A	J	WING COMM		30/03/1943
KEPPIE	E	D	F		SGT		11/01/1945
KER	D	A			SQD LDR	Acting	26/01/1950
KER	M	B			SQD OFF		17/04/1947
KERMODE	A	C			WING COMM	Acting	06/03/1947
KERNAGHAN	S	H			WING COMM	Acting	30/10/1957
KERNER	W	W			FO		07/05/1953
KERR	F	M			SGT		16/01/1957
KERR	G	P			SQD LDR		04/02/1943
KERR	J				SQD LDR		08/05/1947
KERR	N	P			FL		01/03/1945
KERR	T	H			FL		25/06/1953
KERR	W				SGT		07/08/1947

Source	Service No	Force	Notes	Surname
N395 1948	65608	RAFVR		KEEVIL
N59 1944	63792	RAFVR	DFC	KEIR
N1355 1945	86630	RAFVR	DFC	KELLETT
N1638 1942	90082	AAF	DSO,DFC	KELLETT
N298 1953	90082	RAuxAF	DSO,DFC; 1st Clasp	KELLETT
N1355 1945	122985	RAFVR		KELLITT
N722 1962	2650955	RAuxAF	wef 2/9/62	KELLY
N197 1946	142036	RAFVR		KELLY
N653 1943	112545	RAFVR		KELLY
N796 1953	456	WAAF		KELLY
N289 1947	334	WAAF		KELLY
N953 1949	865586	RAuxAF		KELLY
N24 1957	181439	RAFVR		KELLY
N478 1947	2220	WAAF		KELLY
N527 1953	128364	RAFVR		KELLY
N358 1947	866412	AAF		KELLY
N1014 1943	808088	AAF		KELLY
N187 1947	73971	RAFVR		KELLY
N232 1948	870237	RAuxAF		KELSEY
N460 1947	90525	AAF		KELWAY
N805 1948	804289	RAuxAF		KEMBLE
N624 1961	818133	AAF	wef 24/5/44	KEMBLE
N856 1948	812232	RAuxAF		KEMP
N402 1952	818007	RAuxAF		KEMP
N329 1954	810910	RAuxAF		KEMP
N263 1946	84941	RAFVR	DFC	KEMP
N425 1944	740972	RAFVR		KEMP
N621 1951	72093	RAFVR	MB,ChB	KEMP
YB 1997	4232872	RAuxAF		KEMP
N94 1950	752151	RAFVR		KEMP
N70 1945	72078	RAFVR		KEMPE
N596 1944	88655	RAFVR		KEMP-LEWIS
N673 1949	90850	RAuxAF		KEMP-SCRIVEN
YB 1999	n/a	RAuxAF		KENDALL
N915 1952	2686102	RAuxAF		KENDREW
N94 1950	810151	RAuxAF		KENDRICK
N961 1944	84670	RAFVR		KENNAN
N471 1954	142895	RAFVR		KENNAWAY
N197 1946	81358	RAFVR		KENNEDY
YB 1997	n/a	RAuxAF		KENNEDY
N187 1947	81351	RAFVR	DFC	KENNEDY
N953 1949	865054	RAuxAF		KENNEDY
N1232 1945	126030	RAFVR		KENNEDY-FINLAYSON
N456 1957	840978	AAF	wef 16/3/44	KENT
N517 1950	847249	RAuxAF		KENT
N242 1962	198628	RAFVR	wef 12/2/61	KENT
N187 1947	86759	RAFVR	MBE	KENWORTHY
N1014 1943	90003	AAF		KEPPEL
N14 1945	802394	AAF		KEPPIE
N94 1950	72071	RAFVR	MB,ChB	KER
N289 1947	181	WAAF		KER
N187 1947	73262	RAFVR	OBE	KERMODE
N775 1957	37572	RAuxAF	wef 11/8/54	KERNAGHAN
N395 1953	197839	RAFVR		KERNER
N24 1957	728103	RAF	DFM	KERR
N131 1943	90109	AAF		KERR
N358 1947	73263	RAFVR		KERR
N225 1945	115342	RAFVR	DFC,AFC	KERR
N527 1953	163135	RAFVR		KERR
N662 1947	803480	AAF		KERR

Surname	Initials			Rank	Status	Date
KERRIDGE	G	E		SQD LDR	Acting	24/01/1946
KERSHAW	C	D		WING COMM		05/02/1948
KERSHAW	E			FL		05/02/1948
KERSHAW	F	J		SQD LDR	Acting	26/08/1948
KETCHER	W	G		CPL	Temporary	20/01/1960
KETHRO-SCHONBACH	D	M		WO		28/02/1946
KETTLE	C	H		SGT		10/04/1952
KETTLEWELL	H			FLT SGT	Acting	01/10/1953
KEVAN	J			FL	Acting	04/09/1947
KEY	S	J	L	FL		06/03/1947
KEY	V	P		SQD LDR		28/11/1946
KIBBLE	E			FO		08/11/1945
KIDD	E	A		MST GNR		25/06/1953
KIDD	W			SGT		25/05/1950
KIFF	A	W	J	FLT SGT	Acting	19/10/1960
KILBURN	J	G		WO		09/05/1946
KILLICK	P			FO		12/04/1945
KILNER	J	R		FO		17/06/1943
KILROY	J	R		SGT	Acting	31/10/1956
KILROY	J			SGT		14/09/1950
KILSBY	A	E		FL		05/06/1947
KIMBER	T	J		SQD LDR		05/06/1947
KING	A	E		CPL		15/07/1948
KING	A	R		FL	Acting	01/07/1948
KING	A			FL		21/01/1959
KING	A			FL		05/08/1966
KING	C	D		WING COMM	Acting	10/06/1948
KING	C			CPL		01/07/1948
KING	E	A		FL		08/11/1945
KING	E	W		LAC		26/04/1951
KING	E			WO		01/08/1997
KING	F	B		SQD LDR		05/01/1950
KING	G	M		SGT		14/01/1954
KING	H	V		FL		27/04/1944
KING	J	B		SGT		13/12/1945
KING	J	C		LAC		07/08/1947
KING	J	D		FL		21/09/1944
KING	J	E		FL		09/05/1946
KING	J	G		SGT		26/06/1947
KING	J	G		SGT		07/02/1952
KING	K	A		FO		25/05/1944
KING	S	V		LAC		13/01/1949
KING	T	B		SQD LDR		12/06/1947
KING	T	S		FL		09/05/1946
KING	W	J		WO		30/09/1948
KINGABY	D	E		WING COMM	Acting	28/02/1946
KINGS	D	L		WING COMM		15/07/1948
KINGS	R	A		SQD LDR	Acting	10/05/1945
KINGSLEY	S	G		WING COMM		27/04/1950
KINGSTON	R	G		SQD LDR		23/02/1950
KINGTON	E	W	G	SGT		21/01/1959
KIRBY	E	E	H	FL		21/06/1961
KIRBY	L			SGT		20/05/1948
KIRBY	M	D		SAC		06/03/1997
KIRBY	W	F		LAC		01/10/1953
KIRBY-JAMES	F	L		SQD LDR		09/11/1944
KIRK	A	E		CPL		20/05/1948
KIRK	A	E		SGT		01/11/1951
KIRK	G	R	N	FL		26/04/1956
KIRK	G			AC2		18/03/1948

Source	Service No	Force	Notes	Surname
N82 1946	84679	RAFVR		KERRIDGE
N87 1948	90470	RAuxAF	OBE	KERSHAW
N87 1948	83010	RAFVR	MC	KERSHAW
N694 1948	91092	RAuxAF		KERSHAW
N33 1960	752051	RAFVR	wef 17/7/44	KETCHER
N197 1946	745442	RAFVR		KETHRO-SCHONBACH
N278 1952	849210	RAuxAF		KETTLE
N796 1953	799902	RAFVR		KETTLEWELL
N740 1947	124962	RAFVR		KEVAN
N187 1947	169591	RAFVR		KEY
N1004 1946	84929	RAFVR	OBE	KEY
N1232 1945	170894	RAFVR		KIBBLE
N527 1953	804428	RAuxAF		KIDD
N517 1950	803388	RAuxAF		KIDD
N775 1960	2603997	RAFVR	wef 12/6/60	KIFF
N415 1946	755256	RAFVR		KILBURN
N381 1945	144792	RAFVR		KILLICK
N653 1943	63783	RAFVR		KILNER
N761 1956	2699020	RAuxAF		KILROY
N926 1950	753169	RAFVR		KILROY
N460 1947	102086	RAFVR	DFC	KILSBY
N460 1947	72179	RAFVR		KIMBER
N573 1948	845044	RAuxAF		KING
N527 1948	122105	RAFVR		KING
N28 1959	162034	RAFVR	wef 6/8/56	KING
MOD 1967	162034	RAFVR	1st Clasp	KING
N456 1948	72290	RAFVR		KING
N527 1948	743618	RAFVR		KING
N1232 1945	117785	RAFVR		KING
N402 1951	743340	RAFVR		KING
YB 2000	n/a	RAuxAF		KING
N1 1950	103487	RAFVR		KING
N18 1954	2607603	RAFVR		KING
N396 1944	62256	RAFVR		KING
N1355 1945	807111	AAF		KING
N662 1947	811220	AAF		KING
N961 1944	76572	RAFVR		KING
N415 1946	86382	RAFVR		KING
N519 1947	811078	AAF		KING
N103 1952	2688000	RAuxAF	BEM; 1st Clasp	KING
N496 1944	146703	RAFVR	DFC	KING
N30 1949	846003	RAuxAF		KING
N478 1947	72188	RAFVR		KING
N415 1946	149935	RAFVR		KING
N805 1948	759167	RAFVR	DFM	KING
N197 1946	112406	RAFVR	DSO,DFM	KINGABY
N573 1948	72222	RAFVR	GM	KINGS
N485 1945	82953	RAFVR		KINGS
N406 1950	72872	RAFVR	MC	KINGSLEY
N179 1950	72309	RAFVR		KINGSTON
N28 1959	2614238	RAFVR	wef 6/11/58	KINGTON
N478 1961	156863	RAFVR	wef 7/12/60	KIRBY
N395 1948	747045	RAFVR		KIRBY
YB 1997	n/a	RAuxAF		KIRBY
N796 1953	849170	RAuxAF		KIRBY
N1150 1944	72683	RAFVR		KIRBY-JAMES
N395 1948	2681505	RAuxAF		KIRK
N1126 1951	2681505	RAuxAF	1st Clasp	KIRK
N318 1956	183150	RAFVR		KIRK
N232 1948	816111	RAuxAF		KIRK

Surname	Initials			Rank	Status	Date
KIRK	R			CPL		23/02/1950
KIRK	R			SGT		01/10/1996
KIRKHAM	E	R		LAC		07/05/1953
KIRKHAM	E			FLT SGT		20/05/1948
KIRKMAN	N	J		FO		12/06/1947
KIRKPATRICK	A			SGT		08/11/1945
KIRKPATRICK	E	R		WING OFF	Acting	17/04/1947
KIRKPATRICK	I			WING COMM		10/12/1942
KIRKPATRICK	J			FL	Acting	13/01/1949
KIRSTEN	S	S		SQD LDR		23/02/1950
KITCHEN	D			SGT		14/09/1950
KITCHENER	E	W		FL		01/12/1955
KITCHENER	H	H		SQD LDR	Acting	10/08/1944
KITCHING	W	A	C	FO		06/09/1951
KITCHINGHAM	L			CPL		14/10/1948
KITSON	D			FLT SGT		26/01/1950
KIY	J			CPL		01/07/1948
KLEBOE	P	A		SQD LDR	Acting	20/04/1944
KNIGHT	A	I		FLT SGT		03/03/1993
KNIGHT	E	W		LAC		18/11/1948
KNIGHT	H	A	S	WO		04/02/1943
KNIGHT	J	E		CPL		01/07/1948
KNIGHT	J			LAC		05/01/1950
KNIGHT	L	W		SGT		10/02/1949
KNIGHT	P	J		CPL		29/07/1998
KNIGHT	P	W		SGT		28/09/1995
KNIGHT	R	C		FL	Acting	07/08/1947
KNIGHT	R	J		FL		18/04/1946
KNIGHT	R	J		LAC		13/07/1950
KNIGHT	R	J		SGT	Temporary	09/12/1959
KNIGHT	R	W		CPL		04/11/1948
KNIGHT	T	E		FL		13/12/1945
KNIGHT	W	H		CPL		17/03/1949
KNIGHTON	J	W		SGT		18/11/1948
KNIGHTS	J	R		CPL	Acting	30/10/1957
KNOCKER	J	B		SQD LDR		15/07/1948
KNOCKER	W	R	A	WING COMM	Acting	05/01/1950
KNOTT	J			FLT SGT		30/03/1943
KNOWLES	G	C		SQD LDR		21/09/1944
KNOWLES	J	C		PO		26/02/1958
KNOWLES	M	H		SAC		01/12/1993
KNOWLES	V	R		FL		30/10/1957
KNOX	J	G		SGT	Temporary	01/12/1955
KNOX	T	G		FL		06/01/1955
KORNDORFFER	J			SQD LDR		28/06/1945
KRAMER	J	I		SAC		19/04/1996
KRAUTER	F			FL		26/10/1944
KUNSELLA	D	G		FL		05/02/1968
KUTASSY	R			WO		16/03/1998
KYTE	D	I		SQD LDR		24/04/1998
LACEY	A	D		FL		15/07/1948
LACEY	D	G		FL		29/06/1944
LACEY	E	R		SQD LDR	Acting	08/11/1945
LACEY	J	H		SQD LDR	Acting	17/02/1944
LACEY	J	H		SQD LDR		02/08/1951
LACEY	N	S		FL		15/06/1944
LACEY	P	A		FLT SGT		17/10/1995
LACEY	T	R		WO		03/07/1952
LACY	F	S		FL		17/04/1947
LACY	V	S		LAC		19/08/1959

Source	Service No	Force	Notes	Surname
N179 1950	866444	RAuxAF		KIRK
YB 1997	n/a	RAuxAF		KIRK
N395 1953	870899	RAuxAF		KIRKHAM
N395 1948	854608	RAuxAF		KIRKHAM
N478 1947	196306	RAFVR		KIRKMAN
N1232 1945	802417	RAFVR		KIRKPATRICK
N289 1947	273	WAAF		KIRKPATRICK
N1680 1942	90184	AAF		KIRKPATRICK
N30 1949	143805	RAFVR	DFC	KIRKPATRICK
N179 1950	73817	RAFVR		KIRSTEN
N926 1950	750259	RAFVR		KITCHEN
N895 1955	107610	RAFVR		KITCHENER
N812 1944	87029	RAFVR	DFM	KITCHENER
N913 1951	137889	RAFVR		KITCHING
N856 1948	808341	RAuxAF		KITCHINGHAM
N94 1950	868023	RAuxAF		KITSON
N527 1948	749728	RAFVR		KIY
N363 1944	88440	RAFVR	DFC,AFC	KLEBOE
YB 1996	n/a	RAuxAF		KNIGHT
N984 1948	868579	RAuxAF		KNIGHT
N131 1943	804051	AAF		KNIGHT
N527 1948	865043	RAuxAF		KNIGHT
N1 1950	841288	RAuxAF		KNIGHT
N123 1949	850253	RAuxAF		KNIGHT
YB 1999	n/a	RAuxAF		KNIGHT
YB 1996	n/a	RAuxAF		KNIGHT
N662 1947	147821	RAFVR		KNIGHT
N358 1946	86563	RAFVR		KNIGHT
N675 1950	819106	RAuxAF		KNIGHT
N843 1959	2605354	RAFVR	wef 23/10/59	KNIGHT
N934 1948	850293	RAuxAF		KNIGHT
N1355 1945	86651	RAFVR		KNIGHT
N231 1949	855951	RAuxAF		KNIGHT
N984 1948	811147	RAuxAF		KNIGHTON
N775 1957	2694067	RAuxAF	wef 16/7/57	KNIGHTS
N573 1948	73617	RAFVR		KNOCKER
N1 1950	74333	RAFVR	OBE	KNOCKER
N1014 1943	808075	AAF		KNOTT
N961 1944	72000	RAFVR		KNOWLES
N149 1958	2603038	RAFVR	wef 28/9/53	KNOWLES
YB 1996	n/a	RAuxAF		KNOWLES
N775 1957	130305	RAFVR	wef 16/4/54	KNOWLES
N895 1955	807197	RAuxAF		KNOX
N1 1955	162008	RAFVR		KNOX
N706 1945	72444	RAFVR	AFC	KORNDORFFER
YB 1997	n/a	RAuxAF		KRAMER
N1094 1944	88420	RAFVR		KRAUTER
MOD 1968	n/a	RAFVR		KUNSELLA
YB 1999	n/a	RAuxAF		KUTASSY
YB 1999	91469	RAuxAF		KYTE
N573 1948	124582	RAFVR		LACEY
N652 1944	76586	RAFVR		LACEY
N1232 1945	134518	RAFVR	DSO	LACEY
N132 1944	60321	RAFVR	DFM	LACEY
N777 1951	60321	RAFVR	DFM; 1st Clasp	LACEY
N596 1944	72997	RAFVR		LACEY
YB 1997	n/a	RAuxAF		LACEY
N513 1952	846050	RAuxAF		LACEY
N289 1947	134186	RAFVR		LACY
N552 1959	2697577	RAFVR	wef 7/6/59	LACY

Surname	Initials			Rank	Status	Date
LADD	P	H		FLT SGT		12/04/1945
LAIRD	B	C		FL		01/06/1997
LAIRD	D	K		FL		10/04/1952
LAIRD	P	E		FL		06/03/1947
LAKE	A	A	J	FL		08/07/1959
LAKE	E	G	A	FO		03/06/1959
LAKE	G	W	C	FL	Acting	06/03/1947
LAKE	J			FO		29/04/1954
LAKE	M	B	C	SQD LDR		18/11/1948
LAMB	A	A		FL		27/11/1952
LAMB	A	M		WO		28/02/1946
LAMB	D	A		FL		14/10/1948
LAMB	E	M		SEC OFF		07/08/1947
LAMB	G	B		FO		20/04/1944
LAMB	G	E	S	WING COMM		20/07/1960
LAMB	G	M		SEC OFF		07/08/1947
LAMB	P	G		WING COMM	Acting	08/02/1945
LAMB	P	S		FL		20/01/1960
LAMB	R	L		FL		07/01/1943
LAMB	R	L		SQD LDR		23/08/1951
LAMB	W	H	F	FLT SGT		18/02/1954
LAMBERT	A	D		SQD LDR	Acting	20/09/1945
LAMBERT	C	W		SGT		18/03/1948
LAMBERT	D	S		FLT OFF		06/01/1955
LAMBLE	C	E		FL		14/01/1954
LAMBOURNE	A	E		WO		27/03/1947
LAMBOURNE	A	J		CPL		03/07/1952
LAMBSON	O	I	H	FLT OFF	Acting	04/09/1947
LAMPLOUGH	J	S		SQD LDR	Acting	18/04/1946
LANCASTER	E	R		WO		13/07/1950
LANCASTER	E			FL		19/09/1956
LANCASTER	I			PO		12/02/1953
LANCASTER	I			FO		28/02/1962
LAND	C	D		FL		14/01/1954
LAND	F	G		SQD LDR	Acting	12/04/1945
LAND	G			WO		31/05/1945
LAND	J			SGT		28/09/1989
LAND	W	A		FL		24/01/1946
LANDLESS	J	K		FL		20/01/1960
LANDRIDGE	J			CPL		13/01/1949
LANE	A			SGT		24/11/1949
LANE	N	J		SGT		14/10/1948
LANE	R	O		SQD LDR	Acting	13/12/1945
LANE	T	C		CPL		21/09/1995
LANE	W	F		SQD LDR	Acting	24/01/1946
LANG	H			FL		06/01/1955
LANGFORD	E	G		FL		02/04/1953
LANGFORD	H	C		SQD LDR	Acting	31/05/1945
LANGHAM-HOBART	N	C		SQD LDR	Acting	17/04/1947
LANGHORN	E	J	B	SQD LDR	Acting	28/11/1946
LANGHORNE	R	A		FL		23/02/1950
LANGLANDS	C	W		FL		23/02/1950
LANGLEY	W	L		CPL		27/03/1947
LANGMEAD	P			FL		06/03/1947
LANGSTON	S	A	D	FL		05/10/1944
LANGSTON-JONES	P	G		FL		27/11/1997
LANGTON	A	J		FLT SGT		21/10/1959
LANGTON	F			LAC		01/07/1948
LANSDOWNE	A	F		FL		07/04/1955
LANSER	E	H		PO		17/12/1942

Source	Service No	Force	Notes	Surname
N381 1945	812049	AAF		LADD
YB 1999	4230660	RAuxAF		LAIRD
N278 1952	180233	RAFVR		LAIRD
N187 1947	158126	RAFVR	DFC	LAIRD
N454 1959	181005	RAFVR	wef 29/11/52	LAKE
N368 1959	2614503	RAFVR	wef 11/11/58	LAKE
N187 1947	87075	RAFVR		LAKE
N329 1954	184314	RAFVR		LAKE
N984 1948	90739	RAuxAF		LAKE
N915 1952	145223	RAFVR		LAMB
N197 1946	817206	AAF		LAMB
N856 1948	91197	RAuxAF		LAMB
N662 1947	2852	WAAF		LAMB
N363 1944	121332	RAFVR		LAMB
N531 1960	90575	AAF	wef 10/5/44	LAMB
N662 1947	6159	WAAF		LAMB
N139 1945	90349	AAF	AFC	LAMB
N33 1960	196305	RAFVR	wef 14/1/53	LAMB
N3 1943	82718	RAFVR		LAMB
N866 1951	82718	RAFVR	1st Clasp	LAMB
N137 1954	2696017	RAuxAF		LAMB
N1036 1945	80832	RAFVR	DFC	LAMBERT
N232 1948	757021	RAFVR		LAMBERT
N1 1955	6160	WRAAF	MBE	LAMBERT
N18 1954	91195	RAuxAF		LAMBLE
N250 1947	760153	RAFVR		LAMBOURNE
N513 1952	770629	RAFVR		LAMBOURNE
N740 1947	2453	WAAF		LAMBSON
N358 1946	60308	RAFVR		LAMPLOUGH
N675 1950	751565	RAFVR	DFC	LANCASTER
N666 1956	138294	RAFVR		LANCASTER
N115 1953	2602510	RAFVR		LANCASTER
N157 1962	2602510	RAFVR	1st Clasp; wef 7/6/61	LANCASTER
N18 1954	106333	RAFVR		LAND
N381 1945	85686	RAFVR		LAND
N573 1945	741486	RAFVR		LAND
YB 2001	n/a	RAuxAF		LAND
N82 1946	128561	RAFVR		LAND
N33 1960	202751	RAFVR	wef 11/12/52	LANDLESS
N30 1949	812320	RAuxAF		LANDRIDGE
N1193 1949	770278	RAFVR		LANE
N856 1948	840805	RAuxAF		LANE
N1355 1945	82701	RAFVR	DFC	LANE
YB 1996	n/a	RAuxAF		LANE
N82 1946	86652	RAFVR		LANE
N1 1955	85300	RAuxAF		LANG
N298 1953	81955	RAFVR		LANGFORD
N573 1945	86412	RAFVR		LANGFORD
N289 1947	77792	RAFVR		LANGHAM-HOBART
N1004 1946	91169	AAF	MBE	LANGHORN
N179 1950	172884	RAFVR		LANGHORNE
N179 1950	73267	RAFVR		LANGLANDS
N250 1947	860395	AAF		LANGLEY
N187 1947	86629	RAFVR		LANGMEAD
N1007 1944	114119	RAFVR		LANGSTON
YB 1998	4275864	RAuxAF		LANGSTON-JONES
N708 1959	2676429	RAuxAF	wef 29/9/58	LANGTON
N527 1948	857356	RAuxAF		LANGTON
N277 1955	101763	RAFVR		LANSDOWNE
N1708 1942	117251	RAFVR		LANSER

Surname	Initials			Rank	Status	Date
LANSER	E	H		FL		17/01/1952
LAPHAM	R	D		SGT		21/10/1984
LAPHAM	R	D		SGT		21/10/1994
LAPIDGE	E			LAC		03/07/1952
LAPRAIK	D	F		SQD LDR		27/04/1950
LARBALESTIER	B	D		FL		10/12/1942
LARKIN	W	F		CPL		23/06/1955
LARKING	R	A		SGT		10/06/1948
LARMAN	L	G		SQD LDR	Acting	16/10/1947
LASH	E	P		SQD LDR		03/12/1942
LASH	E	P		FL		19/07/1951
LASH	E	P		FL		19/07/1951
LASH	E	P		FL		19/07/1951
LAUDERDALE	V	F		SQD LDR	Acting	06/03/1947
LAUGHTON	A	O	J	SGT		10/06/1948
LAUGHTON	A	O	J	FLT SGT		06/01/1955
LAUGHTON	F	M		FL		13/02/1957
LAUNDER	W	A		FO		07/05/1993
LAUNDON	H	J		SGT		12/06/1947
LAURENCE	A	W		WO		06/03/1947
LAURENCE	W	A		FL	Acting	27/03/1947
LAVELL	R	G		FL		20/01/1960
LAVERS	C	A		SGT	Temporary	25/05/1960
LAVERY	P	J		SGT		01/05/1963
LAVINGTON	G	W		SQD LDR		30/12/1943
LAW	B			CPL	Temporary	19/01/1956
LAW	J	C	S	CPL		20/05/1948
LAW	J	E	F	FL		08/02/1945
LAW	J	P		LAC		14/10/1943
LAW	R	C	E	SQD LDR		13/11/1947
LAW	W			CPL		24/07/1952
LAWDHAM	L	C		FL		05/01/1950
LAWLEDGE	E	C		SQD LDR		05/01/1950
LAWLER	E	S		FO		09/08/1945
LAWLER	R			CPL		05/06/1947
LAWMAN	I	T		FL		01/10/1953
LAWRENCE	E	E		FL		26/04/1961
LAWRENCE	G	H		PO		21/03/1946
LAWRENCE	I	H		SQD LDR		06/03/1947
LAWRENCE	J	K		SQD LDR		30/03/1944
LAWRENCE	J	T		FL		09/08/1945
LAWS	E	S	A	SGT		07/02/1952
LAWSON	D	F		SGT		20/05/1948
LAWSON	I	D	N	WING COMM		04/10/1945
LAWSON	K	J		WING COMM	Acting	17/01/1952
LAWSON	R	D		FO		18/07/1956
LAWTON	P	C	F	WING COMM	Acting	15/04/1943
LAW-WRIGHT	H			WING COMM	Acting	21/09/1944
LAYTON	K			SGT		16/09/1993
LAYTON	P	H		SQD LDR	Acting	15/01/1953
Le BESQUE	E	J	R	SGT		08/07/1959
Le MAY	W	K		WING COMM		22/04/1943
Le MAY	W	K		GRP CAP		08/03/1956
Le SUEUR	F	W		FO		22/04/1943
LEA	B	A		FL		07/01/1954
LEA	R	F	G	SQD LDR	Acting	04/02/1943
LEACH	A	L		FLT SGT		05/09/1946
LEACH	A	M	M	FLT OFF	Acting	05/06/1947
LEACH	C	G		SGT		23/02/1950
LEACH	H	C		SGT	Acting	07/08/1947

Source	Service No	Force	Notes	Surname
N35 1952	117251	RAuxAF	1st Clasp	LANSER
YB 1996	n/a	RAuxAF		LAPHAM
YB 1996	n/a	RAuxAF	1st Clasp	LAPHAM
N513 1952	868699	RAuxAF		LAPIDGE
N406 1950	73496	RAFVR	DFC	LAPRAIK
N1680 1942	84706	RAFVR		LARBALESTIER
N473 1955	2683037	RAuxAF		LARKIN
N456 1948	749665	RAFVR		LARKING
N846 1947	81945	RAFVR		LARMAN
N1638 1942	70383	RAFVR		LASH
N724 1951	70383	RAFVR	AFC; 1st Clasp	LASH
N724 1951	70383	RAFVR	AFC; 2nd Clasp	LASH
N724 1951	70383	RAFVR	AFC; 3rd Clasp	LASH
N187 1947	67045	RAFVR		LAUDERDALE
N456 1948	804321	RAuxAF		LAUGHTON
N1 1955	2684513	RAuxAF	1st Clasp	LAUGHTON
N104 1957	140713	RAFVR	wef 15/1/56	LAUGHTON
YB 1996	2631680	RAuxAF		LAUNDER
N478 1947	801466	AAF		LAUNDON
N187 1947	811034	AAF		LAURENCE
N250 1947	106792	RAFVR		LAURENCE
N33 1960	198284	RAFVR	wef 10/3/53	LAVELL
N375 1960	801389	AAF	wef 28/11/42	LAVERS
N333 1963	2650970	RAuxAF	wef 3/3/63	LAVERY
N1350 1943	70385	RAFVR		LAVINGTON
N43 1956	869337	RAuxAF		LAW
N395 1948	803557	RAuxAF		LAW
N139 1945	108858	RAFVR		LAW
N1065 1943	808097	AAF		LAW
N933 1947	72742	RAFVR	DSO,DFC	LAW
N569 1952	2683001	RAuxAF		LAW
N1 1950	88488	RAFVR		LAWDHAM
N1 1950	90867	RAuxAF		LAWLEDGE
N874 1945	182840	RAFVR		LAWLER
N460 1947	811193	AAF		LAWLER
N796 1953	152593	RAFVR		LAWMAN
N339 1961	193388	RAFVR	wef 3/1/54	LAWRENCE
N263 1946	201847	RAFVR		LAWRENCE
N187 1947	90771	AAF		LAWRENCE
N281 1944	70388	RAFVR		LAWRENCE
N874 1945	104428	RAFVR	AFC	LAWRENCE
N103 1952	2676441	RAFVR		LAWS
N395 1948	752154	RAFVR		LAWSON
N1089 1945	86354	RAFVR	DFC	LAWSON
N35 1952	82728	RAFVR	DSO,DFC; Deceased	LAWSON
N513 1956	2603417	RAFVR		LAWSON
N454 1943	90217	AAF	DFC	LAWTON
N961 1944	79741	RAFVR	DSO,DFC	LAW-WRIGHT
YB 1996	n/a	RAuxAF		LAYTON
N28 1953	76258	RAFVR		LAYTON
N454 1959	2684570	RAuxAF	wef 5/8/58	Le BESQUE
N475 1943	90002	AAF	OBE	Le MAY
N187 1956	90002	RAF	CBE; 1st Clasp	Le MAY
N475 1943	84974	RAFVR		Le SUEUR
MOD 1964	159467	RAFVR		LEA
N131 1943	90116	AAF	OBE	LEA
N751 1946	801396	AAF		LEACH
N460 1947	3487	WAAF		LEACH
N179 1950	755056	RAFVR		LEACH
N662 1947	841534	AAF		LEACH

Surname	Initials			Rank	Status	Date
LEACH	J	B		FL		30/03/1960
LEACH	J	M		FL		05/10/1944
LEACH	K	P		FL		31/10/1956
LEADER	G	H		FLT SGT		11/01/1945
LEAN	G			FO		11/01/1945
LEATHARD	A	C	S	FL		19/01/1956
LEATHEM	E	G	C	FL		28/02/1946
LEATHER	W	J		WING COMM		06/05/1943
LEAVER-POWER	J	A	F	SQD LDR	Acting	24/01/1946
LeCHEMINANT	J			FL		08/11/1945
LECKEY	A			SAC		07/12/1960
LECKIE	J	M		SGT		19/10/1960
LECKIE	T			SGT		03/07/1957
LEDGER	E	C		FLT SGT	Acting	14/10/1948
LEDGER	R	I		SGT		07/12/1960
LEE	C	P		SQD LDR		27/09/1998
LEE	C			SGT		17/01/1952
LEE	D	N		FLT SGT		15/01/1953
LEE	E	H		FL		01/07/1948
LEE	F	G		AC2		13/11/1947
LEE	F	W		SQD LDR		26/01/1950
LEE	K	N	T	SQD LDR	Acting	04/10/1945
LEE	K	T		CPL		14/10/1948
LEE	L	C		FL		28/09/1950
LEE	S	V		AC1		05/06/1947
LEECH	R	F		WO		21/03/1946
LEEK	F	J		LAC		22/04/1943
LEES	A	F	Y	FL		06/03/1947
LEES	A	H		FO		28/02/1946
LEES	A	J		SGT		30/11/1944
LEES	H	L		SGT		13/11/1947
LEES	S	W		LAC		15/09/1949
LEESON	L	G	D	FL		31/10/1956
LEETHAM	L	A		WO		18/04/1946
LEGG	G	A		AC1		13/01/1949
LEGGATT	A	E		CPL		01/05/1952
LEGGE	A	M		FL		24/01/1946
LEGGETT	P	G		FL		08/05/1947
LEGGE-WILLIS	J	W	A	SQD LDR	Acting	08/05/1947
LEIGH	A	C		SQD LDR		01/07/1948
LEIGHTON	A			SGT		05/06/1947
LEIGHTON	J			FO		24/01/1946
LEISHMAN	J			LAC		13/01/1949
LEITHEAD	S			FLT SGT		28/11/1990
LELLIOTT	J	F		FO		08/05/1947
LELLIOTT	J	F		FO		15/01/1953
LELLIOTT	J	F		FL		19/10/1960
LENDON	W	W		FL		27/04/1950
LENG	M	E		FL		08/11/1945
LENNARD	W	A		FL		15/07/1948
LEONARD	R	A		FO		06/03/1947
LERCHE	S			FL		03/07/1957
LERWAY	F	T		FI		28/02/1946
LESLIE	K	A	S	SQD LDR	Acting	12/08/1943
LEST	E	W		FL		08/05/1947
L'ESTRANGE	C	H	N	WING COMM	Acting	18/04/1946
LETCH	A	G	W	LAC		06/01/1955
LETFORD	K	H	F	FL		14/01/1954
LETZER	F	J		WING COMM		05/10/1944
LEVENTHORPE	T	B		FL		17/03/1949

Source	Service No	Force	Notes	Surname
N229 1960	134966	RAFVR	wef 13/2/60	LEACH
N1007 1944	n/a	EX-AAF		LEACH
N761 1956	155174	RAuxAF	DFM	LEACH
N14 1945	856596	AAF		LEADER
N14 1945	145517	RAFVR		LEAN
N43 1956	131183	RAFVR		LEATHARD
N197 1946	78087	RAFVR	DFC	LEATHEM
N526 1943	90355	AAF	DFC	LEATHER
N82 1946	88725	RAFVR	DFC	LEAVER-POWER
N1232 1945	126148	RAFVR	DFC	LeCHEMINANT
N933 1960	2650853	RAuxAF	wef 19/10/60	LECKEY
N775 1960	A.34834	RAAF	wef 22/6/44	LECKIE
N456 1957	2677405	RAuxAF	wef 7/6/54	LECKIE
N856 1948	761035	RAFVR		LEDGER
N933 1960	2691089	RAuxAF	wef 8/10/60	LEDGER
YB 1999	91445	RAuxAF		LEE
N35 1952	808261	RAuxAF		LEE
N28 1953	748633	RAFVR		LEE
N527 1948	90954	RAuxAF		LEE
N933 1947	844707	AAF		LEE
N94 1950	72693	RAFVR		LEE
N1089 1945	72998	RAFVR	DFC	LEE
N856 1948	869281	RAuxAF		LEE
N968 1950	117366	RAFVR		LEE
N460 1947	819096	AAF		LEE
N263 1946	746892	RAFVR		LEECH
N475 1943	805347	AAF		LEEK
N187 1947	81681	RAFVR	DFC	LEES
N197 1946	177360	RAFVR		LEES
N1225 1944	805244	AAF		LEES
N933 1947	815071	AAF		LEES
N953 1949	805460	RAuxAF		LEES
N761 1956	106793	RAFVR		LEESON
N358 1946	751474	RAFVR		LEETHAM
N30 1949	801425	RAuxAF		LEGG
N336 1952	770628	RAFVR		LEGGATT
N82 1946	122760	RAFVR		LEGGE
N358 1947	86329	RAFVR		LEGGETT
N358 1947	72876	RAFVR		LEGGE-WILLIS
N527 1948	111975	RAFVR	DFC,DFM	LEIGH
N460 1947	771035	RAFVR		LEIGHTON
N82 1946	196265	RAFVR		LEIGHTON
N30 1949	874766	RAuxAF		LEISHMAN
YB 1996	n/a	RAuxAF		LEITHEAD
N358 1947	157814	RAFVR		LELLIOTT
N28 1953	157814	RAFVR	1st Clasp	LELLIOTT
N775 1960	157814	RAFVR	2nd Clasp; wef 28/6/60	LELLIOTT
N406 1950	72877	RAFVR		LENDON
N1232 1945	67035	RAFVR		LENG
N573 1948	124735	RAFVR	DFC	LENNARD
N187 1947	134456	RAFVR		LEONARD
N456 1957	176225	RAuxAF	wef 19/5/57	LERCHE
N197 1946	142495	RAFVR		LERWAY
N850 1943	60290	RAFVR		LESLIE
N358 1947	116936	RAFVR		LEST
N358 1946	70394	RAFVR		L'ESTRANGE
N1 1955	846279	RAuxAF		LETCH
N18 1954	132892	RAFVR	DSO,DFC	LETFORD
N1007 1944	70396	RAFVR		LETZER
N231 1949	65070	RAFVR		LEVENTHORPE

Surname	Initials			Rank	Status	Date
LEVER	J			SGT		05/06/1947
LEVIEN	E	D		WING COMM	Acting	03/10/1946
LEVITT	F			FL		08/11/1945
LEVY	D			FO		21/10/1959
LEWARNE	B	R		SQD LDR	Acting	19/09/1956
LEWINGTON	E	G		CPL		22/02/1951
LEWIS	A	D		FL		24/07/1952
LEWIS	A	D		FL		08/04/2000
LEWIS	A	G		LAC		27/03/1947
LEWIS	C	H		WING COMM	Acting	03/12/1942
LEWIS	C	S		FLT SGT		30/03/1944
LEWIS	D	L		FL		21/03/1946
LEWIS	D	M		FL	Acting	01/07/1948
LEWIS	E	W		LAC		02/08/1951
LEWIS	F	R		FL		26/08/1948
LEWIS	H	E		SQD LDR		06/03/1947
LEWIS	H	G		SGT		22/02/1961
LEWIS	L	G		FL		26/04/1951
LEWIS	L	N		LAC		18/11/1948
LEWIS	M	E		WO		17/03/1949
LEWIS	P	E		FO		20/05/1948
LEWIS	R	E	M	FL	Acting	20/09/1961
LEWIS	S			AC1		10/06/1948
LEWIS	T	J		SQD LDR	Acting	12/06/1947
LEWTON	M			WO		07/12/1950
LEYTON	R	N	A	SQD LDR		03/07/1952
LIARDET	V	H		FL		27/11/1952
LIDBURY	R	E	E	FO		24/01/1946
LIDDELL	F	J		SGT		02/04/1953
LIDDELL	F	J		SGT		21/10/1959
LIDDELL	G	H		SGT		20/05/1948
LIDDLE	A	S		SGT		04/02/1943
LIDDLE	I			SGT		09/12/1959
LIGHT	F			FLT SGT		08/03/1951
LIGHTBODY	F	E	H	FL		18/07/1956
LIGHTBODY	H	A		FL		27/07/1944
LIGHTERWOOD	J	H		SQD LDR		20/05/1948
LIGHTERWOOD	J	H		FL		13/10/1955
LIGHTFOOT	F			AC1		30/03/1960
LIGHTFOOT	G	L	S	GRP CAP		06/03/1947
LIMBREY	D	H		FL		28/02/1946
LIMPENNY	E	R		FO		30/08/1945
LIMPITLAW	A			FO		18/02/1954
LINCOLN	M	J		FLT OFF		28/07/1955
LINCOLN	P	L		AIR COMMD		04/12/1947
LINDLEY	J	T		CPL	Temporary	28/07/1955
LINDLEY	R			FL		21/03/1946
LINDSAY	A	W		SQD LDR		30/03/1943
LINDSAY	D			SQD LDR	Acting	26/01/1950
LINDSAY	J	S		PO		21/06/1945
LINES	G	J		SGT		04/10/1945
LININGTON	K	C	G	FL		30/03/1960
LINKLATER	J	A		CPL		19/10/1960
LINKS	J	G		WING COMM		07/08/1947
LINLEY	W			FL		24/01/1946
LINN	R	S		FL		25/03/1954
LINNECAR	D	R		FL		18/04/1946
LINNELL	R	W		FL		20/01/1960
LINTON	J	B		FL		15/06/1944
LIPSKI	J	P		SAC		28/09/1995

Source	Service No	Force	Notes	Surname
N460 1947	811117	AAF		LEVER
N837 1946	79572	RAFVR	DSO,DFC	LEVIEN
N1232 1945	136170	RAFVR		LEVITT
N708 1959	2605032	RAFVR	wef 17/5/53	LEVY
N666 1956	135978	RAuxAF	MBE	LEWARNE
N194 1951	862556	RAuxAF		LEWINGTON
N569 1952	142483	RAFVR		LEWIS
YB 2002	212485	RAuxAF		LEWIS
N250 1947	845501	AAF		LEWIS
N1638 1942	70397	RAFVR		LEWIS
N281 1944	801451	AAF		LEWIS
N263 1946	121473	RAFVR	DFM	LEWIS
N527 1948	159741	RAFVR		LEWIS
N777 1951	859590	RAuxAF		LEWIS
N694 1948	106440	RAFVR		LEWIS
N187 1947	72533	RAFVR		LEWIS
N153 1961	746221	RAFVR	wef 4/7/44	LEWIS
N402 1951	115288	RAFVR	DFC,DFM	LEWIS
N984 1948	743979	RAFVR		LEWIS
N231 1949	881746	WAAF		LEWIS
N395 1948	117920	RAFVR		LEWIS
N720 1961	113133	RAFVR(T)	wef 5/7/60	LEWIS
N456 1948	756753	RAFVR		LEWIS
N478 1947	87469	RAFVR		LEWIS
N1210 1950	755261	RAFVR		LEWTON
N513 1952	137101	RAuxAF	MRCS,LRCP	LEYTON
N915 1952	199389	RAFVR		LIARDET
N82 1946	179232	RAFVR		LIDBURY
N298 1953	866343	RAuxAF		LIDDELL
N708 1959	2699513	RAuxAF	1st Clasp; wef 23/11/56	LIDDELL
N395 1948	807198	RAuxAF		LIDDELL
N131 1943	803298	AAF		LIDDLE
N843 1959	2659272	WRAAF	wef 25/8/59	LIDDLE
N244 1951	799889	RAFVR		LIGHT
N513 1956	153059	RAuxAF	DFC,MD,ChB	LIGHTBODY
N763 1944	116893	RAFVR		LIGHTBODY
N395 1948	85207	RAFVR		LIGHTERWOOD
N762 1955	85207	RAFVR	1st Clasp	LIGHTERWOOD
N229 1960	840373	AAF	wef 28/3/44	LIGHTFOOT
N187 1947	72880	RAFVR	OBE	LIGHTFOOT
N197 1946	102579	RAFVR		LIMBREY
N961 1945	189635	RAFVR		LIMPENNY
N137 1954	204238	RAFVR		LIMPITLAW
N565 1955	5247	WRAAF		LINCOLN
N1007 1947	90416	AAF	CB,DSO,MC	LINCOLN
N565 1955	770346	RAFVR		LINDLEY
N263 1946	116105	RAFVR		LINDLEY
N1014 1943	70401	RAFVR		LINDSAY
N94 1950	72881	RAFVR		LINDSAY
N671 1945	189154	RAFVR		LINDSAY
N1089 1945	805028	AAF		LINES
N229 1960	204467	RAFVR	wef 25/8/56	LININGTON
N775 1960	2691075	RAuxAF	wef 24/7/60	LINKLATER
N662 1947	90439	AAF	OBE	LINKS
N82 1946	88031	RAFVR		LINLEY
N249 1954	170519	RAuxAF		LINN
N358 1946	122813	RAFVR		LINNECAR
N33 1960	146729	RAFVR	wef 6/12/59	LINNELL
N596 1944	81651	RAFVR		LINTON
YB 1996	n/a	RAuxAF		LIPSKI

Surname	Initials			Rank	Status	Date
LISHMAN	J	A		FL		11/01/1945
LISSNER	M			FL		09/08/1945
LISTER	A	F		SGT		20/05/1948
LISTER	D	S		FO		20/09/1945
LISTER	H			SGT	Temporary	26/04/1961
LISTER	J	H		SGT		21/01/1959
LITCHFIELD	H	G		FO		15/04/1943
LITHGOW	H			SGT		09/12/1959
LITTLE	B	W		SQD LDR		07/06/1945
LITTLE	I	M	D	SQD LDR		09/08/1945
LITTLE	J	H		WING COMM		03/12/1942
LITTLE	R	A		WING COMM		01/07/1948
LITTLE	T	W		SQD LDR		05/06/1947
LITTLEFAIR	J	R		FL		12/04/1951
LITTLEWOOD	W	W		CPL		26/08/1948
LIVINGSTONE-BUSSELL	D			FL		11/01/1945
LIZIERI	R	H		FL	Acting	04/09/1947
LLEWELLYN	D	J		SGT		16/03/1950
LLOYD	A	H		AC2		18/03/1948
LLOYD	D			FL		28/02/1946
LLOYD	H	H		FL		30/09/1948
LLOYD	J	E		WING COMM		05/02/1948
LLOYD	M	M		WO		07/07/1949
LLOYD	R	M		SGT		07/07/1949
LLOYD	S			LAC		15/07/1948
LLOYDS	W	H	R	SQD LDR		24/11/1949
LOACH	J	D		PO		30/08/1945
LOCK	A	J		SAC		18/10/1995
LOCK	R	J		WO		28/02/1946
LOCKHART	S	J		SAC		17/03/1999
LOCKHEAD	R	N		WING COMM	Acting	15/09/1949
LOCKSTONE	R	H		FLT SGT	Acting	16/03/1950
LOCKWOOD	E	B	R	SQD LDR	Acting	24/12/1942
LOCKWOOD	E	B	R	FL		06/09/1951
LOCKWOOD	E	B	R	WING COMM		26/04/1956
LOCKYER	C	H		FL		03/10/1946
LODGE	J	D		FL		28/02/1946
LODGE	W	C	A	SQD LDR	Acting	28/02/1946
LOE	H			FL		03/06/1959
LOFTHOUSE	C	W		FL		03/12/1942
LOFTS	K	T		WING COMM	Acting	30/08/1945
LOGAN	P	V		SQD LDR		09/11/1997
LOGSDAIL	J			FL		05/10/1944
LOLE	W	A		SAC		02/07/1958
LOMAS	C	L		FL		15/07/1948
LOMAX	V	A	J	D	FL		05/02/1948
LOMER	H	W		CPL		15/07/1948
LOMER	R	J		CPL		26/01/1950
LONDESBOROUGH	A			FO		22/06/1995
LONG	C	N		SGT		04/09/1947
LONG	J			LAC		24/11/1949
LONG	N	H	C	CPL		23/02/1950
LONG	S	W	J	SGT		23/02/1950
LONG	I	J		SQD LDR	Acting	06/03/1947
LONGHURST	W	M		CPL		06/01/1955
LONGLEY	J	M		FL		03/12/1942
LONGSHAW	C			FL	Acting	16/03/1950
LONGSTAFF	W	M		FL	Acting	26/08/1948
LONGVILLE	H			FL		06/01/1955
LONSDALE	A			FL		13/02/1957

Source	Service No	Force	Notes	Surname
N14 1945	105995	RAFVR		LISHMAN
N874 1945	86413	RAFVR	DFC	LISSNER
N395 1948	804223	RAuxAF		LISTER
N1036 1945	171951	RAFVR		LISTER
N339 1961	855922	AAF	wef 8/6/44	LISTER
N28 1959	2655029	WRAAF	wef 11/11/58	LISTER
N454 1943	89978	RAFVR		LITCHFIELD
N843 1959	2699523	RAuxAF	wef 29/8/55	LITHGOW
N608 1945	90326	AAF		LITTLE
N874 1945	72484	RAFVR	AFC	LITTLE
N1638 1942	90125	AAF	DFC	LITTLE
N527 1948	72142	RAFVR	OBE	LITTLE
N460 1947	73273	RAFVR		LITTLE
N342 1951	125535	RAFVR		LITTLEFAIR
N694 1948	869526	RAuxAF		LITTLEWOOD
N14 1945	72099	RAFVR		LIVINGSTONE-BUSSELL
N740 1947	155690	RAFVR		LIZIERI
N257 1950	818173	RAuxAF		LLEWELLYN
N232 1948	816072	RAuxAF		LLOYD
N197 1946	130461	RAFVR		LLOYD
N805 1948	174028	RAFVR	DFC	LLOYD
N87 1948	91123	RAuxAF	OBE	LLOYD
N673 1949	881144	WAAF		LLOYD
N673 1949	881143	WAAF		LLOYD
N573 1948	848381	RAuxAF		LLOYD
N1193 1949	90688	RAuxAF		LLOYDS
N961 1945	196540	RAFVR		LOACH
YB 2001	n/a	RAuxAF		LOCK
N197 1946	748445	RAFVR	AFC	LOCK
YB 2000	n/a	RAuxAF		LOCKHART
N953 1949	90750	RAuxAF	MC	LOCKHEAD
N257 1950	746243	RAFVR		LOCKSTONE
N1758 1942	89584	RAFVR	MBE	LOCKWOOD
N913 1951	89584	RAFVR	DSO,MBE; 1st Clasp	LOCKWOOD
N318 1956	89584	RAuxAF	DSO;MBE; 2nd Clasp	LOCKWOOD
N837 1946	135483	RAFVR	DFC	LOCKYER
N197 1946	84011	RAFVR		LODGE
N197 1946	77038	RAFVR	DFC	LODGE
N368 1959	156249	RAFVR	wef 14/3/54	LOE
N1638 1942	70405	RAFVR		LOFTHOUSE
N961 1945	90483	AAF	DFC	LOFTS
YB 1998	n/a	RAuxAF		LOGAN
N1007 1944	76587	RAFVR		LOGSDAIL
N452 1958	2689552	RAFVR	wef 21/3/58	LOLE
N573 1948	90559	RAuxAF		LOMAS
N87 1948	91187	RAuxAF		LOMAX
N573 1948	752618	RAFVR		LOMER
N94 1950	752617	RAFVR		LOMER
YB 1996	4232242	RAuxAF		LONDESBOROUGH
N740 1947	813236	AAF		LONG
N1193 1949	860431	RAuxAF	MM	LONG
N179 1950	849951	RAuxAF		LONG
N179 1950	812021	RAuxAF		LONG
N187 1947	133548	RAFVR	DFC	LONG
N1 1955	840862	RAuxAF		LONGHURST
N1638 1942	70407	RAFVR		LONGLEY
N257 1950	138981	RAFVR		LONGSHAW
N694 1948	148010	RAFVR		LONGSTAFF
N1 1955	163420	RAFVR		LONGVILLE
N104 1957	201713	RAFVR	wef 8/5/56	LONSDALE

Surname	Initials			Rank	Status	Date
LONSDALE	E			LAC		20/05/1948
LONSDALE	P	S		CPL		26/08/1948
LONSDALE	R	W		FL		15/01/1953
LOOKER	D	J		FL		07/07/1949
LOOSEMORE	A	I		CPL	Temporary	07/12/1960
LORD	R	C		FO		07/06/1945
LORD	W	A		FO		17/04/1947
LORDEN	P	E	C	CPL		15/09/1949
LORIMER	J	T		FO		16/01/1957
LOTE	E	A		FLT SGT	Temporary	03/06/1959
LOTON	A	G		WING COMM		13/01/1944
LOUCH	H	M		FL	Acting	03/12/1942
LOUDEN	A	L		FL		01/07/1948
LOUDEN	H			FO		19/08/1954
LOUGHAN	J	J		SQD LDR	Acting	10/06/1948
LOUGHERY	A	J		SGT		26/06/1947
LOUGHRAN	J	E		WO		27/04/1944
LOUNSBURY	L	I		SEC OFF		05/06/1947
LOVE	H	A		SQD LDR		30/03/1943
LOVE	P	C		PO		25/03/1954
LOVEITT	R	H		FL	Acting	21/10/1959
LOVEJOY	E	W		FL		05/09/1946
LOVELL	H	C	E	LAC		15/07/1948
LOVELL	L	C		SQD LDR	Acting	06/03/1947
LOVERIDGE	L	S		FL		11/07/1946
LOVERIDGE	L	S		FL		19/08/1954
LOVICK	J	R		LAC		19/08/1954
LOWDEN	G	D		FL		19/01/1956
LOWE	B	J		FL		26/11/1953
LOWE	C			LAC		03/07/1952
LOWE	D	J		FO		16/01/1957
LOWE	F			FL		20/09/1945
LOWE	J	P		SGT	Temporary	26/04/1961
LOWE	L	R		CPL	Temporary	25/07/1962
LOWE	R	E		FL		27/04/1944
LOWERY	J	H		LAC		10/06/1948
LOWERY	J	H		CPL		07/04/1955
LOWNDES	A	P		FL		15/09/1949
LOWNDES	J	K		FL		01/07/1948
LOWRY	J	H		FO		09/05/1946
LOWRY	N	F		FL		21/09/1944
LOWRY	W			SQD LDR	Acting	04/11/1943
LOWTH	F	W		PO		30/11/1944
LOXTON	W	S		FL		06/03/1947
LUCAS	A	J		SGT		14/09/1950
LUCAS	A			LAC		03/07/1957
LUCAS	S	E		FO		08/11/1945
LUCIE-SMITH	H	J		SQD LDR		17/04/1947
LUCK	H	J		FLT SGT		12/02/1953
LUCK	J	F		FO		19/08/1954
LUCKETT	A	P		LAC		30/09/1948
LUDBROOK	L			SGT		11/11/1943
LUDMAN	R	E		FO		30/11/1944
LUGG	W	J		FLT SGT		14/10/1948
LUMLEY	G	E		PO		15/06/1944
LUND-LINTON	W	M		FLT OFF		28/11/1956
LUNN	A	F		FL		21/03/1946
LUNN	W	E	J	FL		03/12/1942
LUNSON	E	A		SQD LDR		14/09/1950
LUNT	C	H		FLT SGT		13/01/1949

Source	Service No	Force	Notes	Surname
N395 1948	841303	RAuxAF		LONSDALE
N694 1948	868581	RAuxAF		LONSDALE
N28 1953	127169	RAFVR	DFC,MM	LONSDALE
N673 1949	90607	RAuxAF		LOOKER
N933 1960	849824	AAF	wef 25/5/44	LOOSEMORE
N608 1945	116730	RAFVR		LORD
N289 1947	178119	RAFVR		LORD
N953 1949	804418	RAuxAF		LORDEN
N24 1957	156855	RAFVR		LORIMER
N368 1959	807173	AAF	wef 27/4/43	LOTE
N11 1944	70408	RAFVR	AFC	LOTON
N1638 1942	84685	RAFVR		LOUCH
N527 1948	171542	RAFVR		LOUDEN
N672 1954	179292	RAuxAF		LOUDEN
N456 1948	81433	RAFVR		LOUGHAN
N519 1947	873296	AAF		LOUGHERY
N396 1944	740051	RAFVR		LOUGHRAN
N460 1947	5323	WAAF		LOUNSBURY
N1014 1943	70409	RAFVR	AFC	LOVE
N249 1954	2601769	RAFVR		LOVE
N708 1959	156488	RAFVR(T)	DFM; wef 16/2/43	LOVEITT
N751 1946	142868	RAFVR	DFC	LOVEJOY
N573 1948	819155	RAuxAF		LOVELL
N187 1947	82766	RAFVR		LOVELL
N594 1946	107936	RAFVR	AFC	LOVERIDGE
N672 1954	107936	RAFVR	AFC; 1st Clasp	LOVERIDGE
N672 1954	75576	RAFVR		LOVICK
N43 1956	189152	RAuxAF		LOWDEN
N952 1953	121591	RAuxAF		LOWE
N513 1952	870013	RAuxAF		LOWE
N24 1957	194256	RAuxAF		LOWE
N1036 1945	118019	RAFVR	DFM	LOWE
N339 1961	742492	RAFVR	wef 2/11/43	LOWE
N549 1962	819124	AAF	wef 28/4/44	LOWE
N396 1944	100029	RAFVR		LOWE
N456 1948	808333	RAuxAF		LOWERY
N277 1955	2686021	RAuxAF	1st Clasp	LOWERY
N953 1949	146114	RAFVR		LOWNDES
N527 1948	155458	RAFVR		LOWNDES
N415 1946	175278	RAFVR		LOWRY
N961 1944	69457	RAFVR		LOWRY
N1144 1943	70411	RAFVR		LOWRY
N1225 1944	130800	RAFVR		LOWTH
N187 1947	79590	RAFVR		LOXTON
N926 1950	756208	RAFVR		LUCAS
N456 1957	870083	AAF	wef 22/7/44	LUCAS
N1232 1945	171647	RAFVR	DFC	LUCAS
N289 1947	72143	RAFVR		LUCIE-SMITH
N115 1953	841729	RAuxAF		LUCK
N672 1954	141632	RAFVR		LUCK
N805 1948	860421	RAuxAF		LUCKETT
N1177 1943	808091	AAF		LUDBROOK
N1225 1944	116108	RAFVR		LUDMAN
N856 1948	818204	RAuxAF		LUGG
N596 1944	171707	RAFVR		LUMLEY
N833 1956	5154	WRAAF		LUND-LINTON
N263 1946	147769	RAFVR		LUNN
N1638 1942	79369	RAFVR		LUNN
N926 1950	73977	RAFVR		LUNSON
N30 1949	810093	RAuxAF		LUNT

The Air Efficiency Award 1942-2005

Surname	Initials				Rank	Status	Date
LUNT	H	G			SQD LDR	Acting	05/02/1948
LUSH	E	G			CPL		30/09/1948
LUTY	W	G			FL		25/07/1962
LUTYENS	R				WING COMM		20/05/1948
LUXFORD	G	H			FLT SGT		20/05/1948
LYALL	G				FL		01/05/1993
LYALL	T	W			SGT	Acting	20/08/1958
LYDDIARD	J	A			FL		21/09/1944
LYE	E	D			FO		25/03/1954
LYNAS	T	R			FLT SGT	Acting	07/12/1960
LYNES	A	H	C		FL		24/01/1946
LYNN	T				WING COMM		22/11/2001
LYON	E				SQD LDR	Acting	06/03/1947
LYON	J	M			SQD OFF	Acting	08/05/1947
LYONS-MONTGOMERY	P				FLT OFF		27/04/1950
LYTHGOE	R	F			FL	Acting	01/03/1945
MABEY	J	E			FO		21/01/1959
MacARTHUR	M	R			SQD LDR		03/12/1942
MacAULAY	I	A			SQD LDR		22/02/1961
MacBEAN	J	I			FL		31/05/1945
MacBEATH	D	H			FL		03/06/1943
MacBEATH	D	H			FL		21/06/1951
MacBETH	E	K			SGT		26/11/1953
MacCARTHY	J	D			FO		07/04/1955
MacDONALD	A	J			SGT		07/01/1943
MacDONALD	A	J			SGT	Temporary	22/09/1955
MacDONALD	D	W	H		FO		28/02/1946
MacDONALD	D				FL		29/04/1954
MacDONALD	H	E			FL		16/10/1947
MacDONALD	H	J			WO		21/02/1952
MacDONALD	I	S			WING COMM		04/12/1947
MacDONALD	J				SQD LDR	Acting	23/02/1950
MacDONALD	J				FL		03/07/1952
MacDONALD	J				REV		18/07/1956
MacDONALD	J				SGT		16/08/1961
MacDONALD	P	R	D	A.C	FL		08/02/1945
MacDONALD	S				FL		26/04/1956
MacDOUGALL	A				GRP CAP		28/11/1946
MacFARLANE	D				FO		03/12/1942
MacFARLANE	D				FO		24/01/1946
MacFIE	C	H			SQD LDR	Acting	28/02/1946
MacGILLIVRAY	L				FL	Acting	06/03/1947
MacGOWAN	T	J	B	A	WING COMM	Acting	26/01/1950
MacGOWN	J	C			WING COMM	Acting	23/02/1950
MacGREGOR	A	N			FO		03/06/1943
MacGREGOR	D				SQD LDR	Acting	06/03/1947
MacGREGOR	I	B	K		SQD LDR		23/02/1950
MACHIN	D				SQD OFF	Acting	05/06/1947
MACHIN	L				CPL		30/09/1948
MACHIN	W				CPL		20/09/1951
MACILWAINE	R				FL		03/10/1946
MacINTOSH	D	R			FO		13/10/1955
MACK	D	B			CPL		01/07/1948
MacKAY	A	M			FLT SGT		28/02/1946
MacKAY	D				FL		28/02/1946
MacKAY	F				FL		29/07/1995
MacKECHNIE	J				AC2		16/03/1950
MacKENZIE	A	A			FL	Acting	13/01/1949
MacKENZIE	D	J			FL		25/03/1959
MacKENZIE	V	O			SGT		20/07/1960

Source	Service No	Force	Notes	Surname
N87 1948	91074	RAuxAF		LUNT
N805 1948	746413	RAFVR		LUSH
N549 1962	109343	RAFVR	wef 22/8/61	LUTY
N395 1948	75290	RAFVR		LUTYENS
N395 1948	869491	RAuxAF		LUXFORD
YB 1996	2626758	RAuxAF		LYALL
N559 1958	2686085	RAuxAF	wef 28/5/58	LYALL
N961 1944	86341	RAFVR		LYDDIARD
N249 1954	89097	RAuxAF		LYE
N933 1960	2650786	RAuxAF	wef 24/1/60	LYNAS
N82 1946	68146	RAFVR		LYNES
YB 2002	4231204	RAuxAF		LYNN
N187 1947	76245	RAFVR		LYON
N358 1947	335	WAAF		LYON
N406 1950	3033	WAAF		LYONS-MONTGOMERY
N225 1945	147216	RAFVR		LYTHGOE
N28 1959	2695022	RAuxAF	wef 30/11/54	MABEY
N1638 1942	70416	RAFVR	DFC	MacARTHUR
N153 1961	65973	AAF	wef 29/7/44	MacAULAY
N573 1945	102580	RAFVR		MacBEAN
N609 1943	73000	RAFVR		MacBEATH
N621 1951	73000	RAFVR	1st Clasp	MacBEATH
N952 1953	743325	RAFVR		MacBETH
N277 1955	2688575	RAuxAF		MacCARTHY
N3 1943	802199	AAF		MacDONALD
N708 1955	802199	RAuxAF	1st Clasp	MacDONALD
N197 1946	188007	RAFVR		MacDONALD
N329 1954	132519	RAuxAF		MacDONALD
N846 1947	90855	AAF		MacDONALD
N134 1952	818088	RAuxAF		MacDONALD
N1007 1947	79587	RAFVR		MacDONALD
N179 1950	116066	RAFVR		MacDONALD
N513 1952	116066	RAFVR	1st Clasp	MacDONALD
N513 1956	n/a	RAFVR		MacDONALD
N624 1961	2696054	RAuxAF	wef 27/5/61	MacDONALD
N139 1945	76589	RAFVR		MacDONALD
N318 1956	186356	RAFVR		MacDONALD
N1004 1946	78441	RAFVR		MacDOUGALL
N1638 1942	102103	RAFVR		MacFARLANE
N82 1946	175409	RAFVR		MacFARLANE
N197 1946	90657	AAF	DFC	MacFIE
N187 1947	149891	RAFVR	MM	MacGILLIVRAY
N94 1950	70425	RAFVR	MB,ChM,FRCS	MacGOWAN
N179 1950	72766	RAFVR	DFC,MB,ChB	MacGOWN
N609 1943	109895	RAFVR		MacGREGOR
N187 1947	61476	RAFVR	DFC	MacGREGOR
N179 1950	72060	RAFVR	MB,ChB	MacGREGOR
N460 1947	2862	WAAF		MACHIN
N805 1948	813150	RAuxAF		MACHIN
N963 1951	858062	RAuxAF		MACHIN
N837 1946	130695	RAFVR	DFC	MACILWAINE
N762 1955	1810144	RAuxAF		MacINTOSH
N527 1948	747334	RAFVR		MACK
N197 1946	1825914	EX-AAF		MacKAY
N197 1946	169708	RAFVR		MacKAY
YB 1997	2676532	RAuxAF		MacKAY
N257 1950	804384	RAuxAF		MacKECHNIE
N30 1949	88504	RAFVR		MacKENZIE
N196 1959	69624	RAFVR	wef 20/6/44	MacKENZIE
N531 1960	803617	AAF	wef 9/5/44	MacKENZIE

Surname	Initials			Rank	Status	Date
MacKENZIE	W	A		SQD LDR	Acting	13/01/1949
MacKIE	A	G		FL		06/01/1955
MacKIE	A	J		FL		13/12/1945
MacKIE	J	A		SGT		18/10/1951
MacKIE	R	J		WO		30/03/1943
MacKIN	J			FO		18/04/1946
MacKINDER	J	H		SQD LDR		15/09/1949
MacKINDER	J	H		FL		02/04/1953
MacKINGTOSH	J			SGT		15/11/1997
MACKLEY	S	G		FL		15/07/1948
MACKMAN	G			FO		12/04/1945
MacLACHLAN	A	M		WING COMM		10/12/1942
MacLAGAN	H	M		LAC		15/09/1949
MacLAREN	M	G		FLT OFF		18/11/1948
MacLEAN	C	H		WING COMM		26/02/1958
MacLEAN	K			FLT SGT	Temporary	01/05/1957
MacLEAN	N	M		FO		05/09/1962
MacLEOD	D	G		FL		16/04/1958
MacLOCHLAINN	P	M		FO		16/01/1957
MacMILLAN	K	G		FL		03/07/1952
MacOWAN	J	R		FL		08/03/1956
MacPHAIL	W	C		FO		26/04/1956
MacPHERSON	C			SGT		13/01/1949
MADDAFORD	W	J		SGT		13/12/1951
MADDICK	P			SAC		09/11/1995
MADDOX	A	G		CPL		01/07/1948
MADDOX	E	H		FL	Acting	26/01/1950
MADERSON	C	W		FL		24/01/1946
MADLE	S	J		FL		28/02/1946
MAFFEY	A	L		FL		13/12/1945
MAGGS	M	H		FL		05/01/1950
MAGGS	P	M		SQD LDR	Acting	28/02/1946
MAGILL	J			CPL		30/09/1948
MAGOR	A	J		LAC		02/07/1958
MAGRATH	H	M		FL		12/08/1943
MAGSON	F			CPL		17/04/1947
MAGUIRE	W	H		SQD LDR	Acting	27/07/1944
MAIDENS	A	L		WING COMM	Acting	14/10/1943
MAIDENS	A	L		FL		30/03/1960
MAIDENS	H			FLT SGT		21/06/1945
MAIN	A			CPL		23/08/1951
MAINWARING	R	N		CPL		13/01/1949
MAINWARING	W	G		SGT		14/10/1948
MAISON	K	J		CPL		01/01/1993
MAITLAND	H	L		SEC OFF		27/03/1947
MAITLAND-WALKER	W	H		WING COMM	Acting	03/12/1942
MAITLAND-WALKER	W	H		WING COMM		19/07/1951
MAKEPIECE	A	M		FL		26/01/1950
MALCOLM	R	A		FL		28/02/1946
MALES	A	J	W	SQD LDR	Acting	10/05/1945
MALEY	M	L		FL		13/12/1951
MALFOY	C	E		WING COMM	Acting	15/04/1943
MALLENDER	F			LAC		30/09/1948
MALLETT	C	F	R	FL		18/04/1946
MALLINSON	R			LAC		20/09/1951
MALLON	M	G		SQD LDR		20/07/2004
MALLORIE	P	W	M	FL		20/09/1945
MALPASS	L			LAC		13/07/1950
MALTBY	D	J		FL		21/10/1959
MALYON	W	W		CPL		13/01/1949

222

Source	Service No	Force	Notes	Surname
N30 1949	76192	RAFVR		MacKENZIE
N1 1955	163994	RAFVR		MacKIE
N1355 1945	79733	RAFVR		MacKIE
N1077 1951	817062	RAuxAF		MacKIE
N1014 1943	803364	AAF		MacKIE
N358 1946	182501	RAFVR		MacKIN
N953 1949	86401	RAFVR		MacKINDER
N298 1953	86401	RAuxAF	1st Clasp	MacKINDER
YB 1998	n/a	RAuxAF		MacKINGTOSH
N573 1948	91057	RAuxAF		MACKLEY
N381 1945	174317	RAFVR		MACKMAN
N1680 1942	90085	AAF		MacLACHLAN
N953 1949	873857	RAuxAF		MacLAGAN
N984 1948	376	WAAF		MacLAREN
N149 1958	90166	RAuxAF	1st Clasp; wef 18/6/48	MacLEAN
N277 1957	800439	RAFVR	wef 14/12/42	MacLEAN
N655 1962	2683142	RAuxAF	wef 8/6/62	MacLEAN
N258 1958	1622913	RAFVR	wef 25/8/54	MacLEOD
N24 1957	2600959	RAuxAF		MacLOCHLAINN
N513 1952	73285	RAFVR	MBE	MacMILLAN
N187 1956	149109	RAFVR		MacOWAN
N318 1956	204229	RAFVR		MacPHAIL
N30 1949	873767	RAuxAF		MacPHERSON
N1271 1951	757413	RAFVR		MADDAFORD
YB 1997	n/a	RAuxAF		MADDICK
N527 1948	750517	RAFVR		MADDOX
N94 1950	102188	RAFVR		MADDOX
N82 1946	155879	RAFVR		MADERSON
N197 1946	86323	RAFVR		MADLE
N1355 1945	90214	AAF		MAFFEY
N1 1950	79357	RAFVR	DFC	MAGGS
N197 1946	79376	RAFVR	MBE	MAGGS
N805 1948	849151	RAuxAF		MAGILL
N452 1958	820204	AAF	wef 21/3/44	MAGOR
N850 1943	72450	RAFVR		MAGRATH
N289 1947	843016	AAF		MAGSON
N763 1944	62249	RAFVR	DFC	MAGUIRE
N1065 1943	70441	RAFVR		MAIDENS
N229 1960	70441	RAFVR	1st Clasp; wef 10/2/55	MAIDENS
N671 1945	808019	AAF		MAIDENS
N866 1951	869501	RAuxAF		MAIN
N30 1949	853765	RAuxAF		MAINWARING
N856 1948	855244	RAuxAF		MAINWARING
YB 1996	n/a	RAuxAF		MAISON
N250 1947	6492	WAAF		MAITLAND
N1638 1942	70707	RAFVR		MAITLAND-WALKER
N724 1951	70707	RAFVR	1st Clasp	MAITLAND-WALKER
N94 1950	91098	RAuxAF		MAKEPIECE
N197 1946	102996	RAFVR		MALCOLM
N485 1945	87642	RAFVR		MALES
N1271 1951	72102	RAFVR	MB,BS,MRCS,LRCP	MALEY
N454 1943	90019	AAF	DFC	MALFOY
N805 1948	868765	RAuxAF		MALLENDER
N358 1946	144634	RAFVR		MALLETT
N963 1951	870038	RAuxAF		MALLINSON
LG 2004	2631426C	RAuxAF		MALLON
N1036 1945	78730	RAFVR		MALLORIE
N675 1950	744274	RAFVR		MALPASS
N708 1959	142414	RAFVR	wef 3/9/59	MALTBY
N30 1949	840046	RAuxAF		MALYON

Surname	Initials			Rank	Status	Date
MANGAN	J	F		FLT SGT		12/10/1950
MANLEY	G	C		WO		21/06/1951
MANLEY	R	H		CPL		17/01/1952
MANN	H	J	G	CPL		05/06/1947
MANN	J			SGT		30/03/1943
MANN	J			FL		12/04/1945
MANNING	E			FL		28/02/1946
MANNING	F	R	C	FL		16/01/1957
MANNING	F			FLT SGT		21/01/1959
MANNING	J	M		FO		05/06/1947
MANNING	J	M		SQD LDR	Acting	01/05/1957
MANNING	L	A		PO		23/02/1950
MANNING	R	J		CPL		13/01/1949
MANSEL	I	C		GRP OFF	Acting	17/04/1947
MANSFIELD	C	H		FL		30/11/1944
MANSFIELD	E	B	H	FO		17/04/1947
MANSFIELD	E	B	H	FL		03/06/1959
MANSFIELD	E			FL		01/07/1948
MANSFIELD	F	W	S	FL		07/07/1949
MANSON	A	K		LAC		05/06/1947
MANTON	G	W		FL		19/08/1954
MAPP	I	P		FO		11/01/1945
MARCH	A			CPL		07/07/1949
MARCH	H	M		FL		24/01/1946
MARDLIN	A			CPL		13/11/1947
MARGARSON	F			WO		18/04/1946
MARIES	P	C		FO		20/04/1944
MARKER	A	T		FL		25/03/1959
MARKHAM	C	E		SGT		12/06/1947
MARKLAND	E	F	B	WING COMM	Acting	24/01/1946
MARKS	H	J		LAC		21/06/1961
MARKS	R	A		FL		08/11/1945
MARLEY	P	M		CPL		28/03/1997
MARLOW	J			SGT	Temporary	13/10/1955
MARLOWE	A	F		SQD LDR		04/05/1944
MARR	J	S		FL		05/01/1950
MARRIOTT	G	S		FLT SGT	Acting	01/05/1952
MARRIOTT	W	J		AC2		06/09/1951
MARRIOTT	W			CPL		15/09/1949
MARRIS	R	Q		FL		17/03/1949
MARRITT	V	H		SGT		07/07/1949
MARSDEN	C	O		FO		31/10/1956
MARSH	A	E		SGT		05/10/1994
MARSH	E	H		FL		16/10/1947
MARSH	K	G		CPL		15/07/1948
MARSH	K	G		SQD LDR	Acting	14/09/1950
MARSH	M	H		SGT		20/03/1957
MARSHALL	A			FLT SGT		19/01/1956
MARSHALL	A			CPL		15/03/1998
MARSHALL	B	W		FL		08/05/1947
MARSHALL	C	G		FL		04/11/1943
MARSHALL	C			SGT		01/07/1948
MARSHALL	G	K		SQD LDR	Acting	26/06/1947
MARSHALL	G	W		FL		20/03/1957
MARSHALL	H	G		FLT SGT		03/06/1959
MARSHALL	J	T		FL		06/03/1947
MARSHALL	J	V		SQD LDR	Acting	28/11/1946
MARSHALL	J			AC2		15/01/1953
MARSHALL	M			WO	Temporary	04/09/1957

Source	Service No	Force	Notes	Surname
N1018 1950	756710	RAFVR		MANGAN
N621 1951	880140	WAAF		MANLEY
N35 1952	849158	RAuxAF		MANLEY
N460 1947	842857	AAF		MANN
N1014 1943	803314	AAF		MANN
N381 1945	127025	RAFVR	DFM	MANN
N197 1946	130218	RAFVR		MANNING
N24 1957	91293	RAuxAF	MRCS,LRCP	MANNING
N28 1959	2676433	RAFVR	wef 1/11/58	MANNING
N460 1947	132103	RAFVR	MBE	MANNING
N277 1957	132103	RAFVR	MBE; 1st Clasp; wef 3/12/53	MANNING
N179 1950	197642	RAFVR		MANNING
N30 1949	843758	RAuxAF		MANNING
N289 1947	162	WAAF		MANSEL
N1225 1944	137560	RAFVR		MANSFIELD
N289 1947	118343	RAFVR		MANSFIELD
N368 1959	118343	RAuxAF	1st Clasp; wef 8/3/59	MANSFIELD
N527 1948	102106	RAFVR		MANSFIELD
N673 1949	122761	RAFVR	Deceased	MANSFIELD
N460 1947	811069	AAF		MANSON
N672 1954	76995	RAFVR		MANTON
N14 1945	155498	RAFVR	DFC	MAPP
N673 1949	865598	RAuxAF		MARCH
N82 1946	82703	RAFVR		MARCH
N933 1947	868716	AAF		MARDLIN
N358 1946	748786	RAFVR		MARGARSON
N363 1944	141972	RAFVR		MARIES
N196 1959	53490	RAuxAF	wef 20/1/59	MARKER
N478 1947	811130	AAF		MARKHAM
N82 1946	72518	RAFVR		MARKLAND
N478 1961	840086	AAF	wef 11/3/44	MARKS
N1232 1945	87018	RAFVR	AFM	MARKS
YB 1998	n/a	RAuxAF		MARLEY
N762 1955	770347	RAFVR		MARLOW
N425 1944	70444	RAFVR		MARLOWE
N1 1950	146320	RAFVR		MARR
N336 1952	2682000	RAuxAF		MARRIOTT
N913 1951	870209	RAuxAF		MARRIOTT
N953 1949	750400	RAFVR		MARRIOTT
N231 1949	159024	RAFVR		MARRIS
N673 1949	847792	RAuxAF		MARRITT
N761 1956	2661301	WRAAF		MARSDEN
YB 1996	n/a	RAuxAF		MARSH
N846 1947	156314	RAFVR		MARSH
N573 1948	818038	RAuxAF		MARSH
N926 1950	90709	RAuxAF		MARSH
N198 1957	2674470	WRAFVR	wef 28/1/55	MARSH
N43 1956	2605767	RAFVR		MARSHALL
YB 1999	n/a	RAuxAF		MARSHALL
N358 1947	104775	RAFVR	DFC	MARSHALL
N1144 1943	61223	RAFVR	DFC	MARSHALL
N527 1948	748884	RAFVR		MARSHALL
N519 1947	90862	AAF		MARSHALL
N198 1957	178542	RAuxAF	wef 8/1/56	MARSHALL
N368 1959	2692521	RAuxAF	wef 31/1/59	MARSHALL
N187 1947	112422	RAFVR		MARSHALL
N1004 1946	83286	RAFVR	DFC	MARSHALL
N28 1953	2683705	RAuxAF		MARSHALL
N611 1957	880576	WAAF	wef 14/5/44	MARSHALL

Surname	Initials			Rank	Status	Date
MARSHALL	N	B		SQD LDR	Acting	14/10/1948
MARSHALL	R	M		SQD LDR		21/06/1945
MARSHALL	R	W		SGT		20/03/1957
MARSHALL	T	B		FL		27/07/1944
MARSHALL	V	E	H	PO	Acting	05/01/1950
MARSHALL	W	A		FO		01/07/1948
MARSHLAND	E			FL		10/05/1945
MARSLAND	A			FL		17/03/1958
MARSLAND	E	I		FLT SGT		07/08/1947
MARSLAND	F	S		SGT		10/02/1949
MARTHEWS	B			FLT OFF		08/05/1947
MARTIN	A	W		SQD LDR	Acting	26/08/1948
MARTIN	A			FL		24/03/1997
MARTIN	B	C		WO		11/07/1946
MARTIN	B	G		SQD OFF	Acting	16/10/1947
MARTIN	C	P	C	FL	Acting	08/05/1947
MARTIN	E	A		CPL		15/07/1948
MARTIN	E	J		FL		27/01/1944
MARTIN	F	H		FL		27/03/1947
MARTIN	H	H		FLT SGT		26/02/1958
MARTIN	H	L		SGT		17/03/1949
MARTIN	H	W		PO		17/12/1942
MARTIN	H			FL		15/07/1948
MARTIN	J	B		FO		21/06/1945
MARTIN	J	H		WO		13/01/1949
MARTIN	J	N		SGT		30/09/1948
MARTIN	J	T		FLT SGT		29/06/1944
MARTIN	N			SGT	Acting	25/03/1954
MARTIN	R	J		CPL		20/07/1960
MARTIN	T	E		WING COMM		10/06/2001
MARTIN	W	F		SGT		01/07/1948
MARTINDALE	A	F		SQD LDR		26/10/1944
MARTINDALE	C	B		SQD LDR		30/12/1943
MARTINSON	J	A		FO		06/03/1947
MARTINSON	J	P		FL		08/11/1945
MARTLEW	P	D		FO		21/01/1959
MASCALL	I	R		SQD LDR		20/05/1948
MASHFORD	R			CPL		13/01/1949
MASON	B			SGT		25/05/1960
MASON	D	B		SGT		07/12/1960
MASON	D	C		FL		28/02/1946
MASON	G	T		LAC		28/11/1956
MASON	H			CPL		28/07/1955
MASON	R	E		LAC		12/04/1951
MASSEY	A	W		FL		21/10/1959
MASSEY	K			FO		30/09/1948
MASSEY	O	N		FL		03/12/1942
MASSEY	P			FL		28/02/1946
MASSEY	W	A		FL		10/02/1949
MASSON	D	G		FL		01/01/2000
MASTERS	H	J	K	FL		06/01/1955
MATHERS	L	A	J	FL		16/03/1950
MATHESON	D			SQD LDR		22/03/1990
MATHESON	D			SQD LDR		22/03/2000
MATHESON	O	R		FL		29/06/1944
MATHIAS	L	F		CPL		16/03/1950
MATHIESON	C	A	C	FO		17/01/1997
MATHIESON	J	M		FO		14/10/1943
MATHIESON	V	P		FLT SGT		12/10/1995
MATTEY	G	L		FL		08/11/1945

Source	Service No	Force	Notes	Surname
N856 1948	73980	RAFVR		MARSHALL
N671 1945	80023	RAFVR		MARSHALL
N198 1957	873977	RAFVR	wef 18/7/44	MARSHALL
N763 1944	102085	RAFVR		MARSHALL
N1 1950	141259	RAFVR		MARSHALL
N527 1948	136603	RAFVR		MARSHALL
N485 1945	84329	RAFVR		MARSHLAND
MOD 1964	2611259	RAFVR		MARSLAND
N662 1947	882422	WAAF		MARSLAND
N123 1949	803574	RAuxAF		MARSLAND
N358 1947	131	WAAF		MARTHEWS
N694 1948	78254	RAFVR	DFC	MARTIN
YB 1998	n/a	RAuxAF		MARTIN
N594 1946	755565	RAFVR		MARTIN
N846 1947	1242	WAAF		MARTIN
N358 1947	127640	RAFVR		MARTIN
N573 1948	847145	RAuxAF		MARTIN
N59 1944	73004	RAFVR		MARTIN
N250 1947	78463	RAFVR	DFC	MARTIN
N149 1958	2601250	RAFVR	wef 6/5/52	MARTIN
N231 1949	842399	RAuxAF		MARTIN
N1708 1942	118819	RAFVR		MARTIN
N573 1948	115727	RAFVR		MARTIN
N671 1945	160767	RAFVR		MARTIN
N30 1949	802578	RAuxAF		MARTIN
N805 1948	819089	RAuxAF		MARTIN
N652 1944	812048	AAF		MARTIN
N249 1954	2676002	RAuxAF		MARTIN
N531 1960	2693031	RAuxAF	wef 23/3/60	MARTIN
YB 2002	5204078	RAuxAF		MARTIN
N527 1948	770469	RAFVR		MARTIN
N1094 1944	73005	RAFVR	AFC	MARTINDALE
N1350 1943	72527	RAFVR		MARTINDALE
N187 1947	177942	RAFVR		MARTINSON
N1232 1945	132897	RAFVR		MARTINSON
N28 1959	131475	RAFVR	wef 3/5/44	MARTLEW
N395 1948	72362	RAFVR		MASCALL
N30 1949	799576	RAFVR		MASHFORD
N375 1960	2661687	WRAAF	wef 14/2/60	MASON
N933 1960	2661044	WRAFVR	wef 15/11/59	MASON
N197 1946	142193	RAFVR		MASON
N833 1956	860360	RAuxAF		MASON
N565 1955	2688766	RAuxAF		MASON
N342 1951	746122	RAFVR		MASON
N708 1959	75954	RAFVR	wef 8/8/59	MASSEY
N805 1948	115600	RAFVR		MASSEY
N1638 1942	83285	RAFVR		MASSEY
N197 1946	147711	RAFVR		MASSEY
N123 1949	147509	RAFVR		MASSEY
YB 2001	212596	RAuxAF		MASSON
N1 1955	203010	RAuxAF		MASTERS
N257 1950	62110	RAFVR		MATHERS
YB 1996	2630060	RAuxAF		MATHESON
YB 2000	2630060	RAuxAF	1st Clasp	MATHESON
N652 1944	84010	RAFVR	DFC	MATHESON
N257 1950	818151	RAuxAF		MATHIAS
YB 1998	2633999	RAuxAF		MATHIESON
N1065 1943	70454	RAFVR		MATHIESON
YB 1996	n/a	RAuxAF		MATHIESON
N1232 1945	81042	RAFVR	DFC	MATTEY

Surname	Initials			Rank	Status	Date
MATTHEW	A	G		FL		15/07/1948
MATTHEW	I	G	S	FL		28/02/1946
MATTHEW	I			FL	Acting	19/10/1960
MATTHEWS	H	E		FL		13/12/1945
MATTHEWS	J	F	P	SQD LDR	Acting	10/08/1944
MATTHEWS	J	T		FL		06/08/2000
MATTHEWS	M	K		PO		12/06/1947
MATTHEWSON	J	H		FL		26/04/1961
MAUD	I	L	I	SQD OFF		26/11/1953
MAUGHAN	R	J		SAC		22/10/1958
MAUNDER	A	J		CPL		01/07/1948
MAURICE	A	D		FLT OFF		12/06/1947
MAW	D	M		SQD LDR		22/04/1943
MAW	J	K		FL		17/06/1954
MAWBY	L			FLT SGT		01/03/1945
MAWE	J	W		FO		02/09/1943
MAWER	E	G	B	FL	Acting	15/07/1948
MAWLAM	E			WO		27/01/1944
MAWLE	N	W	R	GRP CAP	Acting	01/07/1948
MAXFIELD	W	F		FL		16/08/1961
MAXWELL	E	J		SQD LDR		16/01/1995
MAXWELL	G	C		WING COMM		10/04/1952
MAXWELL	G	M	A	SQD LDR	Acting	06/03/1947
MAXWELL	J	K		FL		15/06/1944
MAXWELL	M	W		FL		01/05/1952
MAXWELL	T	C	P	WING COMM	Acting	31/05/1945
MAXWELL	W			CPL		21/01/1959
MAY	C	J		FL		05/01/1950
MAY	R			SGT		06/03/1947
MAYBEY	R	E		LAC		20/05/1948
MAYERS	D			FO		20/08/1958
MAYLAM	J	B		FL		31/05/1945
MAYNARD	E	C		LAC		10/06/1948
MAYNARD	H			SQD LDR	Acting	21/09/1944
MAYNARD	M	M		FL		18/05/1944
MAZZOTTA	J	J		SGT		21/09/1995
McADAM	D	W		PO		13/01/1944
McADAM	E			FL		30/03/1960
McALERY	C	M		GRP OFF		06/03/1947
McALLISTER	A			FO		30/09/1948
McALLISTER	K	H		SGT		17/06/1954
McARTHUR	A	J	D	FL		17/01/1952
McARTHUR	D			WING COMM	Acting	15/06/1944
McAULEY	S			SAC		25/07/1962
McBAIN	J	A		FLT SGT		12/04/1945
McBRIDE	C			FL		19/10/1960
McBRIDE	J	V		WO	Acting	16/01/1957
McCAIG	P	B	K	WO		01/06/1978
McCAIG	P	B	K	WO		01/06/1988
McCAIG	P	B	K	WO		01/06/1998
McCALL	D			FO		06/03/1947
McCALL	M	C		CPL	Acting	30/03/1960
McCALLION	E	E		FL		09/12/1959
McCALLION	E			SACW		27/02/1963
McCALLUM	D	M		CPL	Acting	13/12/1961
McCALLUM	R			FLT SGT	Acting	25/06/1953
McCANDLESS	K			FLT SGT		03/03/1993
McCARTHY	G	C		SQD LDR	Acting	09/05/1946
McCARTHY	R			FL		15/06/1944
McCARTHY	R			SQD LDR		26/02/1958

Source	Service No	Force	Notes	Surname
N573 1948	90611	RAuxAF		MATTHEW
N197 1946	114998	RAFVR	DFC	MATTHEW
N775 1960	2600338	RAuxAF	wef 5/2/58	MATTHEW
N1355 1945	123054	RAFVR		MATTHEWS
N812 1944	60766	RAFVR	AFC	MATTHEWS
YB 2001	91455	RAuxAF		MATTHEWS
N478 1947	203282	RAFVR		MATTHEWS
N339 1961	187658	RAFVR	wef 7/1/60	MATTHEWSON
N952 1953	2372	WRAAF		MAUD
N716 1958	2685514	RAFVR	wef 8/6/57	MAUGHAN
N527 1948	813161	RAuxAF		MAUNDER
N478 1947	218	WAAF		MAURICE
N475 1943	70456	RAFVR	AFC	MAW
N471 1954	78483	RAFVR		MAW
N225 1945	800373	AAF		MAWBY
N918 1943	117038	RAFVR		MAWE
N573 1948	81958	RAFVR		MAWER
N59 1944	808092	AAF		MAWLAM
N527 1948	90635	RAuxAF	DFC	MAWLE
N624 1961	2607619	RAFVR	wef 11/6/61	MAXFIELD
YB 1996	407987	RAuxAF		MAXWELL
N278 1952	90890	RAuxAF	MC,DFC,AFC	MAXWELL
N187 1947	90920	AAF		MAXWELL
N596 1944	62655	RAFVR		MAXWELL
N336 1952	91277	RAuxAF		MAXWELL
N573 1945	76462	RAFVR	DFC	MAXWELL
N28 1959	2681069	WRAAF	wef 9/2/58	MAXWELL
N1 1950	61149	RAFVR		MAY
N187 1947	804247	AAF		MAY
N395 1948	841025	RAuxAF		MAYBEY
N559 1958	191901	RAFVR	wef 14/7/44	MAYERS
N573 1945	85640	RAFVR	DFC	MAYLAM
N456 1948	841340	RAuxAF		MAYNARD
N961 1944	90965	AAF		MAYNARD
N473 1944	72513	RAFVR		MAYNARD
YB 1996	n/a	RAuxAF		MAZZOTTA
N11 1944	155013	RAFVR		McADAM
N229 1960	148462	RAuxAF	wef 24/11/57	McADAM
N187 1947	31	WAAF	OBE	McALERY
N805 1948	136130	RAFVR		McALLISTER
N471 1954	2649503	RAuxAF		McALLISTER
N35 1952	90936	RAuxAF		McARTHUR
N596 1944	72422	RAFVR		McARTHUR
N549 1962	2681145	RAuxAF	wef 18/5/62	McAULEY
N381 1945	803245	AAF		McBAIN
N775 1960	139981	RAFVR	wef 9/9/57	McBRIDE
N24 1957	2650767	RAuxAF		McBRIDE
YB 1996	n/a	RAuxAF		McCAIG
YB 1996	n/a	RAuxAF	1st Clasp	McCAIG
YB 1999	n/a	RAuxAF	2nd Clasp	McCAIG
N187 1947	144876	RAFVR	DFM	McCALL
N229 1960	2659224	WRAAF	wef 13/1/59	McCALL
N843 1959	63150	RAFVR	wef 2/2/55	McCALLION
N155 1963	2663801	WRAAF	wef 13/1/63	McCALLION
N987 1961	5000413	RAFVR	wef 2/11/61	McCALLUM
N527 1953	2677412	RAuxAF		McCALLUM
YB 1996	n/a	RAuxAF		McCANDLESS
N415 1946	134006	RAFVR	AFC	McCARTHY
N596 1944	81666	RAFVR	AFC	McCARTHY
N149 1958	81666	RAuxAF	AFC; 1st Clasp; wef 2/8/49	McCARTHY

Surname	Initials			Rank	Status	Date
McCARTHY	W	H		CPL		07/07/1949
McCARTHY-JONES	C	C		FL		20/09/1945
McCARTHY-JONES	C	C		FL		17/01/1952
McCASH	W			FO		24/05/2005
McCAUGHERTY	G			CPL		29/04/1954
McCAUSLAND	D	E		FO		18/07/1956
McCLEAN	G	E		FO		20/05/1948
McCLEAN	J	S		WO		31/05/1945
McCLEAN	J			SGT		27/11/1952
McCLELLAND	A	G		FL		13/12/1945
McCLELLAND	D			FL		27/03/1963
McCLELLAND	D			SQD LDR		24/10/1964
McCLELLAND	R			LAC		20/08/1958
McCLUNE	C	L		FLT SGT		05/06/1947
McCLUNE	J	A		FL		13/11/1947
McCLURE	C	G	B	SQD LDR		30/03/1943
McCOLM	S	J		CPL		12/06/1996
McCOMB	J	E		WING COMM		04/02/1943
McCONNELL	C	H		FL		29/04/1954
McCONNELL	H			CPL	Acting	12/06/1947
McCONNELL	J			FL		25/05/1950
McCONNELL	L	C		SQD LDR	Acting	18/04/1946
McCONNELL	W			LAC		05/06/1947
McCORMACK	D	L		CPL		01/11/1951
McCORMICK	W	A		FL		17/04/1947
McCOY	R	E		CPL		24/11/1949
McCREANOR	J			CPL		15/07/1948
McCULLOCH	J	W		FL		10/06/1948
McCULLOCH	J			SGT		16/10/1952
McCULLOCH	R	E	A	FL		01/05/1957
McCULLOCH	S			FO		11/01/1945
McCULLOUGH	J			SGT		16/10/1947
McCULLOUGH	J			FLT SGT		29/04/1954
McCURDY	A	K	S	SQD LDR	Acting	08/11/1945
McCUSKER	D			FO		07/07/1949
McDERMOTT	J			FLT SGT		05/06/1947
McDONALD	B	F		FL		18/04/1946
McDONALD	J			LAC		20/03/1957
McDONALD-WEBB	R	N		FO		20/07/1993
McDONALD-WEBB	R	N		FL		16/02/2004
McDONNELL	J	J	R	CPL		19/09/1956
McDOUGALL	R			FL		13/12/1945
McDOUGALL-BLACK	P			FL		01/08/1946
McDOWALL	A			WING COMM	Acting	18/05/1944
McELLIGOTT	G	L	M	GRP CAP		07/04/1955
McEVOY	J			SGT		01/07/1948
McEVOY	L	B		SGT	Temporary	26/04/1956
McFADYEN	T	K		SGT		03/10/1962
McFARLAND	K	A		FL		10/05/1945
McFARLAND	R			LAC		18/07/1956
McFARLANE	A			FO		21/09/1944
McFARLANE	R			WING COMM	Acting	28/02/1946
McGEE	A	D		CPL		09/02/1951
McGEORGE	T	S		FO		06/03/1947
McGIBBON	J	E	G	SQD LDR		17/04/1947
McGIFFEN	W	H		SQD LDR		30/03/1944
McGILL	J	L		FLT SGT		11/10/1995
McGILL	J			FL		13/12/1945
McGILL	J			CPL		28/09/1950
McGINN	M	F		FO		07/04/1955

Source	Service No	Force	Notes	Surname
N673 1949	865509	RAuxAF		McCARTHY
N1036 1945	76590	RAFVR		McCARTHY-JONES
N35 1952	91312	RAuxAF	1st Clasp	McCARTHY-JONES
LG2005	1570997	RAFVR		McCASH
N329 1954	2681001	RAuxAF		McCAUGHERTY
N513 1956	2650774	RAuxAF		McCAUSLAND
N395 1948	179438	RAFVR		McCLEAN
N573 1945	803467	AAF		McCLEAN
N915 1952	746479	RAFVR		McCLEAN
N1355 1945	171075	RAFVR		McCLELLAND
N239 1963	139420	RAuxAF	DFC; wef 24/10/54	McCLELLAND
MOD 1966	139420	RAuxAF	DFC; 1st Clasp	McCLELLAND
N559 1958	816219	AAF	wef 4/7/44	McCLELLAND
N460 1947	811088	AAF		McCLUNE
N933 1947	156550	RAFVR	DFC	McCLUNE
N1014 1943	70418	RAFVR		McCLURE
YB 1997	n/a	RAuxAF		McCOLM
N131 1943	90352	AAF	DFC	McCOMB
N329 1954	103876	RAuxAF		McCONNELL
N478 1947	811055	AAF		McCONNELL
N517 1950	136316	RAFVR		McCONNELL
N358 1946	81938	RAFVR	DFC	McCONNELL
N460 1947	811103	AAF		McCONNELL
N1126 1951	750884	RAFVR		McCORMACK
N289 1947	173606	RAFVR		McCORMICK
N1193 1949	805426	RAuxAF		McCOY
N573 1948	867835	RAuxAF		McCREANOR
N456 1948	143655	RAFVR		McCULLOCH
N798 1952	2683524	RAuxAF		McCULLOCH
N277 1957	164235	RAuxAF	wef 11/11/52	McCULLOCH
N14 1945	174892	RAFVR		McCULLOCH
N846 1947	816213	AAF		McCULLOUGH
N329 1954	2681008	RAuxAF	1st Clasp	McCULLOUGH
N1232 1945	82683	RAFVR	DFC	McCURDY
N673 1949	198058	RAFVR		McCUSKER
N460 1947	807234	AAF		McDERMOTT
N358 1946	155963	RAFVR		McDONALD
N198 1957	867001	RAFVR	wef 19/6/44	McDONALD
YB 1997	8104132E	RAuxAF		McDONALD-WEBB
LG 2004	8104132E	RAuxAF	1st Clasp	McDONALD-WEBB
N666 1956	2612531	RAFVR		McDONNELL
N1355 1945	82706	RAFVR		McDOUGALL
N652 1946	145147	RAFVR		McDOUGALL-BLACK
N473 1944	89229	RAFVR	DFM	McDOWALL
N277 1955	72086	RAFVR	MRCS,LRCP	McELLIGOTT
N527 1948	818048	RAuxAF		McEVOY
N318 1956	2605123	RAFVR		McEVOY
N722 1962	2683694	RAuxAF	wef 17/8/62	McFADYEN
N485 1945	64896	RAFVR		McFARLAND
N513 1956	816219	RAuxAF		McFARLAND
N961 1944	170266	RAFVR	DFM	McFARLANE
N197 1946	111222	RAFVR	DSO,DFC	McFARLANE
N145 1951	874042	RAuxAF		McGEE
N187 1947	118347	RAFVR		McGEORGE
N289 1947	91012	AAF	OBE	McGIBBON
N281 1944	90041	AAF		McGIFFEN
YB 1999	n/a	RAuxAF		McGILL
N1355 1945	64899	RAFVR	DFC	McGILL
N968 1950	853554	RAuxAF		McGILL
N277 1955	199498	RAuxAF		McGINN

Surname	Initials			Rank	Status	Date
McGLEN	D			CPL		07/08/1947
McGOVERN	P			CPL		30/09/1995
McGOWAN	R	A		SQD LDR	Acting	13/01/1944
McGRAIL	B			LAC		22/02/1951
McGRATH	J	K	B	SQD LDR		06/03/1947
McGRATH	P			LAC		25/06/1953
McGREGOR	A	J		SQD LDR	Acting	04/10/1945
McGREGOR	G	A	D	SAC		20/07/1960
McGREGOR	P	R		SQD LDR		13/12/1945
McGREGOR	T	S		SGT		18/05/1944
McGUFFIE	R	A		SGT		12/04/1951
McGUGAN	R			FL		28/02/1946
McGUIGAN	W	H		SGT		09/11/1944
McGUIRE	D			CPL		22/02/1961
McINNES	A			FL		24/01/1946
McINTOSH	A	C		CPL		30/12/1943
McINTOSH	D	W		FL		20/03/1957
McINTOSH	D			FL		16/10/1952
McINTOSH	R			FO		23/02/1950
McINTYRE	D	F		WING COMM		04/02/1943
McINTYRE	N	E		FL		17/03/1949
McINTYRE	P	F		FL		28/02/1946
McIVOR	G	F		SQD LDR		26/06/1947
McKAY	D	A	S	FL		27/07/1944
McKAY	D	A	S	FO		19/07/1951
McKAY	J	D		FLT OFF		18/11/1948
McKAY	M	J		FLT SGT		03/03/1994
McKAY	N	L	A	WO		08/05/1947
McKECHNIE	W			SGT		04/02/1943
McKELVIE	R	F	S	SQD LDR		04/09/1947
McKENZIE	D			FLT SGT		25/01/1945
McKENZIE	D			FLT SGT		04/10/1951
McKENZIE	D			SAC		01/05/1990
McKENZIE	D			SAC		01/05/2000
McKENZIE	J			SQD LDR		05/01/1950
McKENZIE	K	W		FL		24/01/1946
McKEOWN	N	J		CPL		07/07/1949
McKIE	E	J		FL		20/01/1960
McKINLAY	J			LAC		12/06/1947
McKINNON	R	J		CPL		23/02/1950
McKINNON	W	M		FLT SGT		21/03/1946
McKINNON	W	M		FLT SGT		04/10/1951
McKNIGHT	E	J		FLT SGT		01/03/1945
McLACHLAN	J			SGT		12/04/1945
McLANNAHAN	G	G		SQD LDR		03/12/1942
McLAREN	A	C		SQD LDR		06/03/1947
McLAREN	E	B		FLT OFF		07/08/1947
McLAREN	J			LAC		13/07/1950
McLAREN	R	D		SQD LDR	Acting	02/09/1943
McLAUGHLIN	J	W		FL		12/06/1947
McLAUGHLIN	P	G		FLT SGT		04/10/1945
McLAUGHLIN	R			FLT SGT		24/06/1943
McLEAN	A	L		FO		21/10/1959
McLEAN	C	H		SQD LDR		22/04/1943
McLEAN	I			FL		17/04/1947
McLEOD	A	C		SGT		27/07/1944
McLEOD	A			SGT		13/12/1945
McMASTER	A	S		FLT SGT		25/01/1945
McMASTER	J	G		LAC		10/04/1952
McMEEHAN	R	J		CPL		30/03/1960

Source	Service No	Force	Notes	Surname
N662 1947	867015	AAF		McGLEN
YB 1996	n/a	RAuxAF		McGOVERN
N11 1944	73001	RAFVR		McGOWAN
N194 1951	870314	RAuxAF		McGRAIL
N187 1947	90967	AAF	DFC	McGRATH
N527 1953	856633	RAuxAF		McGRATH
N1089 1945	81919	RAFVR		McGREGOR
N531 1960	2683615	RAuxAF	wef 8/2/60	McGREGOR
N1355 1945	73002	RAFVR		McGREGOR
N473 1944	802367	AAF		McGREGOR
N342 1951	859671	RAuxAF		McGUFFIE
N197 1946	115801	RAFVR		McGUGAN
N1150 1944	802249	AAF		McGUIGAN
N153 1961	2661306	WRAAF	wef 14/4/56	McGUIRE
N82 1946	84920	RAFVR		McINNES
N1350 1943	802308	AAF		McINTOSH
N198 1957	163287	RAuxAF	wef 30/12/56	McINTOSH
N798 1952	144397	RAuxAF		McINTOSH
N179 1950	118378	RAFVR		McINTOSH
N131 1943	72248	RAFVR	AFC	McINTYRE
N231 1949	149133	RAFVR		McINTYRE
N197 1946	83295	RAFVR		McINTYRE
N519 1947	90489	AAF		McIVOR
N763 1944	113322	RAFVR	DFM	McKAY
N724 1951	113322	RAFVR	DFM; 1st Clasp	McKAY
N984 1948	377	WAAF		McKAY
YB 1996	n/a	RAuxAF		McKAY
N358 1947	882201	WAAF		McKAY
N131 1943	802069	AAF		McKECHNIE
N740 1947	90260	AAF		McKELVIE
N70 1945	803415	AAF		McKENZIE
N1019 1951	2683504	RAuxAF	1st Clasp	McKENZIE
YB 2000	n/a	RAuxAF		McKENZIE
YB 2000	n/a	RAuxAF	1st Clasp	McKENZIE
N1 1950	83297	RAFVR		McKENZIE
N82 1946	84017	RAFVR	DFC	McKENZIE
N673 1949	855020	RAuxAF		McKEOWN
N33 1960	120195	RAFVR	wef 26/7/44	McKIE
N478 1947	853835	AAF		McKINLAY
N179 1950	743572	RAFVR		McKINNON
N263 1946	802452	AAF		McKINNON
N1019 1951	2683304	RAuxAF	1st Clasp	McKINNON
N225 1945	807079	AAF		McKNIGHT
N381 1945	802372	AAF		McLACHLAN
N1638 1942	70434	RAFVR		McLANNAHAN
N187 1947	90950	AAF		McLAREN
N662 1947	2006	WAAF		McLAREN
N675 1950	873783	RAuxAF		McLAREN
N918 1943	73003	RAFVR		McLAREN
N478 1947	146149	RAFVR		McLAUGHLIN
N1089 1945	811007	AAF		McLAUGHLIN
N677 1943	816044	AAF		McLAUGHLIN
N708 1959	2601757	RAFVR	wef 8/10/52	McLEAN
N475 1943	90166	AAF		McLEAN
N289 1947	68293	RAFVR		McLEAN
N763 1944	853726	AAF		McLEOD
N1355 1945	802435	AAF		McLEOD
N70 1945	802006	EX-AAF		McMASTER
N278 1952	867766	RAuxAF		McMASTER
N229 1960	863470	AAF	wef 29/6/44	McMEEHAN

Surname	Initials			Rank	Status	Date
McMILLAN	J			FO		11/01/1945
McMINN	G	M		SQD LDR	Acting	06/03/1947
McMONAGLE	H	J		FL		24/07/1952
McNAB	A	D		SQD LDR		10/12/1942
McNABB	H	Y		SGT		30/03/1960
McNABB	L	V		WING COMM	Acting	06/03/1947
McNAIR	R	J		SQD LDR	Acting	13/12/1945
McNEILL	P	J		LAC		07/07/1949
McPHEE	J			FL		21/02/1952
McPHEE	T			SQD LDR	Acting	21/06/1945
McPHERSON	C	K		WO		24/06/1943
McQUEEN	D	C		WO		01/05/1995
McQUILLAN	D			FL		19/08/1959
McRINER	G			SQD LDR	Acting	28/11/1946
McROBBIE	W			FL		01/03/1945
McROBERT	J			FL		27/01/1944
McSHEFFREY	A			SAC		03/09/1989
McSHEFFREY	A			SAC		03/09/1990
McSWEIN	D	D		SQD LDR		01/10/1953
McSWEIN	R			FL	Acting	04/11/1948
McVIE	G	M		CPL		20/05/1948
McVIE	G	M		SGT	Acting	04/09/1957
McWHINNIE	D	R		SQD LDR		05/06/1947
MEAD	H	M		FLT SGT		06/03/1947
MEAD	M	P		SGT	Acting	13/01/1949
MEADOWS	F	C		FL	Acting	20/09/1945
MEADOWS	J	P		SQD LDR		30/08/1945
MEADOWS	J	P		SQD LDR		07/02/1952
MEADOWS	J			FLT SGT		31/10/1956
MEADOWS	J			FLT SGT		20/07/1960
MEADS	C	J		CPL		09/02/1951
MEADS	E	R		SQD LDR		17/06/1943
MEAGER	E	W		CPL		13/01/1949
MEAGER	R	W		FLT SGT		17/12/1942
MEAKIN	H	J	W	WING COMM	Acting	05/10/1944
MEALING	R	H	S	WING COMM		06/03/1947
MEASURES	M	B		SQD OFF		05/06/1947
MEDHURST	R	H		FL		19/08/1959
MEDLEY	C	W		FL		30/03/1960
MEDWAY	C	W	T	LAC		07/07/1949
MEDWORTH	J	C	O	FL		13/12/1945
MEECHAN	R			FLT SGT	Acting	25/06/1953
MEEK	B	W		SGT		17/04/1947
MEEK	B	W		SGT		26/11/1958
MEEK	C			FL		20/07/1960
MEGARRY	J	D		LAC		16/04/1958
MEGGINSON	R	R		SQD LDR	Acting	11/01/1945
MELHUISH	A	H		WING OFF		08/05/1947
MELLING	T	J		SQD LDR		30/01/1989
MELLING	T	J		SQD LDR		30/01/1999
MELLOR	T	R		FO		20/02/1940
MELSON	T	II		SGT		13/01/1949
MELVILLE	C			SGT		25/01/1951
MELVILLE	N	R		WING COMM		13/01/1949
MELVILLE-JACKSON	G	H		SQD LDR	Acting	07/06/1945
MELVIN	F	W		LAC		05/01/1950
MELVIN	R	J	McG	FL		19/08/1954
MENHENNET	J	W		SGT		30/03/1943
MENHINICK	L	B	G	FL		07/04/1955

Source	Service No	Force	Notes	Surname
N14 1945	135447	RAFVR		McMILLAN
N187 1947	81700	RAFVR		McMINN
N569 1952	186948	RAFVR		McMONAGLE
N1680 1942	90177	AAF		McNAB
N229 1960	2650805	RAuxAF	wef 23/2/60	McNABB
N187 1947	70438	RAFVR		McNABB
N1355 1945	112522	RAFVR	DFC	McNAIR
N673 1949	840049	RAuxAF		McNEILL
N134 1952	146128	RAFVR	AFC	McPHEE
N671 1945	100603	RAFVR	DFC,DFM	McPHEE
N677 1943	816022	AAF		McPHERSON
YB 1997	n/a	RAuxAF		McQUEEN
N552 1959	189449	RAuxAF	wef 19/12/56	McQUILLAN
N1004 1946	100040	RAFVR	DFC	McRINER
N225 1945	83278	RAFVR		McROBBIE
N59 1944	70439	RAFVR		McROBERT
YB 1996	n/a	RAuxAF		McSHEFFREY
YB 2000	n/a	RAuxAF	1st Clasp	McSHEFFREY
N796 1953	108732	RAuxAF	OBE	McSWEIN
N934 1948	85969	RAFVR		McSWEIN
N395 1948	803527	RAuxAF		McVIE
N611 1957	2683525	RAuxAF	1st Clasp; wef 22/6/57	McVIE
N460 1947	73935	RAFVR		McWHINNIE
N187 1947	760118	RAFVR		MEAD
N30 1949	881708	WAAF		MEAD
N1036 1945	147965	RAFVR		MEADOWS
N961 1945	82672	RAFVR	DFC	MEADOWS
N103 1952	82672	RAuxAF	DFC; 1st Clasp	MEADOWS
N761 1956	2601861	RAFVR		MEADOWS
N531 1960	2601861	RAFVR	DELETED N775 1960; wef 3/4/56	MEADOWS
N145 1951	753071	RAFVR		MEADS
N653 1943	70459	RAFVR		MEADS
N30 1949	843447	RAuxAF		MEAGER
N1708 1942	801160	AAF		MEAGER
N1007 1944	83280	RAFVR	DFC	MEAKIN
N187 1947	73819	RAFVR		MEALING
N460 1947	137	WAAF		MEASURES
N552 1959	157328	RAFVR	wef 29/6/55	MEDHURST
N229 1960	181190	RAFVR	wef 13/7/53	MEDLEY
N673 1949	861019	RAuxAF		MEDWAY
N1355 1945	146294	RAFVR		MEDWORTH
N527 1953	2677404	RAuxAF		MEECHAN
N289 1947	771286	RAFVR		MEEK
N807 1958	2694034	RAuxAF	1st Clasp; wef 6/8/58	MEEK
N531 1960	139260	RAFVR	wef 11/7/44	MEEK
N258 1958	816185	AAF	wef 13/6/44	MEGARRY
N14 1945	84708	RAFVR		MEGGINSON
N358 1947	155	WAAF		MELHUISH
YB 1996	91414	RAuxAF		MELLING
YB 2000	91414	RAuxAF	1st Clasp	MELLING
N197 1946	171744	RAFVR		MELLOR
N30 1949	811137	RAuxAF		MELSON
N93 1951	853653	RAuxAF		MELVILLE
N30 1949	72342	RAFVR		MELVILLE
N608 1945	80842	RAFVR	DFC	MELVILLE-JACKSON
N1 1950	753179	RAFVR		MELVIN
N672 1954	132454	RAuxAF		MELVIN
N1014 1943	808079	AAF		MENHENNET
N277 1955	149026	RAFVR		MENHINICK

Surname	Initials			Rank	Status	Date
MENZIES	M	C		CPL		20/07/1960
MENZIES	S			CPL		23/02/1950
MERCER	A	B		FL	Acting	07/05/1953
MERCER	D			SQD LDR	Acting	28/07/1955
MERCER	E	D	W	SQD LDR		27/03/1947
MERCER	E	G	W	FL		28/08/1963
MERCER	G	S		FL		07/05/1953
MERCER	J			SGT		04/10/1951
MERCHANT	E			FLT OFF		15/09/1949
MERCHANT	H	J		FO		15/04/1943
MEREDITH	A	C		SQD LDR	Acting	04/09/1947
MEREDITH	A	D		FL		12/04/1945
MEREDITH	H			SQD LDR		18/11/1948
MEREDITH	J			CPL		05/06/1947
MEREDITH	P	G		FL	Acting	06/03/1947
MERIFIELD	J	R	H	WING COMM	Acting	28/02/1946
MERRETT	J	C		FL		15/07/1948
MERRICK	C			SQD LDR	Acting	12/07/1945
MERRICK	M	J		FL		13/02/1957
MERRICK	V	E		SQD LDR		07/03/1996
MERRIEN	A	C		FL		19/08/1959
MERRIFIELD	G	M		SQD LDR	Acting	20/04/1944
MERRINGTON	N	B		FL		27/07/1944
MERRYMAN	W	J	C	FL	Acting	04/09/1947
MESSENGER	S	W		SQD LDR	Acting	27/03/1963
METCALF	C			FL		25/07/1962
METCALFE	F			SQD LDR	Acting	09/08/1945
METCALFE	J	B		FLT SGT		30/03/1943
METCALFE	J	W		FL		19/08/2001
METCALFE	R			SEC OFF		07/12/1950
METCALFE	S			SGT		22/06/1997
METHVEN	R	D		FL		24/01/1946
MEYRICK-ROBERTS	J	E		FL		01/05/1957
MICHAEL	D	R	A	SQD LDR	Acting	09/08/1945
MIDDLETON	A	G		FL		13/12/1945
MIDDLETON	A	R		FL		27/07/1944
MIDDLETON	A	R		CPL		22/02/1951
MIDDLETON	D	B		FL		28/11/1956
MIDDLETON	H	R		SQD LDR	Acting	07/08/1947
MIDDLETON	L	F		SQD LDR	Acting	18/04/1946
MIDGLEY	D	A	E	FL		07/07/1949
MIDGLEY	J	E		FL		21/03/1946
MIKKELSEN	J	W		FO		13/07/1950
MILBANK	G	A		SQD LDR		16/03/1950
MILBORROW	J	S		FL	Acting	10/08/1944
MILBURN	R	B		SQD LDR	Acting	13/12/1945
MILES	A	E		FL		05/06/1947
MILES	C	A		CPL		07/08/1947
MILES	J	A	C	FO		16/01/1957
MILES	W	J		LAC		24/07/1963
MILES	W	S		LAC		19/08/1964
MILLAR	D	E		FL		07/06/1945
MILLAR	G			WO		12/04/1945
MILLAR	J	L		SGT		28/10/1943
MILLAR	J	N		FL		13/12/1961
MILLAR	R	G		FL		07/04/1955
MILLARD	J	G		FL		26/10/1944
MILLARD	P	C	G	FL		09/12/1959
MILLER	A	C		SQD LDR		13/12/1945
MILLER	A	G		WING COMM		07/01/1943

Source	Service No	Force	Notes	Surname
N531 1960	2659325	WRAAF	wef 2/3/60	MENZIES
N179 1950	873208	RAuxAF		MENZIES
N395 1953	188288	RAuxAF		MERCER
N565 1955	115079	RAuxAF		MERCER
N250 1947	72317	RAFVR		MERCER
N621 1963	198330	RAFVR	wef 10/7/61	MERCER
N395 1953	199774	RAFVR	DFC	MERCER
N1019 1951	855040	RAuxAF		MERCER
N953 1949	1564	WAAF		MERCHANT
N454 1943	108856	RAFVR		MERCHANT
N740 1947	88257	RAFVR		MEREDITH
N381 1945	62651	RAFVR	AFC	MEREDITH
N984 1948	73538	RAFVR		MEREDITH
N460 1947	810143	AAF		MEREDITH
N187 1947	131247	RAFVR		MEREDITH
N197 1946	74337	RAFVR	DSO,DFC	MERIFIELD
N573 1948	104441	RAFVR		MERRETT
N759 1945	83256	RAFVR	DFC	MERRICK
N104 1957	197134	RAFVR	wef 28/5/55	MERRICK
YB 1998	211641	RAuxAF		MERRICK
N552 1959	193248	RAFVR	wef 12/7/59	MERRIEN
N363 1944	73006	RAFVR		MERRIFIELD
N763 1944	86681	RAFVR		MERRINGTON
N740 1947	108488	RAFVR		MERRYMAN
N239 1963	90863	AAF	wef 10/7/44	MESSENGER
N549 1962	133981	RAFVR	wef 2/8/57	METCALF
N874 1945	79556	RAFVR		METCALFE
N1014 1943	807020	AAF		METCALFE
YB 2002	210330	RAuxAF		METCALFE
N1210 1950	1150	WAAF		METCALFE
YB 1998	n/a	RAuxAF		METCALFE
N82 1946	122758	RAFVR		METHVEN
N277 1957	143814	RAFVR	wef 12/3/54	MEYRICK-ROBERTS
N874 1945	85942	RAFVR		MICHAEL
N1355 1945	78251	RAFVR		MIDDLETON
N763 1944	126543	RAFVR	DFC	MIDDLETON
N194 1951	743378	RAFVR		MIDDLETON
N833 1956	167842	RAuxAF		MIDDLETON
N662 1947	110047	RAFVR		MIDDLETON
N358 1946	82729	RAFVR	DFC	MIDDLETON
N673 1949	60328	RAFVR		MIDGLEY
N263 1946	137258	RAFVR		MIDGLEY
N675 1950	190094	RAFVR		MIKKELSEN
N257 1950	73648	RAFVR		MILBANK
N812 1944	141482	RAFVR		MILBORROW
N1355 1945	63070	RAFVR	DFC	MILBURN
N460 1947	131097	RAFVR		MILES
N662 1947	847038	AAF		MILES
N24 1957	205009	RAuxAF		MILES
N525 1963	2682030	RAuxAF	wef 26/4/63	MILES
N672 1954	865538	RAuxAF		MILES
N608 1945	116730	RAFVR		MILLAR
N381 1945	816061	AAF		MILLAR
N1118 1943	741082	RAFVR		MILLAR
N987 1961	111638	RAuxAF	wef 29/1/59	MILLAR
N277 1955	143417	RAFVR		MILLAR
N1094 1944	83999	RAFVR		MILLARD
N843 1959	163288	RAFVR	wef 11/12/57	MILLARD
N1355 1945	119169	RAFVR	DFC	MILLER
N3 1943	90088	AAF	DFC	MILLER

Surname	Initials			Rank	Status	Date
MILLER	A			LAC		18/11/1948
MILLER	A			CPL		17/03/1949
MILLER	B	R		FL		14/01/1954
MILLER	D	W		FL	Acting	14/09/1950
MILLER	G	S		LAC		07/07/1949
MILLER	G	S		SGT		20/03/1957
MILLER	I			FLT OFF		18/07/1956
MILLER	J	A		CPL		09/11/1944
MILLER	J	A		FL		19/08/1959
MILLER	J			LAC		10/05/1951
MILLER	J			LAC		03/07/1952
MILLER	K	L		CPL	Acting	09/12/1959
MILLER	R	I	M	FL		16/08/1961
MILLER	R	J		FL		27/07/1944
MILLER	W	R		FLT SGT	Acting	01/12/1955
MILLER-WILLIAMS	D	E	A	FL		26/11/1953
MILLETT	C	B		WING COMM		06/03/1947
MILLETT	I	L		FLT OFF		06/03/1947
MILLETT	R	E		SGT		25/10/1997
MILLETT	R	F		FL		26/02/1958
MILLIKEN	R	H		SGT		20/09/1951
MILLS	A	H		SGT		23/06/1955
MILLS	F	R		SGT		20/05/1948
MILLS	H			SQD LDR		12/06/1947
MILLS	I	R	A	FL		13/02/1957
MILLS	J	G		FL		21/03/1946
MILLS	J	H	H	FL		18/04/1946
MILLS	J	P		SQD LDR	Acting	30/08/1945
MILLS	J			CPL		15/01/1953
MILLS	K	F		SQD LDR	Acting	01/07/1948
MILLS	L	A		SGT		01/10/1953
MILLS	R	J		SAC		30/03/1994
MILLS	S			LAC		02/08/1951
MILLS	W	G		FO		06/03/1947
MILNE	C	F		SGT		03/06/1959
MILNE	J	R		WING COMM		15/06/1944
MILNE	W	G	T	FL		05/01/1950
MILNE	W			FL		11/07/1946
MILNER	J	L		FL		12/07/1945
MILNER	J	L		FL		17/01/1952
MILNER	V	J		SGT		20/05/1948
MILNES	A	H		FO		03/06/1943
MILNES	A	H		SQD LDR		21/06/1951
MILSOM	W			CPL		20/05/1948
MILTON	J	E		FL		22/02/1961
MILTON	T	E		CPL		10/06/1948
MINGARD	N	S		SQD LDR	Acting	04/10/1945
MINTER	E			FLT SGT		05/06/1947
MINTER	E			FL	Acting	14/10/1948
MINTERN	E	J		L CPL		30/03/1960
MINTERN	E	J		SAC		16/10/1963
MIRICH	P			SGT		01/03/1996
MISKIN	F			SGT		30/09/1948
MITCHAM	T	D		SQD LDR		01/01/1998
MITCHELL	A			FO		28/07/1955
MITCHELL	B	S		SQD OFF		05/02/1948
MITCHELL	C	R		SQD LDR		05/06/1947
MITCHELL	E	T		SQD LDR		26/08/1948
MITCHELL	F	C		CPL		04/09/1947

Source	Service No	Force	Notes	Surname
N984 1948	854464	RAuxAF		MILLER
N231 1949	874131	RAuxAF		MILLER
N18 1954	89562	RAFVR		MILLER
N926 1950	111751	RAFVR	DFC	MILLER
N673 1949	846074	RAuxAF		MILLER
N198 1957	2683503	RAuxAF	wef 8/12/56	MILLER
N513 1956	7500	WRAFVR		MILLER
N1150 1944	803458	AAF		MILLER
N552 1959	197980	RAFVR	DFC; wef 12/1/58	MILLER
N466 1951	850595	RAuxAF		MILLER
N513 1952	870262	RAuxAF		MILLER
N843 1959	2696027	RAuxAF	wef 25/8/59	MILLER
N624 1961	111576	RAFVR	wef 4/4/59	MILLER
N763 1944	81349	RAFVR		MILLER
N895 1955	2661653	WRAAF		MILLER
N952 1953	163748	RAFVR		MILLER-WILLIAMS
N187 1947	73820	RAFVR		MILLETT
N187 1947	5848	WAAF	BEM	MILLETT
YB 1999	n/a	RAuxAF		MILLETT
N149 1958	171804	RAFVR	wef 12/8/57	MILLETT
N963 1951	816030	RAuxAF		MILLIKEN
N473 1955	2602756	RAFVR		MILLS
N395 1948	840964	RAuxAF		MILLS
N478 1947	90542	AAF		MILLS
N104 1957	80201	RAFVR	wef 18/8/44	MILLS
N263 1946	148002	RAFVR		MILLS
N358 1946	133554	RAFVR	DFM	MILLS
N961 1945	64890	RAFVR	DFC	MILLS
N28 1953	863532	RAuxAF		MILLS
N527 1948	117147	RAFVR		MILLS
N796 1953	746212	RAFVR		MILLS
YB 1996	n/a	RAuxAF		MILLS
N777 1951	854496	RAuxAF		MILLS
N187 1947	114134	RAFVR		MILLS
N368 1959	2683580	RAuxAF	wef 3/1/59	MILNE
N596 1944	90969	AAF	DFC	MILNE
N1 1950	86668	RAFVR		MILNE
N594 1946	170440	RAFVR	DFM	MILNE
N759 1945	86386	RAFVR		MILNER
N35 1952	86386	RAFVR	1st Clasp	MILNER
N395 1948	819038	RAuxAF		MILNER
N609 1943	101001	RAFVR		MILNES
N621 1951	101001	RAFVR	1st Clasp	MILNES
N395 1948	813527	RAuxAF		MILSOM
N153 1961	111833	RAFVR	wef 12/10/60	MILTON
N456 1948	860476	RAuxAF		MILTON
N1089 1945	61944	RAFVR	DSO,DFC	MINGARD
N460 1947	881949	WAAF		MINTER
N856 1948	201177	RAFVR	AFC	MINTER
N229 1960	23662162	TA	Inns of Court Regiment; wef 28/7/53	MINTERN
N756 1963	2684613	RAuxAF	1st Clasp; wef 28/7/63	MINTERN
YB 1997	n/a	RAuxAF		MIRICH
N805 1948	702816	RAFVR		MISKIN
YB 1999	2607677	RAuxAF		MITCHAM
N565 1955	2684019	RAuxAF		MITCHELL
N87 1948	543	WAAF		MITCHELL
N460 1947	72204	RAFVR		MITCHELL
N694 1948	73649	RAFVR		MITCHELL
N740 1947	746072	RAFVR		MITCHELL

Surname	Initials			Rank	Status	Date
MITCHELL	H	M		WING COMM	Acting	10/12/1942
MITCHELL	J	G		FL		28/02/1946
MITCHELL	J	L		SQD LDR	Acting	28/02/1946
MITCHELL	J	N		FL		04/09/1947
MITCHELL	J	R	P	FL		10/06/1948
MITCHELL	L	J	C	SQD LDR		10/12/1942
MITCHELL	M			FLT SGT		24/11/1949
MITCHELL	P	F		FL		24/11/1949
MITCHELL	R	A		FO		07/05/1953
MITCHELL	T			FL		13/11/1947
MITCHELL	W	A		FL		20/01/1960
MITCHENER	E			FLT OFF		27/03/1947
MITTEN	R	J		CPL		08/06/1995
MOABY	S	M		SAC		23/08/1999
MOBSBY	L	G		WING COMM	Acting	08/05/1947
MODLEY	L	W	W	WING COMM		19/09/1956
MOFFAT	C			LAC		09/11/1944
MOFFAT	R	H		FL		29/06/1944
MOGG	A	W		FO		18/04/1946
MOGG	F	G		WING COMM		24/06/1943
MOGG	R	P	L	WO		28/11/1946
MOGG	R	P	L	CAPT		22/02/1961
MOIR	A	G	C	FL		01/01/2003
MOIR	C			FO		13/10/1955
MOIR	R	F		CPL		12/10/1950
MOISEY	M			FLT SGT		12/06/1947
MOLE	G	W	S	FO		26/08/1948
MOLESWORTH	J	H	N	WING COMM	Acting	01/03/1945
MOLL	E	B	D'A	FLT OFF		27/03/1947
MOLONEY	B	H		SQD LDR		04/10/1945
MOLYNEUX	L			SGT		30/12/1943
MONCUR	W	K		FL		05/06/1947
MONCUR	W	K		FL		01/05/1952
MONKMAN	G	B		FL		26/01/1950
MONKMAN	J			SGT		13/01/1949
MONKS	G	A		FO		15/01/1953
MONKS	G	A		FL		01/05/1957
MONNICKENDAM	L			FO		18/03/1948
MONRO	H	S	P	FL		28/07/1955
MONTAGUE	F	S		SQD LDR	Acting	06/03/1947
MONTEITH	J	J		CPL		18/11/1948
MONTGOMERY	M			SGT	Acting	14/11/1962
MONTGOMERY	W	N		WING COMM	Acting	06/03/1947
MOOD	J			FO		06/03/1947
MOODIE	A	M		FL		12/06/1998
MOODIE	W	A		CPL		03/06/1959
MOODY	D	G		FL		11/01/1945
MOODY	G	C	R	FL	Acting	03/12/1942
MOODY	J	H		WO		17/01/1952
MOON	W	B		SQD LDR	Acting	26/01/1950
MOON	W	T		CPL	Temporary	21/06/1961
MOONEY	J			CPL		10/02/1949
MOONEY	M			SGT		26/02/1958
MOONEY	R	E		FL		16/01/1957
MOORE	A	R		FL		28/02/1946
MOORE	A	S		SGT		16/10/1952
MOORE	B	D		SQD OFF	Acting	26/06/1947
MOORE	D	W		FO		31/05/1945
MOORE	E	C		LAC		13/12/1951

Source	Service No	Force	Notes	Surname
N1680 1942	90245	AAF	DFC	MITCHELL
N197 1946	101466	RAFVR		MITCHELL
N197 1946	78986	RAFVR	MVO, DFC	MITCHELL
N740 1947	123159	RAFVR		MITCHELL
N456 1948	91049	RAuxAF		MITCHELL
N1680 1942	70468	RAFVR		MITCHELL
N1193 1949	880415	WAAF		MITCHELL
N1193 1949	88139	RAFVR		MITCHELL
N395 1953	2689007	RAuxAF		MITCHELL
N933 1947	123407	RAFVR		MITCHELL
N33 1960	1710414	RAFVR	wef 9/12/59	MITCHELL
N250 1947	3097	WAAF		MITCHENER
YB 1996	n/a	RAuxAF		MITTEN
YB 2000	n/a	RAuxAF		MOABY
N358 1947	90846	AAF		MOBSBY
N666 1956	24098	RAuxAF	OBE,GM	MODLEY
N1150 1944	803451	AAF		MOFFAT
N652 1944	102994	RAFVR		MOFFAT
N358 1946	175065	RAFVR		MOGG
N677 1943	90028	AAF	GM,MRCS,LRCP	MOGG
N1004 1946	744979	RAFVR	MBE	MOGG
N153 1961	n/a	TA	MBE; 1st Clasp; wef 28/4/53	MOGG
YB 2003	213465	RAuxAF		MOIR
N762 1955	202600	RAuxAF		MOIR
N1018 1950	853645	RAuxAF		MOIR
N478 1947	880907	WAAF		MOISEY
N694 1948	137816	RAFVR		MOLE
N225 1945	80828	RAFVR	DFC,AFC	MOLESWORTH
N250 1947	5803	WAAF		MOLL
N1089 1945	90237	AAF		MOLONEY
N1350 1943	741038	RAFVR		MOLYNEUX
N460 1947	120856	RAFVR		MONCUR
N336 1952	91303	RAuxAF	1st Clasp	MONCUR
N94 1950	91132	RAuxAF		MONKMAN
N30 1949	752777	RAFVR		MONKMAN
N28 1953	165797	RAFVR		MONKS
N277 1957	165797	RAuxAF	1st Clasp; wef 14/2/54	MONKS
N232 1948	77395	RAFVR		MONNICKENDAM
N565 1955	130956	RAFVR		MONRO
N187 1947	86261	RAFVR		MONTAGUE
N984 1948	874074	RAuxAF		MONTEITH
N837 1962	2650956	RAuxAF	wef 16/9/62	MONTGOMERY
N187 1947	90817	AAF		MONTGOMERY
N187 1947	183783	RAFVR		MOOD
YB 1999	2625256	RAuxAF		MOODIE
N368 1959	2691027	RAuxAF	wef 18/3/59	MOODIE
N14 1945	118929	RAFVR		MOODY
N1638 1942	78846	RAFVR		MOODY
N35 1952	858032	RAuxAF		MOODY
N94 1950	86219	RAFVR		MOON
N478 1961	811114	AAF	wef 12/11/43	MOON
N123 1949	802494	RAuxAF		MOONEY
N149 1958	2680527	RAFVR	wef 2/11/57	MOONEY
N24 1957	170478	RAuxAF		MOONEY
N197 1946	102100	RAFVR	DFC	MOORE
N798 1952	799759	RAFVR		MOORE
N519 1947	2966	WAAF	MBE	MOORE
N573 1945	161403	RAFVR	DFC	MOORE
N1271 1951	865122	RAuxAF		MOORE

Surname	Initials			Rank	Status	Date
MOORE	E	J		FLT SGT		09/08/1945
MOORE	E	J		FL		30/09/1954
MOORE	E	U	B	SGT		05/01/1950
MOORE	F	A		FO		10/06/1948
MOORE	J	M	J	FL		26/01/1950
MOORE	L	W		WO		28/02/1946
MOORE	M	A	S	FL		04/02/2000
MOORE	M	J		FLT OFF		12/06/1947
MOORE	R	F		SGT		02/11/1995
MOORE	T	H		FL		24/11/1949
MOORMAN	W	H	S	FL		18/03/1948
MORCOM	B	W		FO		21/06/1996
MORCOM	H	G		SQD LDR	Acting	12/06/1947
MORDAUNT	J	F		FL		25/01/1945
MORE	A			FL		07/08/1947
MOREL	M	P		SQD LDR		19/08/1954
MORELAND	R	G	D	FL	Acting	04/09/1947
MOREN	P	D	J	FL		06/03/1947
MORGAN	A	V		LAC		21/06/1951
MORGAN	D	F		SGT		20/07/1960
MORGAN	F	W		SQD LDR	Acting	06/03/1947
MORGAN	I	D		FL		07/03/1996
MORGAN	J	J		FO		25/03/1954
MORGAN	J	R		WING COMM	Acting	27/03/1947
MORGAN	J			SQD LDR		09/11/1950
MORGAN	J			SGT		30/01/1989
MORGAN	L	A	R	CPL		23/02/1950
MORGAN	L	V	W	FO		23/05/1946
MORGAN	P	H	G	FLT SGT		26/04/1961
MORGAN	R	E		FO		24/07/1963
MORGAN	S	G		WING COMM		08/05/1947
MORGAN	S			SGT	Temporary	14/01/1954
MORGAN	W	H		SGT		06/01/1955
MORGAN	W	T		SGT		30/09/1948
MORGANS	C	E		FL	Acting	01/05/1952
MORGANS	F	W		SGT		12/06/1947
MORHEN	G	J	P	FL		20/09/1945
MORLAND	T	H		CPL		19/08/1954
MORLEY	G	A		FL		31/05/1945
MORLEY	H	D		SQD LDR		30/03/1943
MORLEY	H	F		SQD LDR		17/06/1943
MORLEY	W	G		CPL		16/03/1950
MORLING	H	C		AC1		30/03/1960
MORRELL	S	A		WO		04/11/1948
MORREY	W	G		LAC		04/11/1948
MORRIS	A	D		FL		19/08/1954
MORRIS	A	E	N	FL		07/08/1947
MORRIS	A	S	J	SGT		04/09/1947
MORRIS	G	C		FL		18/02/1954
MORRIS	G	E		SQD LDR	Acting	17/04/1947
MORRIS	G			SGT		01/07/1948
MORRIS	P	A		FLT SGT		04/05/1944
MORRIS	P	A		SGT		24/07/1952
MORRIS	R	E	V	CPL	Temporary	22/09/1955
MORRIS	R	W		FO		26/08/1948
MORRIS	R			FL		04/10/1945
MORRIS	R			FO		21/06/1951
MORRISON	A	W		SAC		15/01/1953
MORRISON	C	W		SGT		01/11/1987
MORRISON	C	W		SGT		01/11/1997

Source	Service No	Force	Notes	Surname
N874 1945	801353	AAF		MOORE
N799 1954	173588	RAFVR		MOORE
N1 1950	864806	RAuxAF		MOORE
N456 1948	171440	RAFVR		MOORE
N94 1950	73468	RAFVR	MC	MOORE
N197 1946	805084	AAF		MOORE
YB 2002	2640445	RAuxAF		MOORE
N478 1947	809	WAAF		MOORE
YB 1997	n/a	RAuxAF		MOORE
N1193 1949	141916	RAFVR	MSM	MOORE
N232 1948	82596	RAFVR	MBE	MOORMAN
YB 1998	2634899	RAuxAF		MORCOM
N478 1947	73303	RAFVR		MORCOM
N70 1945	106680	RAFVR		MORDAUNT
N662 1947	103522	RAFVR	DFC	MORE
N672 1954	72546	RAFVR		MOREL
N740 1947	128356	RAFVR		MORELAND
N187 1947	114336	RAFVR	DFM	MOREN
N621 1951	840063	RAuxAF		MORGAN
N531 1960	843832	AAF	wef 13/3/44	MORGAN
N187 1947	86966	RAFVR	MBE	MORGAN
YB 1998	213228	RAuxAF		MORGAN
N249 1954	195937	RAuxAF		MORGAN
N250 1947	73539	RAFVR	OBE	MORGAN
N1113 1950	82673	RAFVR	DSO	MORGAN
YB 1996	n/a	RAuxAF		MORGAN
N179 1950	863251	RAuxAF		MORGAN
N468 1946	190841	RAFVR		MORGAN
N339 1961	2607552	RAFVR	wef 2/5/56	MORGAN
N525 1963	2689083	RAFVR(T)	wef 20/9/57	MORGAN
N358 1947	72897	RAFVR		MORGAN
N18 1954	813103	RAuxAF		MORGAN
N1 1955	2677602	RAuxAF		MORGAN
N805 1948	855071	RAuxAF		MORGAN
N336 1952	116732	RAFVR		MORGANS
N478 1947	843889	AAF		MORGANS
N1036 1945	113485	RAFVR		MORHEN
N672 1954	2694003	RAuxAF		MORLAND
N573 1945	145118	RAFVR		MORLEY
N1014 1943	70474	RAFVR		MORLEY
N653 1943	70473	RAFVR		MORLEY
N257 1950	770627	RAFVR		MORLEY
N229 1960	843010	AAF	wef 2/3/44	MORLING
N934 1948	815060	RAuxAF		MORRELL
N934 1948	853881	RAuxAF		MORREY
N672 1954	100148	RAFVR		MORRIS
N662 1947	90516	AAF		MORRIS
N740 1947	702229	RAFVR		MORRIS
N137 1954	86833	RAFVR		MORRIS
N289 1947	78464	RAFVR		MORRIS
N527 1948	770899	RAFVR		MORRIS
N425 1944	802370	AAF		MORRIS
N569 1952	2681663	RAuxAF	1st Clasp	MORRIS
N708 1955	860354	RAuxAF		MORRIS
N694 1948	69516	RAFVR		MORRIS
N1089 1945	78773	RAFVR		MORRIS
N621 1951	78733	RAFVR	1st Clasp	MORRIS
N28 1953	3685504	RAuxAF		MORRISON
YB 1999	n/a	RAuxAF		MORRISON
YB 1999	n/a	RAuxAF	1st Clasp	MORRISON

Surname	Initials			Rank	Status	Date
MORRISON	D			FL		01/02/1991
MORRISON	F	J		SGT		13/11/1947
MORRISON	J	L		LAC		07/07/1949
MORRISON	R	D	McD	SQD LDR		30/09/1954
MORRISON	T	R		FL		24/12/1942
MORRISS	E	R		FL		28/02/1946
MORRISSEY	C	B		CPL		05/06/1947
MORROGH-STEWART	R	F	C	FLT OFF		17/04/1947
MORROW	M			CPL		04/09/1947
MORROW	R			LAC		28/02/1962
MORTEMORE	C	F		CPL		18/11/1948
MORTER	P	S	P	WING COMM	Acting	08/03/1951
MORTIMER	J	W	F	SGT		19/01/1956
MORTIMER	M	M		SQD OFF		07/08/1947
MORTON	A	F		LAC		23/02/1950
MORTON	H	V		FO		28/11/1946
MORTON	J	S		WING COMM	Acting	21/09/1944
MORTON	J			LAC		25/06/1953
MORTON	L			FL		30/09/1954
MORTON	P	L	W	SQD LDR		14/01/1954
MORTON	R	D		FL		27/03/1947
MOSEDALE	P	J		FL		13/01/1949
MOSELEY	C	F		SQD LDR		30/03/1943
MOSS	A	E		SGT		01/05/1957
MOSS	F			FO		23/02/1950
MOSS	H	B		FL		06/03/1947
MOSS	R	C		FO		13/12/1945
MOSSOM	C			SGT		01/07/1948
MOSSOP	H	R		FL		18/07/1956
MOTT	A	J		FL		28/02/1946
MOTT	W	H		FL		01/08/1946
MOTTISTONE				FL		07/02/1952
MOTTRAM	W			LAC		26/01/1950
MOUG	A			FL	Acting	05/02/1948
MOULD	R	D		FLT OFF		27/03/1947
MOULD	T	H		CPL	Temporary	03/07/1957
MOULD	V	S		SGT		06/01/1955
MOULSDALE	R	P		FL		10/02/1949
MOULT	C	J		FL		01/01/1998
MOUNTER	D	J		FLT SGT		24/06/1943
MOWAT	M			FLT OFF		26/11/1953
MOXON	C	F		SGT		27/03/1947
MUDIE	R	S		WING COMM	Acting	28/11/1956
MUGGERIDGE	C	H		FL		20/09/1945
MUGGLETON	K	E	S	FL		11/07/1946
MUGGRIDGE	E	J		SQD LDR	Acting	25/05/1950
MUIR	A	M		SQD OFF	Acting	27/03/1947
MUIR	R	W		SQD LDR		01/06/1997
MUIR	R	Y		FL		24/01/1946
MUIRHEAD	J	C		FL		26/04/1961
MULBERRY	S			SGT		12/06/1996
MULHOLLAND	E	S		FL		06/03/1947
MULLAY	G	G		SGT		20/05/1948
MULLEN	W			SQD LDR		28/10/1943
MULLER	C	F		FL		30/08/1945
MULLEY	L	J		SGT		23/08/1951
MULLINS	J	D		CPL		19/07/1951
MULLINS	W	A		FL		26/08/1948
MULVEY	J	M		FL		05/09/1962
MUMFORD	F	J		SGT		26/08/1948

Source	Service No	Force	Notes	Surname
YB 1996	2626701	RAuxAF		MORRISON
N933 1947	816139	AAF		MORRISON
N673 1949	816199	RAuxAF		MORRISON
N799 1954	85611	RAuxAF		MORRISON
N1758 1942	70478	RAFVR		MORRISON
N197 1946	141801	RAFVR		MORRISS
N460 1947	811053	AAF		MORRISSEY
N289 1947	1495	WAAF		MORROGH-STEWART
N740 1947	873823	AAF		MORROW
N157 1962	2650939	RAuxAF	wef 15/1/62	MORROW
N984 1948	846136	RAuxAF		MORTEMORE
N244 1951	90856	RAuxAF		MORTER
N43 1956	867787	RAuxAF		MORTIMER
N662 1947	188	WAAF	MBE	MORTIMER
N179 1950	864764	RAuxAF		MORTON
N1004 1946	195653	RAFVR		MORTON
N961 1944	90727	AAF	DFC	MORTON
N527 1953	853650	RAuxAF		MORTON
N799 1954	119404	RAuxAF		MORTON
N18 1954	90921	RAuxAF		MORTON
N250 1947	84937	RAFVR		MORTON
N30 1949	91081	RAuxAF		MOSEDALE
N1014 1943	72428	RAFVR		MOSELEY
N277 1957	840908	RAuxAF	wef 22/3/44	MOSS
N179 1950	123812	RAFVR		MOSS
N187 1947	61606	RAFVR	MC	MOSS
N1355 1945	174241	RAFVR		MOSS
N527 1948	841528	RAuxAF		MOSSOM
N513 1956	143942	RAFVR	DFC	MOSSOP
N197 1946	120214	RAFVR	MBE	MOTT
N652 1946	139388	RAFVR		MOTT
N103 1952	90527	RAuxAF	Lord	MOTTISTONE
N94 1950	814195	RAuxAF		MOTTRAM
N87 1948	132547	RAFVR		MOUG
N250 1947	760	WAAF		MOULD
N456 1957	870919	AAF	DSM; wef 12/7/44	MOULD
N1 1955	747339	RAFVR		MOULD
N123 1949	81432	RAFVR		MOULSDALE
YB 1999	n/a	RAuxAF		MOULT
N677 1943	813003	AAF		MOUNTER
N952 1953	2110	WRAAF		MOWAT
N250 1947	809194	AAF		MOXON
N833 1956	76429	RAFVR		MUDIE
N1036 1945	65507	RAFVR	DFC	MUGGERIDGE
N594 1946	121536	RAFVR		MUGGLETON
N517 1950	84590	RAFVR		MUGGRIDGE
N250 1947	321	WAAF		MUIR
YB 1999	4231677	RAuxAF		MUIR
N82 1946	104375	RAFVR		MUIR
N339 1961	131991	RAFVR	wef 21/2/61	MUIRHEAD
YB 1997	n/a	RAuxAF		MULBERRY
N187 1947	174229	RAFVR	DFC	MULHOLLAND
N395 1948	803649	RAuxAF		MULLAY
N1118 1943	70484	RAFVR		MULLEN
N961 1945	129251	RAFVR	DFC,DFM	MULLER
N866 1951	744865	RAFVR		MULLEY
N724 1951	846234	RAuxAF		MULLINS
N694 1948	91188	RAuxAF		MULLINS
N655 1962	140270	RAFVR	wef 8/6/462	MULVEY
N694 1948	771207	RAFVR		MUMFORD

Surname	Initials			Rank	Status	Date
MUNDELL	F	G		WING COMM	Acting	07/07/1949
MUNDELL	F			SGT		18/11/1948
MUNDY	A	H		FL		08/11/1945
MUNDY	A			LAC		23/02/1950
MUNFORD	L			CPL		26/06/1947
MUNGHAM	H	H		FL		25/06/1953
MUNNS	S	A	E	SQD LDR	Acting	24/01/1946
MUNRO	A			FL	Acting	23/05/1946
MUNRO	J			SACW		14/11/1962
MUNRO	M			SACW		14/11/1962
MUNSON	K	A	O	FL		31/10/1956
MURCHISON	T	McD		FO		07/04/1955
MURDOCH	D	R	A	PO		01/10/1953
MURISON	D			FLT SGT		16/10/1947
MURLY-GOTTO	H	T		FL		27/03/1952
MURPHY	D	J		LAC		24/11/1949
MURPHY	J	A		FL		24/01/1946
MURPHY	J	T		FO		18/02/1954
MURPHY	N	F	C	FO		06/01/1955
MURRAY	B	A		FL		01/01/1998
MURRAY	F	W		CPL	Temporary	19/10/1960
MURRAY	H	J		LAC		23/02/1950
MURRAY	J	W		SQD LDR	Acting	20/04/1944
MURRAY	J	Y		FO		31/05/1945
MURRAY	M	P		WO		26/08/1948
MURRAY	T	R	M	FLT SGT		03/07/1993
MURRAY	W	J	B	LAC		26/06/1947
MURRELL	J	E		WO		28/02/1946
MURTAGH	W			LAC		18/07/1956
MURTON	R	J		FL		29/04/1954
MUSKER	J	H		SQD LDR	Acting	20/09/1961
MUSKETT	R	W	H	F	FL	21/01/1959
MUSPRATT-WILLIAMS	R	T		SQD LDR	Acting	03/06/1943
MUSSON	F	G	S	SQD LDR	Acting	05/01/1950
MUTTON	A	L		SAC		02/12/1992
MYALL	G	O		FL		26/01/1950
MYALL	J	A		FL		05/01/1950
MYATT	S			FL		30/09/1948
MYCROFT	F	H		FO		28/11/1956
MYERS	L			SGT		27/04/1950
MYERS	W	H		SGT		20/07/1960
NALL	F	J		FO		24/01/1946
NASH	D	F	E	WING COMM		05/01/1950
NASON	J	R		CPL		15/09/1949
NAVA	N	R	A	FO		02/09/1943
NAVIN	J	H		LAC		30/09/1948
NEAL	E	C		SGT		23/02/1950
NEAL	G	B	T	FL		21/03/1946
NEALE	C	L		WO		26/01/1950
NEALE	F	J		FL	Acting	06/03/1947
NEALE	P	A		WO		20/05/1948
NEAVES	E	G	J	WING COMM	Acting	23/02/1950
NEECH	E	W		WING COMM	Acting	05/01/1950
NEEDHAM	E			FL	Acting	05/06/1947
NEEDHAM	E			FLT SGT		16/04/1958
NEELEY	A			CPL		14/11/1998
NEEVES	L	D		WING COMM	Acting	10/06/1948
NEGRAN	R			LAC		30/03/1960
NEIL	J	D		SQD LDR		30/09/1948
NEIL	T	F		SQD LDR	Acting	09/08/1945

Source	Service No	Force	Notes	Surname
N673 1949	72116	RAFVR	MB,BS	MUNDELL
N984 1948	872428	RAuxAF		MUNDELL
N1232 1945	149099	RAFVR		MUNDY
N179 1950	770801	RAFVR		MUNDY
N519 1947	804288	AAF		MUNFORD
N527 1953	193001	RAuxAF		MUNGHAM
N82 1946	116154	RAFVR		MUNNS
N468 1946	144504	RAFVR		MUNRO
N837 1962	2664195	WRAAF	wef 15/9/62	MUNRO
N837 1962	2664196	WRAAF	wef 15/9/62	MUNRO
N761 1956	188691	RAuxAF		MUNSON
N277 1955	176324	RAuxAF		MURCHISON
N796 1953	2658607	WRAAF		MURDOCH
N846 1947	881063	WAAF		MURISON
N224 1952	81436	RAFVR		MURLY-GOTTO
N1193 1949	865691	RAuxAF		MURPHY
N82 1946	118151	RAFVR		MURPHY
N137 1954	2600913	RAFVR		MURPHY
N1 1955	105307	RAFVR		MURPHY
YB 1999	2633448	RAuxAF		MURRAY
N775 1960	852099	AAF	wef 14/6/44	MURRAY
N179 1950	854412	RAuxAF		MURRAY
N363 1944	84977	RAFVR	DFC,DFM	MURRAY
N573 1945	164118	RAFVR		MURRAY
N694 1948	740441	RAFVR		MURRAY
YB 1996	n/a	RAuxAF		MURRAY
N519 1947	840930	AAF		MURRAY
N197 1946	801497	AAF		MURRELL
N513 1956	858916	RAuxAF		MURTAGH
N329 1954	103897	RAuxAF		MURTON
N720 1961	90953	AAF	wef 30/5/44	MUSKER
N28 1959	2692506	RAuxAF	wef 12/1/58	MUSKETT
N609 1943	90169	AAF		MUSPRATT-WILLIAMS
N1 1950	73311	RAFVR		MUSSON
YB 1996	n/a	RAuxAF		MUTTON
N94 1950	127855	RAFVR		MYALL
N1 1950	114388	RAFVR		MYALL
N805 1948	130466	RAFVR		MYATT
N833 1956	2600439	RAuxAF		MYCROFT
N406 1950	771004	RAFVR		MYERS
N531 1960	2691045	RAuxAF	wef 23/11/59	MYERS
N82 1946	199362	RAFVR		NALL
N1 1950	70489	RAFVR	FRCS,LRCP	NASH
N953 1949	743525	RAFVR		NASON
N918 1943	136167	RAFVR		NAVA
N805 1948	857418	RAuxAF		NAVIN
N179 1950	756981	RAFVR		NEAL
N263 1946	174366	RAFVR		NEAL
N94 1950	759228	RAFVR		NEALE
N187 1947	137815	RAFVR		NEALE
N395 1948	801529	RAuxAF		NEALE
N179 1950	72389	RAFVR		NEAVES
N1 1950	72390	RAFVR		NEECH
N460 1947	137049	RAFVR		NEEDHAM
N258 1958	2687020	RAFVR	wef 16/11/57	NEEDHAM
YB 1999	n/a	RAuxAF		NEELEY
N456 1948	72270	RAFVR		NEEVES
N229 1960	816206	AAF	wef 27/6/44	NEGRAN
N805 1948	72324	RAFVR		NEIL
N874 1945	79168	RAFVR	DFC	NEIL

Surname	Initials			Rank	Status	Date
NEILL	A	C		WING COMM	Acting	08/05/1947
NEILL	C	D	A	FL		28/11/1946
NEILL	H	G	W	SGT		01/07/1948
NEILL	H	H		SGT		07/08/1947
NELSON	F	T		SGT		07/06/1951
NELSON	F	V		CPL		04/09/1947
NELSON	J			FL		01/10/1953
NEUSS	R	F		FL		20/08/1958
NEVE	C	A	F	CPL		07/12/1960
NEVILLE	R	T		FL	Acting	07/08/1947
NEVITT	J	E		FL		15/01/1958
NEWALL	G			FL		28/11/1956
NEWBERRY	C	J		SQD LDR	Acting	15/07/1948
NEWBERY	J	F		FL		25/03/1954
NEWBON	F			FL		12/04/1951
NEWBON	J			FL		28/02/1946
NEWBURY	C	W		CPL		26/08/1948
NEWBY	R			LAC		07/07/1949
NEWCASTLE				SQD LDR		10/05/1945
NEWCOMBE	D	C		SAC		08/12/1993
NEWELL	A	G		SGT		21/06/1961
NEWELL	S	W		FL		01/05/1963
NEWEY	M	W		LAC		13/01/1949
NEWEY	R			LAC		21/12/1950
NEWHOUSE	H	C		SQD LDR		11/11/1943
NEWHOUSE	P	S		FL		19/07/1951
NEWITT	A	R		FL		03/10/1946
NEWLAND	P	C	E	CPL		01/07/1948
NEWLING	L	K		SGT		23/02/1950
NEWLYN	G	R		LAC		25/05/1950
NEWMAN	A	E		SGT		10/06/1948
NEWMAN	A	L		WO	Acting	26/06/1947
NEWMAN	C	A		FL	Acting	16/10/1947
NEWMAN	G	E		LAC		30/09/1948
NEWMAN	G	H	A	SQD LDR	Acting	12/06/1947
NEWMAN	J	B		WING COMM		06/03/1947
NEWMAN	P	M	E	FO		30/10/1957
NEWMAN	R	H		LAC		07/12/1960
NEWMAN	T	A		FL		09/11/1950
NEWNHAM	B			FL		18/11/1948
NEWNHAM	E	B		CPL		07/07/1949
NEWSON	G	W		SGT		27/03/1952
NEWSTEAD	H	W		WING COMM		01/07/1948
NEWTON	C	J		SQD LDR	Acting	08/05/1947
NEWTON	D	W		FL		24/01/1946
NEWTON	E	S		FL		13/02/1957
NEWTON	H	M		SQD LDR		24/01/1946
NEWTON	H	S		FL		21/06/1945
NEWTON	H	S		CPL		08/05/1947
NEWTON	J	C		FLT SGT		07/04/1955
NEWTON	J	L	C	SQD LDR	Acting	30/12/1943
NEWTON	J	L		FL	Acting	05/09/1946
NEWTON	K			FO		13/12/1945
NEWTON	P	R		WO		13/12/1945
NEWTON	R	C		FL		21/06/1945
NEWTON	S	H		FL		16/08/1967
NEWTON	T	F		WO		28/02/1946
NEWTON	T	J		FO		13/10/1955
NEWTON	W	E		FL		27/06/1962
NEWZLING	D	A		FLT SGT		07/04/1955

Source	Service No	Force	Notes	Surname
N358 1947	90889	AAF		NEILL
N1004 1946	86353	RAFVR		NEILL
N527 1948	760400	RAFVR		NEILL
N662 1947	811172	AAF		NEILL
N560 1951	802356	RAuxAF		NELSON
N740 1947	811143	AAF		NELSON
N796 1953	185268	RAuxAF		NELSON
N559 1958	116421	RAFVR	DFC; wef 15/6/53	NEUSS
N933 1960	2693557	RAuxAF	wef 15/10/60	NEVE
N662 1947	121843	RAFVR		NEVILLE
N40 1958	21377	RAuxAF	wef 20/8/53	NEVITT
N833 1956	52664	RAuxAF	DFM	NEWALL
N573 1948	91211	RAuxAF		NEWBERRY
N249 1954	141354	RAFVR		NEWBERY
N342 1951	127652	RAFVR		NEWBON
N197 1946	148037	RAFVR		NEWBON
N694 1948	860447	RAuxAF		NEWBURY
N673 1949	856754	RAuxAF		NEWBY
N485 1945	90317	AAF	OBE; The Duke of	NEWCASTLE
YB 1996	n/a	RAuxAF		NEWCOMBE
N478 1961	842789	RAF	wef 21/5/44	NEWELL
N333 1963	155308	RAFVR	wef 20/4/53	NEWELL
N30 1949	849065	RAuxAF		NEWEY
N1270 1950	855932	RAuxAF		NEWEY
N1177 1943	90313	AAFRO		NEWHOUSE
N724 1951	112423	RAFVR	DFC	NEWHOUSE
N837 1946	141408	RAFVR	DFM	NEWITT
N527 1948	750680	RAFVR		NEWLAND
N179 1950	744332	RAFVR		NEWLING
N517 1950	840257	RAuxAF		NEWLYN
N456 1948	746510	RAFVR		NEWMAN
N519 1947	812355	RAFVR		NEWMAN
N846 1947	115424	RAFVR		NEWMAN
N805 1948	846284	RAuxAF		NEWMAN
N478 1947	110043	RAFVR		NEWMAN
N187 1947	73731	RAFVR	OBE	NEWMAN
N775 1957	544	WRAFVR	wef 3/9/44	NEWMAN
N933 1960	2650851	RAuxAF	wef 19/10/60	NEWMAN
N1113 1950	146646	RAFVR	DFC	NEWMAN
N984 1948	127653	RAFVR		NEWNHAM
N673 1949	863509	RAuxAF		NEWNHAM
N224 1952	814153	RAuxAF		NEWSON
N527 1948	72161	RAFVR		NEWSTEAD
N358 1947	72404	RAFVR		NEWTON
N82 1946	130730	RAFVR	AFC	NEWTON
N104 1957	191304	RAFVR	wef 10/2/54	NEWTON
N82 1946	72741	RAFVR		NEWTON
N671 1945	134750	RAFVR	AFC	NEWTON
N358 1947	750785	RAFVR		NEWTON
N277 1955	2686023	RAuxAF		NEWTON
N1350 1943	70500	RAFVR		NEWTON
N751 1946	127780	RAFVR		NEWTON
N1355 1945	187993	RAFVR	DFM	NEWTON
N1355 1945	743058	RAFVR		NEWTON
N671 1945	90032	EX-AAF		NEWTON
MOD 1968	51870	RAFVR		NEWTON
N197 1946	740581	RAFVR		NEWTON
N762 1955	110301	RAFVR		NEWTON
MOD 1967	206944	RAFVR		NEWTON
N277 1955	2602290	RAFVR		NEWZLING

Surname	Initials			Rank	Status	Date
NICE	H	T	W	SGT	Acting	26/01/1950
NICHOLAS	N	J		FO		28/02/1946
NICHOLAS	R	H		SAC		24/10/1997
NICHOLLS	C	H	C	CPL		10/02/1949
NICHOLLS	D	B	F	SQD LDR	Acting	21/06/1945
NICHOLS	C	E		FL		20/11/1957
NICHOLS	C	M		FLT OFF		27/03/1947
NICHOLS	D	H		FL		11/07/1946
NICHOLS	R	E		LAC		15/09/1949
NICHOLSON	A	A	N	SQD LDR		06/03/1947
NICHOLSON	D	A	D L	SQD LDR		27/07/1944
NICHOLSON	H			FL		21/03/1946
NICHOLSON	J	W		SQD OFF	Acting	08/05/1947
NICHOLSON	S	V		FL		15/04/1943
NICHOLSON	S	V		WING COMM	Acting	12/02/1953
NICHOLSON	Y	H		FLT SGT		18/03/1948
NICKELS	E	W		FL		19/10/1960
NICKOLS	P	R		FL		22/04/1943
NICKSON	F			FL		28/02/1946
NICOL	A	S		SGT		27/11/1952
NICOL	E	M		WING COMM	Acting	06/03/1947
NICOL	W	P		SQD LDR	Acting	13/12/1945
NICOL	W	P		FL		30/09/1954
NICOLL	E	S		FL		14/01/1954
NICOLL	J	R		FL		14/11/1962
NICOLL	W			CPL		07/02/1952
NIELD	D	H		FL		26/08/1948
NIVEN	J	B		SQD LDR	Acting	21/03/1946
NIVEN	R			FL		28/02/1946
NIXON	D	B		FL		13/12/1945
NIXON	H	W		CPL		01/07/1948
NIXON	H			FL		21/03/1946
NIXON	J	W		FL		05/09/1946
NIXON	J	W		FL		09/01/1963
NIXON	S	A		CPL		05/01/1950
NOAKES	J	F		CPL		04/11/1948
NOAKS	H			CPL		21/02/1952
NOBES	G	E		FL		21/10/1959
NOBLE	A	W		FO		27/04/1950
NOBLE	B	R		SQD LDR	Acting	09/08/1945
NOBLE	H			SGT		10/02/1949
NODEN	R	F		FL		24/01/1946
NOEL	A	H	P	SQD LDR		15/07/1948
NOKES	R	F		CPL		04/10/1951
NOKES	S	M		FO		12/04/1993
NOLAN	J	A	F	FL		28/11/1956
NOLAN	J	A		LAC		16/01/1957
NORLANDER	C	F		FLT SGT		07/08/1947
NORMAN	A	C	W	SQD LDR	Acting	03/12/1942
NORMAN	C	F	W	FL		21/01/1959
NORMAN	F	P		AC1		01/12/1955
NORMAN	H	N	St. V	GRP CAP		10/12/1942
NORMAN	R	E	J	FL		16/03/2004
NORMINGTON	E			FO		19/08/1954
NORRIE	F	B		LAC		16/08/1961
NORTH	E	W		CPL		23/02/1950
NORTH	R	M		WING COMM	Acting	30/03/1944
NORTH	S			SGT		23/02/1950
NORTH-BOMFORD	D	J		WO		08/11/1945

Source	Service No	Force	Notes	Surname
N94 1950	749826	RAFVR		NICE
N197 1946	199364	RAFVR		NICHOLAS
YB 1998	n/a	RAuxAF		NICHOLAS
N123 1949	859579	RAuxAF		NICHOLLS
N671 1945	114121	RAFVR	DFC	NICHOLLS
N835 1957	144488	RAuxAF	wef 8/12/53	NICHOLS
N250 1947	189	WAAF		NICHOLS
N594 1946	118094	RAFVR		NICHOLS
N953 1949	843925	RAuxAF		NICHOLS
N187 1947	74684	RAFVR		NICHOLSON
N763 1944	72002	RAFVR		NICHOLSON
N263 1946	161684	RAFVR		NICHOLSON
N358 1947	1565	WAAF		NICHOLSON
N454 1943	119527	RAFVR	MBE	NICHOLSON
N115 1953	119527	RAFVR	MBE; 1st Clasp	NICHOLSON
N232 1948	880766	WAAF		NICHOLSON
N775 1960	175682	RAuxAF	wef 13/1/59	NICKELS
N475 1943	90318	AAF		NICKOLS
N197 1946	137337	RAFVR	DFC,DFM	NICKSON
N915 1952	2684006	RAFVR		NICOL
N187 1947	73469	RAFVR	MBE,DSM	NICOL
N1355 1945	119662	RAFVR		NICOL
N799 1954	119662	RAFVR	1st Clasp	NICOL
N18 1954	121954	RAFVR		NICOLL
N837 1962	126060	RAuxAF	1st Clasp; wef 21/4/61 no record of first award	NICOLL
N103 1952	2684012	RAuxAF		NICOLL
N694 1948	90870	RAuxAF		NIELD
N263 1946	109061	RAFVR	DFC	NIVEN
N197 1946	146799	RAFVR	DFM	NIVEN
N1355 1945	141725	RAFVR		NIXON
N527 1948	800523	RAuxAF		NIXON
N263 1946	143370	RAFVR	DFM	NIXON
N751 1946	178866	RAFVR		NIXON
N22 1963	66551	RAFVR	wef 1/12/60	NIXON
N1 1950	743259	RAFVR		NIXON
N934 1948	850507	RAuxAF		NOAKES
N134 1952	800479	RAuxAF		NOAKS
N708 1959	112214	RAFVR	wef 10/8/55	NOBES
N406 1950	198120	RAFVR		NOBLE
N874 1945	81043	RAFVR		NOBLE
N123 1949	858928	RAuxAF		NOBLE
N82 1946	125318	RAFVR	DSO,DFM	NODEN
N573 1948	90604	RAuxAF		NOEL
N1019 1951	819184	RAuxAF		NOKES
YB 1997	213384	RAuxAF		NOKES
N833 1956	205251	RAFVR		NOLAN
N24 1957	857512	RAuxAF		NOLAN
N662 1947	841225	AAF		NORLANDER
N1638 1942	90121	AAF		NORMAN
N28 1959	144701	RAuxAF	wef 20/5/57	NORMAN
N895 1955	858962	RAuxAF		NORMAN
N1680 1942	90147	AAF	Bart; Sir	NORMAN
LG 2004	0300872Y	RAuxAF		NORMAN
N672 1954	160864	RAuxAF		NORMINGTON
N624 1961	750854	RAFVR	wef 19/7/44	NORRIE
N179 1950	743246	RAFVR		NORTH
N281 1944	76453	RAFVR	DFC,AFC	NORTH
N179 1950	880300	WAAF		NORTH
N1232 1945	742833	RAFVR		NORTH-BOMFORD

Surname	Initials			Rank	Status	Date
NORTHOVER	F	E		FL		02/07/1958
NORTON	B			SGT		18/11/1948
NORTON	C			FL		01/05/1952
NORTON	F	A		AC1		16/01/1957
NORTON	G	R		SGT	Temporary	28/07/1955
NORTON	H	A	M	SQD LDR		17/04/1947
NORTON	R	C		FL		20/05/1948
NORTON	W	R		WING COMM	Acting	15/11/1951
NORVAL	G	G		SQD LDR	Acting	17/04/1947
NORWELL	J	K		FO		20/04/1944
NORWELL	J	K		FL		02/04/1953
NORWOOD	R	K	C	FL		21/06/1945
NOWELL	G	L		PO		30/03/1944
NOXON	E	H		FL		18/03/1948
NUNN	C	R		SGT		14/10/1948
NUNN	G			CPL		28/06/1945
NUNN	J	L		FL		21/03/1946
NUNN	S	G		SQD LDR	Acting	05/06/1947
NUNN	W	A	C	CPL		31/10/1956
NUTT	M			FLT SGT		07/08/1947
NUTTALL	F			FL		15/07/1948
NUTTER	A	C		FO		29/04/1954
NUTTER	R	C		FL		13/12/1945
OAKLEY	R	G	W	FO		10/12/1942
OATES	R	J		WO		04/05/1944
O'BRIEN	B			SGT	Temporary	27/02/1963
O'BRIEN	E	C		SGT		01/07/1948
O'BRIEN	F	W		FL		07/04/1955
O'BRIEN	F			SQD LDR	Acting	20/09/1951
O'BRIEN	H	W		CPL		05/06/1947
O'BRIEN	H			FL		19/01/1956
O'BRIEN	N	J	S	FL		08/03/1956
O'BRIEN	P	J		SGT		19/07/1997
O'BRIEN	V	J		LAC		12/02/1953
O'BYRNE	P			WO		04/05/1944
O'CONNELL	J	E		SGT	Acting	16/03/1950
O'CONNOR	F	P		SGT		06/03/1947
O'CONNOR	H	S	H	WING COMM	Acting	18/11/1948
ODLING	E	R		FL		21/01/1959
O'DONNELL	T	P		SAC		14/11/1962
ODY	O	A		CPL		15/09/1949
OGBORNE	W	A		WO	Temporary	09/01/1963
OGDEN	F			FL		26/01/1950
OGILVIE	J	N		SQD LDR		09/08/1945
OGILVIE	W	K	N	CPL		11/04/1997
OGILVIE-FORBES	M	F		FL		27/01/1944
OGLEY	G	A		LAC		10/02/1949
O'HAGAN	J	D		FL	Acting	21/06/1951
O'HARA	H	F		FL		16/04/1958
OLDFIELD	E			FL		06/03/1947
OLDMAN	C	W		FO		07/12/1960
OLGIVIE	J	M		FLT OFF		07/08/1947
OLIVER	A	J		FL		15/04/1996
OLIVER	C	W		WO		11/05/1944
OLIVER	F			FO		24/11/1949
OLIVER	G	L	J	FL		12/08/1943
OLIVER	H	W		FL		14/01/1954
OLIVER	I	D		FL		19/08/1954
OLIVER	I	H		FO		23/06/1955
OLIVER	R	G		FO		07/08/1947

Source	Service No	Force	Notes	Surname
N452 1958	166042	RAFVR	wef 10/10/52	NORTHOVER
N984 1948	857274	RAuxAF		NORTON
N336 1952	82755	RAFVR		NORTON
N24 1957	869496	RAuxAF		NORTON
N565 1955	844077	RAuxAF		NORTON
N289 1947	84830	RAFVR		NORTON
N395 1948	85962	RAFVR		NORTON
N1185 1951	72162	RAFVR		NORTON
N289 1947	129686	RAFVR		NORVAL
N363 1944	129717	RAFVR		NORWELL
N298 1953	129717	RAFVR	AFC; 1st Clasp	NORWELL
N671 1945	85232	RAFVR		NORWOOD
N281 1944	146389	RAFVR	DFM	NOWELL
N232 1948	169916	RAFVR	DFC	NOXON
N856 1948	862021	RAuxAF		NUNN
N706 1945	800380	AAF		NUNN
N263 1946	74685	RAFVR	DFC	NUNN
N460 1947	81935	RAFVR	DFC	NUNN
N761 1956	847049	RAuxAF		NUNN
N662 1947	883410	WAAF		NUTT
N573 1948	125927	RAFVR		NUTTALL
N329 1954	2604812	RAuxAF		NUTTER
N1355 1945	108855	RAFVR	DFC	NUTTER
N1680 1942	61011	RAFVR	DFM	OAKLEY
N425 1944	740013	RAFVR		OATES
N155 1963	761027	RAFVR	wef 1/9/44	O'BRIEN
N527 1948	752564	RAFVR		O'BRIEN
N277 1955	174334	RAFVR		O'BRIEN
N963 1951	91025	RAuxAF		O'BRIEN
N460 1947	811145	AAF		O'BRIEN
N43 1956	146016	RAFVR		O'BRIEN
N187 1956	82120	RAFVR		O'BRIEN
YB 1998	n/a	RAuxAF		O'BRIEN
N115 1953	843442	RAuxAF		O'BRIEN
N425 1944	740334	RAFVR		O'BYRNE
N257 1950	744630	RAFVR		O'CONNELL
N187 1947	801427	AAF		O'CONNOR
N984 1948	72096	RAFVR		O'CONNOR
N28 1959	85496	RAFVR	wef 10/9/53	ODLING
N837 1962	2684655	RAuxAF	wef 27/9/62	O'DONNELL
N953 1949	865643	RAuxAF		ODY
N22 1963	850034	AAF	wef 4/4/44	OGBORNE
N94 1950	126858	RAFVR		OGDEN
N874 1945	70508	RAFVR		OGILVIE
YB 1998	n/a	RAuxAF		OGILVIE
N59 1944	70509	RAFVR		OGILVIE-FORBES
N123 1949	870213	RAuxAF		OGLEY
N621 1951	89612	RAFVR		O'HAGAN
N258 1958	195482	RAFVR	wef 10/4/55	O'HARA
N187 1947	145843	RAFVR		OLDFIELD
N933 1960	2693558	RAuxAF	wef 15/10/60	OLDMAN
N662 1947	1614	WAAF		OLGIVIE
YB 1998	474340	RAuxAF		OLIVER
N444 1944	740921	RAFVR		OLIVER
N1193 1949	110039	RAFVR		OLIVER
N850 1943	89056	RAFVR		OLIVER
N18 1954	91252	RAuxAF		OLIVER
N672 1954	84308	RAFVR	DFC	OLIVER
N473 1955	2659557	WRAAF		OLIVER
N662 1947	105277	RAFVR		OLIVER

Surname	Initials			Rank	Status	Date
OLIVER	R	M		FL		16/01/1957
OLIVER	R			SGT		17/04/1947
OLLIS	G	S	A	WING COMM		07/07/1949
OLSSON	J	F		FL		17/06/1954
OLVER	P			SQD LDR		28/11/1946
O'MALLEY	A			SQD LDR	Acting	07/07/1949
OMAND	J	M	W	SQD LDR		14/10/1948
O'NEILL	J	J		FLT SGT		25/11/1992
O'NEILL	S	W	S	FL		04/09/1957
ONION	R	E	S	FLT SGT		15/09/1949
ONLEY	S	D		FO		20/04/1944
ORANGE	W			FL		08/05/1947
ORCHARD	E			SGT		26/11/1953
ORCHARD	O	R		SQD LDR		05/02/1948
ORMISTON	J	D	R	SAC		22/02/1961
ORMISTON	S	D		FL		06/03/1947
ORMONROYD	F			SQD LDR	Acting	08/11/1945
ORR	W	D		CPL		26/08/1948
ORR	W	G		SGT		24/06/1943
ORR-EWING	C	I		WING COMM		13/12/1961
ORRINGE	E	T		SQD LDR	Acting	23/05/1946
ORRY	D	G		FL		15/01/1958
ORTON	T			CPL	Acting	09/12/1959
ORZEL	M	N	F	SQD LDR		02/09/1998
OSBALDISTON	W			CPL		20/05/1948
OSBON	L	A	E	SQD LDR	Acting	18/04/1946
OSBORN	D	W		FLT SGT	Acting	22/02/1961
OSBORNE	A	H		LAC		04/09/1947
OSBORNE	R	K		FL		09/05/1946
OSBORNE	T	W		FLT SGT		15/09/1949
OSBOURNE	A	H		LAC		05/06/1947
OSCROFT	C	A		FL		07/07/1949
O'SHAUGHNESSY	K	M	P	SQD LDR		14/12/1987
O'SHAUGHNESSY	K	M	P	SQD LDR		07/06/1996
OSMOND	L	W		FLT SGT		01/10/1953
OSWALD	J			FO		21/10/1959
OTTER	E	J	S	FL		09/05/1946
OTTEWELL	F			CPL		26/01/1950
OUGHTON	J	H		SQD LDR	Acting	20/04/1944
OULTON	G			FL		30/09/1954
OUSTON	P	A		FL		28/03/1962
OVERTON	W	M		SQD OFF		14/10/1948
OWEN	C	M		SQD LDR	Acting	27/03/1947
OWEN	C	M		FL		01/01/2002
OWEN	D	A		FL		01/07/1948
OWEN	D			FL		05/01/1950
OWEN	E	C		WO		10/02/1949
OWEN	E	I		FL		04/09/1957
OWEN	O	K		WING COMM		01/07/1948
OWEN	R	S		FO		01/07/1948
OWEN	R	T		FL		08/03/1956
OWEN	W	G		FL		24/01/1946
OWENS	J	H		SGT		30/09/1948
OXBY	C			LAC		16/10/1952
OXTOBY	J	C		CPL		30/09/1948
PADDEN	R	T		FO		24/01/1946
PADLEY	J	T		PO		08/11/1945
PAGE	C	R	A	SQD LDR		29/06/1944
PAGE	C	R		CPL		07/08/1947
PAGE	E	E		FLT SGT		20/05/1948

Source	Service No	Force	Notes	Surname
N24 1957	64924	RAuxAF		OLIVER
N289 1947	743925	RAFVR		OLIVER
N673 1949	72205	RAFVR		OLLIS
N471 1954	158485	RAuxAF		OLSSON
N1004 1946	84963	RAFVR	DFC	OLVER
N673 1949	133849	RAFVR		O'MALLEY
N856 1948	90448	RAuxAF		OMAND
YB 1996	n/a	RAuxAF		O'NEILL
N611 1957	190131	RAFVR	wef 23/11/54	O'NEILL
N953 1949	812020	RAuxAF		ONION
N363 1944	143671	RAFVR		ONLEY
N358 1947	84049	RAFVR		ORANGE
N952 1953	2698119	RAuxAF		ORCHARD
N87 1948	73321	RAFVR		ORCHARD
N153 1961	2651102	RAuxAF	wef 28/10/60	ORMISTON
N187 1947	161248	RAFVR		ORMISTON
N1232 1945	115569	RAFVR	DFM	ORMONROYD
N694 1948	874680	RAuxAF		ORR
N677 1943	816018	AAF		ORR
N987 1961	73548	RAFVR	OBE; wef 26/6/44	ORR-EWING
N468 1946	127170	RAFVR	DFC	ORRINGE
N40 1958	190405	RAFVR	wef 16/5/52	ORRY
N843 1959	2699522	RAuxAF	wef 2/10/54	ORTON
YB 1999	n/a	RAuxAF		ORZEL
N395 1948	870183	RAuxAF		OSBALDISTON
N358 1946	82725	RAFVR	DFC	OSBON
N153 1961	2651001	RAuxAF	wef 19/10/60	OSBORN
N740 1947	840071	AAF		OSBORNE
N415 1946	129139	RAFVR		OSBORNE
N953 1949	844555	RAuxAF		OSBORNE
N460 1947	840071	AAF		OSBOURNE
N673 1949	67043	RAFVR		OSCROFT
YB 1996	4231081	RAuxAF		O'SHAUGHNESSY
YB 1997	4231081	RAuxAF	1st Clasp	O'SHAUGHNESSY
N796 1953	818032	RAuxAF		OSMOND
N708 1959	2601752	RAFVR	wef 2/7/52	OSWALD
N415 1946	172942	RAFVR		OTTER
N94 1950	756756	RAFVR		OTTEWELL
N363 1944	102586	RAFVR		OUGHTON
N799 1954	117350	RAuxAF		OULTON
N242 1962	187459	RAFVR	wef 6/3/59	OUSTON
N856 1948	232	WAAF		OVERTON
N250 1947	86642	RAFVR	DFC	OWEN
YB 2003	91465	RAuxAF		OWEN
N527 1948	142136	RAFVR		OWEN
N1 1950	118421	RAFVR	DFC	OWEN
N123 1949	813122	RAuxAF		OWEN
N611 1957	197662	RAF	wef 20/4/52	OWEN
N527 1948	76435	RAFVR		OWEN
N527 1948	147252	RAFVR		OWEN
N187 1956	91176	RAuxAF		OWEN
N82 1946	103386	RAFVR	DFC	OWEN
N805 1948	818049	RAuxAF		OWENS
N798 1952	869518	RAuxAF		OXBY
N805 1948	872403	RAuxAF		OXTOBY
N82 1946	176113	RAFVR	DFC	PADDEN
N1232 1945	187727	RAFVR		PADLEY
N652 1944	70517	RAFVR		PAGE
N662 1947	746616	RAFVR		PAGE
N395 1948	815131	RAuxAF		PAGE

Surname	Initials		Rank	Status	Date
PAGE	E E		SGT		13/12/1951
PAGE	F L	A	CPL		30/09/1948
PAGE	F O		FLT SGT		01/07/1948
PAGE	G S		SGT		06/09/1951
PAGE	H C		WING COMM		12/04/1951
PAGE	J T		FL		20/03/1957
PAGE	V D		SQD LDR	Acting	04/10/1945
PAGET	M Y		SQD LDR		31/10/1956
PAIN	C D		SQD LDR		23/04/1944
PAINTER	C J		WO	Acting	15/07/1948
PAIRMAN	J R		SQD LDR	Acting	17/03/1949
PAIRMAN	J R		FL		28/07/1955
PALIN	R		SGT		13/07/1950
PALIN	R		FLT SGT		07/05/1953
PALLISER	G C	C	FL		28/02/1946
PALLOT	A G		SQD LDR		25/01/1945
PALMER	A E	J	PO		14/01/1954
PALMER	C J		FO		07/07/1949
PALMER	G F		LAC		10/02/1949
PALMER	G M		SQD LDR	Acting	26/08/1948
PALMER	N S		FLT OFF	Acting	08/05/1947
PALMER	R A		CPL		18/11/1948
PALMER	R A	M	SQD LDR	Acting	10/02/1949
PALMER	R G		CPL		19/02/1993
PALMER	R H	F	FL		12/02/1953
PALMER	S G		CPL	Temporary	21/10/1959
PALMER	S G		CPL	Temporary	21/10/1959
PALMER	S J		WING COMM		26/01/1950
PALMER	S W	F	SQD LDR	Acting	28/02/1946
PALMER	T		AC1		13/01/1949
PALMER	W S		FLT SGT	Acting	08/11/1945
PALSER	C H	F	FL		15/07/1948
PANNELL	E G		SQD LDR	Acting	09/08/1945
PARIS	G P		CPL		20/03/1957
PARIS	M V		CPL		24/07/1963
PARISH	K G	H	WO		26/08/1948
PARK	J S		FL		08/03/1956
PARK	J		WING COMM	Acting	16/10/1947
PARK	M M		FL		15/01/1953
PARKER	B B		SQD OFF	Acting	17/04/1947
PARKER	D A		FO		08/11/1945
PARKER	D A		FL		26/08/1948
PARKER	D K		FL		24/01/1946
PARKER	E J		LAC		21/06/1951
PARKER	E		SGT		26/01/1950
PARKER	G V		FO		28/07/1955
PARKER	H D		FL		06/01/1955
PARKER	H E		SGT		13/07/1950
PARKER	H J		CPL		06/09/1951
PARKER	I R		WING COMM		10/12/1942
PARKER	J G		FL		08/07/1959
PARKER	J J		FL		13/02/1957
PARKER	J P		SQD LDR	Acting	30/09/1948
PARKER	M W		FLT OFF		26/06/1947
PARKER	P F		SGT		20/03/1957
PARKER	R A		SGT		03/10/2000
PARKER	R D		FL		13/12/1945
PARKER	R I	P	SQD LDR	Acting	26/01/1950
PARKER	T G		FO		03/07/1957
PARKER	W E		FL		30/03/1960

Source	Service No	Force	Notes	Surname
N1271 1951	2681504	RAuxAF	1st Clasp	PAGE
N805 1948	845499	RAuxAF		PAGE
N527 1948	747341	RAFVR		PAGE
N913 1951	845709	RAuxAF		PAGE
N342 1951	72909	RAFVR		PAGE
N198 1957	174068	RAuxAF	DFM; wef 29/4/55	PAGE
N1089 1945	65502	RAFVR	DFC	PAGE
N761 1956	133921	RAuxAF	MRCS,LRCP,DRCOG	PAGET
MOD 1964	90971	AAF		PAIN
N573 1948	701207	RAFVR		PAINTER
N231 1949	72910	RAFVR		PAIRMAN
N565 1955	72910	RAFVR	1st Clasp	PAIRMAN
N675 1950	810124	RAuxAF		PALIN
N395 1953	2687502	RAuxAF	1st Clasp	PALIN
N197 1946	64891	RAFVR	DFC	PALLISER
N70 1945	90389	AAF		PALLOT
N18 1954	2602514	RAFVR		PALMER
N673 1949	190981	RAFVR		PALMER
N123 1949	770495	RAFVR		PALMER
N694 1948	90710	RAuxAF		PALMER
N358 1947	833	WAAF		PALMER
N984 1948	860424	RAuxAF		PALMER
N123 1949	115772	RAFVR	VC,DFC; Deceased	PALMER
YB 1996	n/a	RAuxAF		PALMER
N115 1953	180101	RAFVR		PALMER
N708 1959	2605160	RAFVR	wef 21/6/44	PALMER
N708 1959	2605160	RAFVR	1st Clasp; wef 5/7/59	PALMER
N94 1950	90598	RAuxAF		PALMER
N197 1946	78466	RAFVR	DFC	PALMER
N30 1949	847014	RAuxAF		PALMER
N1232 1945	801377	AAF		PALMER
N573 1948	91023	RAuxAF		PALSER
N874 1945	86680	RAFVR		PANNELL
N198 1957	2655858	WRAAF	wef 21/7/56	PARIS
N525 1963	5006746	RAuxAF	wef 4/5/63	PARIS
N694 1948	755846	RAFVR		PARISH
N187 1956	194523	RAuxAF		PARK
N846 1947	72594	RAFVR		PARK
N28 1953	153700	RAFVR		PARK
N289 1947	1738	WAAF		PARKER
N1232 1945	172777	RAFVR	DFC	PARKER
N694 1948	73719	RAFVR		PARKER
N82 1946	128987	RAFVR		PARKER
N621 1951	818066	RAuxAF		PARKER
N94 1950	813078	RAuxAF		PARKER
N565 1955	2688604	RAuxAF		PARKER
N1 1955	125756	RAFVR		PARKER
N675 1950	842411	RAuxAF		PARKER
N913 1951	853794	RAuxAF		PARKER
N1680 1942	90335	AAF		PARKER
N454 1959	157127	RAFVR	wef 19/2/58	PARKER
N104 1957	187906	RAFVR	wef 14/7/56	PARKER
N805 1948	90695	RAuxAF		PARKER
N519 1947	530	WAAF		PARKER
N198 1957	2674901	WRAFVR	wef 11/8/54	PARKER
YB 2001	n/a	RAuxAF		PARKER
N1355 1945	120875	RAFVR		PARKER
N94 1950	73564	RAFVR		PARKER
N456 1957	104765	RAuxAF	wef 24/5/57	PARKER
N229 1960	1485107	RAFVR	wef 22/10/54	PARKER

Surname	Initials			Rank	Status	Date
PARKES	J	J		FL		24/01/1946
PARKES	K	E		WING COMM	Acting	26/08/1948
PARKHOUSE	W	R		WING COMM		28/10/1943
PARKIN	A	B	H	FL		24/01/1946
PARKIN	E	G		FL		07/06/1945
PARKINSON	D	R		SQD LDR		27/03/1947
PARKINSON	W	E		SGT	Acting	05/06/1947
PARKS	F	J		LAC		19/07/1951
PARKS	J	L		CPL		15/09/1949
PARLETT	A	G	B	FL		06/01/1955
PARMEE	C			SQD LDR		13/01/1949
PARR	L	A		SQD LDR	Acting	01/03/1945
PARR	L	C		FL		08/11/1945
PARRATT	N	S		SQD LDR	Acting	27/07/1944
PARREN	L	A	J	SGT		15/03/1990
PARRISH	H	T		SGT		26/08/1948
PARROTT	F	G		WO		05/01/1950
PARROTT	J	T		CPL		13/07/1950
PARRY	D	A	G	SQD LDR	Acting	26/08/1948
PARRY	N	R		SQD LDR	Acting	23/02/1950
PARRY	O	A		FL		15/01/1958
PARRY	T	W		CPL	Temporary	23/06/1955
PARSONS	A	G		WO		27/03/1947
PARSONS	A			WO		08/11/1945
PARSONS	B	K		WING COMM	Acting	10/02/1949
PARSONS	C	H		WING COMM		23/02/1950
PARSONS	C	H		SQD LDR		21/01/1959
PARSONS	F	W		LAC		19/09/1956
PARSONS	H	J		FL		14/09/1950
PARSONS	H	M		WING COMM	Acting	05/06/1947
PARSONS	H			SQD LDR	Acting	27/03/1947
PARSONS	J	G		FL		11/07/1946
PARSONS	J	L		SGT		20/05/1948
PARSONS	J	L		FLT SGT		20/03/1957
PARSONS	R	V	G	FLT SGT		20/05/1948
PARTRIDGE	E	F		WING COMM		17/02/1997
PARTRIDGE	J	E		WING COMM	Acting	20/09/1945
PARVIN	D			SGT		25/03/1954
PASCOE	J	G		FO		30/08/1945
PASFIELD	E	J		SGT		17/06/1954
PASHLEY	C	L		FL		04/10/1945
PASKE	C	H		CPL		10/02/1949
PASQUALL	R	L		PO		08/05/1947
PASSY	C	W		WING COMM	Acting	11/01/1945
PATERSON	A	J		FL		05/01/1950
PATERSON	H	D		WING COMM	Acting	10/06/1948
PATERSON	J			FL	Acting	30/10/1957
PATERSON	M			FL		27/07/1944
PATERSON	W			FL		07/04/1955
PATIENCE	J	L		LAC		13/01/1949
PATMORE	F	W		FL		28/11/1956
PATON	J	R		FO		08/11/1945
PATON	R	A		FO		04/05/1999
PATRICK	L	F		PO		04/02/1943
PATRICK	L	F		FL		21/02/1952
PATRICK	L	F		FL		22/02/1961
PATTENDEN	S	M		SQD LDR		12/06/1999
PATTERSON	F	J		CPL		09/12/1959
PATTERSON	H	R	P	SQD LDR		07/08/1947
PATTERSON	J	E		LAC		07/08/1947

Source	Service No	Force	Notes	Surname
N82 1946	90418	EX-AAFRO		PARKES
N694 1948	90940	RAuxAF	MBE	PARKES
N1118 1943	70520	RAFVR		PARKHOUSE
N82 1946	133348	RAFVR	DFC	PARKIN
N608 1945	79734	RAFVR		PARKIN
N250 1947	72137	RAFVR		PARKINSON
N460 1947	747412	RAFVR		PARKINSON
N724 1951	846065	RAuxAF		PARKS
N953 1949	800535	RAuxAF		PARKS
N1 1955	79613	RAuxAF		PARLETT
N30 1949	73604	RAFVR		PARMEE
N225 1945	67605	RAFVR		PARR
N1232 1945	135468	RAFVR	DFM	PARR
N763 1944	77681	RAFVR		PARRATT
YB 1998	n/a	RAuxAF	BEM	PARREN
N694 1948	812259	RAuxAF		PARRISH
N1 1950	746549	RAFVR		PARROTT
N675 1950	810187	RAuxAF		PARROTT
N694 1948	79162	RAFVR	DSO,DFC	PARRY
N179 1950	61517	RAFVR		PARRY
N40 1958	134128	RAuxAF	DFC; wef 10/11/57	PARRY
N473 1955	844790	RAuxAF		PARRY
N250 1947	749370	RAFVR		PARSONS
N1232 1945	807125	AAF		PARSONS
N123 1949	90646	RAuxAF		PARSONS
N179 1950	72911	RAFVR		PARSONS
N28 1959	72911	RAuxAF	1st Clasp; wef 29/8/55	PARSONS
N666 1956	844131	RAuxAF		PARSONS
N926 1950	119978	RAFVR		PARSONS
N460 1947	72912	RAFVR	MBE	PARSONS
N250 1947	72710	RAFVR		PARSONS
N594 1946	143773	RAFVR		PARSONS
N395 1948	819107	RAuxAF		PARSONS
N198 1957	2689501	RAuxAF	1st Clasp; wef 8/3/53	PARSONS
N395 1948	842280	RAuxAF		PARSONS
YB 1998	9159	RAuxAF		PARTRIDGE
N1036 1945	121910	RAFVR	DSO,DFC	PARTRIDGE
N249 1954	2659553	WRAAF		PARVIN
N961 1945	171010	RAFVR	DFM	PASCOE
N471 1954	2602546	RAFVR		PASFIELD
N1089 1945	89305	RAFVR	AFC	PASHLEY
N123 1949	860452	RAuxAF		PASKE
N358 1947	203287	RAFVR		PASQUALL
N14 1945	72028	RAFVR	DFC	PASSY
N1 1950	73330	RAFVR	MA	PATERSON
N456 1948	72343	RAFVR		PATERSON
N775 1957	56671	RAuxAF	wef 22/11/54	PATERSON
N763 1944	61232	RAFVR	DFC	PATERSON
N277 1955	73895	RAFVR		PATERSON
N30 1949	816116	RAuxAF	BEM	PATIENCE
N833 1956	61111	RAFVR		PATMORE
N1232 1945	163243	RAFVR		PATON
YB 2000	2638741	RAuxAF		PATON
N131 1943	123303	RAFVR		PATRICK
N134 1952	123303	RAFVR	1st Clasp	PATRICK
N153 1961	123303	RAFVR	2nd Clasp; wef 16/12/52	PATRICK
YB 2000	408859	RAuxAF		PATTENDEN
N843 1959	2614806	RAFVR	wef 28/4/59	PATTERSON
N662 1947	73332	RAFVR		PATTERSON
N662 1947	816184	AAF		PATTERSON

Surname	Initials					Rank	Status	Date
PATTERSON	J	J	L			FL		17/04/1947
PATTERSON	J					SGT	Temporary	28/07/1955
PATTERSON	W	A				LAC		20/05/1948
PATTINSON	J	H	G			FL		10/06/1948
PATTISON	J	E				CPL		16/03/1950
PATTISSON	J	R				SQD LDR		06/03/1947
PAUL	D	J				WO		13/12/1945
PAULEY	E	S				SGT		01/07/1948
PAULL	D	F	C	G	A	SGT		21/12/1950
PAULSEN	E					SGT		20/09/1951
PAULSEN	E					SGT		30/09/1954
PAULSEN	K	A				FO		20/05/1948
PAVEY	R	B				WING COMM	Acting	10/06/1948
PAXTON	J	McC.				FL		24/01/1946
PAYLOR	A	L				FLT SGT		31/10/1956
PAYNE	A	C	V			LAC		15/07/1948
PAYNE	C	J				FLT SGT		27/07/1944
PAYNE	E	D	B			WING COMM	Acting	24/11/1949
PAYNE	G					WO		01/03/1945
PAYNE	H	F				SQD LDR		30/08/1945
PAYNE	L	W				LAC		22/02/1961
PAYNE	S	R				SGT		26/06/1947
PEACE	A	J	M			FO		26/08/1948
PEACHEY	C	L				FL	Acting	03/07/1952
PEACHEY	D	J				CPL		15/02/1996
PEACOCK	C					LAC		04/09/1947
PEACOCK	G	P				FO		06/01/1955
PEACOCK	M	H				FL		07/06/1945
PEACOCK	O	W				SQD LDR	Acting	27/03/1947
PEAKE	A					FL		26/08/1948
PEAKE	H					AIR COMMD	Acting	03/12/1942
PEARCE	A	A				LAC		10/06/1948
PEARCE	A	C	R			WING COMM	Acting	24/11/1949
PEARCE	A	C	R			FL		03/10/1962
PEARCE	B					FLT SGT	Acting	16/10/1947
PEARCE	D					FO		26/11/1958
PEARCE	F	W				WO		12/08/1943
PEARCE	G	E				SQD LDR	Acting	08/05/1947
PEARCE	H	B				FL		29/04/1954
PEARCE	H	E				SQD LDR	Acting	05/01/1950
PEARCE	R	S				SQD LDR	Acting	30/03/1944
PEARCE	T	H	W			SQD LDR	Acting	10/06/1948
PEARL	D	W				FLT SGT	Acting	27/04/1950
PEARL	G	E	R			SQD LDR	Acting	09/08/1945
PEARMAN	R	H				FO		20/09/1945
PEARN	C	F				SGT		15/07/1948
PEARN	J	E				CPL		27/03/1952
PEARSE	F	D				SGT		19/10/1960
PEARSE	R	J	B			SQD LDR		10/05/1945
PEARSON	C	E				FL		09/08/1945
PEARSON	D	H				FL	Acting	25/06/1953
PEARSON	H					LAC		03/06/1959
PEARSON	K	R				FL		05/09/1962
PEARSON	L	H				SGT		07/08/1947
PEARSON	M	G				WING COMM	Acting	25/06/1953
PEARSON	P	G	H			SQD LDR		01/08/1946
PEARSON	R	E				FL		20/11/1957
PEATFIELD	J	E				FL		20/09/1945
PECK	E	R				LAC		22/02/1961
PECKETT	J	F				LAC		13/07/1950

Source	Service No	Force	Notes	Surname
N289 1947	73653	RAFVR		PATTERSON
N565 1955	811083	RAuxAF		PATTERSON
N395 1948	816177	RAuxAF		PATTERSON
N456 1948	90812	RAuxAF		PATTINSON
N257 1950	804312	RAuxAF		PATTISON
N187 1947	72914	RAFVR		PATTISSON
N1355 1945	745094	RAFVR		PAUL
N527 1948	747279	RAFVR		PAULEY
N1270 1950	756171	RAFVR		PAULL
N963 1951	804369	RAuxAF		PAULSEN
N799 1954	2684509	RAuxAF		PAULSEN
N395 1948	187740	RAFVR		PAULSEN
N456 1948	72399	RAFVR		PAVEY
N82 1946	81063	RAFVR		PAXTON
N761 1956	2677221	RAuxAF		PAYLOR
N573 1948	844321	RAuxAF		PAYNE
N763 1944	808032	AAF		PAYNE
N1193 1949	72306	RAFVR		PAYNE
N225 1945	741103	RAFVR		PAYNE
N961 1945	75731	RAFVR		PAYNE
N153 1961	846517	AAF	wef 21/5/44	PAYNE
N519 1947	846033	AAF		PAYNE
N694 1948	84964	RAFVR		PEACE
N513 1952	67734	RAFVR		PEACHEY
YB 1997	n/a	RAuxAF		PEACHEY
N740 1947	870075	AAF		PEACOCK
N1 1955	2603553	RAFVR		PEACOCK
N608 1945	124960	RAFVR		PEACOCK
N250 1947	130354	RAFVR		PEACOCK
N694 1948	176424	RAFVR		PEAKE
N1638 1942	90316	AAF		PEAKE
N456 1948	841356	RAuxAF		PEARCE
N1193 1949	72206	RAFVR		PEARCE
N722 1962	72206	RAFVR	1st Clasp; wef 5/12/60	PEARCE
N846 1947	771055	AAF		PEARCE
N807 1958	153859	RAFVR	wef 27/4/58	PEARCE
N850 1943	740980	RAFVR		PEARCE
N358 1947	135841	RAFVR		PEARCE
N329 1954	106036	RAFVR		PEARCE
N1 1950	81692	RAFVR		PEARCE
N281 1944	90713	AAF		PEARCE
N456 1948	72712	RAFVR		PEARCE
N406 1950	753281	RAFVR		PEARL
N874 1945	84303	RAFVR		PEARL
N1036 1945	162512	RAFVR		PEARMAN
N573 1948	846135	RAuxAF		PEARN
N224 1952	848328	RAuxAF		PEARN
N775 1960	810130	RAF	wef 1/12/43	PEARSE
N485 1945	78521	RAFVR	AFC	PEARSE
N874 1945	82950	RAFVR		PEARSON
N527 1953	107976	RAFVR		PEARSON
N368 1959	857434	AAF	wef 9/6/44	PEARSON
N655 1962	169199	RAFVR	wef 12/6/62	PEARSON
N662 1947	750435	RAFVR		PEARSON
N527 1953	60927	RAuxAF		PEARSON
N652 1946	91068	AAF	DFC	PEARSON
N835 1957	117640	RAFVR	wef 16/2/54	PEARSON
N1036 1945	137204	RAFVR	AFC	PEATFIELD
N153 1961	2693566	RAuxAF	wef 19/11/60	PECK
N675 1950	754571	RAFVR		PECKETT

Surname	Initials			Rank	Status	Date
PECOVER	H	A		FL		23/02/1950
PEDGE	S	R		SAC		24/11/1993
PEET	E	A		AC1		26/01/1950
PEET	E	P	F	FO		18/02/1954
PEET	W	V		PO		30/12/1943
PEGLER	D	E	G	FLT SGT		07/08/1947
PEGLER	S	W		FO		05/06/1947
PEIRCE	A	F		FL		08/11/1945
PELL	R	A		FO		07/07/1949
PELLATT	O	W		SQD LDR	Acting	06/03/1947
PENDLEBURY	C	L		GRP CAP		20/05/1948
PENFOLD	W	D		WING COMM	Acting	23/05/1946
PENGALLEY	J	I		SQD LDR	Acting	21/03/1946
PENGELLY	A	T		SAC		19/02/1993
PENIKETT	E	J	K	FL		27/04/1944
PENNELL	C	F		SQD LDR		18/03/1948
PENNINGTON	C			AC2		25/06/1953
PENNINGTON	I			SGT		05/02/1948
PENNINGTON	J			CPL		04/11/1948
PENNY	A	A		FL		13/02/1957
PENNY	A	A		FL		23/03/1966
PENNYCUICK	B			FL		23/05/1946
PENRIDGE	F	B		SGT		07/12/1960
PEOCK	W	D		WING COMM		28/11/1946
PEPPER	F	A		LAC		29/04/1954
PEPPRELL	H	E		CPL		27/04/1950
PERCIVAL	E	A		FL		07/07/1949
PERCY	J	T		FO		28/11/1946
PERCY-SMITH	K	E		SQD OFF		07/08/1947
PERFECT	W	T		WING COMM	Acting	28/10/1943
PERFECT	W	T		FL		02/08/1951
PERKES	W	C		CPL	Acting	16/03/1950
PERKIN	F	S		FL		24/01/1946
PERKIN	K	A		SQD LDR	Acting	30/11/1944
PERKINS	C			SGT		04/09/1947
PERKINS	F	H		FL		03/10/1946
PERKINS	H	E		FL		27/03/1952
PERKINS	P	J		CPL		30/09/1948
PERRENS	D	F		SQD LDR		06/01/1955
PERRING	H			SQD LDR		30/03/1943
PERRY	A	A		CPL		30/09/1948
PERRY	B	F		CPL	Temporary	20/08/1958
PERRY	E	R		FL		26/11/1953
PERRY	F	C		FLT SGT	Acting	01/07/1948
PERRY	G	N	H	SQD LDR	Acting	08/05/1947
PERRY	H	V		FL		10/02/1949
PERRY	I	R		FL		13/11/1947
PERRY	R			LAC		10/06/1948
PERRY	T	A		FLT SGT		30/03/1944
PERRY	W	K		FO		13/01/1949
PERRY	W	M		LAC		20/05/1948
PERRYMAN	C	A		CPL		09/08/1945
PERT	E			FL		21/03/1946
PETERS	J	H		CPL		20/05/1948
PETFORD	E	H		FLT SGT		27/03/1947
PETTETT	C	W		SGT		10/06/1948
PETTMAN	F	W		SGT		24/07/1963
PETTY	J	R	N	CPL		07/12/1950
PETTY	R	L		FL		25/05/1960
PEXTON	R	D		SQD LDR		17/06/1943

Source	Service No	Force	Notes	Surname
N179 1950	121848	RAFVR		PECOVER
YB 1996	n/a	RAuxAF		PEDGE
N94 1950	870186	RAuxAF		PEET
N137 1954	2603508	RAFVR		PEET
N1350 1943	108169	EX-RAFVR		PEET
N662 1947	881400	WAAF		PEGLER
N460 1947	122786	RAFVR		PEGLER
N1232 1945	122969	RAFVR		PEIRCE
N673 1949	191760	RAFVR		PELL
N187 1947	72916	RAFVR		PELLATT
N395 1948	90669	RAuxAF	MC,TD	PENDLEBURY
N468 1946	115753	RAFVR		PENFOLD
N263 1946	60295	RAFVR	DFC	PENGALLEY
YB 1996	n/a	RAuxAF		PENGELLY
N396 1944	102957	RAFVR		PENIKETT
N232 1948	91502	RAuxAF		PENNELL
N527 1953	2691601	RAuxAF		PENNINGTON
N87 1948	815112	RAuxAF		PENNINGTON
N934 1948	868756	RAuxAF		PENNINGTON
N104 1957	140250	RAFVR	wef 23/3/56	PENNY
MOD 1967	140250	RAFVR	1st Clasp	PENNY
N468 1946	122315	RAFVR		PENNYCUICK
N933 1960	2650412	RAuxAF	wef 22/10/60	PENRIDGE
N1004 1946	90365	AAF		PEOCK
N329 1954	846020	RAuxAF		PEPPER
N406 1950	863375	RAuxAF		PEPPRELL
N673 1949	115883	RAFVR		PERCIVAL
N1004 1946	70535	RAFVR		PERCY
N662 1947	123	WAAF		PERCY-SMITH
N1118 1943	78532	RAFVR		PERFECT
N777 1951	78532	RAFVR	1st Clasp	PERFECT
N257 1950	842656	RAuxAF		PERKES
N82 1946	104446	RAFVR		PERKIN
N1225 1944	84703	RAFVR		PERKIN
N740 1947	868528	AAF		PERKINS
N837 1946	156320	RAFVR		PERKINS
N224 1952	164176	RAFVR		PERKINS
N805 1948	859548	RAuxAF		PERKINS
N1 1955	47493	RAuxAF	DSO,DFC	PERRENS
N1014 1943	70536	RAFVR		PERRING
N805 1948	841113	RAuxAF		PERRY
N559 1958	752651	RAFVR	wef 28/7/44	PERRY
N952 1953	174337	RAFVR		PERRY
N527 1948	845380	RAuxAF		PERRY
N358 1947	74384	RAFVR		PERRY
N123 1949	122234	RAFVR		PERRY
N933 1947	75291	RAFVR		PERRY
N456 1948	808291	RAuxAF		PERRY
N281 1944	802342	AAF		PERRY
N30 1949	134889	RAFVR		PERRY
N395 1948	856655	RAuxAF		PERRY
N874 1945	861005	AAF		PERRYMAN
N263 1946	170014	RAFVR		PERT
N395 1948	841217	RAuxAF		PETERS
N250 1947	741888	RAFVR		PETFORD
N456 1948	812145	RAuxAF		PETTETT
N525 1963	756274	RAF	wef 20/7/44	PETTMAN
N1210 1950	855165	RAuxAF		PETTY
N375 1960	205910	RAFVR	TD; wef 12/3/60	PETTY
N653 1943	72150	RAFVR		PEXTON

Surname	Initials			Rank	Status	Date
PHAIR	E	H		CPL		20/05/1948
PHELAN	J	T		WING COMM		07/08/1947
PHIBBS	R	K	D	FL		07/08/1947
PHILIP	J			SGT		21/01/1959
PHILIP	M	S		SQD OFF	Acting	27/03/1947
PHILIPS	D	G		FL		14/08/1966
PHILIPS	E	C	M	SGT		15/09/1949
PHILIPS	J	S		LAC		24/12/1942
PHILLIPPS	R	M		FL		28/02/1946
PHILLIPS	A	J		SQD LDR		26/01/1950
PHILLIPS	A	W		FLT SGT		24/06/1943
PHILLIPS	H	G		FL		30/10/1957
PHILLIPS	H			SGT	Acting	20/11/1957
PHILLIPS	J	S	P	SQD LDR		09/11/1944
PHILLIPS	K	O		FL		21/10/1959
PHILLIPS	P	A	G	SQD LDR		27/03/1947
PHILLIPS	R	F	P	FL		25/05/1944
PHILLIPS	R			SGT		12/06/1947
PHILLIPS	S			LAC		01/07/1948
PHILLPOT	J	E		FO		19/08/1954
PHILPOT	L	F		FLT SGT		25/06/1953
PHILPOTT	O	C		SQD LDR		14/01/1954
PHIPPS	G	F		SQD LDR		13/12/1945
PHIPPS	L	T		FL		07/05/1953
PICK	S	M		FLT SGT	Acting	19/08/1954
PICK	T	C		SQD LDR	Acting	27/07/1944
PICKARD	D	J		SGT		04/10/1997
PICKARD	S	G		FL		28/06/1945
PICKERING	A			FL		13/12/1945
PICKERING	J			FO		17/02/1944
PICKERING	J			FO		26/04/1961
PICKERING	T	G		FL		01/07/1948
PICKETT	A	R		WO		20/04/1995
PICKFORD	J	B		SGT		02/08/1951
PICKFORD	J	T		FO		10/06/1948
PICKUP	G	J		LAC		05/06/1947
PIERCE	F	H		FO		24/07/1963
PIERCY	W			SAC		22/02/1961
PIERPOINT	J	E	G	FO		25/03/1959
PIGG	P			FLT OFF		05/06/1947
PIKE	C	A		WING COMM		04/11/1943
PIKE	R	L		FO		12/02/1953
PIKE	S	G		FL		30/09/1954
PILKINGTON	V	T		CPL		10/04/1952
PILKINGTON	V	W		FL	Acting	13/07/1950
PILLERS	R	K		WING COMM		06/03/1947
PILLING	F			SQD LDR	Acting	10/08/1944
PILLING	T			LAC		01/12/1955
PIM	A	W		WING COMM	Acting	08/05/1947
PINDER	J			FO		30/09/1948
PINKERTON	G	C		WING COMM		03/12/1942
PINNELL	D	A		CPL		07/07/1949
PINNER	W	F		SGT		07/08/1947
PINNEY	E	F	W	FO		09/11/1944
PINNOCK	S	G		FO		25/06/1953
PINSENT	J			FL		03/10/1962
PIPER	A	H		FL		13/12/1945
PIPER	F	W		FL		06/01/1965
PIPER	H	R		FO		21/06/1945
PIPER	J	D		SQD LDR		17/04/1947

Source	Service No	Force	Notes	Surname
N395 1948	819117	RAuxAF		PHAIR
N662 1947	90424	AAF		PHELAN
N662 1947	109444	RAFVR		PHIBBS
N28 1959	2683569	RAuxAF	wef 6/10/58	PHILIP
N250 1947	1277	WAAF		PHILIP
MOD 1967	123606	RAFVR		PHILIPS
N953 1949	859514	RAuxAF		PHILIPS
N1758 1942	808024	AAF		PHILIPS
N197 1946	80563	RAFVR	DFC	PHILLIPPS
N94 1950	91161	RAuxAF		PHILLIPS
N677 1943	813029	AAF		PHILLIPS
N775 1957	142416	RAFVR	wef 4/9/55	PHILLIPS
N835 1957	2660968	WRAAF	wef 4/5/55	PHILLIPS
N1150 1944	90311	AAF	DFC	PHILLIPS
N708 1959	103927	RAFVR	wef 20/5/59	PHILLIPS
N250 1947	72918	RAFVR		PHILLIPS
N496 1944	116899	RAFVR		PHILLIPS
N478 1947	811048	AAF		PHILLIPS
N527 1948	844866	RAuxAF		PHILLIPS
N672 1954	186797	RAuxAF		PHILLPOT
N527 1953	2605028	RAFVR		PHILPOT
N18 1954	131184	RAuxAF		PHILPOTT
N1355 1945	70541	RAFVR	AFC	PHIPPS
N395 1953	116895	RAFVR		PHIPPS
N672 1954	2658073	WRAAF		PICK
N763 1944	73009	RAFVR	AFC	PICK
YB 1998	n/a	RAuxAF		PICKARD
N706 1945	143865	RAFVR	DFC	PICKARD
N1355 1945	109382	RAFVR		PICKERING
N132 1944	117397	RAFVR		PICKERING
N339 1961	117397	RAFVR	1st Clasp; wef 11/5/49	PICKERING
N527 1948	114471	RAFVR		PICKERING
YB 1996	n/a	RAuxAF		PICKETT
N777 1951	770574	RAFVR		PICKFORD
N456 1948	175421	RAFVR	DFM	PICKFORD
N460 1947	811004	AAF		PICKUP
N525 1963	2600086	RAFVR(T)	wef 19/9/57	PIERCE
N153 1961	2691095	RAuxAF	wef 5/11/60	PIERCY
N196 1959	2607286	RAFVR	wef 20/9/58	PIERPOINT
N460 1947	1963	WAAF		PIGG
N1144 1943	70543	RAFVR	AFC	PIKE
N115 1953	190670	RAFVR		PIKE
N799 1954	90785	RAuxAF		PIKE
N278 1952	868655	RAuxAF		PILKINGTON
N675 1950	109260	RAFVR		PILKINGTON
N187 1947	73720	RAFVR	OBE	PILLERS
N812 1944	91014	AAF		PILLING
N895 1955	858813	RAuxAF		PILLING
N358 1947	72453	RAFVR		PIM
N805 1948	191044	RAFVR		PINDER
N1638 1942	90160	AAF	DFC	PINKERTON
N673 1949	865541	RAuxAF		PINNELL
N662 1947	842495	AAF		PINNER
N1150 1944	137451	RAFVR	DFC	PINNEY
N527 1953	195468	RAFVR		PINNOCK
N722 1962	150362	RAFVR	wef 27/2/57	PINSENT
N1355 1945	68149	RAFVR	DFC	PIPER
MOD 1965	195907	RAFVR		PIPER
N671 1945	179245	RAFVR		PIPER
N289 1947	73341	RAFVR		PIPER

Surname	Initials			Rank	Status	Date
PIRIE	W	J		FLT SGT		19/10/1960
PITMAN	R	H		FL		11/07/1946
PITT	A	M	D	FL		18/03/1948
PITT	D	L		WING COMM	Acting	13/12/1945
PITT	D	M	B	SQD LDR		16/10/1947
PITTAM	E	C		FO		30/09/1948
PITTAWAY	R			SGT		01/07/1948
PITTER	E	M		WING OFF	Acting	08/05/1947
PITTS	T	B		FL		07/08/1947
PIZZEY	D	A		PO		28/07/1955
PLAISTOWE	G	W	B	SQD LDR		10/06/1948
PLANK	C			SAC		01/12/1973
PLANK	C			SAC		01/12/1983
PLANK	C			SAC		01/12/1993
PLASKITT	R	A		SQD LDR	Acting	15/07/1948
PLASKITT	R	H	J	CPL		20/05/1948
PLATER	T	H		CHIEF TECH		14/01/1954
PLATT	A	M		FLT OFF		27/03/1947
PLAYFAIR	R			LAC		10/06/1948
PLENDERLEITH	R			FL	Acting	21/09/1944
PLOWMAN	T			SAC		22/02/1961
PLUMB	G			FO		07/04/1955
PLUMMER	C	E		SQD LDR		24/11/1949
PLUMMER	T	S		CPL		05/06/1947
PLUMMER	W	L		WO		07/07/1949
PLUMRIDGE	D	L		FL		07/08/1996
PLUMSTEAD	L	J		FL		10/08/1944
POCOCK	C	H		FL		12/06/1947
POCOCK	J	R		WING COMM	Acting	06/03/1947
POCOCK	W	T		FL		23/02/1950
PODMORE	G	P		FL		12/02/1953
POINTON	R	A		SQD LDR	Acting	22/02/1951
POLDEN	A	E		FL		04/05/1944
POLE	D	R		FL		05/01/1950
POLE	E	G		FL		20/05/1948
POLLARD	J	F		SGT		22/06/1995
POLLARD	L	T		LAC		13/11/1947
POLLARD	L	V		SQD LDR	Acting	06/03/1947
POLLARD	W	H	C	CPL		09/02/1951
POLLITT	K	T		SGT	Acting	28/08/1963
POLLOCK	D	J		FO		25/05/1960
POLLOCK	G	S		SQD LDR	Acting	15/07/1948
POLLOCK	J	T		SAC		16/04/1958
POLLOCK	S	J		SQD LDR	Acting	13/02/1957
POLSON	E	L		CPL		26/11/1953
POMEROY	J	F	B	SQD LDR	Acting	05/01/1950
PONTEFRACT	R	M		FL		25/03/1959
POOL	D	C		FL		13/02/1957
POOL	J	H		WING COMM		15/07/1948
POOLE	A	R		SQD LDR	Acting	23/05/1946
POOLE	A	R		SQD LDR		26/11/1953
POOLE	D	F		SQD LDR		05/01/1950
POOLE	F	J		FL		06/03/1947
POOLE	E	W		FL		23/05/1946
POOLE	F	F	F	SGT		15/01/1958
POOLE	M	C		LAC		01/12/1955
POPE	E	W		WO		24/01/1946
POPE	O	A		CPL		20/05/1948
POPE	R	E		FL		15/06/1944
POPE	V	D		FL		28/02/1946

Source	Service No	Force	Notes	Surname
N775 1960	2649722	RAuxAF	wef 8/1/60	PIRIE
N594 1946	148452	RAFVR	DSO	PITMAN
N232 1948	91146	RAuxAF		PITT
N1355 1945	83986	RAFVR	DSO,DFC,AFC	PITT
N846 1947	73342	RAFVR	MBE	PITT
N805 1948	62102	RAFVR	DCM	PITTAM
N527 1948	851450	RAuxAF		PITTAWAY
N358 1947	708	WAAF		PITTER
N662 1947	72754	RAFVR		PITTS
N565 1955	2601258	RAFVR		PIZZEY
N456 1948	72173	RAFVR		PLAISTOWE
YB 1996	n/a	RAuxAF		PLANK
YB 1996	n/a	RAuxAF	1st Clasp	PLANK
YB 1996	n/a	RAuxAF	2nd Clasp	PLANK
N573 1948	110031	RAFVR		PLASKITT
N395 1948	840803	RAuxAF		PLASKITT
N18 1954	2694000	RAuxAF		PLATER
N250 1947	3490	WAAF		PLATT
N456 1948	816138	RAuxAF		PLAYFAIR
N961 1944	139411	RAFVR	DFC	PLENDERLEITH
N153 1961	2686171	RAFVR	wef 1/10/60	PLOWMAN
N277 1955	195729	RAuxAF		PLUMB
N1193 1949	72498	RAFVR		PLUMMER
N460 1947	811200	AAF		PLUMMER
N673 1949	751284	RAFVR		PLUMMER
YB 1997	n/a	RAuxAF		PLUMRIDGE
N812 1944	101505	RAFVR		PLUMSTEAD
N478 1947	62782	RAFVR		POCOCK
N187 1947	72163	RAFVR		POCOCK
N179 1950	144241	RAFVR		POCOCK
N115 1953	176023	RAFVR		PODMORE
N194 1951	85293	RAFVR		POINTON
N425 1944	116131	RAFVR		POLDEN
N1 1950	142323	RAFVR		POLE
N395 1948	75293	RAFVR	MBE	POLE
YB 1996	n/a	RAuxAF		POLLARD
N933 1947	865207	AAF		POLLARD
N187 1947	118899	RAFVR	DFC	POLLARD
N145 1951	770857	RAFVR		POLLARD
N621 1963	2614260	RAuxAF	wef 1/7/63	POLLITT
N375 1960	2606609	RAFVR	wef 28/11/59	POLLOCK
N573 1948	91028	RAuxAF	MBE	POLLOCK
N258 1958	2686057	RAFVR	wef 20/9/57	POLLOCK
N104 1957	87457	RAFVR	DFC; wef 12/7/45	POLLOCK
N952 1953	2685041	RAuxAF		POLSON
N1 1950	82679	RAFVR		POMEROY
N196 1959	197902	RAuxAF	wef 26/7/53	PONTEFRACT
N104 1957	140706	RAFVR	wef 29/1/56	POOL
N573 1948	90789	RAuxAF		POOL
N468 1946	129193	RAFVR	DSO,DFC	POOLE
N952 1953	129193	RAFVR	1st Clasp	POOLE
N1 1950	72192	RAFVR		POOLE
N187 1947	85240	RAFVR		POOLE
N468 1946	142578	RAFVR		POOLE
N40 1958	841789	AAF	wef 18/4/44	POOLE
N895 1955	850004	RAuxAF		POOLE
N82 1946	742557	RAFVR		POPE
N395 1948	744404	RAFVR		POPE
N596 1944	81360	RAFVR		POPE
N197 1946	120980	RAFVR		POPE

Surname	Initials			Rank	Status	Date
PORRITT	E			LAC		15/07/1948
PORTEOUS	J	S		FL		05/09/1962
PORTER	E	F		FL		22/02/1961
PORTER	G	R		WO	Temporary	20/01/1960
PORTER	R			FO		28/11/1946
POTE	J			SQD LDR		09/05/1998
POTT	D	S		FO		10/05/1951
POTTER	C	A		SQD LDR		05/01/1950
POTTER	E	A		CPL	Temporary	03/07/1957
POTTER	F	W		LAC		20/09/1951
POTTER	G	B		FL		11/01/1945
POTTER	K	D		SQD LDR		22/02/1951
POTTER	M	A	G	FLT OFF	Acting	26/06/1947
POTTER	M	F	W	CPL		24/11/1949
POTTER	P	S		SQD LDR		10/06/1948
POTTER	R	H		LAC		26/06/1947
POTTER	R	K		FL		21/09/1944
POTTER-IRWIN	M	H		FL		06/03/1947
POTTS	A			FL	Acting	13/01/1949
POTTS	A			SAC		01/05/1995
POTTS	E			FLT OFF		21/02/1952
POTTS	E			FLT OFF		22/02/1961
POTTS	J	E		FL		26/06/1947
POTTS	J			FL		07/07/1949
POULTER	M	F		SQD LDR		26/01/1950
POULTON	H	R	G	FL		21/03/1946
POVEY	A	R		FL		27/04/1944
POWELL	A	S		FL		11/07/1946
POWELL	C	T		LAC		15/11/1951
POWELL	D			SQD LDR		29/04/1954
POWELL	D			SQD LDR	Acting	20/03/1957
POWELL	F	H		FLT SGT		26/11/1953
POWELL	F	J		FL		27/04/1950
POWELL	F			LAC		20/05/1948
POWELL	G	F		FL		06/05/1943
POWELL	H			SGT		22/02/1951
POWELL	J	F		FL		17/04/1947
POWELL	J	H		SGT		20/05/1948
POWELL	K	B		FL		25/06/1953
POWELL	K	R		FL		01/07/1948
POWELL	L			LAC		03/07/1952
POWELL	S	W	M	FL		13/01/1949
POWELL	W	G		FL		24/01/1946
POWER	D	C		FL		28/11/1956
POWER	E	S	F	AC1		01/10/1953
POWLING	R	H	C	FL		04/11/1948
PRATER	L	R		WO		24/06/1943
PRATER	W	J		FLT SGT		18/04/1946
PRATLEY	F			FL		01/07/1948
PRATT	J	C		SGT		07/06/1951
PREECE	M	L		FL		20/07/2004
PREECE	T	W		SGT		26/01/1950
PREEDY	J	R		WING COMM	Acting	21/03/1946
PRENTICE	J			WO		30/03/1943
PRENTICE	T	D		LAC		20/05/1948
PRESCOTT	G	H		CPL		21/06/1951
PRESCOTT-DENNIS	W	F		FL		18/03/1948
PRESS	L	G		FL		09/08/1945
PRESTON	H			WING COMM		18/04/1946
PRESTWICH	M	E		FLT OFF		12/06/1947

Source	Service No	Force	Notes	Surname
N573 1948	873903	RAuxAF		PORRITT
N655 1962	83038	RAuxAF	wef 30/7/62	PORTEOUS
N153 1961	144962	RAFVR	wef 5/12/60	PORTER
N33 1960	751583	RAFVR	wef 21/7/44	PORTER
N1004 1946	179882	RAFVR		PORTER
YB 1999	2623289	RAuxAF		POTE
N466 1951	159286	RAFVR		POTT
N1 1950	90647	RAuxAF		POTTER
N456 1957	800522	AAF	wef 27/9/43	POTTER
N963 1951	864157	RAuxAF		POTTER
N14 1945	81932	RAFVR		POTTER
N194 1951	72454	RAFVR		POTTER
N519 1947	2746	WAAF		POTTER
N1193 1949	857631	RAuxAF		POTTER
N456 1948	90394	RAuxAF		POTTER
N519 1947	811128	AAF		POTTER
N961 1944	81364	RAFVR		POTTER
N187 1947	157205	RAFVR		POTTER-IRWIN
N30 1949	118295	RAFVR		POTTS
YB 1996	n/a	RAuxAF		POTTS
N134 1952	184	WAAF		POTTS
N153 1961	184	WRAAF	1st Clasp; wef 20/10/60	POTTS
N519 1947	76204	RAFVR		POTTS
N673 1949	146352	RAFVR		POTTS
N94 1950	73655	RAFVR		POULTER
N263 1946	84925	RAFVR	DFC	POULTON
N396 1944	78712	RAFVR		POVEY
N594 1946	86631	RAFVR		POWELL
N1185 1951	868605	RAuxAF		POWELL
N329 1954	136120	RAuxAF		POWELL
N198 1957	136120	RAuxAF	1st Clasp; wef 4/9/56	POWELL
N952 1953	749194	RAFVR		POWELL
N406 1950	60348	RAFVR		POWELL
N395 1948	801537	RAuxAF		POWELL
N526 1943	70550	RAFVR		POWELL
N194 1951	849062	RAuxAF		POWELL
N289 1947	73348	RAFVR		POWELL
N395 1948	818087	RAuxAF		POWELL
N527 1953	188291	RAFVR		POWELL
N527 1948	106019	RAFVR		POWELL
N513 1952	855927	RAuxAF		POWELL
N30 1949	68136	RAFVR		POWELL
N82 1946	106361	RAFVR	DFC	POWELL
N833 1956	168452	RAFVR		POWER
N796 1953	755045	RAFVR		POWER
N934 1948	85682	RAFVR		POWLING
N677 1943	813004	AAF		PRATER
N358 1946	808157	AAF		PRATER
N527 1948	162853	RAFVR		PRATLEY
N560 1951	770192	RAFVR		PRATT
LG 2004	0091516N	RAuxAF		PREECE
N94 1950	849941	RAuxAF		PREECE
N263 1946	79183	RAFVR	DSO	PREEDY
N1014 1943	803093	AAF		PRENTICE
N395 1948	803585	RAuxAF		PRENTICE
N621 1951	855042	RAuxAF		PRESCOTT
N232 1948	115315	RAFVR		PRESCOTT-DENNIS
N874 1945	117153	RAFVR	AFC	PRESS
N358 1946	70554	RAFVR		PRESTON
N478 1947	515	WAAF		PRESTWICH

Surname	Initials			Rank	Status	Date
PREW	J	I		SGT		09/12/1959
PRICE	A			SQD LDR	Acting	09/05/1946
PRICE	B	L		FL		21/09/1944
PRICE	C	M		FO		30/09/1954
PRICE	F	C		LAC		23/02/1950
PRICE	H	W		FL		20/01/1960
PRICE	J			FLT SGT		28/02/1946
PRICE	S	L		SGT		12/06/1947
PRICE	U	M		FLT SGT		08/05/1947
PRICE	W	C		SGT		07/08/1947
PRICE	W	J	R	FO		13/02/1957
PRIESTLEY	A	R		SGT	Temporary	03/06/1959
PRIESTLEY	G			SGT		03/07/1952
PRIESTLEY	H			FL	Acting	30/03/1943
PRIESTLEY	K	S		FL		14/10/1943
PRIESTLY	G			LAC		14/01/1954
PRIME	J	C		SQD LDR	Acting	12/04/1945
PRIMER	R			FO		23/05/1946
PRINCE	N	S		FO		05/06/1947
PRINCE	S			LAC		05/06/1947
PRINGLE	H			CPL		13/01/1949
PRINGLE	R			WO	Acting	05/06/1947
PRIOR	R	W	A	FL		07/12/1960
PRITCHARD	A	S		SQD OFF		05/06/1947
PRITCHARD	C	A		WING COMM	Acting	22/04/1943
PRITCHARD	E	W		LAC		01/10/1953
PRITCHARD	J	W		AC2		02/08/1951
PRITCHARD	L			SGT		18/03/1948
PRITCHARD	W			FL	Acting	04/11/1948
PRITCHETT	I	P		SEC OFF		12/06/1947
PROCTOR	A	C		SGT		10/02/1949
PROCTOR	A	C		SGT		20/08/1958
PROCTOR	A			SAC		03/09/1989
PROCTOR	A			SAC		03/09/1990
PROCTOR	E	H		WING COMM	Acting	27/04/1950
PROCTOR	H	W		CPL		15/09/1949
PROCTOR	J	W		FL		01/08/1946
PRYER	F	J	M	SQD LDR		13/11/1947
PRYER	W	E		CPL		05/06/1947
PUCKLE	A	M		SQD LDR	Acting	04/09/1947
PUGH	F	C		SGT		07/05/1953
PUGH	J	S		FL		28/11/1946
PULHAM	D	A		FLT SGT		17/04/1947
PULVERMACHER	K			FL		04/10/1945
PUMMELL	E	W		FLT SGT		26/08/1948
PUMMERY	C	F		WO		09/05/1946
PUNSHEON	R	G		CPL		07/05/1953
PURCELL	A	E		CPL		12/06/1947
PURCER-SMITH	H	C		SQD LDR		06/03/1947
PURSER	J	W	G	FL	Acting	10/02/1949
PURSEY	E	F		CPL		16/03/1950
PURVIS	A			CPL		10/06/1948
PURVIS	G	C		FL		08/07/1959
PUTLEY	A	O		SAC		20/07/1960
PUXLEY	W	G	V	FL		24/01/1946
PYGALL	A	S		CPL		19/10/1960
PYKETT	F	A		CPL		10/04/1952
PYNE	T	S		LAC		25/06/1953
PYPER	A	H	D	FLT SGT	Acting	03/06/1959
QUARRINGTON	C	A	G	WING COMM		15/07/1948

Source	Service No	Force	Notes	Surname
N843 1959	2655769	WRAAF	wef 20/2/59	PREW
N415 1946	83299	RAFVR	DFC	PRICE
N961 1944	70814	RAFVR		PRICE
N799 1954	2657512	WRAAF		PRICE
N179 1950	865690	RAuxAF		PRICE
N33 1960	168303	RAFVR	wef 25/2/58	PRICE
N197 1946	805044	AAF		PRICE
N478 1947	882730	WAAF		PRICE
N358 1947	880429	WAAF		PRICE
N662 1947	744898	RAFVR		PRICE
N104 1957	196189	RAFVR	wef 21/5/54	PRICE
N368 1959	812144	AAF	wef 6/4/44	PRIESTLEY
N513 1952	2688538	RAuxAF		PRIESTLEY
N1014 1943	122115	RAFVR		PRIESTLEY
N1065 1943	84326	RAFVR		PRIESTLEY
N18 1954	870934	RAuxAF		PRIESTLY
N381 1945	85264	RAFVR		PRIME
N468 1946	160831	RAFVR		PRIMER
N460 1947	107291	RAFVR		PRINCE
N460 1947	811093	AAF		PRINCE
N30 1949	803546	RAuxAF		PRINGLE
N460 1947	852227	AAF		PRINGLE
N933 1960	150438	RAuxAF	wef 3/10/59	PRIOR
N460 1947	214	WAAF		PRITCHARD
N475 1943	90092	AAF	DFC	PRITCHARD
N796 1953	803488	RAuxAF		PRITCHARD
N777 1951	856031	RAuxAF		PRITCHARD
N232 1948	811146	RAuxAF		PRITCHARD
N934 1948	88892	RAFVR		PRITCHARD
N478 1947	244	WAAF		PRITCHETT
N123 1949	749620	RAFVR		PROCTOR
N559 1958	2602376	RAFVR	1st Clasp; wef 8/9/55	PROCTOR
YB 1996	n/a	RAuxAF		PROCTOR
YB 2000	n/a	RAuxAF	1st Clasp	PROCTOR
N406 1950	72310	RAFVR		PROCTOR
N953 1949	848410	RAuxAF		PROCTOR
N652 1946	169093	RAFVR		PROCTOR
N933 1947	73354	RAFVR		PRYER
N460 1947	841546	AAF		PRYER
N740 1947	78103	RAFVR	MBE	PUCKLE
N395 1953	840067	RAuxAF		PUGH
N1004 1946	120329	RAFVR	DFC	PUGH
N289 1947	881084	WAAF		PULHAM
N1089 1945	146897	RAFVR		PULVERMACHER
N694 1948	859516	RAuxAF	DCM	PUMMELL
N415 1946	755709	RAFVR		PUMMERY
N395 1953	2698581	RAuxAF		PUNSHEON
N478 1947	811194	AAF		PURCELL
N187 1947	70633	RAFVR		PURCER-SMITH
N123 1949	91107	RAuxAF		PURSER
N257 1950	861205	RAuxAF		PURSEY
N456 1948	808388	RAuxAF		PURVIS
N454 1959	150375	RAFVR	wef 6/6/55	PURVIS
N531 1960	2663924	WRAAF	wef 1/5/60	PUTLEY
N82 1946	117933	RAFVR		PUXLEY
N775 1960	750942	RAFVR	wef 23/7/44	PYGALL
N278 1952	799707	RAFVR		PYKETT
N527 1953	850606	RAuxAF		PYNE
N368 1959	803609	AAF	wef 19/3/44	PYPER
N573 1948	72921	RAFVR		QUARRINGTON

Surname	Initials			Rank	Status	Date
QUARTERMAINE	A	H		FL		13/01/1944
QUAYE	W	J		LAC		13/07/1950
QUELCH	B	H		FL		10/05/1945
QUELCH	B	H		FL		22/09/1955
QUICK	L	W		FL		11/07/1946
QUICK	S	P		WING COMM	Acting	05/06/1947
QUIN	J			CPL		26/08/1948
QUINE	J	T		CPL		30/09/1948
QUINN	D	R	S P	SQD LDR		30/03/1943
QUINN	F	E	A	FL		24/01/1946
QUINN	J			WO		24/11/1949
QUINTON	A	E		FL		05/09/1946
QUIRK	J	C		FO		17/06/1954
QUITZOW	J	A	U	FL		10/05/1945
RABBITT	A	J		LAC		15/07/1948
RABONE	J	H	M	FL		13/12/1945
RACHER	K	W		LAC		27/03/1952
RACKHAM	R	J		WING COMM		04/09/1947
RADCLIFFE	A			LAC		12/10/1950
RADDEN	A	H		FLT SGT		24/11/1949
RADFORD	E	A	B	FL		26/04/1961
RADFORD	H	D		FL		30/09/1948
RADFORD	P	I		FL		03/10/1946
RADFORD	W			CPL		13/01/1949
RADLEY	R	S		SQD LDR	Acting	24/01/1946
RADLEY	W	S		SGT	Acting	07/04/1955
RADNEDGE	J	L		SAC		04/10/1996
RAE	G			WO		01/07/1948
RAE	W			FL		01/12/1955
RAFFAN	J	T		FL		06/03/1947
RAFTER	R			CPL		24/07/1952
RAGAN	J	H		SGT		07/07/1949
RAHILLY	S	D		PO		07/12/1960
RAILTON	W	H		SGT		26/08/1948
RAINFORTH	W	O		FL		26/08/1948
RALPH	R	J		SQD LDR	Acting	17/02/1944
RAMPLING	E	F		FL		28/02/1946
RAMSAY	G	M	I	FLT OFF		12/06/1947
RAMSAY	G	R	L	FLT SGT		04/09/1957
RAMSAY	I	L		WO		18/04/1946
RAMSAY	I	McT		SQD LDR	Acting	11/01/1945
RAMSAY	I	McT		WING COMM		14/01/1954
RAMSAY	J	D		SQD LDR		13/02/1944
RAMSAY	N	H	D	FL		03/10/1946
RAMSAY-SMITH	D	L		FL		12/04/1945
RAMSBOTTOM	B	R	T	FL		19/08/1959
RAMSDEN	O	A		WING OFF	Acting	06/03/1947
RAMSEY	C	M		FO		12/08/1943
RAMSHAW	H			CPL		16/03/1950
RANDALL	A	H	C	SQD LDR		17/04/1947
RANDALL	C	H	C	FL		13/12/1945
RANDALL	P			SQD LDR		19/07/1951
RANDERSON	Γ			FLT SGT		14/10/1948
RANKIN	A			SGT		16/07/1948
RANKIN	H	D		FL		23/02/1950
RANKIN	R	M		WING COMM		26/01/1950
RANKIN	W			CPL	Temporary	21/01/1959
RANN	K	V		WO		28/02/1946
RANSLEY	L	F		SGT		18/03/1948
RANSON	P	L		FO		19/01/1956

Source	Service No	Force	Notes	Surname
N11 1944	78848	RAFVR		QUARTERMAINE
N675 1950	817084	RAuxAF		QUAYE
N485 1945	115130	RAFVR	DFC	QUELCH
N708 1955	115130	RAFVR	DFC; 1st Clasp	QUELCH
N594 1946	146028	RAFVR	DFC	QUICK
N460 1947	72175	RAFVR		QUICK
N694 1948	743812	RAFVR		QUIN
N805 1948	855152	RAuxAF		QUINE
N1014 1943	72344	RAFVR		QUINN
N82 1946	84721	RAFVR	MBE	QUINN
N1193 1949	759133	RAFVR		QUINN
N751 1946	113836	RAFVR	DFM	QUINTON
N471 1954	172031	RAuxAF		QUIRK
N485 1945	122724	RAFVR		QUITZOW
N573 1948	743470	RAFVR		RABBITT
N1355 1945	90226	AAF		RABONE
N224 1952	743288	RAFVR		RACHER
N740 1947	73357	RAFVR		RACKHAM
N1018 1950	814196	RAuxAF		RADCLIFFE
N1193 1949	743764	RAFVR		RADDEN
N339 1961	87888	RAFVR	wef 9/11/60	RADFORD
N805 1948	147926	RAFVR		RADFORD
N837 1946	174982	RAFVR		RADFORD
N30 1949	849863	RAuxAF		RADFORD
N82 1946	83263	RAFVR	DFC	RADLEY
N277 1955	2689513	RAuxAF		RADLEY
YB 1998	n/a	RAuxAF		RADNEDGE
N527 1948	802568	RAuxAF		RAE
N895 1955	165452	RAuxAF		RAE
N187 1947	90938	AAF		RAFFAN
N569 1952	2688535	RAuxAF		RAFTER
N673 1949	804348	RAuxAF		RAGAN
N933 1960	2650849	RAuxAF	wef 17/10/60	RAHILLY
N694 1948	770028	RAFVR		RAILTON
N694 1948	82754	RAFVR		RAINFORTH
N132 1944	79526	RAFVR	AFC	RALPH
N197 1946	156332	RAFVR	DFC	RAMPLING
N478 1947	445	WAAF		RAMSAY
N611 1957	799979	RAFVR	wef 30/5/44	RAMSAY
N358 1946	747915	RAFVR		RAMSAY
N14 1945	85667	RAFVR		RAMSAY
N18 1954	85667	RAuxAF	1st Clasp	RAMSAY
MOD 1964	72367	RAFVR		RAMSAY
N837 1946	62658	RAFVR	DFC	RAMSAY
N381 1945	82175	RAFVR		RAMSAY-SMITH
N552 1959	128926	RAFVR	wef 13/10/52	RAMSBOTTOM
N187 1947	286	WAAF		RAMSDEN
N850 1943	106078	RAFVR	DFC	RAMSEY
N257 1950	750233	RAFVR		RAMSHAW
N289 1947	72922	RAFVR		RANDALL
N1355 1945	131137	RAFVR		RANDALL
N724 1951	73626	RAFVR		RANDALL
N856 1948	814116	RAuxAF		RANDERSON
N573 1948	802542	RAuxAF		RANKIN
N179 1950	83710	RAFVR		RANKIN
N94 1950	72923	RAFVR		RANKIN
N28 1959	873074	RAFVR	wef 8/6/44	RANKIN
N197 1946	746989	RAFVR		RANN
N232 1948	860287	RAuxAF		RANSLEY
N43 1956	2601336	RAFVR		RANSON

Surname	Initials				Rank	Status	Date
RASMUSSEN	I	O			FL		20/01/1960
RASON	A				CPL		30/09/1948
RATCLIFF	T	G			WING COMM	Acting	26/08/1948
RATCLIFFE	L	F			WING COMM		01/07/1948
RATCLIFFE	R	G			WO		03/10/1946
RATCLIFFE	W	A			SGT		30/09/1954
RATHBONE	W	F			FL		01/03/1945
RATHBONE	W				FL		17/01/1952
RATHBONE	W				FL		22/02/1961
RATHKEY	K	W			LAC		07/07/1949
RAVEN	F	H	G	R	FL		04/09/1957
RAVENHILL	A	W			SQD LDR	Acting	06/03/1947
RAVENSCROFT	E				SGT		11/07/1946
RAW	J	H			FL		20/01/1960
RAWBONE	A				SGT		30/03/1943
RAWLINS	C	G	C		SQD LDR	Acting	09/05/1946
RAWLINS	D	F			FO		19/08/1954
RAWNSLEY	C	F			FL		11/11/1943
RAWSON	F	W	B		SGT		15/07/1948
RAWSON	R	F	W		SGT	Temporary	22/10/1958
RAY	C	R			SQD LDR	Acting	14/11/1962
RAYNER	E	B			SQD LDR		26/02/1958
RAYNER	G	J			WING COMM		09/08/1945
RAYNER	J	L			LAC		27/03/1952
RAYNER	J	W			SQD LDR	Acting	30/03/1943
RAYNER	S				CPL		30/09/1948
RAYNES	C				FL		04/11/1948
RAYNOR	R				PO		07/04/1955
REA	A	T			AC1		27/04/1950
REA	H	S			CPL	Temporary	19/10/1960
REA	K	R			FL		26/04/1961
REACROFT	F	D			FL		01/05/1957
READ	F	I			SGT		16/10/1947
READ	G	B			WING COMM	Acting	13/01/1949
READ	J				WO		16/10/1947
READ	R	G			FO		31/05/1945
READ	W	C			LAC		06/01/1955
READ	W	J			CPL		25/06/1953
READING	F	C			CPL		18/11/1948
READING	R				SGT		23/02/1950
READINGS	P				FL		10/06/1948
REAH	B	M			AC2		25/05/1960
REAKS	K	E			FL		26/08/1948
REAM	T	E	J		FO		21/06/1945
REASON	S	J			FL		28/11/1956
REAVELL-CARTER	L				SQD LDR	Acting	24/01/1946
RECKETT	J	C	T		FL		20/03/1957
REDDING	G	W			CPL		20/05/1948
REDDING	J	E			GRP CAP	Acting	27/03/1947
REDDY	L	W			LAC		26/08/1948
REDFEARN	G	B			FL		21/06/1951
REDFERN	G	F			SQD LDR		20/07/1960
REDFERN	T				CPL		14/10/1948
REDGRAVE	C	F			FO		15/07/1948
REDHEAD	A				FLT SGT	Acting	26/01/1950
REDMAN	E	W	G		FO		27/04/1950
REDMAN	W	C	A		CPL		30/09/1948
REDMAYNE	R	P			PO		04/01/2005

Source	Service No	Force	Notes	Surname
N33 1960	202158	RAFVR	wef 3/12/59	RASMUSSEN
N805 1948	846990	RAuxAF		RASON
N694 1948	72278	RAFVR		RATCLIFF
N527 1948	87022	RAFVR	DSO,DFC,AFC	RATCLIFFE
N837 1946	745200	RAFVR		RATCLIFFE
N799 1954	2688621	RAuxAF		RATCLIFFE
N225 1945	88422	RAFVR		RATHBONE
N35 1952	90544	RAuxAF		RATHBONE
N153 1961	90544	RAFVR	1st Clasp; wef 10/11/60	RATHBONE
N673 1949	865627	RAuxAF		RATHKEY
N611 1957	127029	RAFVR	wef 16/5/56	RAVEN
N187 1947	115773	RAFVR	DFC	RAVENHILL
N594 1946	805343	AAF		RAVENSCROFT
N33 1960	199051	RAFVR	wef 9/9/57	RAW
N1014 1943	805051	AAF		RAWBONE
N415 1946	72510	RAFVR	DFC	RAWLINS
N672 1954	140240	RAFVR		RAWLINS
N1177 1943	102089	RAFVR	DSO,DFC,DFM	RAWNSLEY
N573 1948	846106	RAuxAF		RAWSON
N716 1958	746369	RAFVR	wef 23/6/44	RAWSON
N837 1962	103950	RAFVR	wef 1/9/62	RAY
N149 1958	72035	RAFVR	MB,ChB; 1st Clasp; wef 19/12/56; no record of first award	RAYNER
N874 1945	90436	AAF		RAYNER
N224 1952	2688523	RAuxAF		RAYNER
N1014 1943	70565	RAFVR		RAYNER
N805 1948	840431	RAuxAF		RAYNER
N934 1948	79602	RAFVR		RAYNES
N277 1955	2605373	RAFVR		RAYNOR
N406 1950	849960	RAuxAF		REA
N775 1960	865075	AAF	wef 8/7/44	REA
N339 1961	183778	RAFVR	DFM; wef 10/3/59	REA
N277 1957	199968	RAuxAF	wef 2/10/56	REACROFT
N846 1947	843610	RAFVR		READ
N30 1949	77209	RAFVR	DFC	READ
N846 1947	810154	AAF		READ
N573 1945	149516	RAFVR		READ
N1 1955	840374	RAuxAF		READ
N527 1953	853555	RAuxAF		READ
N984 1948	844226	RAuxAF		READING
N179 1950	748828	RAFVR		READING
N456 1948	124002	RAFVR		READINGS
N375 1960	2661813	WRAAF	wef 8/5/59	REAH
N694 1948	128706	RAFVR		REAKS
N671 1945	182002	RAFVR		REAM
N833 1956	83388	RAFVR		REASON
N82 1946	76017	RAFVR		REAVELL-CARTER
N198 1957	180442	RAuxAF	wef 15/9/56	RECKETT
N395 1948	800572	RAuxAF		REDDING
N250 1947	90530	AAF		REDDING
N694 1948	75961	RAFVR		REDDY
N621 1951	142992	RAFVR	DFM	REDFEARN
N531 1960	78446	RAFVR	wef 7/7/44	REDFERN
N856 1948	856723	RAuxAF		REDFERN
N573 1948	156429	RAFVR	DFC	REDGRAVE
N94 1950	750390	RAFVR		REDHEAD
N406 1950	136236	RAFVR		REDMAN
N805 1948	744011	RAFVR		REDMAN
LG 2005	5007156	RAuxAF		REDMAYNE

Surname	Initials			Rank	Status	Date
REDMOND	B	J		FL		17/02/1944
REED	F	A	O	WO		18/04/1946
REED	F	E		SQD LDR	Acting	21/03/1946
REED	H	J		CPL		13/01/1949
REED	H	T		SGT		30/09/1948
REED	H			FL		26/06/1947
REED	W	A	A	FL		01/03/1945
REES	E	T		WING COMM		17/04/1947
REES	F	A		CPL		26/08/1948
REES	M	B	J	CPL		26/06/1947
REES	O	W		FL		21/03/1946
REES	W	E		SGT		04/11/1948
REES	W	L		SQD LDR		07/08/1947
REES-WILLIAMS	A	E		SAC		22/02/1995
REEVE	C	G		WING COMM	Acting	13/02/1957
REEVE	F	H		CPL	Temporary	20/01/1960
REEVE	J	H		SQD LDR	Acting	28/06/1945
REEVES	A	C		FL		21/07/1994
REEVES	G	L		CPL		30/09/1948
REEVES	J	W		LAC		25/05/1950
REEVES	N	E		WING COMM	Acting	06/03/1947
REEVES	R	S		CPL		16/09/1993
REEVES	V	E		SGT		17/04/1947
REEVES	W	L		FO		17/03/1949
REGAN	R	J	W	LAC		26/06/1947
REGAN	R	J	W	SGT		01/05/1952
REID	A			SQD LDR	Acting	18/04/1946
REID	B	K	J	CPL		19/05/1984
REID	B	K	J	CPL		19/05/1994
REID	D			FO		25/05/1960
REID	D			FO		16/08/1961
REID	J	C	A	SQD LDR		15/07/1948
REID	J	C		FL		01/05/1997
REID	J	S		SGT		31/05/1945
REID	J	W		LAC		01/10/1953
REID	M	D		SGT		01/07/1948
REID	R	H		FL		04/09/1957
REID	R	S		FO		26/04/1961
REID	R			FL		30/12/1943
REID	R			CPL		26/08/1948
REID	W			FLT SGT		27/03/1947
REIDY	R	M		WO		30/12/1943
REILLY	P			SGT		13/12/1945
RELF	H	L		FLT SGT		05/02/1948
RENAUT	J	K		FL		19/10/1960
RENDELL	J	H		SQD LDR		17/03/1949
RENDELL	J	H		FL		20/11/1957
RENDLE	R			SQD LDR		15/04/1943
RENGERT	E	J	K	FL		14/01/1954
RENNIE	J	L		CPL		13/01/1949
RENNIE	J	L		CPL		07/04/1955
RENNIE	J			SGT		07/07/1949
RENNIE	J			OOT		16/03/1950
RENNIE	W	M	B	CPL		20/05/1948
RENNIE	W	M	B	SGT		16/10/1952
RENSEN	L	W		SQD OFF	Acting	06/03/1947
RENSHAW	E	S		WO		17/03/1949
RENTALL	C	H	F	CPL		07/08/1947
RENVOIZE	J	V		FL		28/02/1946
RENWICK	A	E		SQD LDR		13/11/1947

Source	Service No	Force	Notes	Surname
N132 1944	68143	RAFVR	DFC	REDMOND
N358 1946	801374	AAF		REED
N263 1946	100599	RAFVR		REED
N30 1949	750998	RAFVR		REED
N805 1948	743265	RAFVR		REED
N519 1947	129933	RAFVR		REED
N225 1945	80822	RAFVR	AFC	REED
N289 1947	73363	RAFVR		REES
N694 1948	857268	RAuxAF		REES
N519 1947	756792	RAFVR		REES
N263 1946	122061	RAFVR	DSO,DFC	REES
N934 1948	870918	RAuxAF		REES
N662 1947	72452	RAFVR		REES
YB 1996	n/a	RAuxAF		REES-WILLIAMS
N104 1957	82965	RAF	MBE; wef 17/8/44	REEVE
N33 1960	846224	AAF	wef 11/4/44	REEVE
N706 1945	79527	RAFVR	DFC	REEVE
YB 1999	91496	RAuxAF		REEVES
N805 1948	749313	RAFVR		REEVES
N517 1950	846044	RAuxAF		REEVES
N187 1947	110797	RAFVR	DSO,DFC	REEVES
YB 1996	n/a	RAuxAF		REEVES
N289 1947	844351	AAF		REEVES
N231 1949	190737	RAFVR	DFC	REEVES
N519 1947	813180	AAF		REGAN
N336 1952	2680500	RAuxAF	1st Clasp	REGAN
N358 1946	79184	RAFVR	DFC	REID
YB 1996	n/a	RAuxAF		REID
YB 1996	n/a	RAuxAF	1st Clasp	REID
N375 1960	2649788	RAuxAF	wef 12/3/60	REID
N624 1961	2696046	RAuxAF	wef 1/2/61	REID
N573 1948	90998	RAuxAF		REID
YB 1999	8024882	RAuxAF		REID
N573 1945	873046	EX-AAF		REID
N796 1953	873162	RAuxAF		REID
N527 1948	749995	RAFVR		REID
N611 1957	171758	RAFVR	wef 8/4/57	REID
N339 1961	2601925	RAuxAF	wef 31/5/58	REID
N1350 1943	80836	RAFVR		REID
N694 1948	743843	RAFVR		REID
N250 1947	873238	AAF		REID
N1350 1943	740024	RAFVR		REIDY
N1355 1945	803333	AAF		REILLY
N87 1948	812068	RAuxAF		RELF
N775 1960	117645	RAFVR	wef 18/3/58	RENAUT
N231 1949	72241	RAFVR		RENDELL
N835 1957	72241	RAFVR	1st Clasp; wef 11/4/55	RENDELL
N454 1943	72209	RAFVR		RENDLE
N18 1954	86565	RAuxAF		RENGERT
N30 1949	803594	RAuxAF		RENNIE
N277 1955	2683502	RAuxAF	1st Clasp	RENNIE
N673 1949	803485	RAuxAF		RENNIE
N257 1950	803352	RAuxAF		RENNIE
N395 1948	803557	RAuxAF		RENNIE
N798 1952	2683506	RAuxAF	1st Clasp	RENNIE
N187 1947	4000	WAAF		RENSEN
N231 1949	820045	RAuxAF		RENSHAW
N662 1947	842511	AAF		RENTALL
N197 1946	110352	RAFVR		RENVOIZE
N933 1947	90583	AAF	MC	RENWICK

Surname	Initials			Rank	Status	Date
RENWICK	J	E		LAC		22/05/1952
RENWICK	J	F		LAC		14/11/1962
RESTALL	J			CPL		30/09/1948
REYNOLDS	A	R		SAC		09/12/1992
REYNOLDS	H	S		SGT		01/11/1951
REYNOLDS	J	C		WING COMM	Acting	05/10/1944
REYNOLDS	J	H		SGT		08/11/1945
REYNOLDS	J	H		FL		23/02/1950
REYNOLDS	M	E		SGT		09/11/1950
REYNOLDS	S	G		FL		03/11/1989
REYNOLDS	S	G		FL		03/11/1989
REYNOLDS	V	J		WO		23/05/1946
RHODES	C	F		FL		21/01/1959
RHODES	E	F		SQD LDR		06/03/1947
RHODES	H	A		SGT		17/04/1947
RHYS	R	O		WING COMM		08/11/1945
RICE	D	J		WO		13/12/1945
RICE	R	C		FL	Acting	10/06/1948
RICE	W	H		CPL		24/11/1949
RICH	A	E	B	SQD LDR		16/10/1947
RICHARDS	C	G		WO		15/04/1943
RICHARDS	D	J		FL		01/12/1955
RICHARDS	G	H	B	FL		06/03/1947
RICHARDS	I	J		FO		06/01/1955
RICHARDS	J	M		FL		24/07/1952
RICHARDS	J	W		SQD LDR		06/03/1947
RICHARDS	K	F		FO		08/02/1945
RICHARDS	M	D		SGT		04/04/1997
RICHARDS	P	J		CPL		12/10/1979
RICHARDS	P	J		CPL		12/10/1989
RICHARDS	T	W		SGT		27/03/1947
RICHARDSON	A	E		FLT SGT		15/07/1948
RICHARDSON	A	F		SGT		25/06/1953
RICHARDSON	A			FL		25/06/1953
RICHARDSON	B	M		CPL		19/08/1959
RICHARDSON	D	B		WO		30/08/1945
RICHARDSON	E			WO		27/07/1944
RICHARDSON	G	E		SGT	Temporary	28/07/1955
RICHARDSON	H	D		FL		30/12/1943
RICHARDSON	N	H		SQD LDR	Acting	27/03/1947
RICHARDSON	R	J		LAC		15/07/1948
RICHARDSON	R	W		FL		21/03/1946
RICHARDSON	R			CPL		04/10/1951
RICHARDSON	T	N		FO		01/12/1955
RICHDALE	E	B		GRP OFF	Acting	06/03/1947
RICHES	E	R		FL	Acting	14/10/1943
RICHES	G	R		CPL		15/07/1948
RICHES	R	W	P	FL	Acting	17/04/1947
RICHES	S			CPL		29/10/1990
RICHMOND	G	A		FL		07/02/1952
RICHMOND	R	L		WING COMM	Acting	15/07/1948
RICHMOND-WATSON	E	O		WING COMM	Acting	03/10/1946
RICKABY	F	A		FL		11/01/1945
RICKABY	G	B		FLT SGT		16/01/1957
RICKARDS	E	S		SQD LDR		26/01/1950
RICKARDS	F	J		CPL		26/11/1953
RICKETT	G	E		LAC		15/11/1951
RICKETTS	A	V		FL		29/04/1954
RICKETTS	R	A		CPL		15/12/1994
RICKUS	Z	A		FLT OFF	Acting	10/05/1951

Source	Service No	Force	Notes	Surname
N402 1952	854389	RAuxAF		RENWICK
N837 1962	2605941	RAuxAF	wef 19/8/61	RENWICK
N805 1948	863383	RAuxAF		RESTALL
YB 1996	n/a	RAuxAF		REYNOLDS
N1126 1951	750485	RAFVR		REYNOLDS
N1007 1944	90064	AAF	AFC	REYNOLDS
N1232 1945	816047	AAF	GM	REYNOLDS
N179 1950	73722	RAFVR	MM	REYNOLDS
N1113 1950	753002	RAFVR		REYNOLDS
YB 1996	2630118	RAuxAF		REYNOLDS
YB 2000	2630118	RAuxAF	1st Clasp	REYNOLDS
N468 1946	755865	RAFVR		REYNOLDS
N28 1959	104903	RAFVR	wef 12/11/53	RHODES
N187 1947	74654	RAFVR	AFC	RHODES
N289 1947	809067	AAF		RHODES
N1232 1945	90386	AAF		RHYS
N1355 1945	741422	RAFVR		RICE
N456 1948	159268	RAFVR		RICE
N1193 1949	842774	RAuxAF		RICE
N846 1947	72314	RAFVR		RICH
N454 1943	740276	RAFVR		RICHARDS
N895 1955	150992	RAFVR		RICHARDS
N187 1947	73658	RAFVR		RICHARDS
N1 1955	2604345	RAFVR		RICHARDS
N569 1952	141978	RAFVR		RICHARDS
N187 1947	73622	RAFVR		RICHARDS
N139 1945	141708	RAFVR		RICHARDS
YB 1999	n/a	RAuxAF		RICHARDS
YB 1996	n/a	RAuxAF		RICHARDS
YB 1996	n/a	RAuxAF	1st Clasp	RICHARDS
N250 1947	857300	AAF		RICHARDS
N573 1948	801511	RAuxAF		RICHARDSON
N527 1953	814158	RAuxAF		RICHARDSON
N527 1953	91288	RAFVR		RICHARDSON
N552 1959	2655851	WRAAF	wef 23/6/59	RICHARDSON
N961 1945	742416	RAFVR		RICHARDSON
N763 1944	808117	AAF		RICHARDSON
N565 1955	800481	RAuxAF		RICHARDSON
N1350 1943	60322	RAFVR		RICHARDSON
N250 1947	72717	RAFVR		RICHARDSON
N573 1948	807042	RAuxAF		RICHARDSON
N263 1946	113339	RAFVR		RICHARDSON
N1019 1951	867264	RAuxAF		RICHARDSON
N895 1955	2699514	RAuxAF		RICHARDSON
N187 1947	190	WAAF		RICHDALE
N1065 1943	108980	RAFVR		RICHES
N573 1948	860369	RAuxAF		RICHES
N289 1947	143641	RAFVR		RICHES
YB 1996	n/a	RAuxAF		RICHES
N103 1952	91070	RAuxAF		RICHMOND
N573 1948	90747	RAuxAF	OBE	RICHMOND
N837 1946	84927	RAFVR		RICHMOND-WATSON
N14 1945	100041	RAFVR	AFC	RICKABY
N24 1957	2682125	RAuxAF		RICKABY
N94 1950	90918	RAuxAF		RICKARDS
N952 1953	743972	RAFVR		RICKARDS
N1185 1951	869324	RAuxAF		RICKETT
N329 1954	161732	RAuxAF		RICKETTS
YB 1996	n/a	RAuxAF		RICKETTS
N466 1951	3521	WAAF		RICKUS

Surname	Initials			Rank	Status	Date
RICKWOOD	A	L		SGT		01/07/1948
RIDDELL	R	N		WING COMM		14/10/1943
RIDDLE	C	J	H	WING COMM	Acting	21/06/1945
RIDDLE	H	J		WING COMM	Acting	20/09/1945
RIDGE	A	C		WING COMM	Acting	06/03/1947
RIDGE	M	C		SQD LDR		23/08/1997
RIDGEWAY	E	V		WO		08/05/1947
RIDGWAY	A			CPL		26/01/1950
RIDLEY	H	G		FL		20/09/1945
RIDLEY	K	C		FL		04/09/1947
RIGBY	J			SQD LDR		10/02/1949
RIGBY	T	P		FO		10/04/1952
RIGGS	A	F		FL		26/04/1961
RIGGS	P	H		WING COMM	Acting	05/06/1947
RIGGS	W	C		SGT		26/01/1950
RILEY	D			FO		16/01/1957
RILEY	G	A		FL		15/07/1948
RILEY	H	J		SQD LDR		23/02/1950
RILEY	J			LAC		26/02/1958
RILEY	L	M		SQD LDR		27/03/1947
RIMMER	L	T		FLT SGT	Temporary	25/03/1959
RINTOUL	A			SQD LDR		03/12/1942
RIPPON	T	W		WO		13/10/1955
RISELEY	A	H		PO		04/02/1943
RISK	T	N		FL		07/05/1953
RITCHIE	J			FO		26/11/1953
RITCHIE	L	S		SQD LDR		10/05/1945
RIVERS	R	M		FO		06/03/1947
RIXON	W	F		LAC		26/08/1948
ROACH	C	T		CPL		24/11/1949
ROACH	J	G		CPL		25/06/1953
ROACH	J	S		CPL		20/05/1948
ROAKE	H	T		CPL		18/03/1948
ROAKE	H	T		FLT SGT	Acting	25/03/1954
ROBB	A	G		CPL		24/11/1949
ROBB	A	G		CPL		28/11/1956
ROBBINS	C	J		SQD LDR	Acting	07/08/1947
ROBBINS	G	A		LAC		22/02/1951
ROBBINS	J	T	W	FO		20/04/1944
ROBBINS	R	H		FL		24/01/1946
ROBBINS	R	W	F	CPL		04/09/1947
ROBERTS	A	W	J	FL		20/05/1948
ROBERTS	D	W		SQD LDR	Acting	23/02/1950
ROBERTS	E	C		FL		16/04/1958
ROBERTS	E	M		LAC		26/08/1948
ROBERTS	E			FL		25/05/1960
ROBERTS	F	D		PO		04/02/1943
ROBERTS	F	H		CPL	Temporary	26/04/1961
ROBERTS	H	W		FL		15/06/1944
ROBERTS	J	C		SGT		01/07/1948
ROBERTS	J	G		FL		06/03/1947
ROBERTS	J			FO		20/04/1944
ROBERTS	K			FLT OFF		20/05/1948
ROBERTS	M	M		FLT OFF		05/06/1947
ROBERTS	O			CPL		24/11/1949
ROBERTS	R	J		AC1		15/09/1949
ROBERTS	R			FL		28/11/1956
ROBERTS	S	G		SQD LDR		28/09/1992
ROBERTS	S	J		FL		13/10/1955
ROBERTS	T	R	P	SGT		07/02/1952

Source	Service No	Force	Notes	Surname
N527 1948	844326	RAuxAF		RICKWOOD
N1065 1943	70572	RAFVR		RIDDELL
N671 1945	90143	AAF		RIDDLE
N1036 1945	90141	AAF		RIDDLE
N187 1947	72345	RAFVR		RIDGE
YB 1998	593529	RAuxAF		RIDGE
N358 1947	810048	AAF		RIDGEWAY
N94 1950	810175	RAuxAF		RIDGWAY
N1036 1945	141278	RAFVR		RIDLEY
N740 1947	175155	RAFVR		RIDLEY
N123 1949	90779	RAuxAF		RIGBY
N278 1952	2688513	RAuxAF		RIGBY
N339 1961	205647	RAFVR	wef 23/1/61	RIGGS
N460 1947	72301	RAFVR		RIGGS
N94 1950	752615	RAFVR		RIGGS
N24 1957	2600927	RAuxAF		RILEY
N573 1948	145479	RAFVR	DFC,DFM	RILEY
N179 1950	91179	RAuxAF	MB,ChB,MRCS,LRCP	RILEY
N149 1958	2688738	RAuxAF	wef 19/2/54	RILEY
N250 1947	73366	RAFVR		RILEY
N196 1959	757004	RAFVR	wef 2/9/44	RIMMER
N1638 1942	90178	AAF		RINTOUL
N762 1955	841506	RAuxAF		RIPPON
N131 1943	106080	RAFVR		RISELEY
N395 1953	130959	RAFVR		RISK
N952 1953	177240	RAFVR		RITCHIE
N485 1945	90198	AAF		RITCHIE
N187 1947	179124	RAFVR		RIVERS
N694 1948	859659	RAuxAF		RIXON
N1193 1949	865153	RAuxAF		ROACH
N527 1953	865649	RAuxAF		ROACH
N395 1948	865148	RAuxAF		ROACH
N232 1948	840918	RAuxAF		ROAKE
N249 1954	2692503	RAuxAF	1st Clasp	ROAKE
N1193 1949	817123	RAuxAF		ROBB
N833 1956	2684002	RAuxAF	1st Clasp	ROBB
N662 1947	68840	RAFVR	MBE	ROBBINS
N194 1951	864161	RAuxAF		ROBBINS
N363 1944	141266	RAFVR		ROBBINS
N82 1946	161726	RAFVR		ROBBINS
N740 1947	805326	AAF		ROBBINS
N395 1948	174466	RAFVR		ROBERTS
N179 1950	72164	RAFVR		ROBERTS
N258 1958	90857	AAF	wef 2/6/44	ROBERTS
N694 1948	818060	RAuxAF		ROBERTS
N375 1960	123972	RAFVR	wef 20/4/60	ROBERTS
N131 1943	122972	RAFVR		ROBERTS
N339 1961	840085	AAF	wef 11/3/44	ROBERTS
N596 1944	70576	RAFVR		ROBERTS
N527 1948	849046	RAuxAF		ROBERTS
N187 1947	129108	RAFVR	DFC,DFM	ROBERTS
N363 1944	156926	RAFVR		ROBERTS
N395 1948	579	WAAF		ROBERTS
N460 1947	1693	WAAF		ROBERTS
N1193 1949	865692	RAuxAF		ROBERTS
N953 1949	853621	RAuxAF		ROBERTS
N833 1956	91351	RAuxAF		ROBERTS
YB 1997	9228	RAuxAF		ROBERTS
N762 1955	114488	RAFVR		ROBERTS
N103 1952	816084	RAuxAF		ROBERTS

Surname	Initials			Rank	Status	Date
ROBERTS	V			CPL		05/06/1947
ROBERTS	W	B		SGT		03/07/1952
ROBERTSON	A	M		SGT		12/11/1998
ROBERTSON	A			FLT SGT		01/03/1945
ROBERTSON	A			SGT		01/03/1945
ROBERTSON	C	A		FO		09/12/1959
ROBERTSON	C	B		FL		20/06/1944
ROBERTSON	D	C	B	FL		21/03/1946
ROBERTSON	D	H		FO		25/05/1960
ROBERTSON	D	M	G	FL		15/07/1948
ROBERTSON	G	M		SQD LDR	Acting	21/03/1946
ROBERTSON	G			CPL		20/05/1948
ROBERTSON	G			SGT		15/01/1953
ROBERTSON	I	A		SQD LDR	Acting	24/11/1949
ROBERTSON	I	M		FO		21/09/1944
ROBERTSON	R	G		FL		30/09/1954
ROBERTSON	S	M		FO		04/05/1944
ROBERTSON	V	A		PO		20/04/1944
ROBERTSON	W	L		CPL		10/08/1950
ROBINS	L	E		FO		22/02/1961
ROBINSON	A	E		WO		14/09/1950
ROBINSON	A	E		SGT		13/12/1951
ROBINSON	A	W		FL		03/07/1952
ROBINSON	B			WING COMM	Acting	06/03/1947
ROBINSON	D	G		SQD LDR	Acting	17/04/1947
ROBINSON	D	N		FL		11/01/1945
ROBINSON	E	A		WO		07/07/1949
ROBINSON	E			FLT SGT		18/03/1948
ROBINSON	G	P		SGT		26/01/1950
ROBINSON	H	O		SGT		07/08/1947
ROBINSON	H			CPL		13/01/1949
ROBINSON	J	E		FO		09/08/1945
ROBINSON	J	M		CPL		07/06/1951
ROBINSON	J	R		FL		28/02/1946
ROBINSON	M			WING COMM		03/12/1942
ROBINSON	M			SQD LDR		13/12/1951
ROBINSON	P	B		WING COMM	Acting	30/12/1943
ROBINSON	R	F		SQD LDR	Acting	12/07/1945
ROBINSON	R	R		SGT		30/09/1948
ROBINSON	S	H		FL	Acting	24/11/1949
ROBINSON	T	A	F	SQD LDR		26/01/1950
ROBINSON	T	W	C	WING COMM	Acting	04/11/1943
ROBINSON	V			WO		17/04/1947
ROBOTHAM	A	W		FL		05/09/1946
ROBSON	J	C		FL		08/11/1945
ROBSON	J			CPL		10/02/1949
ROBSON	R	W		CHIEF TECH		06/06/1962
RODGER	J	K		FL		17/06/1954
RODGER	W	L		FL		12/04/1945
RODGERS	H			LAC		05/01/1950
RODGERS	J	W		CPL		08/03/1951
RODGERS	J	W		CPL		19/06/1963
RODGERS	R	J		CPL	Acting	29/04/1954
RODLEY	E	E		SQD LDR	Acting	30/03/1944
ROE	E	C	P	SGT		22/10/1958
ROE	S	H		FL		17/04/1947
ROE	S			SGT		20/07/1997
ROGERS	C	C		FL		19/01/1956
ROGERS	C	J		FL		21/03/1946
ROGERS	D	G		FL		06/03/1947

Source	Service No	Force	Notes	Surname
N460 1947	811205	AAF		ROBERTS
N513 1952	799924	RAFVR		ROBERTS
YB 2002	n/a	RAuxAF		ROBERTSON
N225 1945	803404	AAF		ROBERTSON
N225 1945	803379	AAF		ROBERTSON
N843 1959	2600543	RAF	wef 8/7/57	ROBERTSON
MOD 1965	85688	RAFVR		ROBERTSON
N263 1946	85688	RAFVR		ROBERTSON
N375 1960	8781	WRAAF	wef 24/3/59	ROBERTSON
N573 1948	86375	RAFVR		ROBERTSON
N263 1946	86657	RAFVR	DFC	ROBERTSON
N395 1948	873800	RAuxAF		ROBERTSON
N28 1953	874563	RAuxAF		ROBERTSON
N1193 1949	86636	RAFVR	DFC	ROBERTSON
N961 1944	131645	RAFVR		ROBERTSON
N799 1954	134090	RAuxAF		ROBERTSON
N425 1944	126652	RAFVR		ROBERTSON
N363 1944	157995	RAFVR		ROBERTSON
N780 1950	874103	RAuxAF		ROBERTSON
N153 1961	2651000	RAuxAF	wef 19/8/56	ROBINS
N926 1950	748199	RAFVR		ROBINSON
N1271 1951	843300	RAuxAF		ROBINSON
N513 1952	91256	RAuxAF	DFC	ROBINSON
N187 1947	73823	RAFVR		ROBINSON
N289 1947	72928	RAFVR		ROBINSON
N14 1945	60515	RAFVR		ROBINSON
N673 1949	882185	WAAF		ROBINSON
N232 1948	820014	RAuxAF		ROBINSON
N94 1950	770280	RAFVR		ROBINSON
N662 1947	770265	RAFVR		ROBINSON
N30 1949	855070	RAuxAF		ROBINSON
N874 1945	160676	RAFVR		ROBINSON
N560 1951	820002	RAuxAF		ROBINSON
N197 1946	110565	RAFVR		ROBINSON
N1638 1942	90161	AAF	AFC	ROBINSON
N1271 1951	90161	RAuxAF	AFC; 1st Clasp	ROBINSON
N1350 1943	90462	RAFRO		ROBINSON
N759 1945	78849	RAFVR	DFC	ROBINSON
N805 1948	800237	RAuxAF		ROBINSON
N1193 1949	123528	RAFVR		ROBINSON
N94 1950	90996	RAuxAF		ROBINSON
N1144 1943	72492	RAFVR		ROBINSON
N289 1947	880556	WAAF		ROBINSON
N751 1946	171371	RAFVR		ROBOTHAM
N1232 1945	85683	RAFVR	DFC	ROBSON
N123 1949	873214	RAuxAF		ROBSON
N435 1962	748831	RAFVR	wef 3/7/44	ROBSON
N471 1954	90895	RAuxAF		RODGER
N381 1945	76464	RAFVR		RODGER
N1 1950	869480	RAuxAF		RODGERS
N244 1951	852761	RAuxAF		RODGERS
N434 1963	2680164	RAuxAF	wef 15/3/63	RODGERS
N329 1954	2681012	RAuxAF		RODGERS
N281 1944	61472	RAFVR	DSO,DFC	RODLEY
N716 1958	2692013	RAuxAF	wef 29/7/58	ROE
N289 1947	169207	RAFVR		ROE
YB 1998	n/a	RAuxAF		ROE
N43 1956	156096	RAuxAF	DFC	ROGERS
N263 1946	126529	RAFVR	DFC	ROGERS
N187 1947	86748	RAFVR		ROGERS

Surname	Initials			Rank	Status	Date
ROGERS	E	B		FL		11/05/1944
ROGERS	E	C		CPL	Temporary	21/10/1959
ROGERS	F			CPL		07/07/1949
ROGERS	G	D	W	SQD LDR	Acting	13/12/1945
ROGERS	G	R		FO		04/10/1951
ROGERS	H	A		FO		31/10/1956
ROGERS	H	D		WO		30/08/1945
ROGERS	J	E		FO		20/04/1944
ROGERS	J	R		CPL		18/11/1948
ROGERS	J	S		FLT SGT		16/10/1947
ROGERS	R			SGT		05/01/1950
ROGERS	S			WO	Acting	07/07/1949
ROGERS	W	T		FL	Acting	26/06/1947
ROGERSON	G			FL		26/01/1950
ROLFS	R	A		FO		28/02/1946
ROLLASON	W	A		WING COMM	Acting	14/10/1943
ROLLETT	J	H	B	FL		05/01/1950
ROLLINGS	H			FL		19/09/1956
ROLLS	W	T	E	FL		30/09/1948
ROOK	A	H		WING COMM	Acting	30/03/1944
ROOM	R	E		WO		27/04/1944
ROOT	R			FL		22/09/1955
ROPER	H	S		LAC		15/09/1949
ROPER	W			FL		26/01/1950
ROSE	A	D		GRP CAP	Acting	01/07/1948
ROSE	A	D		GRP CAP		03/07/1957
ROSE	B	D		LAC		15/01/1953
ROSE	D	M		SQD OFF	Acting	04/09/1947
ROSE	D			AC1		01/07/1948
ROSE	D			CPL		17/06/1954
ROSE	G	M		WING COMM		10/06/1948
ROSE	H			FO		08/05/1947
ROSE	L	G		FL		18/03/1948
ROSE	R	H		FL		28/02/1946
ROSE	S	N		SQD LDR	Acting	04/10/1945
ROSEVEAR	R	C	G	FL		01/08/1946
ROSOMAN	R	F		FL		30/09/1948
ROSS	A	A		CPL		02/09/1998
ROSS	A			FO		27/04/1950
ROSS	C	M		FO		13/01/1949
ROSS	D			FO		22/10/1958
ROSS	F	F		CPL		01/05/1957
ROSS	G	R		FL		08/11/1945
ROSS	I	H	F	SGT		08/11/1945
ROSS	R	M		SAC		17/03/1999
ROSS	S	A		LAC		07/08/1947
ROSSER	L	V		FO		06/03/1947
ROSSER	W	J		FL		04/10/1945
ROSSER	W	J		FL		25/03/1959
ROSSINGTON	E	H		FL		28/10/1943
ROSSITER	H	C		SGT	Acting	12/06/1947
ROSTANCE	J	H		FL	Acting	01/07/1948
ROTHWELL	P	F		FL		08/11/1945
ROUND	H	B		SGT		26/11/1953
ROUPELL	C	F		FL		18/07/1956
ROURKE	F	D		LAC		21/06/1951
ROUSE	A	W		FL		18/10/1951
ROUSE	C	A		CPL		20/05/1948

Source	Service No	Force	Notes	Surname
N444 1944	81373	RAFVR		ROGERS
N708 1959	841694	AAF	wef 26/3/44	ROGERS
N673 1949	865529	RAuxAF		ROGERS
N1355 1945	86646	RAFVR		ROGERS
N1019 1951	173879	RAFVR	DFM	ROGERS
N761 1956	2698513	RAuxAF		ROGERS
N961 1945	742155	RAFVR		ROGERS
N363 1944	157081	RAFVR		ROGERS
N984 1948	866481	RAuxAF		ROGERS
N846 1947	810005	AAF		ROGERS
N1 1950	771184	RAFVR		ROGERS
N673 1949	750388	RAFVR		ROGERS
N519 1947	117228	RAFVR		ROGERS
N94 1950	73375	RAFVR		ROGERSON
N197 1946	186911	RAFVR		ROLFS
N1065 1943	70586	RAFVR		ROLLASON
N1 1950	85229	RAFVR	DFC	ROLLETT
N666 1956	84631	RAFVR		ROLLINGS
N805 1948	116492	RAFVR	DFC,DFM	ROLLS
N281 1944	90071	AAF	DFC	ROOK
N396 1944	740242	RAFVR		ROOM
N708 1955	199525	RAFVR		ROOT
N953 1949	865658	RAuxAF		ROPER
N94 1950	90925	RAuxAF	DFM	ROPER
N527 1948	72193	RAFVR	OBE	ROSE
N456 1957	72193	RAFVR	OBE; 1st Clasp; wef 30/10/54	ROSE
N28 1953	864249	RAuxAF		ROSE
N740 1947	5989	WAAF		ROSE
N527 1948	805511	RAuxAF		ROSE
N471 1954	2685000	RAuxAF	1st Clasp	ROSE
N456 1948	72074	RAFVR		ROSE
N358 1947	130757	RAFVR		ROSE
N232 1948	68841	RAFVR		ROSE
N197 1946	81652	RAFVR	DSO	ROSE
N1089 1945	81920	RAFVR		ROSE
N652 1946	67048	RAFVR	DFC,DFM	ROSEVEAR
N805 1948	142272	RAFVR		ROSOMAN
YB 1999	n/a	RAuxAF		ROSS
N406 1950	157769	RAFVR		ROSS
N30 1949	109273	RAFVR		ROSS
N716 1958	2604506	RAFVR	wef 31/12/54	ROSS
N277 1957	752459	RAFVR	wef 15/7/44	ROSS
N1232 1945	80830	RAFVR	DFC	ROSS
N1232 1945	817017	AAF		ROSS
YB 2001	n/a	RAuxAF		ROSS
N662 1947	840928	AAF		ROSS
N187 1947	185963	RAFVR	DFC	ROSSER
N1089 1945	102991	RAFVR	DFC	ROSSER
N196 1959	102991	RAF	DFC; 1st Clasp; wef 24/6/50	ROSSER
N1118 1943	70589	RAFVR		ROSSINGTON
N478 1947	840796	AAF		ROSSITER
N527 1948	113287	RAFVR		ROSTANCE
N1232 1945	86647	RAFVR		ROTHWELL
N952 1953	2685007	RAuxAF		ROUND
N513 1956	73377	RAFVR		ROUPELL
N621 1951	845837	RAuxAF		ROURKE
N1077 1951	72720	RAFVR		ROUSE
N395 1948	756226	RAFVR		ROUSE

Surname	Initials			Rank	Status	Date
ROUSE	G	W		FL		28/02/1946
ROUTELEDGE	G			FLT SGT		29/07/1999
ROUTLEDGE	P	V		FLT SGT		23/02/1950
ROUTLEDGE	W	J		LAC		13/01/1949
ROWE	E	D	E	FL		11/07/1946
ROWE	J	H	C	FL		04/10/1945
ROWELL	P	A		FL		21/06/1945
ROWLAND	E			CPL		24/11/1949
ROWLAND	T			SGT		17/04/1947
ROWLANDS	D	H		FL	Acting	07/07/1949
ROWLEY	F			FL		31/10/1956
ROXBURGH	A	J	J	WO	Acting	24/01/1946
ROXBURGH	A			SGT		01/09/1970
ROXBURGH	A			SGT		01/09/1980
ROXBURGH	A			SGT		01/09/1990
ROXBURGH	D			FL		01/07/1948
ROXBURGH	I	D		FL		10/08/1944
ROY	J	W		FLT SGT		12/04/1945
ROYCE	W	B		WING COMM	Acting	02/09/1943
ROYDEN	E	A		SGT		16/03/1950
ROYLE	E			LAC		17/03/1949
ROYLE	W	R		LAC		21/06/1961
RUDD	P	R		FL		20/01/1960
RUDDLE	R	C		LAC		15/07/1948
RUDGE	H			SGT		01/07/1948
RUDLAND	C	P		WING COMM	Acting	24/01/1946
RUMBELOW	N	J		FO		19/10/1960
RUMFITT	E	W		FL		25/03/1954
RUNCIMAN	W	L		WING COMM		17/12/1942
RUNDELL	H	R		SGT		14/09/1950
RUNDLE	D	E		SGT		03/03/1994
RUSHTON	A	H		FLT SGT		26/08/1948
RUSSELL	A	F		FO		01/07/1948
RUSSELL	A	G		FL		13/12/1945
RUSSELL	A	P		FO		14/10/1948
RUSSELL	D	M		FL		12/07/1945
RUSSELL	E	L		SGT	Temporary	20/07/1960
RUSSELL	F	P		FO		28/02/1946
RUSSELL	H			FL		18/04/1946
RUSSELL	H			LAC		16/10/1947
RUSSELL	J	A	S	SQD LDR	Acting	08/11/1945
RUSSELL	K	C		WO		22/04/1943
RUSSELL	M	A	V	WING COMM	Acting	07/07/1949
RUSSELL	M	H		SEC OFF		01/07/1948
RUSSELL	R	R		WING COMM	Acting	10/08/1944
RUSSELL	S	P		SQD LDR	Acting	13/01/1944
RUSSELL	T	K		FLT SGT	Acting	23/06/1955
RUSSELL	T	R		SQD LDR	Acting	25/01/1945
RUSSELL	T	R		FL		27/03/1952
RUSSON	G			CPL		24/12/1942
RUST	T	P		FL		19/08/1954
RUTHERFORD	E	O		SQD LDR	Acting	26/01/1950
RUTHERFORD-JONES	M			SQD OFF		17/04/1947
RUTLAND	H	W	J	FO		23/06/1955
RUTLEDGE	C	P		LAC		05/06/1947
RYALL	L	J		SQD LDR		10/06/1948
RYAN	G	A	E	FL		21/09/1995
RYAN	J	C		FL		12/07/1945
RYAN	J	P		SGT		26/08/1948
RYDER	F	G		SGT		04/11/1943

Source	Service No	Force	Notes	Surname
N197 1946	132337	RAFVR	DFC	ROUSE
YB 2000	n/a	RAuxAF		ROUTELEDGE
N179 1950	809132	RAuxAF		ROUTLEDGE
N30 1949	853576	RAuxAF		ROUTLEDGE
N594 1946	147235	RAFVR		ROWE
N1089 1945	90089	AAF		ROWE
N671 1945	115449	RAFVR		ROWELL
N1193 1949	744551	RAFVR		ROWLAND
N289 1947	871751	AAF		ROWLAND
N673 1949	122998	RAFVR	DFC; Deceased	ROWLANDS
N761 1956	188118	RAFVR		ROWLEY
N82 1946	802388	AAF		ROXBURGH
YB 1996	n/a	RAuxAF		ROXBURGH
YB 1996	n/a	RAuxAF	1st Clasp	ROXBURGH
YB 1997	n/a	RAuxAF	2nd Clasp	ROXBURGH
N527 1948	147186	RAFVR		ROXBURGH
N812 1944	63424	RAFVR		ROXBURGH
N381 1945	808138	AAF		ROY
N918 1943	90062	AAF	DFC	ROYCE
N257 1950	811135	RAuxAF		ROYDEN
N231 1949	857635	RAuxAF		ROYLE
N478 1961	854603	AAF	wef 26/6/44	ROYLE
N33 1960	147777	RAFVR	wef 1/1/53	RUDD
N573 1948	816215	RAuxAF		RUDDLE
N527 1948	805402	RAuxAF		RUDGE
N82 1946	65998	RAFVR	DFC	RUDLAND
N775 1960	3119438	RAFVR	wef 27/6/58	RUMBELOW
N249 1954	117246	RAFVR		RUMFITT
N1708 1942	90272	AAF	AFC; The Honorable	RUNCIMAN
N926 1950	748832	RAFVR		RUNDELL
YB 1996	n/a	RAuxAF		RUNDLE
N694 1948	810095	RAuxAF		RUSHTON
N527 1948	110397	RAFVR		RUSSELL
N1355 1945	120491	RAFVR		RUSSELL
N856 1948	84693	RAFVR	AFC	RUSSELL
N759 1945	102955	RAFVR		RUSSELL
N531 1960	804213	AAF	wef 7/12/42	RUSSELL
N197 1946	177117	RAFVR		RUSSELL
N358 1946	134753	RAFVR		RUSSELL
N846 1947	869475	AAF		RUSSELL
N1232 1945	81352	RAFVR	DFC	RUSSELL
N475 1943	740065	RAFVR		RUSSELL
N673 1949	79015	RAFVR		RUSSELL
N527 1948	1185	WAAF		RUSSELL
N812 1944	90375	AAF		RUSSELL
N11 1944	73011	RAFVR	DFC	RUSSELL
N473 1955	2650753	RAuxAF		RUSSELL
N70 1945	100614	RAFVR	DFC,AFC	RUSSELL
N224 1952	100614	RAuxAF	DFC,AFC; 1st Clasp	RUSSELL
N1758 1942	805161	AAF		RUSSON
N672 1954	152327	RAFVR		RUST
N94 1950	72721	RAFVR		RUTHERFORD
N289 1947	710	WAAF	MBE	RUTHERFORD-JONES
N473 1955	168527	RAFVR		RUTLAND
N460 1947	842258	AAF		RUTLEDGE
N456 1948	72443	RAFVR		RYALL
YB 1998	2634105	RAuxAF		RYAN
N759 1945	120728	RAFVR		RYAN
N694 1948	756905	RAFVR		RYAN
N1144 1943	808086	AAF		RYDER

Surname	Initials			Rank	Status	Date
RYE	L	T		FL		26/01/1950
RYMILLS	F	E		SQD LDR		21/06/1961
SACK	B	G		FO		09/12/1959
SADD	H	J		FL		21/03/1946
SAGAR	L	H		SQD LDR	Acting	09/11/1944
SAINSBURY	E	M		FLT OFF	Acting	08/05/1947
SAINT	D	E	J	FL		06/05/1943
SALANDIN	J	R		FL	Acting	15/01/1953
SALISBURY-HUGHES	K	H		SQD LDR		04/02/1943
SALMON	K	L		FO		24/01/1946
SALMON	N	M		WING OFF		06/03/1947
SALMOND	N			WO		26/01/1950
SALSBURY	A	W		FLT SGT		12/06/1947
SALTER	A	A		FL		14/12/1944
SALTER	A	A		FL		05/09/1946
SALTER	A	C		CPL		01/07/1948
SALTER	F	J		SGT		04/11/1948
SALTHOUSE	E	D		WING COMM	Acting	10/06/1948
SALTMARSH	R	V		SQD LDR		07/07/1949
SAMBROOK	R	D		FL		20/01/1960
SAMPHIER	L	P		FO		30/09/1948
SAMS	J	S		FL		13/12/1945
SAMSON	A	J		SQD LDR	Acting	03/06/1943
SAMSON	G	H		FO		04/10/1945
SAMUELS	F	A		SQD LDR		05/06/1947
SAMWAYS	A	C		SGT		27/03/1947
SANDELL	R	E		FLT SGT	Acting	14/01/1954
SANDEMAN	D	C		WING COMM	Acting	20/09/1945
SANDERMAN	F	A		FL		26/10/1944
SANDERS	A	J		FO		20/05/1948
SANDERS	H			SQD LDR		10/05/1945
SANDERS	J	C		FL		07/04/1955
SANDERS	J	S		SGT		22/06/1995
SANDERS	M			FLT OFF	Acting	13/01/1949
SANDERS	P	F		SAC		02/12/1992
SANDERS	V	H		SGT		01/07/1948
SANDERSON	A	E		CPL		26/06/1947
SANDERSON	R			SNR TECH		08/07/1959
SANDERSON	W	E		CPL		18/03/1948
SANDES	T	L		WING COMM	Acting	28/02/1946
SANDFORD	A	J		WING COMM		27/04/1950
SANDFORD	L	A		FL	Acting	07/06/1951
SANDIFER	A	K		FLT SGT		04/02/1943
SANDYS	J	L		CPL		21/02/1952
SANFORD	W	A		CPL		22/02/1951
SANKEY	H			WO		02/05/1962
SANKEY	W	K		SQD LDR	Acting	25/01/1951
SAPSTED	F	G		FO		13/01/1949
SAREL	C	R		SQD LDR	Acting	03/06/1943
SARGEANT	F	C		FLT SGT		15/07/1948
SARGENT	R	E	B	FL		09/08/1945
SARRE	A	R		FO		01/08/1946
SAUNDERS	A	E		SAC		16/10/1963
SAUNDERS	E	J		FL		25/01/1945
SAUNDERS	F	A		FL		20/05/1948
SAUNDERS	H	R		FLT SGT		25/06/1953
SAUNDERS	J	F		FL		05/06/1947
SAUNDERSON	E	J		SQD LDR	Acting	31/05/1945
SAVAGE	A	J		SQD LDR	Acting	28/02/1946
SAVAGE	A	R	J	FL		30/03/1944

Source	Service No	Force	Notes	Surname
N94 1950	91011	RAuxAF		RYE
N478 1961	115338	RAF	DFC,DFM; wef 9/6/44	RYMILLS
N843 1959	2603253	RAFVR	wef 29/5/52	SACK
N263 1946	137201	RAFVR		SADD
N1150 1944	90320	AAF		SAGAR
N358 1947	3680	WAAF		SAINSBURY
N526 1943	82713	RAFVR		SAINT
N28 1953	2684604	RAuxAF		SALANDIN
N131 1943	70597	RAFVR	DFC	SALISBURY-HUGHES
N82 1946	196213	RAFVR		SALMON
N187 1947	49	WAAF		SALMON
N94 1950	880020	WAAF		SALMOND
N478 1947	604202	AAF		SALSBURY
MOD 1965	81660	RAFVR	1st Clasp	SALTER
N751 1946	81660	RAFVR		SALTER
N527 1948	744682	RAFVR		SALTER
N934 1948	743207	RAFVR		SALTER
N456 1948	90436	RAuxAF		SALTHOUSE
N673 1949	90999	RAuxAF	Deceased	SALTMARSH
N33 1960	2605120	RAFVR	wef 25/2/54	SAMBROOK
N805 1948	195699	RAFVR		SAMPHIER
N1355 1945	82705	RAFVR	AFC	SAMS
N609 1943	78850	RAFVR	DFC	SAMSON
N1089 1945	182086	RAFVR		SAMSON
N460 1947	73386	RAFVR		SAMUELS
N250 1947	862594	AAF		SAMWAYS
N18 1954	2694508	RAuxAF		SANDELL
N1036 1945	81052	RAFVR	DFC	SANDEMAN
N1094 1944	86686	RAFVR	AFC	SANDERMAN
N395 1948	89099	RAFVR		SANDERS
N485 1945	70599	RAFVR		SANDERS
N277 1955	178066	RAFVR		SANDERS
YB 1996	n/a	RAuxAF		SANDERS
N30 1949	729	WAAF		SANDERS
YB 1996	n/a	RAuxAF		SANDERS
N527 1948	744480	RAFVR		SANDERS
N519 1947	841114	AAF		SANDERSON
N454 1959	2686586	RAuxAF	wef 28/3/59	SANDERSON
N232 1948	870284	RAuxAF		SANDERSON
N197 1946	72486	RAFVR	DFC	SANDES
N406 1950	72242	RAFVR		SANDFORD
N560 1951	842542	RAuxAF		SANDFORD
N131 1943	804127	AAF		SANDIFER
N134 1952	2682510	RAuxAF		SANDYS
N194 1951	847044	RAuxAF		SANFORD
N314 1962	2614804	RAFVR	wef 24/2/62	SANKEY
N93 1951	62117	RAFVR		SANKEY
N30 1949	145580	RAFVR		SAPSTED
N609 1943	90091	AAF		SAREL
N573 1948	846775	RAuxAF		SARGEANT
N874 1945	122024	RAFVR	AFC	SARGENT
N652 1946	197053	RAFVR		SARRE
N756 1963	2683728	RAuxAF	wef 23/8/63	SAUNDERS
N70 1945	115510	RAFVR		SAUNDERS
N395 1948	130914	RAFVR		SAUNDERS
N527 1953	745944	RAFVR		SAUNDERS
N460 1947	73387	RAFVR		SAUNDERS
N573 1945	84319	RAFVR	DSO,DFC	SAUNDERSON
N197 1946	113754	RAFVR	DFC	SAVAGE
N281 1944	70603	RAFVR		SAVAGE

Surname	Initials			Rank	Status	Date
SAVAGE	A			CPL		13/01/1949
SAVAGE	H	C	A	SGT		19/06/1963
SAVAGE	L	W		CPL		21/06/1951
SAVERY	R	A		CPL		01/07/1995
SAVERY	R	J	H	SAC		20/01/1960
SAVIGAR	W	J	C	CPL		30/09/1948
SAVILL	J	E		WO		20/09/1945
SAWDON	H			FO	Acting	03/12/1942
SAWYER	E	A		FL		15/06/1944
SAWYER	P	C		FL		31/05/1945
SAWYER	R	W		FL		06/06/1962
SAYER	A	J		FL		10/12/1942
SAYER	A	L		FL		10/02/1949
SCAIFF	C	C		LAC		15/07/1948
SCANDRETT	D	P		FLT OFF		15/07/1948
SCARBOROUGH	J	E		FL		20/05/1948
SCARBOROUGH	J	R		SAC		20/05/1991
SCARLETT	J	D		SGT		04/09/1947
SCHOFIELD	C	R	B	FL		13/01/1944
SCHOFIELD	J			FL		21/10/1959
SCHOON	G	W		FL		12/04/1945
SCHREIBER	J	S		WING COMM	Acting	30/09/1948
SCHREUDER	S	G		FL		07/02/1962
SCHROEDER	R	J	H	LAC		10/02/1949
SCHUMM	M	W		PO		26/11/1953
SCORER	G	R		SQD LDR	Acting	28/02/1946
SCORER	I			FL		21/03/1946
SCORGIE	D			FL		28/11/1946
SCORGIE	I	A		FL		30/10/1957
SCOTNEY	M	S		FL		06/03/1947
SCOTT	A	G		LAC		30/09/1948
SCOTT	A	H		SGT		06/01/1955
SCOTT	A			FL		27/11/1952
SCOTT	E	R	H	SQD LDR	Acting	16/10/1947
SCOTT	E			FO		12/10/1950
SCOTT	G	H		FL		28/02/1946
SCOTT	G	J	W	FL	Acting	17/03/1949
SCOTT	G	W		SQD LDR	Acting	20/04/1944
SCOTT	H			SQD OFF		16/10/1947
SCOTT	I	E		FLT SGT	Acting	04/11/1948
SCOTT	J	H		WO		08/11/1945
SCOTT	J			FO		24/12/1942
SCOTT	J			FL		31/10/1956
SCOTT	L	R		FLT SGT		24/06/1943
SCOTT	M	D		FLT SGT		02/09/1998
SCOTT	N	T	H	FL		24/01/1946
SCOTT	P	D		SQD LDR	Acting	11/07/1946
SCOTT	P	L	H	FL	Acting	26/01/1950
SCOTT	R	H		GRP CAP	Acting	04/10/1945
SCOTT	R	J		SGT	Acting	29/04/1954
SCOTT	W	J		FL	Acting	03/07/1952
SCOTT	W			CPL		21/06/1951
SCOTT-MALDEN	F	D	S	WING COMM		06/03/1947
SCOULAR	L	G		WING COMM	Acting	26/08/1948
SCRIMGEOUR	J			WING COMM		20/05/1948
SCRIVENER	N	H		SQD LDR		06/03/1947
SCRIVENER	R	V		WING COMM		04/12/1947
SCROGGS	R	B	H	FL		23/05/1946
SCULL	F	H		CPL		05/06/1947
SCULL	F	M		FL		26/08/1948

Source	Service No	Force	Notes	Surname
N30 1949	873039	RAuxAF		SAVAGE
N434 1963	819042	RAF	wef 17/6/44	SAVAGE
N621 1951	848426	RAuxAF		SAVAGE
YB 1996	n/a	RAuxAF		SAVERY
N33 1960	2609333	RAFVR	wef 21/2/54	SAVERY
N805 1948	841064	RAuxAF		SAVIGAR
N1036 1945	740971	RAFVR		SAVILL
N1638 1942	64278	RAFVR		SAWDON
N596 1944	83728	RAFVR		SAWYER
N573 1945	61235	RAFVR	DFC	SAWYER
N435 1962	78626	RAuxAF	wef 11/12/61	SAWYER
N1680 1942	70604	RAFVR		SAYER
N123 1949	81427	RAFVR		SAYER
N573 1948	855056	RAuxAF		SCAIFF
N573 1948	2530	WAAF		SCANDRETT
N395 1948	102198	RAFVR		SCARBOROUGH
YB 1996	n/a	RAuxAF		SCARBOROUGH
N740 1947	746263	RAFVR		SCARLETT
N11 1944	63786	RAFVR		SCHOFIELD
N708 1959	85792	RAFVR	wef 22/2/54	SCHOFIELD
N381 1945	141463	RAFVR	DFC	SCHOON
N805 1948	90577	RAuxAF		SCHREIBER
N102 1962	167791	RAFVR	wef 15/6/59	SCHREUDER
N123 1949	843309	RAuxAF		SCHROEDER
N952 1953	2659571	WRAAF		SCHUMM
N197 1946	87023	RAFVR	DFC	SCORER
N263 1946	119879	RAFVR	DFC	SCORER
N1004 1946	115082	RAFVR		SCORGIE
N775 1957	187869	RAuxAF	wef 3/4/57	SCORGIE
N187 1947	170701	RAFVR		SCOTNEY
N805 1948	843205	RAuxAF		SCOTT
N1 1955	2692001	RAuxAF		SCOTT
N915 1952	66616	RAFVR		SCOTT
N846 1947	87970	RAFVR		SCOTT
N1018 1950	158682	RAFVR		SCOTT
N197 1946	128709	RAFVR		SCOTT
N231 1949	146193	RAFVR		SCOTT
N363 1944	62257	RAFVR		SCOTT
N846 1947	112	WAAF		SCOTT
N934 1948	812119	RAuxAF		SCOTT
N1232 1945	741721	RAFVR		SCOTT
N1758 1942	108068	RAFVR		SCOTT
N761 1956	170134	RAF		SCOTT
N677 1943	812050	AAF		SCOTT
YB 1999	n/a	RAuxAF		SCOTT
N82 1946	130065	RAFVR	AFC	SCOTT
N594 1946	104442	RAFVR		SCOTT
N94 1950	137729	RAFVR		SCOTT
N1089 1945	90221	AAF		SCOTT
N329 1954	2680004	RAuxAF		SCOTT
N513 1952	138573	RAFVR		SCOTT
N621 1951	867323	RAuxAF		SCOTT
N187 1947	74690	RAFVR	DSO,DFC	SCOTT-MALDEN
N694 1948	73037	RAFVR		SCOULAR
N395 1948	90444	RAuxAF	OBE	SCRIMGEOUR
N187 1947	102596	RAFVR	DSO,DFC	SCRIVENER
N1007 1947	72199	RAFVR		SCRIVENER
N468 1946	130387	RAFVR		SCROGGS
N460 1947	813168	AAF		SCULL
N694 1948	134300	RAFVR		SCULL

Surname	Initials			Rank	Status	Date
SCULL	R	S		CPL		04/11/1948
SEABOURNE	E	W		FO	Acting	17/12/1942
SEABOURNE	E	W		SQD LDR		20/09/1951
SEABROOK	S			LAC		14/10/1948
SEADEN-JONES	H	F		SGT		23/02/1950
SEAFORD	H	P		FL		27/11/1992
SEAL	J			LAC		18/03/1948
SEAMAN	J	R		SGT		01/07/1948
SEARLE	W	J		SQD LDR	Acting	02/04/1953
SEARLES	B			SGT		18/11/1948
SEARLES	G	E		SGT		31/03/1994
SEARLES	J	A		SGT		01/01/1997
SEAZELL	F	W		FL		28/09/1964
SEDGLEY	L	N		SGT		26/08/1948
SEED	F	G		FL		10/06/1948
SEELEY	T			FL		19/08/1959
SEELY	N	R	W	SQD LDR		17/12/1942
SELDON	F	C		SQD LDR		06/01/1955
SELLARS	K	E		FL		24/01/1946
SELLARS	W	H	W	SQD LDR		17/04/1947
SELLERS	R	F		FL		01/07/1948
SELLEY	W	T		SQD LDR	Acting	16/10/1947
SELLORS	J	F		FO		23/05/1946
SELLS	G	de G		FL		12/08/1943
SELVES	L	A		WO		15/07/1948
SELWAY	J	B		SQD LDR		21/06/1945
SERGEANT	C	F	H	SQD LDR		26/08/1948
SERGEANT	R	H		FL		15/09/1949
SERLE	P	P	O	FL		21/03/1946
SETCHFIELD	W	F		FL		25/03/1959
SETCHFIELD	W	F		FL		24/02/1964
SEVERN	P	J		SQD LDR		05/01/1994
SEWELL	H	S		SQD LDR	Acting	24/01/1946
SEYMOUR	A	H		SQD LDR		16/10/1947
SEYMOUR	C	W		SGT		30/09/1948
SHACKLADY	H			CPL		16/03/1950
SHADBOLT	D	G		SGT		07/12/1960
SHALLCROSS	P	E		FO		15/07/1948
SHAND	E	E		FL		27/04/1944
SHANKS	F			SQD LDR	Acting	18/02/1954
SHANNON	E	M		FL		07/01/1943
SHANNON	G			LAC		15/09/1949
SHANNON	H			SQD LDR		25/03/1954
SHANNON	J	E	E	SAC		07/06/1951
SHARDLOW	K	E		FL	Acting	28/11/1956
SHARMAN	H	R		SQD LDR	Acting	13/12/1945
SHARP	G	R		FL		28/02/1946
SHARP	H	F	H	FL		26/10/1944
SHARPE	C	M	W	LAC		05/06/1947
SHARPE	E	J		FO		16/01/1957
SHARPE	J	B		FL	Acting	03/07/1952
SHARPE	J	J		FLT SGT		14/09/1950
SHAW	A	W		CPL		17/03/1949
SHAW	A			FLT SGT		19/07/1951
SHAW	C			FLT SGT		17/03/1949
SHAW	E	C		SQD LDR		05/06/1947
SHAW	G			WING COMM		03/12/1942
SHAW	G			FL		09/09/1996
SHAW	J	H		SQD OFF	Acting	12/04/1951
SHAW	J	T		WING COMM	Acting	24/01/1946

Source	Service No	Force	Notes	Surname
N934 1948	861055	RAuxAF		SCULL
N1708 1942	105162	RAFVR	DFC	SEABOURNE
N963 1951	105162	RAFVR	DFC; 1st Clasp	SEABOURNE
N856 1948	844647	RAuxAF		SEABROOK
N179 1950	752380	RAFVR		SEADEN-JONES
YB 1996	n/a	RAuxAF		SEAFORD
N232 1948	860302	RAuxAF		SEAL
N527 1948	819054	RAuxAF		SEAMAN
N298 1953	73038	RAFVR		SEARLE
N984 1948	840030	RAuxAF		SEARLES
YB 2001	n/a	RAuxAF		SEARLES
YB 2003	n/a	RAuxAF		SEARLES
MOD 1967	185983	RAFVR		SEAZELL
N694 1948	805492	RAuxAF		SEDGLEY
N456 1948	89736	RAFVR		SEED
N552 1959	140290	RAFVR	wef 22/2/57	SEELEY
N1708 1942	90150	AAF		SEELY
N1 1955	76259	RAFVR		SELDON
N82 1946	169730	RAFVR		SELLARS
N289 1947	90854	AAF		SELLARS
N527 1948	111224	RAFVR	AFC	SELLERS
N846 1947	79713	RAFVR		SELLEY
N468 1946	177020	RAFVR		SELLORS
N850 1943	84002	RAFVR		SELLS
N573 1948	804332	RAuxAF		SELVES
N671 1945	90219	AAF	DFC	SELWAY
N694 1948	72491	RAFVR		SERGEANT
N953 1949	111777	RAFVR		SERGEANT
N263 1946	87358	RAFVR		SERLE
N196 1959	170765	RAuxAF	wef 24/2/54	SETCHFIELD
MOD 1964	170765	RAuxAF	1st Clasp	SETCHFIELD
YB 1997	4231169	RAuxAF		SEVERN
N82 1946	86667	RAFVR	DFC,AFC	SEWELL
N846 1947	72369	RAFVR		SEYMOUR
N805 1948	752002	RAFVR		SEYMOUR
N257 1950	853610	RAuxAF		SHACKLADY
N933 1960	2602649	RAFVR	wef 3/10/60	SHADBOLT
N573 1948	176891	RAFVR		SHALLCROSS
N396 1944	81053	RAFVR		SHAND
N137 1954	118313	RAuxAF		SHANKS
N3 1943	86419	RAFVR		SHANNON
N953 1949	854325	RAuxAF		SHANNON
N249 1954	83254	RAFVR		SHANNON
N560 1951	863448	RAuxAF		SHANNON
N833 1956	174967	RAFVR		SHARDLOW
N1355 1945	78257	RAFVR	AFC	SHARMAN
N197 1946	130470	RAFVR		SHARP
N1094 1944	112555	RAFVR		SHARP
N460 1947	842705	AAF		SHARPE
N24 1957	128384	RAuxAF		SHARPE
N513 1952	91280	RAuxAF		SHARPE
N926 1950	807004	RAuxAF		SHARPE
N231 1949	770717	RAFVR		SHAW
N724 1951	802572	RAuxAF		SHAW
N231 1949	810083	RAuxAF		SHAW
N460 1947	72165	RAFVR		SHAW
N1638 1942	90297	AAF	DFC	SHAW
YB 1997	2636997	RAuxAF		SHAW
N342 1951	1197	WAAF		SHAW
N82 1946	108962	RAFVR	DSO,DFC	SHAW

Surname	Initials			Rank	Status	Date
SHAW	L	S		SGT		25/06/1953
SHAW	N	D		LAC		02/04/1953
SHAW	P	A	T	FL		19/07/1993
SHAW	R	G		GRP CAP		24/12/1942
SHAW	R	J		FL		17/07/1990
SHAW	R	P	A	FL		25/06/1953
SHEAD	H	F	W	FL		13/12/1945
SHEDDEN	G	M		SQD OFF	Acting	27/03/1947
SHEERAN	G	H		WO		10/02/1949
SHEFFIELD	J	A		FL		24/01/1946
SHELLARD	D	B		SAC		19/04/1996
SHELLARD	G	T		FL		12/07/1945
SHELLEY	J	A		FO		20/05/1948
SHELTON	L	B		FL		09/05/1946
SHEPHERD	A	G		FL		27/04/1944
SHEPHERD	F	G		LAC		20/05/1948
SHEPHERD	J	B		FL		10/05/1945
SHEPHERD	M	W		SAC		02/12/1992
SHEPHERD	O			SGT		10/06/1948
SHEPHERD	O			FLT SGT		16/04/1958
SHEPHERD	R	H		FO		17/06/1954
SHEPHERDSON	L	W		WING COMM	Acting	10/06/1948
SHEPHERDSON	W			CPL		19/07/1951
SHEPPARD	A	G		FL		15/01/1953
SHEPPARD	D	G		SQD LDR	Acting	28/02/1946
SHEPPARD	D	W		LAC		15/09/1949
SHEPPARD	F	A		SEC OFF		20/05/1948
SHEPPARD	G	D'O		FL	Acting	15/01/1953
SHEPPARD	J	S		FO		30/12/1943
SHEPPARD	W	J	P	WO		15/09/1949
SHERMAN	M	D		FLT OFF		26/02/1958
SHERRY	M	D		FLT OFF	Acting	05/06/1947
SHERWEN	D	W		SQD LDR	Acting	07/08/1947
SHIEL	L	A		FL		09/05/1946
SHIELD	J	T		SGT		01/10/1953
SHIELDS	M	T	S	FL		05/09/1946
SHIELS	J	P		FL		20/05/1990
SHILLING	C	W		SQD LDR	Acting	06/03/1947
SHIMMONS	W	B		FL		07/04/1955
SHINNIE	G	M		FL		23/05/1946
SHIPWRIGHT	D	E	B K	SQD LDR	Acting	25/05/1944
SHIRCORE	J			FO		07/06/1945
SHORT	C	H	T	FL		17/06/1943
SHORT	J	H		WO		12/04/1945
SHOWELL	J			LAC		04/11/1948
SHRIVES	A	H		FO		07/02/1952
SHRIVES	L	E		FLT SGT		12/10/1950
SHRUBSOLE	A	T		CPL		26/01/1950
SHUTER	D	C		SQD LDR		05/02/1948
SHUTTE	R	V	N	SQD LDR		30/09/1948
SIBLEY	G	W	G	SGT		14/11/1962
SIBLEY	J	G		CPL		09/01/1963
SIBLEY	R	A	G	FL		09/01/1963
SILABON	G	F		LAC		20/07/1960
SILK	E	J	F	LAC		05/01/1950
SILK	F	H		FL		03/10/1946
SILLARS	M			SGT		04/09/1947
SILVER	A			CPL		05/06/1947
SILVERMAN	S	G		FL		09/12/1959
SILVERMAN	S	G		FL		25/05/1965

Source	Service No	Force	Notes	Surname
N527 1953	2680002	RAuxAF		SHAW
N298 1953	873125	RAuxAF		SHAW
YB 1996	586591	RAuxAF		SHAW
N1758 1942	90146	AAF	DFC	SHAW
YB 1997	2630813	RAuxAF		SHAW
N527 1953	112243	RAFVR		SHAW
N1355 1945	147535	RAFVR	DFC	SHEAD
N250 1947	41	WAAF		SHEDDEN
N123 1949	751175	RAFVR		SHEERAN
N82 1946	149306	RAFVR		SHEFFIELD
YB 1997	n/a	RAuxAF		SHELLARD
N759 1945	147953	RAFVR		SHELLARD
N395 1948	163846	RAFVR		SHELLEY
N415 1946	172931	RAFVR		SHELTON
N396 1944	81056	RAFVR		SHEPHERD
N395 1948	840926	RAuxAF		SHEPHERD
N485 1945	104447	RAFVR		SHEPHERD
YB 1996	n/a	RAuxAF		SHEPHERD
N456 1948	808404	RAuxAF		SHEPHERD
N258 1958	2686048	RAuxAF	1st Clasp; wef 29/9/57	SHEPHERD
N471 1954	185276	RAuxAF		SHEPHERD
N456 1948	90859	RAuxAF	MBE	SHEPHERDSON
N724 1951	857465	RAuxAF		SHEPHERDSON
N28 1953	78447	RAFVR		SHEPPARD
N197 1946	118951	RAFVR	DFC	SHEPPARD
N953 1949	812327	RAuxAF		SHEPPARD
N395 1948	2941	WAAF		SHEPPARD
N28 1953	90452	RAuxAF		SHEPPARD
N1350 1943	70620	RAFVR		SHEPPARD
N953 1949	747895	RAFVR		SHEPPARD
N149 1958	8707	WRAAF	wef 12/8/57	SHERMAN
N460 1947	1452	WAAF		SHERRY
N662 1947	62719	RAFVR		SHERWEN
N415 1946	139655	RAFVR	DFM	SHIEL
N796 1953	2602519	RAFVR		SHIELD
N751 1946	128371	RAFVR		SHIELDS
YB 1997	4200496	RAuxAF		SHIELS
N187 1947	138171	RAFVR		SHILLING
N277 1955	85687	RAFVR		SHIMMONS
N468 1946	118664	RAFVR	DFC	SHINNIE
N496 1944	73519	RAFVR		SHIPWRIGHT
N608 1945	161401	RAFVR		SHIRCORE
N653 1943	81361	RAFVR		SHORT
N381 1945	741459	RAFVR		SHORT
N934 1948	805087	RAuxAF		SHOWELL
N103 1952	125351	RAFVR		SHRIVES
N1018 1950	746084	RAFVR		SHRIVES
N94 1950	846959	RAuxAF		SHRUBSOLE
N87 1948	90976	RAuxAF		SHUTER
N805 1948	90462	RAuxAF		SHUTTE
N837 1962	2651058	RAFVR	wef 19/8/62	SIBLEY
N22 1963	2664212	WRAFVR	wef 5/11/62	SIBLEY
N22 1963	166311	RAFVR	wef 17/4/62	SIBLEY
N531 1960	872344	AAF	wef 5/6/44	SILABON
N1 1950	859638	RAuxAF		SILK
N837 1946	111979	RAFVR	DFC	SILK
N740 1947	742144	RAFVR		SILLARS
N460 1947	811170	AAF		SILVER
N843 1959	185993	RAFVR	wef 25/5/55	SILVERMAN
MOD 1967	185993	RAFVR	1st Clasp	SILVERMAN

Surname	Initials			Rank	Status	Date
SILVERWOOD	E	H		FL	Acting	30/09/1948
SILVESTER	G	F		SQD LDR	Acting	24/01/1946
SIM	A	A		FL		06/03/1947
SIM	J	W		WING COMM	Acting	17/04/1947
SIMMONDS	A	J		CPL		07/07/1949
SIMMONDS	C			SGT		28/11/1946
SIMMONDS	C			WO		07/08/1947
SIMMONDS	D	C	V	PO		01/10/1953
SIMMONS	W	J		FLT SGT		15/07/1948
SIMON	T	Ap		SQD LDR		26/01/1950
SIMONS	H	A	B	FL		18/04/1946
SIMPKIN	R	A		FL		19/10/1960
SIMPKIN	R	A		FL		17/02/1968
SIMPSON	A			CPL		10/04/1952
SIMPSON	F	W		FL		13/12/1961
SIMPSON	G	E		CPL	Temporary	28/07/1955
SIMPSON	H	A		SQD LDR	Acting	30/03/1943
SIMPSON	H	F		WO		08/11/1945
SIMPSON	H	G		FLT SGT		21/01/1992
SIMPSON	J	H		WING COMM	Acting	08/11/1945
SIMPSON	J	H		WO	Acting	03/07/1952
SIMPSON	L	A		SQD LDR		17/06/1943
SIMPSON	L	J		WING COMM		09/12/1959
SIMPSON	L			WO		10/05/1945
SIMPSON	P	A		FL		13/01/1944
SIMPSON	W	L		CPL		22/02/1951
SIMPSON	W			SQD LDR		10/06/1948
SIMPSON-FEATHERSTONE	E	R		FO		15/11/1961
SIMS	J	E		AC1		20/05/1948
SIMS	L	G	C	FL		13/01/1949
SINCLAIR	C	P		CPL		10/10/1995
SINCLAIR	I			SAC		06/09/1994
SINCLAIR	J	D		SQD LDR		30/03/1943
SINCLAIR	J	McI		SGT		21/12/1950
SINCLAIR	J			CPL		27/03/1947
SINGER	R	M		SQD LDR	Acting	26/01/1950
SISMORE	E	B		SQD LDR	Acting	06/03/1947
SIZEN	H			FLT SGT		31/10/1956
SKEGGS	A	E		LAC		26/08/1948
SKELLY	J	S		SQD LDR	Acting	06/03/1947
SKELLY	R			LAC		15/07/1948
SKELTON	D	B		FLT SGT		15/09/1949
SKELTON	G			SQD LDR		06/03/1947
SKENE	A			SQD LDR		25/06/1953
SKERRETT	W			CPL		10/08/1950
SKILLMAN	D	W		FL		25/05/1944
SKINNER	A	E	L	SQD LDR		26/01/1950
SKINNER	C	D	E	SQD LDR		13/01/1944
SKINNER	D	H		FL		10/02/1949
SKINNER	J	H		CPL		30/09/1948
SKINNER	L	H		SQD LDR	Acting	05/10/1944
SKINNER	T	G		SGT		07/12/1950
SKINNER	W	M		FL		30/08/1945
SKUES	R	K		SQD LDR		20/07/1994
SKUSE	E	F		FL		23/06/1955
SKUSE	R	J		WO		24/11/1949
SLACK	E	A	M	FL		18/07/1956
SLACK	L			FL		23/03/1966
SLACK	W	C	H	LAC		01/07/1948
SLADE	F	W		CPL		14/01/1954

Source	Service No	Force	Notes	Surname
N805 1948	142331	RAFVR		SILVERWOOD
N82 1946	113838	RAFVR	DFC	SILVESTER
N187 1947	144196	RAFVR		SIM
N289 1947	73397	RAFVR	GM	SIM
N673 1949	865667	RAuxAF		SIMMONDS
N1004 1946	810015	AAF		SIMMONDS
N662 1947	746320	RAFVR		SIMMONDS
N796 1953	2602512	RAFVR		SIMMONDS
N573 1948	841525	RAuxAF		SIMMONS
N94 1950	72430	RAFVR		SIMON
N358 1946	86362	RAFVR		SIMONS
N775 1960	201080	RAFVR	DFM; wef 17/1/58	SIMPKIN
MOD 1968	201080	RAFVR	DFM; 1st Clasp	SIMPKIN
N278 1952	857285	RAuxAF		SIMPSON
N987 1961	178129	RAFVR	wef 30/8/61	SIMPSON
N565 1955	747024	RAFVR		SIMPSON
N1014 1943	84939	RAFVR		SIMPSON
N1232 1945	742744	RAFVR		SIMPSON
YB 1996	n/a	RAuxAF		SIMPSON
N1232 1945	117583	RAFVR	DFC	SIMPSON
N513 1952	803495	RAuxAF		SIMPSON
N653 1943	70624	RAFVR		SIMPSON
N843 1959	91350	RAuxAF	DFC; wef 3/10/59	SIMPSON
N485 1945	808139	AAF		SIMPSON
N11 1944	70625	RAFVR		SIMPSON
N194 1951	743385	RAFVR		SIMPSON
N456 1948	72062	RAFVR		SIMPSON
N902 1961	2606742	RAFVR	wef 31/7/61	SIMPSON-FEATHERSTONE
N395 1948	869362	RAuxAF		SIMS
N30 1949	81959	RAFVR		SIMS
YB 1997	n/a	RAuxAF		SINCLAIR
YB 1996	n/a	RAuxAF		SINCLAIR
N1014 1943	70627	RAFVR		SINCLAIR
N1270 1950	817004	RAuxAF		SINCLAIR
N250 1947	803479	AAF		SINCLAIR
N94 1950	123819	RAFVR		SINGER
N187 1947	130208	RAFVR	DSO,DFC	SISMORE
N761 1956	754686	RAF		SIZEN
N694 1948	845315	RAuxAF		SKEGGS
N187 1947	120172	RAFVR		SKELLY
N573 1948	816226	RAuxAF		SKELLY
N953 1949	816102	RAuxAF		SKELTON
N187 1947	90796	AAF		SKELTON
N527 1953	91274	RAuxAF	MBE	SKENE
N780 1950	856613	RAuxAF		SKERRETT
N496 1944	62250	RAFVR		SKILLMAN
N94 1950	72940	RAFVR	MC	SKINNER
N11 1944	90225	AAF		SKINNER
N123 1949	83265	RAFVR		SKINNER
N805 1948	755041	RAFVR		SKINNER
N1007 1944	85652	RAFVR		SKINNER
N1210 1950	745226	RAFVR		SKINNER
N961 1945	68722	RAFVR	DFM	SKINNER
YB 1997	5059829	RAuxAF		SKUES
N473 1955	128859	RAuxAF		SKUSE
N1193 1949	748628	RAFVR		SKUSE
N513 1956	77126	RAFVR		SLACK
MOD 1967	121097	RAFVR		SLACK
N527 1948	840396	RAuxAF		SLACK
N18 1954	864055	RAuxAF		SLADE

Surname	Initials			Rank	Status	Date
SLADEN	A	I		FL		20/04/1944
SLATER	A	W		WING COMM		24/11/1949
SLATER	G	H		CPL		14/10/1948
SLATER	I	H	F	CPL		15/09/1949
SLATER	J	R		WING COMM	Acting	16/10/1947
SLATER	P	A		CPL		27/03/1952
SLATTER	A	P	A	FO		10/08/1950
SLEEP	R	M		SQD LDR	Acting	28/02/1946
SLEIGHT	A			FL		28/02/1946
SLEIGHT	R			SQD LDR	Acting	30/08/1945
SLINN	W	A		FL		13/02/1957
SMALE	A	R		FL		29/04/1954
SMALLTHWAITE	N	J		FL		26/08/1948
SMALLWOOD	A	R		WING COMM	Acting	08/07/1959
SMALLWOOD	O	D		AIR COMMD	Acting	10/05/1945
SMALLY	E	B		LAC		16/03/1950
SMART	A	C		FL	Acting	08/02/1945
SMART	A			FLT SGT		28/06/1945
SMART	C	G		LAC		25/05/1950
SMART	E	J		GRP CAP		06/03/1947
SMART	F	H		SGT		07/04/1955
SMART	J	M		FO		16/10/1947
SMART	R	C		FL		05/09/1946
SMART	S	L		FL		30/10/1957
SMEE	C	N		FO		14/01/1954
SMETHAM	B	J		FL		21/10/1959
SMILES	H	J	D	FL		18/04/1946
SMITH	A	B	A	FL		21/06/1945
SMITH	A	B		SQD LDR	Acting	21/03/1946
SMITH	A	C		FO		26/02/1958
SMITH	A	E		FO		18/04/1946
SMITH	A	E		CPL		07/08/1947
SMITH	A	S		SGT		01/07/1948
SMITH	A	T		SQD LDR	Acting	08/05/1947
SMITH	A	W		FL		21/03/1946
SMITH	A			PO		24/12/1942
SMITH	A			FLT SGT		10/06/1948
SMITH	A			LAC		10/02/1949
SMITH	A			CPL		18/10/1951
SMITH	A			SGT		07/02/1952
SMITH	B	J		FLT SGT	Acting	18/11/1948
SMITH	B	S		SGT		16/03/1950
SMITH	C	E	A	FL		16/04/1958
SMITH	C	H	P	FO		24/01/1946
SMITH	C			SGT		10/05/1945
SMITH	C			SGT		06/09/1951
SMITH	D	A		WING COMM	Acting	12/08/1943
SMITH	D	B		WING COMM	Acting	30/03/1943
SMITH	D	E		FL		26/01/1950
SMITH	D	J	M	SGT	Acting	25/06/1953
SMITH	D	M		FL		07/05/1953
SMITH	D			FL		18/04/1946
SMITH	D			CPL		17/03/1949
SMITH	E	B	B	WING COMM	Acting	27/01/1944
SMITH	E	B		SQD LDR	Acting	23/05/1946
SMITH	E	C		PO		04/10/1945
SMITH	E	C		FO		17/01/1952
SMITH	E	E	F	G	SGT	20/05/1948
SMITH	E	E		CPL	Temporary	02/05/1962
SMITH	E	G		FL		30/09/1954

Source	Service No	Force	Notes	Surname
N363 1944	102578	RAFVR	DSO	SLADEN
N1193 1949	73827	RAFVR	MBE	SLATER
N856 1948	852833	RAuxAF		SLATER
N953 1949	800631	RAuxAF		SLATER
N846 1947	72373	RAFVR		SLATER
N224 1952	747354	RAFVR		SLATER
N780 1950	134815	RAFVR		SLATTER
N197 1946	104434	RAFVR	DFC	SLEEP
N197 1946	146645	RAFVR		SLEIGHT
N961 1945	114942	RAFVR	DFM	SLEIGHT
N104 1957	83589	RAFVR	wef 15/1/54	SLINN
N329 1954	103997	RAuxAF		SMALE
N694 1948	90813	RAuxAF		SMALLTHWAITE
N454 1959	68966	RAuxAF	wef 10/5/59	SMALLWOOD
N485 1945	90579	AAF	CBE,TD,DL	SMALLWOOD
N257 1950	857587	RAuxAF		SMALLY
N139 1945	161078	RAFVR	DFC	SMART
N706 1945	800384	AAF		SMART
N517 1950	805484	RAuxAF		SMART
N187 1947	90406	AAF		SMART
N277 1955	841286	RAuxAF		SMART
N846 1947	174288	RAFVR		SMART
N751 1946	177202	RAFVR		SMART
N775 1957	124548	RAFVR	DSO,DFC; wef 17/9/56	SMART
N18 1954	141293	RAFVR		SMEE
N708 1959	201643	RAFVR	wef 8/8/54	SMETHAM
N358 1946	103535	RAFVR	DFM	SMILES
N671 1945	126104	RAFVR		SMITH
N263 1946	115767	RAFVR		SMITH
N149 1958	2691002	RAuxAF	wef 3/1/58	SMITH
N358 1946	186571	RAFVR		SMITH
N662 1947	841850	AAF		SMITH
N527 1948	770718	RAFVR		SMITH
N358 1947	72724	RAFVR		SMITH
N263 1946	113497	RAFVR		SMITH
N1758 1942	120028	RAFVR		SMITH
N456 1948	869392	RAuxAF		SMITH
N123 1949	868500	RAuxAF		SMITH
N1077 1951	744277	RAFVR		SMITH
N103 1952	2684008	RAuxAF		SMITH
N984 1948	770821	RAFVR		SMITH
N257 1950	800489	RAuxAF		SMITH
N258 1958	196754	RAFVR	wef 29/1/58	SMITH
N82 1946	182088	RAFVR		SMITH
N485 1945	741069	RAFVR		SMITH
N913 1951	817015	RAuxAF		SMITH
N850 1943	90292	AAF	MBE,MB,ChB,MRCP(E)	SMITH
N1014 1943	70630	RAFVR		SMITH
N94 1950	78110	RAFVR		SMITH
N527 1953	2677408	RAuxAF		SMITH
N395 1953	154656	RAFVR		SMITH
N358 1946	127802	RAFVR		SMITH
N231 1949	866547	RAuxAF		SMITH
N59 1944	90340	AAF	DFC	SMITH
N468 1946	78448	RAFVR		SMITH
N1089 1945	197580	RAFVR		SMITH
N35 1952	197580	RAFVR	1st Clasp	SMITH
N395 1948	841323	RAuxAF		SMITH
N314 1962	812077	AAF	wef 27/9/43	SMITH
N799 1954	150968	RAuxAF		SMITH

Surname	Initials			Rank	Status	Date
SMITH	E	P		FL		01/11/1951
SMITH	E	S		WING COMM	Acting	22/04/1943
SMITH	E	V		SQD LDR	Acting	08/11/1945
SMITH	E	W		FO		08/11/1945
SMITH	E			CPL		12/02/1953
SMITH	F	D		LAC		24/07/1952
SMITH	F	J		SGT		26/04/1961
SMITH	F	W		WING COMM		08/05/1947
SMITH	F	W		LAC		26/08/1948
SMITH	G	F		LAC		01/07/1948
SMITH	G	I		FO		08/03/1956
SMITH	G	J	E	SQD LDR		01/07/1948
SMITH	G	M		FL		05/09/1962
SMITH	G	P		SQD LDR		12/08/1943
SMITH	G	P		FL		01/05/1957
SMITH	G	P		FL		20/09/1963
SMITH	G	R		SGT		09/12/1992
SMITH	G			LAC		08/03/1956
SMITH	H	J	E	SQD LDR		02/07/1958
SMITH	H	J		SGT	Temporary	25/07/1962
SMITH	H	O		SGT		27/03/1952
SMITH	H			LAC		07/02/1952
SMITH	H			FL		18/02/1954
SMITH	H			PO		28/11/1956
SMITH	J	A		CPL		23/08/1951
SMITH	J	C		LAC		24/07/1952
SMITH	J	H	M	SQD LDR		06/05/1943
SMITH	J	H		SQD LDR	Acting	24/06/1943
SMITH	J	H		WING COMM	Acting	04/10/1951
SMITH	J	H		FLT SGT		08/03/1951
SMITH	J	P		SGT		23/02/1950
SMITH	J	P	R	SGT		17/06/1954
SMITH	J	R	A	FL		25/06/1953
SMITH	J	W		FLT SGT		22/09/1955
SMITH	J			CPL		18/11/1948
SMITH	K	G	S	FL		08/11/1945
SMITH	L	E		FL		04/10/1945
SMITH	L	E		FO		25/03/1954
SMITH	L	F		CPL		02/04/1953
SMITH	L	G		SQD LDR	Acting	17/02/1944
SMITH	L	G		FL		04/05/1944
SMITH	L	V		SGT		07/01/1943
SMITH	L			FL		30/03/1960
SMITH	M	A		WING COMM		28/10/1943
SMITH	M	A		SQD LDR		21/09/1944
SMITH	M	A		SQD LDR	Acting	04/10/1945
SMITH	M	A	B	SGT		07/07/1949
SMITH	M	G	B	FO		18/07/1956
SMITH	N			LAC		26/11/1953
SMITH	P	G		FLT SGT		02/06/1995
SMITH	P	H		SGT		15/07/1948
SMITH	P	J		SAC		30/03/1994
SMITH	P	W		FL		19/08/1954
SMITH	P			FL		28/02/1946
SMITH	R	A		SQD LDR		27/03/1947
SMITH	R	B		FL		07/12/1960
SMITH	R	H		FL		21/01/1959
SMITH	R	K		FL	Acting	07/07/1949
SMITH	R	L		FL		08/11/1945
SMITH	R	R		FL		15/07/1948

Source	Service No	Force	Notes	Surname
N1126 1951	87079	RAFVR		SMITH
N475 1943	90093	AAF		SMITH
N1232 1945	82172	RAFVR	DSO,DFC,AFC	SMITH
N1232 1945	174520	RAFVR	DFC	SMITH
N115 1953	850148	RAuxAF		SMITH
N569 1952	843959	RAuxAF		SMITH
N339 1961	2607254	RAFVR	wef 29/11/59	SMITH
N358 1947	72224	RAFVR		SMITH
N694 1948	840032	RAuxAF		SMITH
N527 1948	869472	RAuxAF		SMITH
N187 1956	2689557	RAuxAF		SMITH
N527 1948	73664	RAFVR		SMITH
N655 1962	2614060	RAFVR	wef 20/4/62	SMITH
N850 1943	72008	RAFVR	OBE,MB,ChB	SMITH
N277 1957	166902	RAFVR	wef 20/9/53	SMITH
MOD 1966	166902	RAFVR	1st Clasp	SMITH
YB 1996	n/a	RAuxAF		SMITH
N187 1956	850687	RAuxAF		SMITH
N452 1958	90554	AAF	MBE; wef 14/3/44	SMITH
N549 1962	819021	AAF	wef 6/12/43	SMITH
N224 1952	840546	RAuxAF		SMITH
N103 1952	868773	RAuxAF		SMITH
N137 1954	161734	RAuxAF		SMITH
N833 1956	2676600	RAuxAF		SMITH
N866 1951	818211	RAuxAF		SMITH
N569 1952	818035	RAuxAF		SMITH
N526 1943	73828	RAFVR		SMITH
N677 1943	90049	AAF		SMITH
N1019 1951	90049	RAuxAF	1st Clasp	SMITH
N244 1951	820032	RAuxAF		SMITH
N179 1950	808285	RAuxAF		SMITH
N471 1954	2650021	RAuxAF		SMITH
N527 1953	130791	RAFVR		SMITH
N708 1955	2687025	RAuxAF		SMITH
N984 1948	853640	RAuxAF		SMITH
N1232 1945	72055	RAFVR		SMITH
N1089 1945	112251	RAFVR		SMITH
N249 1954	112251	RAFVR	1st Clasp	SMITH
N298 1953	804318	RAuxAF		SMITH
N132 1944	103090	RAFVR		SMITH
N425 1944	119816	RAFVR	DFC,DFM	SMITH
N3 1943	808081	AAF		SMITH
N229 1960	172404	RAFVR	DFC; wef 23/11/56	SMITH
N1118 1943	70776	RAFVR		SMITH
N961 1944	72029	RAFVR		SMITH
N1089 1945	88658	RAFVR	DFC	SMITH
N673 1949	771181	RAFVR		SMITH
N513 1956	2658602	WRAAF		SMITH
N952 1953	2695511	RAuxAF		SMITH
YB 1996	n/a	RAuxAF		SMITH
N573 1948	743958	RAFVR		SMITH
YB 1996	n/a	RAuxAF		SMITH
N672 1954	62314	RAFVR	DFC	SMITH
N197 1946	159536	RAFVR		SMITH
N250 1947	73724	RAFVR		SMITH
N933 1960	105681	RAFVR	wef 14/8/60	SMITH
N28 1959	120727	RAFVR	wef 5/10/53	SMITH
N673 1949	117833	RAFVR		SMITH
N1232 1945	129958	RAFVR	DFC,DFM	SMITH
N573 1948	78141	RAFVR		SMITH

Surname	Initials			Rank	Status	Date
SMITH	R	W		FL		18/04/1946
SMITH	R	W	E	SGT		18/03/1948
SMITH	S	C		FL		31/05/1945
SMITH	S	G		FO		17/06/1943
SMITH	S	J	M	FL		05/10/1944
SMITH	S	L		FLT SGT		10/05/1945
SMITH	T	G		FL		28/11/1956
SMITH	V	G		SGT	Acting	18/11/1948
SMITH	V	P		AC1		24/07/1952
SMITH	W	A		FL		20/01/1960
SMITH	W	J		SGT		12/10/1950
SMITH	W	L		SGT		14/10/1948
SMITH	W			SGT		18/07/1956
SMITHER	E	J	C	FL		17/04/1947
SMULIAN	P	K		SQD LDR	Acting	26/10/1944
SMYTH	J	C		SQD LDR		02/09/1943
SMYTH	R	N		FLT SGT		27/04/1950
SMYTH	T	G		SGT	Acting	18/02/1954
SNAPE	J	N		FO		02/04/1953
SNARR	J			CPL	Temporary	03/06/1959
SNEE	J	J		FL		06/03/1947
SNEESBY	R	J	H	FLT SGT		13/07/1950
SNELL	A	G		LAC		07/07/1949
SNELL	G	F		FL		13/02/1957
SNOW	S	J	M	JNR TECH		28/07/1955
SNOW	T			SGT		26/08/1948
SNOWDEN	E	G		FL		13/12/1945
SOAR	T	E		LAC		07/07/1949
SOARS	H	J		FL		20/09/1945
SOARS	H	J		FO		02/08/1951
SOLE	J	L		WING COMM	Acting	22/10/1958
SOLLARS	R	G		FL		09/05/1946
SOLLOM	H	W		FO		09/03/1944
SOLOMAN	A	F	H	LAC		30/09/1948
SOLOMAN	D	A		FLT SGT		07/07/1949
SOLOMON	J	W		FO		23/02/1950
SOMERVILLE	H	G		CPL		07/08/1947
SOMERVILLE	J	C		MST SIG		28/08/1963
SONES	L	C		PO		10/12/1942
SOPER	C	H	G	SQD LDR	Acting	05/06/1947
SOPPITT	G	E		FL		01/07/1948
SORBY	C	D	V	FL		24/01/1946
SOUTAR	J	M		SGT		01/10/1996
SOUTER	C	E		SAC		01/09/1996
SOUTER	K	P		FL		21/03/1946
SOUTHALL	E			FL		30/10/1957
SOUTHALL	L	H		FO		30/08/1945
SOUTHAN	K			FL	Acting	30/03/1943
SOUTHEN	R	G		FO		07/12/1960
SOUTHERDEN	N	G		CPL		21/02/1952
SOUTHERN	J	M	M	SGT		25/03/1959
SOUTHEY	E	A		FL		04/10/1945
SOUTHWARD	D	F		SGT		30/05/1998
SOUTHWELL	K	A		FL		06/03/1947
SOUTOU	N			SAC		11/10/1995
SOWNEY	J	T		FL	Acting	07/08/1947
SOWTER	R	S		FL		13/02/1957
SPALDING	R	C	M	SAC		01/09/1999
SPALDING	R	E	F	SGT	Acting	07/12/1960
SPARKES	G	H		FO		25/05/1950

Source	Service No	Force	Notes	Surname
N358 1946	131882	RAFVR	DFM	SMITH
N232 1948	746379	RAFVR		SMITH
N573 1945	139821	RAFVR	DFC	SMITH
N653 1943	119502	RAFVR		SMITH
N1007 1944	118423	RAFVR		SMITH
N485 1945	815033	AAF		SMITH
N833 1956	166754	RAuxAF		SMITH
N984 1948	752717	RAFVR		SMITH
N569 1952	842295	RAuxAF		SMITH
N33 1960	167054	RAFVR	wef 31/12/52	SMITH
N1018 1950	845035	RAuxAF		SMITH
N856 1948	770095	RAFVR		SMITH
N513 1956	756145	RAF		SMITH
N289 1947	78113	RAFVR		SMITHER
N1094 1944	102533	RAFVR	AFC	SMULIAN
N918 1943	70635	RAFVR		SMYTH
N406 1950	743796	RAFVR		SMYTH
N137 1954	2676818	RAuxAF		SMYTH
N298 1953	105461	RAFVR		SNAPE
N368 1959	2605511	RAFVR	wef 26/11/58	SNARR
N187 1947	173858	RAFVR		SNEE
N675 1950	756004	RAFVR		SNEESBY
N673 1949	845827	RAuxAF		SNELL
N104 1957	177968	RAFVR	DFC; wef 15/6/53	SNELL
N565 1955	2603875	RAFVR		SNOW
N694 1948	847727	RAuxAF		SNOW
N1355 1945	101013	RAFVR		SNOWDEN
N673 1949	815259	RAuxAF		SOAR
N1036 1945	134228	RAFVR		SOARS
N777 1951	134228	RAFVR	1st Clasp	SOARS
N716 1958	136660	RAuxAF	MBE; wef 1/9/58	SOLE
N415 1946	72381	RAFVR		SOLLARS
MOD 1964	110934	RAFVR		SOLLOM
N805 1948	847294	RAuxAF		SOLOMAN
N673 1949	744530	RAFVR		SOLOMAN
N179 1950	140989	RAFVR	MBE	SOLOMON
N662 1947	841753	AAF		SOMERVILLE
N621 1963	751232	RAF	wef 24/5/44	SOMERVILLE
N1680 1942	127803	RAFVR		SONES
N460 1947	73407	RAFVR		SOPER
N527 1948	122792	RAFVR		SOPPITT
N82 1946	68135	RAFVR		SORBY
YB 1997	n/a	RAuxAF		SOUTAR
YB 1997	n/a	RAuxAF		SOUTER
N263 1946	129001	RAFVR		SOUTER
N775 1957	171323	RAFVR	wef 18/6/54	SOUTHALL
N961 1945	169927	RAFVR		SOUTHALL
N1014 1943	103485	RAFVR		SOUTHAN
N933 1960	2699056	RAuxAF	wef 3/10/60	SOUTHEN
N134 1952	799774	RAFVR		SOUTHERDEN
N196 1959	2657102	WRAAF	wef 28/6/58	SOUTHERN
N1089 1945	110344	RAFVR	AFC	SOUTHEY
YB 1999	n/a	RAuxAF		SOUTHWARD
N187 1947	80565	RAFVR		SOUTHWELL
YB 1996	n/a	RAuxAF		SOUTOU
N662 1947	125923	RAFVR		SOWNEY
N104 1957	153023	RAFVR	wef 16/5/52	SOWTER
YB 2000	n/a	RAuxAF		SPALDING
N933 1960	2693554	RAuxAF	wef 15/10/60	SPALDING
N517 1950	145218	RAFVR		SPARKES

Surname	Initials			Rank	Status	Date
SPARKES	P	J		SQD LDR		16/12/1995
SPARKS	J	P		FL		02/04/1953
SPARLING	T	A	G	FL		20/09/1945
SPARREY	L	E		CPL		16/04/1958
SPARROW	G	F		FL		20/09/1945
SPARROW	H	F		CPL		18/11/1948
SPARROW	J	W		SQD LDR	Acting	27/03/1947
SPEAR	N	A		LAC		13/01/1949
SPEARING	B	J	H	FL		25/03/1959
SPEARS	R	F		FL		06/03/1947
SPEED	H	V		FO		20/05/1948
SPEEDY	R	F		FL		08/11/1945
SPENCE	C	S		FL		22/10/1958
SPENCER	A	H	G	FL		06/01/1955
SPENCER	C	F	A	PO		04/09/1947
SPENCER	D	C		FL		30/12/1943
SPENCER	E			FL		08/03/1956
SPENCER	J	F		SGT		19/09/1956
SPENCER	J			SQD OFF	Acting	13/11/1947
SPENCER	R	G		FL		26/04/1961
SPENCER	S			SGT		01/10/1953
SPENCER-EDWARDS	E	A		SQD LDR	Acting	29/06/1944
SPEYER	E	P	K	SQD LDR		07/04/1955
SPIERS	A	J		FL		10/02/1949
SPIERS	A			WO		26/08/1948
SPIERS	G	W		SQD LDR		27/03/1947
SPIERS	J	A		SGT		03/07/1993
SPINK	P	J		FL		26/10/1944
SPINK	R			WING OFF		14/10/1948
SPINKS	H			CPL		27/03/1947
SPINKS	J	A		FL		20/03/1957
SPIRES	J	H		FL		15/09/1949
SPLUDE	D	P		CPL		22/10/1994
SPONG	L	G		SQD LDR	Acting	01/03/1945
SPONG	L	G		FL		22/10/1958
SPONG	L	G		FL		22/10/1958
SPONTON	T			LAC		19/10/1960
SPOONER	A			SQD LDR	Acting	21/09/1944
SPOONER	E			LAC		05/01/1950
SPOONER	I	D		FL		07/08/1947
SPOONER	R	L		SQD LDR	Acting	17/04/1947
SPRAGET	H	J		CPL		07/08/1947
SPRAGG	F	E		CPL	Acting	02/04/1953
SPRATT	J			LAC		13/01/1949
SPRIGG	C	J		FL		06/03/1947
SPROXTON	A			FL		15/09/1949
SPURGEON	J	H		SQD LDR	Acting	19/07/1951
SQUIER	J	W	C	FL		24/01/1946
SQUIRE	H	F		SQD LDR		26/01/1950
SQUIRE	J			FLT SGT		01/05/1957
SQUIRE	P	G		CPL		30/09/1948
SQUIRES	W	J	W	FL		15/01/1958
St JOHN CHAPPELL	H	M		FLT OFF		03/07/1957
STABLER	J	T		LAC		07/04/1955
STABLES	J	M		FL		09/02/1951
STACEY	I	G	R	FL		19/10/1960
STACEY	P	M		FO		18/02/1954
STACEY	R	C		WO		10/08/1950

Source	Service No	Force	Notes	Surname
YB 1997	9934	RAuxAF		SPARKES
N298 1953	191973	RAFVR		SPARKS
N1036 1945	84932	RAFVR		SPARLING
N258 1958	2692004	RAuxAF	wef 4/3/58	SPARREY
N1036 1945	132338	RAFVR		SPARROW
N984 1948	760302	RAFVR		SPARROW
N250 1947	72944	RAFVR	MM	SPARROW
N30 1949	861022	RAuxAF		SPEAR
N196 1959	105359	RAFVR	wef 27/10/54	SPEARING
N187 1947	121244	RAFVR		SPEARS
N395 1948	162709	RAFVR		SPEED
N1232 1945	63422	RAFVR		SPEEDY
N716 1958	196090	RAFVR	wef 13/4/55	SPENCE
N1 1955	145359	RAFVR		SPENCER
N740 1947	204088	RAFVR		SPENCER
N1350 1943	85638	RAFVR	DFM	SPENCER
N187 1956	132758	RAuxAF		SPENCER
N666 1956	844691	RAF		SPENCER
N933 1947	1834	WAAF		SPENCER
N339 1961	1868800	RAuxAF	wef 15/3/61	SPENCER
N796 1953	799856	RAFVR		SPENCER
N652 1944	91113	AAF		SPENCER-EDWARDS
N277 1955	73535	RAFVR		SPEYER
N123 1949	72945	RAFVR		SPIERS
N694 1948	749478	RAFVR		SPIERS
N250 1947	78266	RAFVR		SPIERS
YB 1996	n/a	RAuxAF		SPIERS
N1094 1944	81925	RAFVR		SPINK
N856 1948	104	WAAF		SPINK
N250 1947	842793	AAF		SPINKS
N198 1957	172109	RAuxAF	wef 29/4/55	SPINKS
N953 1949	121239	RAFVR	DFC,DFM	SPIRES
YB 1996	n/a	RAuxAF		SPLUDE
N225 1945	114062	RAFVR	DFC	SPONG
N716 1958	114062	RAFVR	DFC; 1st Clasp; wef 25/11/49	SPONG
N716 1958	114062	RAFVR	DFC; 2nd Clasp; wef 27/2/58	SPONG
N775 1960	2699554	RAuxAF	wef 23/4/60	SPONTON
N961 1944	82948	RAFVR	DSO,DFC	SPOONER
N1 1950	800671	RAuxAF		SPOONER
N662 1947	90698	AAF		SPOONER
N289 1947	86761	RAFVR		SPOONER
N662 1947	840869	AAF		SPRAGET
N298 1953	841784	RAuxAF		SPRAGG
N30 1949	816128	RAuxAF		SPRATT
N187 1947	134994	RAFVR		SPRIGG
N953 1949	115524	RAFVR		SPROXTON
N724 1951	128360	RAFVR	DFC,AFC	SPURGEON
N82 1946	125762	RAFVR		SQUIER
N94 1950	72075	RAFVR	MRCS,LRCP	SQUIRE
N277 1957	800427	RAuxAF	wef 4/12/42	SQUIRE
N805 1948	864812	RAuxAF		SQUIRE
N40 1958	136561	RAFVR	wef 4/12/53	SQUIRES
N456 1957	2968	WAAF	wef 25/7/44	St JOHN CHAPPELL
N277 1955	842886	RAuxAF		STABLER
N145 1951	79766	RAFVR		STABLES
N775 1960	182891	RAFVR	wef 3/7/44	STACEY
N137 1954	198075	RAFVR		STACEY
N780 1950	755302	RAFVR		STACEY

Surname	Initials			Rank	Status	Date
STAFF	C	S		SGT		26/06/1947
STAINER	A	E		FL		25/03/1954
STAINES	W	P		SGT	Temporary	01/05/1957
STAINFORTH	A	P		FO		08/03/1956
STAINSBY	S	J		SGT		19/07/1997
STAIT	L	J	E	SGT		16/03/1950
STAKEMIRE	D	A		WO		24/01/1946
STAMMERS	K	S		SQD LDR	Acting	20/04/1944
STAMP	R	A		SGT		10/06/1948
STANANOUGHT	A			SGT		10/04/1952
STANBROOK	J	A		SQD LDR	Acting	17/04/1947
STANDEN	A	C		FL		29/04/1954
STANDEN	H	A		FL		06/03/1947
STANFIELD	H			CPL		16/10/1947
STANFIELD	H			CPL		29/04/1954
STANFORD	P	A		FL		28/11/1946
STANLEY	C	R		FL		24/01/1946
STANLEY	H	B		FLT SGT		14/10/1948
STANLEY	J			SQD LDR	Acting	25/05/1944
STANLEY	T	A		FL		31/05/1945
STANNARD	C	R		FO		03/12/1942
STANTON	W	J		FL		26/04/1961
STAPLES	L			FL		30/08/1945
STAPLES	T	N		SQD LDR	Acting	12/04/1945
STARK	R			LAC		05/01/1950
STARKEY	E	M		SQD OFF	Acting	17/04/1947
STATON	W	R		FO		25/06/1953
STEAD	P	K		WING COMM		10/12/1942
STEADMAN	D	J		FL		18/04/1946
STEARN	F	J	E	FL		30/11/1944
STEED	M	A		SGT		17/04/1947
STEEDMAN	W	F		WO		25/01/1945
STEEL	J	G		FO		20/04/1944
STEEL	W	M		FL		03/06/1959
STEELE	A	S		FL		01/03/1945
STEELE	R	P		SGT		15/01/1958
STEER	E	R	A	FO		16/01/1957
STEFF	B	G		FL	Acting	10/06/1948
STENHOUSE	J			FO		21/06/1945
STEPHEN	C			SGT		28/11/1956
STEPHEN	E			SGT		15/07/1948
STEPHEN	H	M		WING COMM	Acting	27/01/1944
STEPHEN	R	J		WING COMM		05/06/1947
STEPHENS	A			WING OFF	Acting	17/04/1947
STEPHENS	B	C		SQD LDR	Acting	21/06/1961
STEPHENS	E	M		FL		18/09/1998
STEPHENS	F	F		SGT		30/09/1948
STEPHENS	I	R		FL		20/09/1945
STEPHENS	J	C		FLT SGT		09/12/1992
STEPHENSON	A			FO		01/07/1948
STEPHENSON	G	C		SGT		07/12/1960
STEPHENSON	G	L		WO	Acting	26/04/1961
STEPHENSON	P	J	T	WING COMM	Acting	04/10/1945
STEVEN	J	S		WING COMM	Acting	26/08/1948
STEVENS	A	H	K	SQD LDR		12/06/1947
STEVENS	C	H		WO		28/02/1946
STEVENS	E	J		SQD LDR	Acting	13/12/1945
STEVENS	F	E		LAC		15/07/1948
STEVENS	G			FL		27/04/1950
STEVENS	G			FL		19/07/1951

Source	Service No	Force	Notes	Surname
N519 1947	841039	AAF		STAFF
N249 1954	66102	RAFVR		STAINER
N277 1957	846292	RAuxAF	wef 10/4/44	STAINES
N187 1956	2600562	RAuxAF		STAINFORTH
YB 1998	n/a	RAuxAF		STAINSBY
N257 1950	743338	RAFVR		STAIT
N82 1946	746768	RAFVR		STAKEMIRE
N363 1944	65989	RAFVR	DFC,DFM	STAMMERS
N456 1948	747360	RAFVR		STAMP
N278 1952	810177	RAuxAF		STANANOUGHT
N289 1947	73408	RAFVR		STANBROOK
N329 1954	189425	RAFVR		STANDEN
N187 1947	171924	RAFVR		STANDEN
N846 1947	816067	AAF		STANFIELD
N329 1954	2681000	RAuxAF	1st Clasp	STANFIELD
N1004 1946	130712	RAFVR	DFC	STANFORD
N82 1946	143119	RAFVR		STANLEY
N856 1948	810085	RAuxAF		STANLEY
N496 1944	70640	RAFVR		STANLEY
N573 1945	115740	RAFVR		STANLEY
N1638 1942	89297	RAFVR		STANNARD
N339 1961	188541	RAFVR	wef 28/11/59	STANTON
N961 1945	117410	RAFVR		STAPLES
N381 1945	100604	RAFVR	DFM	STAPLES
N1 1950	840477	RAuxAF		STARK
N289 1947	1542	WAAF		STARKEY
N527 1953	201886	RAuxAF		STATON
N1680 1942	90306	AAF		STEAD
N358 1946	62261	RAFVR		STEADMAN
N1225 1944	63067	RAFVR	AFC	STEARN
N289 1947	749613	RAFVR		STEED
N70 1945	803122	AAF		STEEDMAN
N363 1944	155005	RAFVR		STEEL
N368 1959	171687	RAFVR	wef 30/1/55	STEEL
N225 1945	79732	RAFVR		STEELE
N40 1958	2689527	RAFVR	wef 27/5/57	STEELE
N24 1957	2603600	RAFVR		STEER
N456 1948	161929	RAFVR		STEFF
N671 1945	156124	RAFVR		STENHOUSE
N833 1956	2684029	RAuxAF		STEPHEN
N573 1948	810139	RAuxAF		STEPHEN
N59 1944	78851	RAFVR	DSO,DFC	STEPHEN
N460 1947	73410	RAFVR		STEPHEN
N289 1947	86	WAAF	MBE	STEPHENS
N478 1961	199554	RAuxAF	wef 19/4/61	STEPHENS
YB 1999	9945	RAuxAF		STEPHENS
N805 1948	749741	RAFVR		STEPHENS
N1036 1945	81915	RAFVR	DFC	STEPHENS
YB 1996	n/a	RAuxAF		STEPHENS
N527 1948	139968	RAFVR		STEPHENSON
N933 1960	2699542	RAuxAF	wef 17/9/60	STEPHENSON
N339 1961	2699502	RAuxAF	wef 23/4/54	STEPHENSON
N1089 1945	81343	RAFVR	DFC	STEPHENSON
N694 1948	90977	RAuxAF	MBE	STEVEN
N478 1947	72316	RAFVR		STEVENS
N197 1946	747763	RAFVR	DFM	STEVENS
N1355 1945	82660	RAFVR		STEVENS
N573 1948	845027	RAuxAF		STEVENS
N406 1950	137305	RAFVR		STEVENS
N724 1951	137305	RAFVR	1st Clasp	STEVENS

Surname	Initials			Rank	Status	Date
STEVENS	J	E		FL		22/02/1961
STEVENS	J	H		SQD LDR		07/08/1947
STEVENS	K	C		FL		21/09/1944
STEVENS	L	B		SGT		28/11/1946
STEVENS	P	M		SGT		17/04/1947
STEVENS	P	N	J	SGT		09/11/1950
STEVENS	R	F		FL		23/02/1950
STEVENS	S	G		FL		14/01/1954
STEVENS	T	A		SQD LDR	Acting	28/02/1946
STEVENS	V	A		FL		30/08/1945
STEVENS	W	N		AC2		16/10/1947
STEVENSON	A	A		FL		19/10/1960
STEVENSON	G	A		FO		06/03/1947
STEVENSON	G			SGT		22/02/1961
STEVENSON	T			FL		20/04/1944
STEVENTON	D	K		CPL		19/04/1996
STEWARD	W	A		SQD LDR		19/07/1951
STEWART	A	D		SQD LDR	Acting	24/01/1946
STEWART	D	J		SQD LDR		17/02/1944
STEWART	E	H		FL		26/06/1947
STEWART	G	C		CPL		16/10/1947
STEWART	J	B		SQD LDR		27/03/1947
STEWART	J	B		FL		01/05/1957
STEWART	J	W		SQD LDR		18/05/1944
STEWART	L	A	R	CPL		13/01/1949
STEWART	L	C		SGT		18/03/1948
STEWART	N	L	McN	SQD OFF		27/04/1950
STEWART	P	G		GRP CAP		03/12/1942
STEWART	P	G		GRP CAP		20/03/1957
STEWART	P	G		GRP CAP		20/03/1957
STEWART	R	W		SQD LDR		06/03/1947
STEWART	R	W		FO		07/05/1953
STEWART	R			SGT		26/08/1948
STEWART	S	L		FLT SGT		22/04/1943
STEWART	T	F		SQD LDR		03/07/1952
STEWART	W	K		AIR COMMD		11/08/1944
STILES	H			SQD LDR	Acting	24/01/1946
STILES	R	H		FLT SGT		15/06/1944
STILLWELL	R	W		SQD LDR	Acting	06/03/1947
STINSON	F	G		CPL		05/01/1950
STIRK	C	W		FLT SGT		20/05/1948
STOBART	R	D		SGT	Temporary	25/05/1960
STOCK	F			SGT		05/01/1950
STOCK	K	C	M	WING COMM	Acting	07/07/1949
STOCK	R	A		FL		03/07/1952
STOCKDALE	G	G		WING COMM		15/06/1944
STOCKEN	H	E	W	WING COMM	Acting	15/07/1948
STOCKER	W	P		SGT		07/07/1949
STOCKS	G	F		SGT	Temporary	16/04/1958
STOCKS	P	J		FO		26/01/1950
STOCKS	W			FL	Acting	05/06/1947
STOCKTON	A	G		CPL		23/02/1950
STOCKWELL	R	H		SQD LDR	Acting	17/04/1947
STODDART	J	V		WO	Acting	21/12/1950
STODDART	K	M		SQD LDR		30/03/1943
STODHART	A	H		FL		27/11/1952
STOKES	B	W	B	CPL		16/03/1950
STOKES	E	V		SQD LDR		05/02/1948
STOKES	R	T		FL		12/08/1943
STOKES	W	M		SGT		02/07/1958

Source	Service No	Force	Notes	Surname
N153 1961	179126	RAFVR	wef 6/7/44	STEVENS
N662 1947	73412	RAFVR		STEVENS
N961 1944	110605	RAFVR		STEVENS
N1004 1946	815051	AAF		STEVENS
N289 1947	880530	WAAF		STEVENS
N1113 1950	813100	RAuxAF		STEVENS
N179 1950	157961	RAFVR		STEVENS
N18 1954	149614	RAFVR		STEVENS
N197 1946	86669	RAFVR		STEVENS
N961 1945	86632	RAFVR		STEVENS
N846 1947	850012	AAF		STEVENS
N775 1960	168134	RAFVR	wef 30/5/59	STEVENSON
N187 1947	131178	RAFVR		STEVENSON
N153 1961	2691096	RAuxAF	wef 3/12/60	STEVENSON
N363 1944	82957	RAFVR		STEVENSON
YB 1997	n/a	RAuxAF		STEVENTON
N724 1951	72392	RAFVR		STEWARD
N82 1946	79583	RAFVR		STEWART
N132 1944	70649	RAFVR		STEWART
N519 1947	90980	AAF		STEWART
N846 1947	816200	AAF		STEWART
N250 1947	72393	RAFVR		STEWART
N277 1957	72393	RAFVR	1st Clasp; wef 25/10/53	STEWART
N473 1944	70651	RAFVR		STEWART
N30 1949	846409	RAuxAF		STEWART
N232 1948	771202	RAFVR		STEWART
N406 1950	241	WAAF	MBE	STEWART
N1638 1942	90107	AAF		STEWART
N198 1957	90107	RAuxAF	1st Clasp; wef 18/9/40	STEWART
N198 1957	90107	RAuxAF	2nd Clasp; wef 18/9/45	STEWART
N187 1947	72079	RAFVR		STEWART
N395 1953	157865	RAFVR		STEWART
N694 1948	844524	RAuxAF		STEWART
N475 1943	802420	AAF		STEWART
N513 1952	91275	RAuxAF	MB,ChB,FRFPS	STEWART
MOD 1965	73558	RAF	CB,CBE,AFC	STEWART
N82 1946	103500	RAFVR	DFC	STILES
N596 1944	801315	AAF		STILES
N187 1947	126839	RAFVR	DFC,DFM	STILLWELL
N1 1950	853759	RAuxAF		STINSON
N395 1948	870000	RAuxAF		STIRK
N375 1960	807218	AAF	wef 28/6/43	STOBART
N1 1950	813181	RAuxAF		STOCK
N673 1949	90960	RAuxAF		STOCK
N513 1952	190897	RAFVR		STOCK
N596 1944	90006	AAF		STOCKDALE
N573 1948	72243	RAFVR		STOCKEN
N673 1949	842821	RAuxAF		STOCKER
N258 1958	810087	AAF	wef 14/7/43	STOCKS
N94 1950	159825	RAFVR		STOCKS
N460 1947	146821	RAFVR		STOCKS
N179 1950	857408	RAuxAF		STOCKTON
N289 1947	131727	RAFVR		STOCKWELL
N1270 1950	752872	RAFVR		STODDART
N1014 1943	90358	AAF		STODDART
N915 1952	171754	RAFVR		STODHART
N257 1950	744556	RAFVR		STOKES
N87 1948	73413	RAFVR		STOKES
N850 1943	87031	RAFVR		STOKES
N452 1958	2655404	WRAAF	wef 11/10/57	STOKES

Surname	Initials				Rank	Status	Date
STOKOE	J				SQD LDR	Acting	23/05/1946
STONE	A				SGT		09/12/1959
STONE	C	J			FO		23/02/1950
STONE	C	P			FL		05/01/1950
STONE	F	E			CPL		07/07/1949
STONE	G	S			FL		20/04/1944
STONE	K	E	T	V	SAC		27/11/1952
STONE	K	T			FL		07/08/1947
STONE	P	D			FL		20/05/1948
STONEBRIDGE	B	B			FL		05/01/1950
STONESTREET	J	T			FL		03/07/1952
STOPHER	W	G			SGT		16/10/1947
STOREY	A	T			CPL		15/07/1948
STOREY	C	B			FL		01/01/1998
STOREY	R	T			FO		21/12/1950
STOREY	R	T			FL		25/05/1960
STOREY	R	W			WO		23/05/1946
STORRAR	J	E			SQD LDR		19/09/1956
STORRY	G	D			WO		01/07/1948
STORRY	G	D			WO		19/10/1960
STOTT	A				CPL		06/03/1947
STOTT	F				FL		24/01/1946
STOTT	G	M			SGT		01/07/1948
STOTT	M	R			FO		01/07/1948
STOTTER	L	W			WING COMM		12/06/1947
STOW	F	G			CPL		10/02/1949
STRADLING	P				SQD LDR	Acting	06/03/1947
STRAIGHT	W	W			AIR COMMD	Acting	13/12/1945
STRAIN	J				FLT SGT		22/04/1943
STRATHDOE	R	W			CPL		16/03/1950
STRAW	C	W			FL		16/01/1957
STREET	D	E			SQD LDR	Acting	08/11/1945
STRETCH	R	R			WO		15/07/1948
STRETTON	H				FL		22/10/1958
STRINGER	F	R	H		AC1		02/04/1953
STRONG	F	M			FL		10/06/1948
STRONG	F				FO		27/11/1952
STRONG	R	C			FL		12/07/1945
STRONGE	R	J			CPL		18/12/1963
STROUD	G	A			FL		24/01/1946
STROUD	G	A			FL		15/01/1953
STROWBRIDGE	D	W			CPL		01/07/1948
STRUTHERS	C	B			LAC		20/08/1958
STRUTHERS	J				FO		18/05/1944
STUART	C	F			CPL		07/02/1952
STUART	D				SAC		30/11/1993
STUART	G	H			CPL		25/05/1950
STUART	G	V			CPL		21/12/1950
STUART	J	A	G		WING COMM	Acting	27/03/1947
STUART-SMITH	L	P			FL		06/01/1955
STUBBS	D	R			WING COMM	Acting	24/01/1946
STUBBS	J	F	O		FL		07/11/1960
STUBBS	J				SGT		30/03/1960
STURMAN	E				WING COMM	Acting	07/08/1947
STURMEY	F	W			SQD LDR	Acting	30/09/1948
STUTELY	C	G			FO		03/10/1946
STYCHE	R	C			PO		08/11/1945
SUCH	E	B			FLT SGT		03/07/1952
SUDDABY	J	L			FL		09/12/1959
SUDWORTH	A	G			SQD LDR		15/07/1948

Source	Service No	Force	Notes	Surname
N468 1946	60512	RAFVR	DFC	STOKOE
N843 1959	2674953	WRAFVR	wef 15/10/59	STONE
N179 1950	171586	RAFVR		STONE
N1 1950	108961	RAFVR	DFM	STONE
N673 1949	848322	RAuxAF		STONE
N363 1944	108574	RAFVR		STONE
N915 1952	2680611	RAuxAF		STONE
N662 1947	85032	RAFVR		STONE
N395 1948	67525	RAFVR		STONE
N1 1950	78114	RAFVR	MBE	STONEBRIDGE
N513 1952	91239	RAuxAF		STONESTREET
N846 1947	743290	RAFVR		STOPHER
N573 1948	819122	RAuxAF		STOREY
YB 1999	212212	RAuxAF		STOREY
N1270 1950	138574	RAFVR		STOREY
N375 1960	138574	RAFVR	1st Clasp; wef 1/9/59	STOREY
N468 1946	759022	RAFVR	DFM	STOREY
N666 1956	41881	RAuxAF	DFC,AFC	STORRAR
N527 1948	810009	RAuxAF		STORRY
N775 1960	2697037	RAFVR	1st Clasp; wef 8/9/59	STORRY
N187 1947	743899	RAFVR		STOTT
N82 1946	142143	RAFVR	DFM	STOTT
N527 1948	749469	RAFVR		STOTT
N527 1948	170801	RAFVR	DCM	STOTT
N478 1947	72225	RAFVR	OBE	STOTTER
N123 1949	841796	RAuxAF		STOW
N187 1947	109292	RAFVR		STRADLING
N1355 1945	90680	AAF	CBE,MC,DFC	STRAIGHT
N475 1943	740003	RAFVR		STRAIN
N257 1950	817063	RAuxAF		STRATHDOE
N24 1957	161124	RAFVR		STRAW
N1232 1945	79557	RAFVR	DFC	STREET
N573 1948	747953	RAFVR		STRETCH
N716 1958	193479	RAFVR	wef 27/12/55	STRETTON
N298 1953	760312	RAFVR		STRINGER
N456 1948	66655	RAFVR		STRONG
N915 1952	110528	RAFVR		STRONG
N759 1945	108242	RAFVR		STRONG
N917 1963	2650986	RAuxAF	wef 8/10/63	STRONGE
N82 1946	141736	RAFVR		STROUD
N28 1953	141736	RAFVR	1st Clasp	STROUD
N527 1948	747363	RAFVR		STROWBRIDGE
N559 1958	803591	AAF	wef 6/11/43	STRUTHERS
N473 1944	143643	RAFVR		STRUTHERS
N103 1952	2684013	RAuxAF		STUART
YB 1996	n/a	RAuxAF		STUART
N517 1950	815058	RAuxAF		STUART
N1270 1950	752873	RAFVR		STUART
N250 1947	73415	RAFVR		STUART
N1 1955	147588	RAFVR		STUART-SMITH
N82 1946	87017	RAFVR	DSO,DFC	STUBBS
MOD 1964	147046	RAFVR		STUBBS
N229 1960	2655466	WRAAF	wef 11/12/59	STUBBS
N662 1947	72950	RAFVR	OBE	STURMAN
N805 1948	101517	RAFVR		STURMEY
N837 1946	198288	RAFVR		STUTELY
N1232 1945	191057	RAFVR		STYCHE
N513 1952	843776	RAuxAF		SUCH
N843 1959	122517	RAFVR	wef 24/4/54	SUDDABY
N573 1948	90487	RAuxAF		SUDWORTH

Surname	Initials			Rank	Status	Date
SULLIVAN	R	P	P	CPL		20/08/1958
SULLIVAN	W	P		CPL		25/05/1960
SUMMERELL	B			CPL		07/07/1949
SUMMERS	A	R	G	LAC		20/01/1960
SUMMERS	L	M		FO		15/09/1949
SUMMERS	R	N		CPL		27/03/1947
SUMNER	G	S		FL		01/08/1946
SUNDERLAND	G	P		CPL		15/07/1948
SUNDERLAND	J			FLT SGT		28/03/1997
SURGEON	W			FL		07/12/1960
SUTER	H	E		SQD LDR		11/01/1945
SUTHERLAND	F	G		CPL	Acting	06/01/1955
SUTHERLAND	G			FO		24/01/1946
SUTHERLAND	J	T		FL		23/05/1946
SUTHERLAND	L	A		SQD LDR		03/12/1942
SUTHERLAND	T	L		LAC		28/09/1950
SUTHERLAND	W	S		LAC		25/05/1960
SUTTON	E	W	A	FL		01/08/1946
SUTTON	H	I		FL		08/11/1945
SUTTON	J	N		WO		24/01/1946
SUTTON	T			FO		26/01/1950
SUTTON	T			FL		25/05/1960
SWABEY	F	P		LAC		12/06/1947
SWAFFIN	J	P		FL		14/10/1943
SWAINSON	W	N		FO		26/11/1953
SWAN	A	D		FO		22/10/1958
SWAN	F	C		CPL		22/10/1958
SWAN	F	C		CPL		08/07/1959
SWAN	F	W		LAC		03/06/1959
SWAN	R			CPL		02/05/1962
SWANN	L	J		SQD LDR		27/03/1947
SWANSON	A			FO		08/07/1959
SWANSON	G			CPL		07/07/1949
SWANSON	W	F		CPL		30/09/1948
SWANSTON	J			FL		25/05/1960
SWANWICK	G	W		FL	Acting	02/09/1943
SWANWICK	G	W		SQD LDR		16/01/1957
SWATTON	L	E		WING OFF	Acting	17/01/1952
SWAYNE	R	C		SQD LDR	Acting	12/06/1947
SWEENEY	C	A		SAC		23/04/1993
SWEETING	R	F		SQD LDR	Acting	03/10/1946
SWIFT	J			FO		06/01/1955
SWIFT	J			FL		18/01/1963
SWIFT	R	H		CPL		01/07/1948
SWINBURNE	S	M		SQD OFF	Acting	18/03/1948
SWINDELL	W	S		CPL	Temporary	20/08/1958
SWINDELLS	G	H		SGT		05/06/1947
SWINDELLS	R			LAC		07/08/1947
SWINNERTON	J	W		SGT		04/09/1947
SWORDER	C	N		FL		24/01/1946
SYKENS	W	F		MST PIL		26/11/1953
SYKES	G	D		FL		16/10/1952
SYKES	G	F		SGT		16/04/1958
SYME	M	S		SGT		08/07/1959
SYMMONS	F			FL		13/12/1945
SYMONDS	J	E		SQD LDR	Acting	01/08/1946
SYMONDS	M	L		SQD LDR		02/12/1992
SYNDERCOMBE	H	A		FL		18/12/1963
SYRETT	L			FL		13/12/1945
TACKLEY	R	E		LAC		10/05/1951

Source	Service No	Force	Notes	Surname
N559 1958	2689157	RAF	wef 23/5/54	SULLIVAN
N375 1960	799833	RAFVR	wef 28/5/44	SULLIVAN
N673 1949	860450	RAuxAF		SUMMERELL
N33 1960	865094	AAF	wef 12/7/44	SUMMERS
N953 1949	178544	RAFVR	DFC	SUMMERS
N250 1947	747424	RAFVR		SUMMERS
N652 1946	85641	RAFVR		SUMNER
N573 1948	809205	RAuxAF		SUNDERLAND
YB 1997	n/a	RAuxAF		SUNDERLAND
N933 1960	194377	RAuxAF	wef 4/3/60	SURGEON
N14 1945	90912	AAF		SUTER
N1 1955	2692002	RAuxAF		SUTHERLAND
N82 1946	172983	RAFVR	DFC	SUTHERLAND
N468 1946	169148	RAFVR		SUTHERLAND
N1638 1942	90174	AAF		SUTHERLAND
N968 1950	874715	RAuxAF		SUTHERLAND
N375 1960	2684141	RAuxAF	wef 12/3/60	SUTHERLAND
N652 1946	178354	RAFVR		SUTTON
N1232 1945	111099	RAFVR		SUTTON
N82 1946	746728	RAFVR		SUTTON
N94 1950	172158	RAFVR		SUTTON
N375 1960	172158	RAuxAF	1st Clasp; wef 15/5/57	SUTTON
N478 1947	800571	AAF		SWABEY
N1065 1943	88019	RAFVR		SWAFFIN
N952 1953	2686030	RAuxAF		SWAINSON
N716 1958	2605049	RAFVR	wef 16/12/56	SWAN
N716 1958	2684611	RAuxAF	wef 28/6/44	SWAN
N454 1959	2684611	RAuxAF	1st Clasp; wef 24/2/59	SWAN
N368 1959	2693527	AAF	wef 28/9/43	SWAN
N314 1962	2659516	WRAAF	wef 2/3/62	SWAN
N250 1947	72951	RAFVR		SWANN
N454 1959	2693531	RAuxAF	wef 2/5/59	SWANSON
N673 1949	809207	RAuxAF		SWANSON
N805 1948	857431	RAuxAF		SWANSON
N375 1960	194410	RAFVR	wef 6/7/59	SWANSTON
N918 1943	118533	RAFVR		SWANWICK
N24 1957	118533	RAF	1st Clasp	SWANWICK
N35 1952	11	WAAF		SWATTON
N478 1947	84822	RAFVR	MBE	SWAYNE
YB 1996	n/a	RAuxAF		SWEENEY
N837 1946	67027	RAFVR	DFC	SWEETING
N1 1955	2603539	RAFVR	DFC	SWIFT
MOD 1966	2603539	RAFVR	DFC; 1st Clasp	SWIFT
N527 1948	851272	RAuxAF		SWIFT
N232 1948	855	WAAF		SWINBURNE
N559 1958	2605115	RAFVR	wef 17/2/54	SWINDELL
N460 1947	811020	AAF		SWINDELLS
N662 1947	811232	AAF		SWINDELLS
N740 1947	810006	AAF		SWINNERTON
N82 1946	84301	RAFVR		SWORDER
N952 1953	2686013	RAuxAF		SYKENS
N798 1952	51061	RAFVR		SYKES
N258 1958	2688581	RAFVR	wef 21/2/58	SYKES
N454 1959	2659181	WRAAF	wef 4/11/58	SYME
N1355 1945	61469	RAFVR	DFC,DFM	SYMMONS
N652 1946	103539	RAFVR		SYMONDS
YB 1996	2631438	RAuxAF		SYMONDS
N917 1963	185639	RAFVR	wef 15/1/59	SYNDERCOMBE
N1355 1945	156924	RAFVR		SYRETT
N466 1951	750313	RAFVR		TACKLEY

Surname	Initials				Rank	Status	Date
TAIT	L	J	B		LAC		18/03/1948
TAIT	R	L			FL	Acting	01/10/1953
TALTON	S	J	S		FL		01/01/2002
TAMBLIN	D	V			FL		13/02/1957
TAMBLIN	D	V			FL		26/05/1966
TAMBLYN-WATTS	T	M			FLT SGT		01/10/1953
TANGYE	N	T			WING COMM	Acting	10/12/1942
TANNER	E	F			WING COMM	Acting	05/06/1947
TANNER	P	H			WING COMM		23/02/1950
TANSLEY	A	L			SGT		13/01/1949
TANSLEY	F	A			CPL		16/03/1950
TAPHOUSE	B	M			SEC OFF		12/06/1947
TAPLIN	P				SQD LDR	Acting	04/05/1944
TAPPIN	H	E			SQD LDR	Acting	17/02/1944
TAPPIN	H	G	L		SGT		30/09/1948
TARGETT	H				FO		26/04/1951
TARLTON	H	C			FL		16/03/1950
TARPLEE	L	E			FL		01/03/1945
TARPLETT	E	W			FL		07/07/1949
TARRANT	H	H			WING COMM	Acting	06/03/1947
TARRANT	K	W	J		SQD LDR	Acting	06/03/1947
TARRATT	L	H			FL	Acting	02/08/1951
TATHAM	A	L			CPL		12/06/1947
TATMAN	A	C			FL		07/04/1955
TATTERSALL	A	E			FL	Acting	22/02/1951
TATTON	B				FLT SGT		11/11/1943
TAVENNER	A				FO		13/10/1955
TAYLOR	A	F			SGT	Acting	10/04/1952
TAYLOR	A	P			SQD LDR		17/06/1954
TAYLOR	A				CPL		22/02/1951
TAYLOR	B	A			WING COMM		01/07/1948
TAYLOR	B	E	G	W	FL		21/01/1959
TAYLOR	C				FL	Acting	12/07/1945
TAYLOR	D	F			CPL		01/05/1952
TAYLOR	D	J			FL		28/02/1946
TAYLOR	D	P			FL		21/03/1946
TAYLOR	E	F			FL		23/02/1950
TAYLOR	E	M	R		FLT SGT		20/08/1958
TAYLOR	E				SQD LDR		06/03/1947
TAYLOR	F	C	H		FO		10/12/1942
TAYLOR	F	C	H		FL		14/01/1954
TAYLOR	F	H			FL		16/10/1947
TAYLOR	F	J			CPL		10/02/1949
TAYLOR	F	J	O		SQD LDR		30/03/1960
TAYLOR	G	A			FO		18/04/1946
TAYLOR	G	E			SGT		04/11/1948
TAYLOR	G	N			FL		24/01/1946
TAYLOR	H	P			FL		07/08/1947
TAYLOR	H	W	H		FO		06/03/1947
TAYLOR	J	B			WING COMM	Acting	27/03/1947
TAYLOR	J	C			FO		15/07/1948
TAYLOR	J	F			FO		20/09/1945
TAYLOR	J	F			SQD LDR	Acting	16/03/1950
TAYLOR	J	L			FLT SGT		10/05/1945
TAYLOR	J	T	R		FL		05/09/1946
TAYLOR	J				SGT		07/08/1947
TAYLOR	J				FO		26/08/1948
TAYLOR	L	M			FL		28/11/1956
TAYLOR	L	V			FO		13/07/1950
TAYLOR	M	E	U		FLT OFF		10/08/1950

Source	Service No	Force	Notes	Surname
N232 1948	844072	RAuxAF		TAIT
N796 1953	143768	RAuxAF		TAIT
YB 2003	213209	RAuxAF		TALTON
N104 1957	140116	RAFVR	wef 26/5/56	TAMBLIN
MOD 1967	140116	RAFVR	1st Clasp	TAMBLIN
N796 1953	2603558	RAFVR		TAMBLYN-WATTS
N1680 1942	70661	RAFVR		TANGYE
N460 1947	90848	AAF		TANNER
N179 1950	90464	RAuxAF	MBE,MC,TD	TANNER
N30 1949	749039	RAFVR		TANSLEY
N257 1950	843970	RAuxAF		TANSLEY
N478 1947	3879	WAAF		TAPHOUSE
N425 1944	102101	RAFVR		TAPLIN
N132 1944	89304	RAFVR	DFC	TAPPIN
N805 1948	843503	RAuxAF		TAPPIN
N402 1951	158464	RAFVR		TARGETT
N257 1950	125895	RAFVR		TARLTON
N225 1945	85252	RAFVR		TARPLEE
N673 1949	78137	RAFVR		TARPLETT
N187 1947	91076	AAF		TARRANT
N187 1947	83732	RAFVR		TARRANT
N777 1951	108443	RAFVR		TARRATT
N478 1947	811167	AAF		TATHAM
N277 1955	149006	RAFVR		TATMAN
N194 1951	131736	RAFVR		TATTERSALL
N1177 1943	805154	AAF	MM	TATTON
N762 1955	2692502	RAuxAF		TAVENNER
N278 1952	2683320	RAuxAF		TAYLOR
N471 1954	170068	RAuxAF		TAYLOR
N194 1951	858762	RAuxAF		TAYLOR
N527 1948	72325	RAFVR		TAYLOR
N28 1959	148364	RAFVR	wef 4/2/56	TAYLOR
N759 1945	182540	RAFVR		TAYLOR
N336 1952	799738	RAFVR		TAYLOR
N197 1946	142219	RAFVR		TAYLOR
N263 1946	145913	RAFVR	DFC	TAYLOR
N179 1950	90889	RAuxAF		TAYLOR
N559 1958	881953	WAAF	wef 7/8/44	TAYLOR
N187 1947	72376	RAFVR		TAYLOR
N1680 1942	111531	RAFVR		TAYLOR
N18 1954	111531	RAFVR	1st Clasp	TAYLOR
N846 1947	174057	RAFVR		TAYLOR
N123 1949	813096	RAuxAF		TAYLOR
N229 1960	73419	RAF	wef 29/6/44	TAYLOR
N358 1946	198218	RAFVR		TAYLOR
N934 1948	814163	RAuxAF		TAYLOR
N82 1946	61013	RAFVR	AFC	TAYLOR
N662 1947	149563	RAFVR		TAYLOR
N187 1947	161512	RAFVR		TAYLOR
N250 1947	72177	RAFVR		TAYLOR
N573 1948	137978	RAFVR		TAYLOR
N1036 1945	174904	RAFVR		TAYLOR
N257 1950	78122	RAFVR		TAYLOR
N485 1945	811003	AAF		TAYLOR
N751 1946	143232	RAFVR	DFC	TAYLOR
N662 1947	863283	AAF		TAYLOR
N694 1948	173111	RAFVR	DFM	TAYLOR
N833 1956	135197	RAuxAF		TAYLOR
N675 1950	112259	RAFVR		TAYLOR
N780 1950	346	WAAF		TAYLOR

Surname	Initials			Rank	Status	Date
TAYLOR	N	S		FO		30/09/1954
TAYLOR	N			FL		09/05/1946
TAYLOR	P	A		FL		16/03/2004
TAYLOR	P	E		FO		16/01/1957
TAYLOR	R	A		FL		09/02/1951
TAYLOR	R	A		FL		21/06/1961
TAYLOR	R	D		FL		21/03/1946
TAYLOR	R			FL		27/03/1947
TAYLOR	R			FO		07/12/1960
TAYLOR	S			FLT SGT		04/09/1947
TAYLOR	S			CPL	Acting	24/07/1952
TAYLOR	T	B		FL		09/05/1946
TAYLOR	T	B		FL	Acting	17/04/1947
TAYLOR	T	W		LAC		30/09/1948
TAYLOR	W	A		FL	Acting	05/01/1950
TAYLOR	W	C		LAC		06/01/1955
TAYLOR	W	T		LAC		05/01/1950
TEAGUE	P	B	P	CPL		04/11/1948
TEARLE	F	G		FL		21/06/1945
TEDDER	C	L		FL	Acting	10/12/1942
TEDDER	E	C		FL		22/04/1943
TEER	R	J		CPL		01/07/1948
TEMPLE	H	F	R	SQD LDR		07/07/1949
TEMPLE	H	F	R	FL		01/05/1957
TEMPLE	J			SGT		23/04/1993
TEMPLE	V	B		FL		30/10/1957
TEMPLE	W			SGT		15/09/1949
TEMPLEMAN	A			SGT		03/07/1952
TEMPLETON	F	A		CPL		07/06/1951
TENNANT	J	M		CPL		11/04/1995
TESORIERE	J	H	A	FO		06/03/1947
TESTER	R	W		SQD LDR	Acting	11/05/1944
TEVENDALE	T			SAC		06/01/1955
TEW	K	G		FL		26/08/1948
TEWSON	F	W		FLT SGT		10/02/1949
THACKER	F	W		WO		24/12/1942
THACKERAY	J			FL		23/02/1950
THACKERY	S	J		FL	Acting	30/03/1943
THEOBALD	L	G		FL		09/12/1959
THIRLWELL	T			FLT SGT	Acting	14/09/1950
THOM	A	H		FL		06/03/1947
THOM	W	G		FL		03/07/1957
THOMAS	A	C		FL		20/01/1960
THOMAS	A	E		SGT		17/03/1949
THOMAS	A	H		FL		20/03/1957
THOMAS	A	T		LAC		30/09/1948
THOMAS	C	M		SGT		19/07/1996
THOMAS	D	J		FO		02/07/1944
THOMAS	D	O		FL		27/04/1950
THOMAS	E	T	M	FL		26/04/1961
THOMAS	F	V		FL		13/12/1945
THOMAS	G	G		SGT		05/06/1947
THOMAS	G	H		FO		19/01/1956
THOMAS	G	S		WO		30/03/1944
THOMAS	H	D		CPL		24/10/1997
THOMAS	H	E		FL		13/12/1961
THOMAS	I	C		FL	Acting	15/07/1948
THOMAS	J	E		WO		28/02/1946
THOMAS	J	E		FO		23/06/1955
THOMAS	J	J		WO	Acting	26/08/1948

Source	Service No	Force	Notes	Surname
N799 1954	121824	RAFVR		TAYLOR
N415 1946	101500	RAFVR	DFC	TAYLOR
LG 2004	0091536R	RAuxAF		TAYLOR
N24 1957	172007	RAFVR		TAYLOR
N145 1951	79701	RAFVR		TAYLOR
N478 1961	79701	RAFVR	1st Clasp; wef 21/12/59	TAYLOR
N263 1946	74341	RAFVR		TAYLOR
N250 1947	82733	RAFVR		TAYLOR
N933 1960	2603337	RAFVR	wef 5/9/53	TAYLOR
N740 1947	750128	RAFVR		TAYLOR
N569 1952	2688531	RAuxAF		TAYLOR
N415 1946	128412	RAFVR		TAYLOR
N289 1947	172677	RAFVR		TAYLOR
N805 1948	847287	RAuxAF		TAYLOR
N1 1950	141259	RAFVR		TAYLOR
N1 1955	840094	RAuxAF		TAYLOR
N1 1950	849908	RAuxAF		TAYLOR
N934 1948	840552	RAuxAF		TEAGUE
N671 1945	123198	RAFVR	DFC	TEARLE
N1680 1942	61338	RAFVR		TEDDER
N475 1943	82675	RAFVR		TEDDER
N527 1948	812116	RAuxAF		TEER
N673 1949	72420	RAFVR		TEMPLE
N277 1957	72420	RAFVR	1st Clasp; wef 13/6/56	TEMPLE
YB 1996	n/a	RAuxAF		TEMPLE
N775 1957	177848	RAFVR	DFC; wef 29/3/57	TEMPLE
N953 1949	843244	RAuxAF		TEMPLE
N513 1952	804166	RAuxAF		TEMPLEMAN
N560 1951	856657	RAuxAF		TEMPLETON
YB 1997	n/a	RAuxAF		TENNANT
N187 1947	146146	RAFVR		TESORIERE
N444 1944	78983	RAFVR		TESTER
N1 1955	2605788	RAFVR		TEVENDALE
N694 1948	84692	RAFVR	DFC	TEW
N123 1949	748035	RAFVR		TEWSON
N1758 1942	805121	AAF		THACKER
N179 1950	86561	RAFVR		THACKERAY
N1014 1943	111982	RAFVR		THACKERY
N843 1959	197247	RAFVR	wef 11/11/52	THEOBALD
N926 1950	807246	RAuxAF		THIRLWELL
N187 1947	114075	RAFVR	DFC	THOM
N456 1957	187289	RAuxAF	wef 5/1/57	THOM
N33 1960	183077	RAFVR	wef 14/12/59	THOMAS
N231 1949	860137	RAuxAF		THOMAS
N198 1957	188296	RAuxAF	wef 25/2/54	THOMAS
N805 1948	860383	RAuxAF		THOMAS
YB 1997	n/a	RAuxAF		THOMAS
MOD 1964	161926	RAFVR		THOMAS
N406 1950	72729	RAFVR		THOMAS
N339 1961	156228	RAFVR	wef 10/7/60	THOMAS
N1355 1945	80845	RAFVR		THOMAS
N460 1947	813210	AAF		THOMAS
N43 1956	2692501	RAuxAF		THOMAS
N281 1944	804256	AAF		THOMAS
YB 1997	n/a	RAuxAF		THOMAS
N987 1961	142777	RAFVR	wef 5/11/61	THOMAS
N573 1948	146196	RAFVR		THOMAS
N197 1946	751969	RAFVR		THOMAS
N473 1955	2601772	RAuxAF	1st Clasp	THOMAS
N694 1948	808191	RAuxAF		THOMAS

Surname	Initials			Rank	Status	Date
THOMAS	J	S		SQD LDR		15/09/1949
THOMAS	J	W		FL		06/01/1955
THOMAS	J	W		FL		13/02/1957
THOMAS	L	C		FL	Acting	04/02/1943
THOMAS	L			FO		30/09/1948
THOMAS	N	F		SGT		05/06/1947
THOMAS	P	E	H	FL		06/03/1947
THOMAS	R	A	B	FL		07/07/1949
THOMAS	R	R	J	FL		18/04/1946
THOMPSON	A	J		FL		09/12/1959
THOMPSON	A	M		GRP OFF		08/05/1947
THOMPSON	A	P		FO		09/01/1963
THOMPSON	A			CPL	Acting	03/06/1959
THOMPSON	A			CPL		22/10/1994
THOMPSON	B	V		FL		15/01/1958
THOMPSON	C	B		SQD LDR		22/04/1943
THOMPSON	C	W		SQD LDR		07/08/1947
THOMPSON	C			FO		07/04/1955
THOMPSON	D	S		FL	Acting	07/04/1955
THOMPSON	E	A		FL		17/06/1954
THOMPSON	E	M		FLT OFF	Acting	07/12/1950
THOMPSON	E			FL		08/11/1945
THOMPSON	E			FLT SGT		01/07/1948
THOMPSON	F	N		FL		13/12/1945
THOMPSON	G	G		FL		06/03/1947
THOMPSON	G	V		SGT		30/09/1948
THOMPSON	G			FO		28/10/1943
THOMPSON	H	D		SQD LDR	Acting	26/01/1950
THOMPSON	J	B	J	FL		26/10/1944
THOMPSON	J	C		FLT SGT		23/02/1950
THOMPSON	J	E	P	FL		06/03/1947
THOMPSON	J	E		FL		07/07/1949
THOMPSON	J	H	G	LAC		23/02/1950
THOMPSON	J	M	M	FO		07/06/1945
THOMPSON	J	McM		FLT SGT		01/10/1953
THOMPSON	J	R		FO		24/07/1952
THOMPSON	J			FL		25/06/1953
THOMPSON	L	C		SQD LDR	Acting	06/03/1947
THOMPSON	M	O		FO		08/05/1947
THOMPSON	M	P		WING COMM	Acting	18/04/1946
THOMPSON	P	D		SQD LDR	Acting	24/01/1946
THOMPSON	R	G		SGT		13/01/1949
THOMPSON	R	J	W	LAC		05/06/1947
THOMPSON	R	R		SQD LDR		31/07/1944
THOMPSON	R	T		PO		17/04/1947
THOMPSON	R	W		CPL		25/01/1951
THOMPSON	R	W		SGT		26/04/1961
THOMPSON	S	G		SQD LDR		07/07/1949
THOMPSON	S			LAC		15/09/1949
THOMPSON	T	H		LAC		10/06/1948
THOMPSON	W	B		SQD LDR		03/12/1942
THOMPSON	W	B		WING OFF	Acting	27/03/1947
THOMSETT	A	E	L	CPL		15/07/1948
THOMSON	F			FLT SGT		07/06/1951
THOMSON	H	R		FL	Acting	17/03/1949
THOMSON	H			WO		10/08/1944
THOMSON	H			CPL		24/11/1949
THOMSON	J	L		LAC		07/12/1960
THOMSON	J			CPL		12/06/1947
THOMSON	P	H		WING COMM		12/06/1947

Source	Service No	Force	Notes	Surname
N953 1949	72307	RAFVR		THOMAS
N1 1955	120795	RAuxAF		THOMAS
N104 1957	140420	RAFVR	wef 1/2/56	THOMAS
N131 1943	108177	RAFVR		THOMAS
N805 1948	121810	RAFVR		THOMAS
N460 1947	811012	AAF		THOMAS
N187 1947	60297	RAFVR		THOMAS
N673 1949	85954	RAFVR		THOMAS
N358 1946	132334	RAFVR		THOMAS
N843 1959	167844	RAFVR	wef 23/2/57	THOMPSON
N358 1947	192	WAAF		THOMPSON
N22 1963	744744	RAFVR(T)	wef 1/6/44	THOMPSON
N368 1959	2693529	RAuxAF	wef 21/3/59	THOMPSON
YB 1996	n/a	RAuxAF		THOMPSON
N40 1958	187347	RAFVR	wef 27/5/52	THOMPSON
N475 1943	90244	AAF		THOMPSON
N662 1947	73427	RAFVR		THOMPSON
N277 1955	165345	RAFVR		THOMPSON
N277 1955	161511	RAFVR		THOMPSON
N471 1954	181175	RAFVR		THOMPSON
N1210 1950	1120	WAAF		THOMPSON
N1232 1945	86671	RAFVR		THOMPSON
N527 1948	874510	RAuxAF		THOMPSON
N1355 1945	78258	RAFVR		THOMPSON
N187 1947	100982	RAFVR	DFM	THOMPSON
N805 1948	805441	RAuxAF		THOMPSON
N1118 1943	115906	RAFVR		THOMPSON
N94 1950	91084	RAuxAF	MC	THOMPSON
N1094 1944	126008	RAFVR		THOMPSON
N179 1950	812019	RAuxAF		THOMPSON
N187 1947	77794	RAFVR	AFC	THOMPSON
N673 1949	78536	RAFVR	MBE	THOMPSON
N179 1950	845565	RAuxAF		THOMPSON
N608 1945	161679	RAFVR		THOMPSON
N796 1953	803525	RAuxAF		THOMPSON
N569 1952	110395	RAuxAF		THOMPSON
N527 1953	177947	RAFVR		THOMPSON
N187 1947	110009	RAFVR		THOMPSON
N358 1947	203031	RAFVR		THOMPSON
N358 1946	77274	RAFVR		THOMPSON
N82 1946	84697	RAFVR	DFC	THOMPSON
N30 1949	770403	RAFVR		THOMPSON
N460 1947	747119	RAFVR		THOMPSON
MOD 1966	73428	RAFVR		THOMPSON
N289 1947	198137	RAFVR		THOMPSON
N93 1951	808406	RAuxAF		THOMPSON
N339 1961	2696537	RAFVR	1st Clasp; wef 21/8/59	THOMPSON
N673 1949	84595	RAFVR		THOMPSON
N953 1949	870202	RAuxAF		THOMPSON
N456 1948	808302	RAuxAF		THOMPSON
N1638 1942	70673	RAFVR		THOMPSON
N250 1947	270	WAAF		THOMPSON
N573 1948	750676	RAFVR		THOMSETT
N560 1951	809079	RAuxAF		THOMSON
N231 1949	86746	RAFVR		THOMSON
N812 1944	816038	AAF		THOMSON
N1193 1949	817072	RAuxAF		THOMSON
N933 1960	2676234	RAuxAF	wef 19/8/58	THOMSON
N478 1947	816167	AAF		THOMSON
N478 1947	72244	RAFVR	OBE	THOMSON

Surname	Initials			Rank	Status	Date
THOMSON	R	B		WING COMM	Acting	11/05/1944
THOMSON	W	R		SQD LDR		17/04/1947
THORN	L	E	A	FL		20/11/1957
THORNABY	G	T		CPL	Temporary	19/08/1959
THORNE	H			WO		15/07/1948
THORNE	K	L		SQD LDR		20/05/1948
THORNE	P	R		CPL		30/09/1948
THORNLEY	H	E		FO		28/02/1946
THORNTON	F	K		FL		24/01/1946
THOROGOOD	L	A		SQD LDR	Acting	04/10/1945
THOROGOOD	L	A		FL		24/07/1952
THORPE	A			FL		13/01/1949
THORPE-TRACEY	R	S	J	SGT		30/09/1948
THURRELL	C	C		SQD LDR		30/03/1943
THWAITE	J	R		LAC		05/06/1947
THYNNE	B	S		GRP CAP		03/12/1942
TICE	J	E		CPL		05/06/1947
TICKLE	G			LAC		23/06/1955
TIDMARSH	E	A		FO		15/09/1949
TIERNEY	D	J		WO		10/02/1949
TILEY	H	S		CPL		23/02/1950
TILEY	W	M	S	CPL		25/05/1950
TILLEY	J	N	G	FL		11/07/1946
TILSTON	J	E		FL		27/07/1944
TILSTON	J	E		FL		26/11/1953
TILSTON	J			FL		25/06/1953
TIMINGS	B	A		FL		10/05/1945
TIMMINS	D	L		FL		10/06/1948
TIMMINS	E			LAC		27/03/1963
TIMMS	A	S		FL		13/12/1945
TIMMS	A	T		SGT		13/12/1951
TIMMS	K	G		WO		05/10/1997
TINGLE	D			FL		06/03/1990
TINGLE	D			FL		06/03/2000
TINKLER	A			SGT		24/11/1949
TIPPETT	E	J		SQD LDR		01/07/1948
TIPPLE	E	W		FL		30/08/1945
TISDALE	H			CPL		04/09/1947
TITLER	E	A		CPL		30/09/1948
TITLEY	E	A		SQD LDR		06/03/1947
TITLOW	M	R		PO		12/08/1943
TITMAN	R	A		FLT SGT		12/04/1945
TITMAN	T	H		FLT SGT		12/06/1947
TIVNEY	E			FO		08/03/1956
TODD	A	G		SQD LDR	Acting	06/03/1947
TODD	D	J	A	FL		04/05/1968
TODD	E	W		FO		25/01/1945
TODD	J	M		WING COMM		21/09/1944
TODD	J	P		FL		01/05/1957
TODD	J	W		SGT		20/05/1948
TODD	R	D		FL		01/03/1945
TODD	R	G		FL		03/06/1959
TODD	R	G		FL		19/08/1959
TODD	W			SGT		12/04/1945
TODMAN	C	F		LAC		01/07/1948
TOFT	R	W		SQD LDR		09/11/1944
TOFTS	R	A		FO		21/03/1946
TOLLERTON	D			SQD LDR		17/04/1947
TOMKINS	L	M		FL		18/05/1944
TOMLIN	E	J		SAC		30/03/1960

Source	Service No	Force	Notes	Surname
N444 1944	90370	AAF	DSO,DFC	THOMSON
N289 1947	73431	RAFVR		THOMSON
N835 1957	131543	RAFVR	wef 10/1/54	THORN
N552 1959	812211	AAF	wef 21/5/44	THORNABY
N573 1948	755793	RAFVR		THORNE
N395 1948	81948	RAFVR		THORNE
N805 1948	842500	RAuxAF		THORNE
N197 1946	195551	RAFVR		THORNLEY
N82 1946	86676	RAFVR		THORNTON
N1089 1945	107939	RAFVR		THOROGOOD
N569 1952	107939	RAFVR	DFC; 1st Clasp	THOROGOOD
N30 1949	131044	RAFVR		THORPE
N805 1948	770554	RAFVR		THORPE-TRACEY
N1014 1943	70677	RAFVR		THURRELL
N460 1947	811233	AAF		THWAITE
N1638 1942	90119	AAF	AFC	THYNNE
N460 1947	861935	AAF		TICE
N473 1955	856661	RAuxAF		TICKLE
N953 1949	101236	RAFVR		TIDMARSH
N123 1949	805466	RAuxAF		TIERNEY
N179 1950	813134	RAuxAF		TILEY
N517 1950	861030	RAuxAF		TILEY
N594 1946	124124	RAFVR	DFC	TILLEY
N763 1944	117402	RAFVR		TILSTON
N952 1953	117402	RAFVR	1st Clasp	TILSTON
N527 1953	174222	RAuxAF		TILSTON
N485 1945	129932	RAFVR		TIMINGS
N456 1948	134659	RAFVR		TIMMINS
N239 1963	856504	AAF	wef 18/6/44	TIMMINS
N1355 1945	124626	RAFVR		TIMMS
N1271 1951	818006	RAuxAF		TIMMS
YB 1999	n/a	RAuxAF		TIMMS
YB 1997	210299	RAuxAF		TINGLE
YB 2001	210299	RAuxAF	1st Clasp	TINGLE
N1193 1949	815189	RAuxAF		TINKLER
N527 1948	73435	RAFVR		TIPPETT
N961 1945	64277	RAFVR		TIPPLE
N740 1947	849201	AAF		TISDALE
N805 1948	840498	RAuxAF		TITLER
N187 1947	90271	AAF		TITLEY
N850 1943	144390	RAFVR		TITLOW
N381 1945	801351	AAF		TITMAN
N478 1947	749021	RAFVR		TITMAN
N187 1956	2650002	RAuxAF		TIVNEY
N187 1947	119873	RAFVR	DFC	TODD
MOD 1968	1603802	RAFVR		TODD
N70 1945	147140	RAFVR	DFC	TODD
N961 1944	90700	AAF		TODD
N277 1957	87080	RAFVR	wef 28/7/44	TODD
N395 1948	756735	RAFVR		TODD
N225 1945	138788	RAFVR		TODD
N368 1959	170240	RAFVR	wef 28/7/44	TODD
N552 1959	170240	RAFVR	1st Clasp; wef 12/6/59	TODD
N381 1945	808053	AAF		TODD
N527 1948	841341	RAuxAF		TODMAN
N1150 1944	90968	AAF		TOFT
N263 1946	174512	RAFVR		TOFTS
N289 1947	72138	RAFVR		TOLLERTON
N473 1944	118690	RAFVR		TOMKINS
N229 1960	2691556	RAuxAF	wef 26/11/59	TOMLIN

Surname	Initials			Rank	Status	Date
TOMLIN	P	A		CPL		01/07/1948
TOMLINSON	F	G		FLT SGT	Acting	07/07/1949
TOMLINSON	F			PO		02/07/1958
TOMLINSON	M			CPL		20/01/1960
TOMPKINS	B	C		SGT		09/11/1950
TONKIN	P	W		CPL		03/03/1994
TOOKE	E	E		FL		10/06/1948
TOOMBS	J	R		FO		24/01/1946
TOOP	K	J	F	FL		27/03/1947
TOOTELL	T			LAC		17/06/1954
TOOZE	R	J	W	FL		01/01/1990
TOPHAM	J			FL		08/05/1947
TOPPING	F	G		WO		05/06/1947
TOPSFIELD	E	F		FL		14/11/1962
TOSELAND	J	N		FL	Acting	24/11/1949
TOTTLE	T	E	A	CPL		23/02/1950
TOUCH	D	F		SQD LDR	Acting	04/10/1945
TOULMIN	J			SQD LDR		27/04/1950
TOVEY	H	R		SGT		04/09/1947
TOWE	S	E		SGT	Acting	07/02/1962
TOWERS	K			FLT SGT	Acting	19/01/1956
TOWERS	P	M	F	SGT		01/07/1948
TOWNEND	C	W		FLT SGT		15/11/1951
TOWNHILL	S	A	H	FO		27/11/1952
TOWNSEND	A	R		FO		19/10/1960
TOWNSEND	C	A		LAC		10/02/1949
TOWNSEND	E	J		FL		11/07/1946
TOWNSEND	F	H		FL		22/02/1961
TOWNSEND	H	G		CPL		01/07/1948
TOWNSEND	N	G		SQD LDR	Acting	03/10/1946
TOWNSEND	V	J		FL		01/10/1953
TOWNSEND	W	F	A	WO		21/10/1959
TOZER	J			FLT SGT	Acting	17/01/1952
TOZER	P	D		SGT		14/11/1998
TRACY	W	L		FL		23/06/1955
TRAVELL	W	R		FL		13/12/1945
TRAVIS	D	J		FL		02/07/1958
TREADGOLD	W	H		LAC		19/08/1954
TREANOR	H			SGT		07/08/1947
TREFUSIS-FORBES	K	J		AIR CHIEF COMM		06/03/1947
TREGUNNA	G	G		SAC		08/01/1994
TRETHEWAY	E	J		SGT		13/07/1950
TRETHEWAY	E	J		FLT SGT	Acting	28/11/1956
TREVELYAN	C	d'E		FL		18/11/1948
TREVOR	H			CPL		07/04/1955
TRICKETT	W	H		FL		30/09/1954
TRICKEY	R	S		FLT SGT		06/03/1947
TRICKLEBANK	F	A		FL		09/05/1946
TRIGGS	H	W	G	SQD LDR		10/06/1948
TRIM	R	F		SQD LDR	Acting	09/08/1945
TRIMBLE	W	J		CPL		29/04/1954
TRIMMER	R			SGT		01/06/1998
TRINDER	A	F		WING COMM		17/04/1947
TRIPP	R	C	H	SQD LDR		21/01/1959
TRIPTREE	F	J		FL		19/09/1956
TROTMAN	S	E		LAC		25/06/1953
TROUGHTON-ROBERTS	G	E		FL		06/03/1947
TROWBRIDGE	S	C		SGT		14/09/1950

Source	Service No	Force	Notes	Surname
N527 1948	842533	RAuxAF		TOMLIN
N673 1949	745828	RAFVR		TOMLINSON
N452 1958	2612035	RAFVR	wef 29/12/55	TOMLINSON
N33 1960	2691041	RAuxAF	wef 16/10/59	TOMLINSON
N1113 1950	804425	RAuxAF		TOMPKINS
YB 1996	n/a	RAuxAF		TONKIN
N456 1948	153729	RAFVR		TOOKE
N82 1946	179612	RAFVR		TOOMBS
N250 1947	78880	RAFVR		TOOP
N471 1954	857283	RAuxAF		TOOTELL
YB 1996	2626676	RAuxAF		TOOZE
N358 1947	116892	RAFVR		TOPHAM
N460 1947	800553	AAF		TOPPING
N837 1962	102766	RAFVR	wef 13/6/62	TOPSFIELD
N1193 1949	162628	RAFVR		TOSELAND
N179 1950	861180	RAuxAF		TOTTLE
N1089 1945	116513	RAFVR	AFC	TOUCH
N406 1950	90488	RAuxAF		TOULMIN
N740 1947	842462	AAF		TOVEY
N102 1962	2650934	RAuxAF	wef 8/1/62	TOWE
N43 1956	2676606	RAuxAF		TOWERS
N527 1948	817092	RAuxAF		TOWERS
N1185 1951	749460	RAFVR		TOWNEND
N915 1952	49912	RAFVR		TOWNHILL
N775 1960	178259	RAFVR	wef 23/6/44	TOWNSEND
N123 1949	755559	RAFVR		TOWNSEND
N594 1946	124177	RAFVR		TOWNSEND
N153 1961	180524	RAFVR	wef 15/3/60	TOWNSEND
N527 1948	755560	RAFVR		TOWNSEND
N837 1946	108958	RAFVR	DFC	TOWNSEND
N796 1953	185733	RAuxAF	DFC	TOWNSEND
N708 1959	2601124	RAFVR	wef 13/8/59	TOWNSEND
N35 1952	801502	RAuxAF		TOZER
YB 1999	n/a	RAuxAF		TOZER
N473 1955	105611	RAFVR		TRACY
N1355 1945	116983	RAFVR	DFC	TRAVELL
N452 1958	172834	RAFVR	wef 26/10/57	TRAVIS
N672 1954	845718	RAuxAF		TREADGOLD
N662 1947	811021	AAF		TREANOR
N187 1947	226	WAAF	DBE; Dame	TREFUSIS-FORBES
YB 1996	n/a	RAuxAF		TREGUNNA
N675 1950	744266	RAFVR		TRETHEWAY
N833 1956	2695002	RAuxAF	1st Clasp	TRETHEWAY
N984 1948	78096	RAFVR		TREVELYAN
N277 1955	2686014	RAuxAF		TREVOR
N799 1954	80596	RAFVR		TRICKETT
N187 1947	811040	AAF		TRICKEY
N415 1946	129229	RAFVR	DFM	TRICKLEBANK
N456 1948	72196	RAFVR		TRIGGS
N874 1945	82664	RAFVR		TRIM
N329 1954	2681016	RAuxAF		TRIMBLE
YB 1999	n/a	RAuxAF		TRIMMER
N289 1947	72959	RAFVR	OBE	TRINDER
N28 1959	23174	RAuxAF	MA,MB,MRCS,LRCP; wef 8/9/58	TRIPP
N666 1956	159636	RAuxAF	1st Clasp; no record of first award	TRIPTREE
N527 1953	850243	RAuxAF		TROTMAN
N187 1947	90930	AAF		TROUGHTON-ROBERTS
N926 1950	846043	RAuxAF		TROWBRIDGE

Surname	Initials			Rank	Status	Date
TRUBODY	W	W		SGT		18/03/1948
TRUDGEN	B	J		FLT SGT		25/11/1992
TRUEFELDT	E			AC1		10/02/1949
TRUMAN	R	G		FL		14/11/1962
TRUNDLE	G	M		SQD LDR		26/01/1950
TRUNDLE	R	M		WO		24/01/1946
TRURAN	J	W	J	SQD LDR		10/12/1942
TRUSLER	K	A	E	FL		26/11/1958
TUBBS	D	F		SQD LDR		13/11/1947
TUCK	J			FL		01/01/2000
TUCKER	A	B		FL		28/10/1943
TUCKER	A	P		CPL		15/09/1949
TUCKER	B	E		FL		30/08/1945
TUCKER	J	D		SQD LDR		06/05/1943
TUCKERMAN	S			SGT		20/09/1945
TUGWELL	D	H		FO		15/09/1949
TULIP	A	E	Y	FLT SGT		24/12/1942
TULL	T	S		WING COMM	Acting	23/05/1946
TURNBALL	J	D		FO		07/04/1955
TURNBULL	A			LAC		22/10/1958
TURNBULL	J	R		WING COMM	Acting	05/06/1947
TURNBULL	T	P		SQD LDR	Acting	26/11/1953
TURNBULL	W	C		LAC		07/07/1949
TURNBULL	W	D		SGT	Temporary	13/10/1955
TURNER	A	E		FO		20/05/1948
TURNER	A			AC2		04/09/1947
TURNER	C	G		CPL	Temporary	09/01/1963
TURNER	G	A		SGT		13/01/1949
TURNER	G	A		FL		22/09/1955
TURNER	G	W		FL		24/01/1946
TURNER	H	R		SQD LDR		30/03/1944
TURNER	J	C	S	FL		20/04/1944
TURNER	J	G		FLT SGT		04/02/1943
TURNER	J	G		CPL		15/11/1951
TURNER	J	M		FO		20/08/1958
TURNER	K	W	A	FLT OFF		12/06/1947
TURNER	L	G		FL	Acting	30/03/1943
TURNER	L	G		FL		16/10/1952
TURNER	L	M		PO		26/01/1950
TURNER	O	C		FO		20/03/1957
TURNER	R	A		SAC		14/11/1962
TURNER	R	F	W	FL		11/07/1946
TURNER	S	J		LAC		16/10/1947
TURNER	S	J		CPL		28/07/1955
TURNER	S			LAC		06/09/1951
TURNER	W	E		FO		14/09/1950
TURNHAM	F	C		SQD LDR	Acting	10/06/1948
TURPIN	A	R		SQD LDR		30/03/1943
TURPIN	G	H		SQD LDR		10/06/1948
TURTON	R	T		SQD LDR	Acting	11/01/1945
TURTON	W	A		SGT		17/01/1952
TUSLER	G			FL		09/05/1946
TWEED	L	J		FL		16/03/1950
TWEEDALE	B	H		SQD LDR	Acting	24/01/1946
TWEEDIE	J	H		WO		26/08/1948
TWIGG	J	M		FL	Acting	10/02/1949
TWIST	K	M		SEC OFF		15/09/1949
TWITCHETT	F	J		FL		06/03/1947
TWYMAN	D	A		FO		27/07/1944
TYE	L	C		WO		04/02/1943

Source	Service No	Force	Notes	Surname
N232 1948	851338	RAuxAF		TRUBODY
YB 1996	n/a	RAuxAF		TRUDGEN
N123 1949	844234	RAuxAF		TRUEFELDT
N837 1962	171816	RAFVR	wef 27/8/56	TRUMAN
N94 1950	90441	RAuxAF		TRUNDLE
N82 1946	745529	RAFVR		TRUNDLE
N1680 1942	70682	RAFVR		TRURAN
N807 1958	137619	RAFVR	wef 1/3/57	TRUSLER
N933 1947	72400	RAFVR		TUBBS
YB 2003	n/a	RAuxAF		TUCK
N1118 1943	70683	EX-RAFVR		TUCKER
N953 1949	865171	RAuxAF		TUCKER
N961 1945	86349	RAFVR		TUCKER
N526 1943	70684	RAFVR		TUCKER
N1036 1945	807089	AAF		TUCKERMAN
N953 1949	102753	RAFVR		TUGWELL
N1758 1942	808082	AAF		TULIP
N468 1946	70686	RAFVR	OBE	TULL
N277 1955	91269	RAF		TURNBALL
N716 1958	868001	AAF	wef 4/7/44	TURNBULL
N460 1947	72226	RAFVR	MBE	TURNBULL
N952 1953	147990	RAuxAF		TURNBULL
N673 1949	846463	RAuxAF		TURNBULL
N762 1955	866612	RAuxAF		TURNBULL
N395 1948	163228	RAFVR		TURNER
N740 1947	810208	AAF		TURNER
N22 1963	753297	RAFVR	wef 14/8/44	TURNER
N30 1949	810123	RAuxAF		TURNER
N708 1955	115613	RAuxAF	DFC	TURNER
N82 1946	127490	RAFVR	DFM	TURNER
N281 1944	70688	RAFVR		TURNER
N363 1944	77789	RAFVR		TURNER
N131 1943	808105	AAF		TURNER
N1185 1951	808105	RAuxAF	1st Clasp	TURNER
N559 1958	2657710	WRAAF	wef 28/5/58	TURNER
N478 1947	462	WAAF		TURNER
N1014 1943	113837	RAFVR		TURNER
N798 1952	113837	RAFVR	DFC,AFC; 1st Clasp	TURNER
N94 1950	162713	RAFVR		TURNER
N198 1957	2605595	RAuxAF	wef 17/6/54	TURNER
N837 1962	2692177	RAuxAF	wef 20/10/62	TURNER
N594 1946	126809	RAFVR	DFC	TURNER
N846 1947	816230	AAF		TURNER
N565 1955	2681004	RAuxAF	1st Clasp	TURNER
N913 1951	810172	RAuxAF		TURNER
N926 1950	121823	RAFVR		TURNER
N456 1948	106023	RAFVR		TURNHAM
N1014 1943	70690	RAFVR		TURPIN
N456 1948	72146	RAFVR		TURPIN
N14 1945	83247	RAFVR		TURTON
N35 1952	818130	RAuxAF		TURTON
N415 1946	139466	RAFVR		TUSLER
N257 1950	138195	RAFVR		TWEED
N82 1946	86352	RAFVR		TWEEDALE
N694 1948	816176	RAuxAF		TWEEDIE
N123 1949	68211	RAFVR		TWIGG
N953 1949	3987	WAAF		TWIST
N187 1947	115346	RAFVR		TWITCHETT
N763 1944	145866	RAFVR		TWYMAN
N131 1943	805000	AAF		TYE

Surname	Initials			Rank	Status	Date
TYLER	C	J		FLT SGT	Temporary	25/05/1960
TYLER	G	H		SGT		25/05/1950
TYNDALL	F	E		WING COMM		17/06/1954
TYNE	H	L		SQD LDR	Acting	17/04/1947
TYRER	F	H		CPL		13/11/1947
TYRIE	J	S	B	FL		28/02/1946
TYSON	E	H		FL		13/12/1945
UEZZELL	H	W		SQD LDR	Acting	01/07/1948
ULYET	M	W		CPL		30/09/1948
UNDERHILL	D	T		FL		22/02/1961
UNDERHILL	H	J		FL		02/07/1958
UNDERTOWN	F	G		CPL		07/12/1950
UNDERWOOD	G	E		CPL		26/08/1948
UNDERWOOD	L	E		CPL		15/07/1948
UNKLES	R	H		FL		18/07/1956
UNWIN	A			WO		24/01/1946
UNWINS	C	F		FL		04/11/1943
UPTON	W			CPL		18/11/1948
URCH	M	E		SGT		28/09/1950
URE	W	A		FO		25/03/1959
URIE	J	D		WING COMM		22/04/1943
URIE	J	D		FL		17/01/1952
URQUHART	W	A		FL		10/08/1944
URQUHART	W			SGT		09/08/1945
URQUHART-MacKAY	J	M		FLT OFF		06/03/1947
USHER	C	W		SQD LDR		27/04/1950
USHER	H	T		FLT SGT		11/01/1945
USHER	V	H		FL		18/02/1954
USMAR	F			FL		09/08/1945
VACQUIER	A	T		FL		03/10/1946
VALE	E	S		FL		13/01/1949
VALENTINE	S	T		SGT		25/05/1950
VALENTINE	S	T		SGT		20/07/1960
VALENTINE	V	D		FL		20/08/1958
VAN GEENE	R	G		FL		29/03/1998
VANN	J	H		WO	Acting	27/04/1950
VANNECK	P	B	R	FL		19/09/1956
VANNECK	P	B	R	GRP CAP		27/03/1963
VANNECK	P	B	R	GRP CAP		27/03/1963
VANSTONE	E	H		LAC		26/08/1948
VANT	R	J		SGT		28/02/1962
VARCOE	F	H	L	WING COMM	Acting	06/03/1947
VARCOE	V	C		WING COMM	Acting	12/06/1947
VARLEY	G	A		FL	Acting	30/09/1948
VARLEY	G	W		FL		21/03/1946
VARLEY	R	M	G	FL		01/10/1953
VARNEY	E	L		FO		06/03/1947
VAUGHAN	A	G		SGT		28/02/1962
VAUGHAN	C	H		WING COMM	Acting	17/04/1947
VAUGHAN	J			WO		04/10/1945
VAUGHAN	R	C		GRP CAP	Acting	17/04/1947
VAUGHAN	W	C		SQD LDR		07/08/1947
VAUGHAN	W	E	G	SQD LDR		23/02/1950
VAUX	P	D	O	WING COMM		03/12/1942
VEALE	A	L	W	FL		28/11/1946
VEAR	W	P	T	FL		29/04/1954

Source	Service No	Force	Notes	Surname
N375 1960	846365	AAF	wef 13/5/44	TYLER
N517 1950	845645	RAuxAF		TYLER
N471 1954	73442	RAFVR		TYNDALL
N289 1947	120680	RAFVR		TYNE
N933 1947	750812	RAFVR		TYRER
N197 1946	87636	RAFVR		TYRIE
N1355 1945	119893	RAFVR		TYSON
N527 1948	146145	RAFVR		UEZZELL
N805 1948	843174	RAuxAF		ULYET
N153 1961	186858	RAuxAF	wef 1/6/60	UNDERHILL
N452 1958	90891	AAF	wef 23/3/44	UNDERHILL
N1210 1950	863454	RAuxAF		UNDERTOWN
N694 1948	842599	RAuxAF		UNDERWOOD
N573 1948	842598	RAuxAF		UNDERWOOD
N513 1956	125404	RAFVR		UNKLES
N82 1946	745936	RAFVR		UNWIN
N1144 1943	70963	RAFVR	AFC	UNWINS
N984 1948	850653	RAuxAF		UPTON
N968 1950	881136	WAAF		URCH
N196 1959	2691527	RAuxAF	wef 11/11/58	URE
N475 1943	90164	AAF		URIE
N35 1952	90164	RAuxAF	1st Clasp	URIE
N812 1944	117151	RAFVR		URQUHART
N874 1945	802414	AAF		URQUHART
N187 1947	338	WAAF		URQUHART-MacKAY
N406 1950	73443	RAFVR		USHER
N14 1945	807055	AAF		USHER
N137 1954	202847	RAFVR		USHER
N874 1945	115914	RAFVR		USMAR
N837 1946	127524	RAFVR		VACQUIER
N30 1949	101470	RAFVR	Deceased	VALE
N517 1950	846097	RAuxAF		VALENTINE
N531 1960	2601416	RAFVR	1st Clasp; wef 19/11/59	VALENTINE
N559 1958	186163	RAFVR	wef 24/4/58	VALENTINE
YB 1999	8024936	RAuxAF		VAN GEENE
N406 1950	750238	RAFVR		VANN
N666 1956	205378	RAuxAF	AFC; The Honorable	VANNECK
N239 1963	205378	RAuxAF	OBE,AFC,ADC,MA; The Honorable; 1st Clasp; wef 17/11/54	VANNECK
N239 1963	205378	RAuxAF	OBE,AFC,ADC,MA; The Honorable; 2nd Clasp; wef 19/7/61	VANNECK
N694 1948	849893	RAuxAF		VANSTONE
N157 1962	2614231	RAFVR	wef 21/10/61	VANT
N187 1947	73667	RAFVR	MC	VARCOE
N478 1947	72961	RAFVR	OBE	VARCOE
N805 1948	133895	RAFVR		VARLEY
N263 1946	85670	RAFVR	DFC	VARLEY
N796 1953	179686	RAFVR		VARLEY
N187 1947	109693	RAFVR		VARNEY
N157 1962	2607433	RAFVR	wef 8/1/61	VAUGHAN
N289 1947	60967	RAFVR		VAUGHAN
N1089 1945	813071	AAF		VAUGHAN
N289 1947	74262	RAFVR	OBE,MC	VAUGHAN
N662 1947	73444	RAFVR	MBE	VAUGHAN
N179 1950	72291	RAFVR		VAUGHAN
N1638 1942	90315	AAF		VAUX
N1004 1946	106558	RAFVR		VEALE
N329 1954	83798	RAFVR		VEAR

Surname	Initials			Rank	Status	Date
VEGLIO	A	J		SGT		16/03/1950
VENN	J	V		FL		21/09/1944
VENNER	J	F		GRP CAP	Acting	13/01/1949
VENTON	C	L	L	LAC		22/05/1952
VERE	R	H	M	FL		19/08/1954
VERITY	A	M		SQD LDR	Acting	30/12/1943
VERITY	F			CPL	Acting	20/08/1958
VERITY	H	B		WING COMM	Acting	20/09/1945
VERLING	W			FO		07/04/1955
VERNEY-CAVE	A	J		FL		21/09/1944
VERNON	A	G	B	SQD LDR	Acting	04/10/1945
VEYSEY	E	W		SQD LDR		06/03/1947
VICK	J	A		WING COMM	Acting	17/12/1942
VICK	N	S		FL		15/10/1966
VICKERMAN	J	W		FLT SGT		13/07/1950
VICKERS	C	W	S	FLT SGT		01/07/1948
VICKERS	G			SGT		08/02/1995
VICKERS	V			FL		30/03/1943
VICKERY	B	H		WO		18/04/1946
VICKERY	C			FLT SGT		26/08/1948
VILES	J	E		FL		22/05/1952
VINCE	E			WO		05/02/1948
VINCENT	B	S	C	LAC		14/10/1948
VINCENT	C	L		FO		05/01/1950
VINCENT	W	G		FL		26/11/1958
VINEY	C	T		FL		26/10/1944
VINEY	C	T		FL		24/07/1952
VIVASH	W	H	J	LAC		22/02/1961
VIVERS	T	H	A	SQD LDR		05/06/1947
von DONOP	D	B		FLT OFF	Acting	23/02/1950
VOSPER	F	C		CPL		10/02/1949
VOTIER	B	E		SACW		09/12/1959
VOUSDEN	F	C		SAC		03/06/1959
VOUT	T			SGT		19/07/1997
WADDELL	C			FLT SGT		21/09/1944
WADDINGTON	F	O		FL		18/04/1946
WADDINGTON	F	R		CPL		18/03/1948
WADDINGTON	W	D		FL		06/01/1955
WADDINGTON	W	D		FL		06/01/1955
WADE	G	H		FO		24/01/1946
WADE	H	F		SGT		26/08/1948
WADE	M	M		SQD OFF	Acting	06/03/1947
WADE	T	S		SQD LDR	Acting	11/01/1945
WADE	W	C		CPL	Temporary	21/06/1961
WADESON	G	W		FL	Acting	15/07/1948
WAGNER	A	D		FL		15/06/1944
WAILES	P	J		SAC		19/01/1994
WAINWRIGHT	C	M		SGT		04/09/1947
WAINWRIGHT	S	A	G	SGT		27/03/1947
WAITE	K	E		FLT SGT		13/11/1947
WAITE	W	J		LAC		20/05/1948
WAKE	J	A		FL		05/08/1947
WAKEFIELD	E	R		FO		28/07/1955
WAKEFIELD	H	K		SQD LDR	Acting	05/01/1950
WAKELIN	D	E		FL		26/11/1953
WAKELIN	D	E		SQD LDR	Acting	18/12/1963
WAKLEY	D	F		FL		13/01/1949
WALBOURN	D	M		SQD LDR		17/06/1943
WALBY	B	J		FLT SGT		04/03/1998
WALDEN	H			CPL		01/10/1953

Source	Service No	Force	Notes	Surname
N257 1950	801318	RAuxAF		VEGLIO
N961 1944	88653	RAFVR		VENN
N30 1949	90872	RAuxAF		VENNER
N402 1952	2698077	RAuxAF		VENTON
N672 1954	186944	RAFVR		VERE
N1350 1943	70697	RAFVR		VERITY
N559 1958	2605307	RAuxAF	DFM; wef 20/2/54	VERITY
N1036 1945	72507	RAFVR	DSO,DFC	VERITY
N277 1955	2602274	RAFVR		VERLING
N961 1944	100031	RAFVR	The Honorable	VERNEY-CAVE
N1089 1945	108830	RAFVR		VERNON
N187 1947	72246	RAFVR		VEYSEY
N1708 1942	90274	AAF		VICK
MOD 1967	140306	RAFVR		VICK
N675 1950	752676	RAFVR		VICKERMAN
N527 1948	771113	RAFVR		VICKERS
YB 1996	n/a	RAuxAF		VICKERS
N1014 1943	70698	RAFVR		VICKERS
N358 1946	748625	RAFVR		VICKERY
N694 1948	752017	RAFVR		VICKERY
N402 1952	91286	RAuxAF		VILES
N87 1948	819047	RAuxAF		VINCE
N856 1948	860363	RAuxAF		VINCENT
N1 1950	166658	RAFVR		VINCENT
N807 1958	165349	RAFVR	wef 15/6/53	VINCENT
N1094 1944	100028	RAFVR		VINEY
N569 1952	100028	RAFVR	1st Clasp	VINEY
N153 1961	842377	AAF	wef 24/3/44	VIVASH
N460 1947	73738	RAFVR		VIVERS
N179 1950	475	WAAF		von DONOP
N123 1949	750053	RAFVR		VOSPER
N843 1959	2656346	WRAAF	wef 10/10/59	VOTIER
N368 1959	2692525	RAuxAF	wef 14/2/59	VOUSDEN
YB 1998	n/a	RAuxAF		VOUT
N961 1944	816029	AAF		WADDELL
N358 1946	129211	RAFVR	DFM	WADDINGTON
N232 1948	747055	RAFVR		WADDINGTON
N1 1955	174802	RAFVR		WADDINGTON
N1 1955	174802	RAFVR	1st Clasp	WADDINGTON
N82 1946	195251	RAFVR		WADE
N694 1948	742814	RAFVR		WADE
N187 1947	48	WAAF	MBE	WADE
N14 1945	78984	RAFVR	DFC,AFC	WADE
N478 1961	801575	AAF	wef 9/7/44	WADE
N573 1948	147972	RAFVR		WADESON
N596 1944	65993	RAFVR	DFC	WAGNER
YB 1997	n/a	RAuxAF		WAILES
N740 1947	805175	AAF		WAINWRIGHT
N250 1947	771150	RAFVR		WAINWRIGHT
N933 1947	809017	AAF		WAITE
N395 1948	849100	RAuxAF		WAITE
N460 1947	160960	RAFVR	DFC	WAKE
N565 1955	2689082	RAuxAF		WAKEFIELD
N1 1950	78267	RAFVR	DFC	WAKEFIELD
N952 1953	110000	RAuxAF		WAKELIN
N917 1963	110000	RAuxAF	1st Clasp; wef 23/10/63	WAKELIN
N30 1949	87561	RAFVR		WAKLEY
N653 1943	72004	RAFVR		WALBOURN
YB 1999	n/a	RAuxAF		WALBY
N796 1953	812065	RAuxAF		WALDEN

Surname	Initials				Rank	Status	Date
WALES	K	C			FL		05/01/1950
WALKER	A	A			CPL		01/03/1970
WALKER	A	A			CPL		01/03/1980
WALKER	A	A			CPL		01/03/1990
WALKER	A	D			FL		15/09/1949
WALKER	A	E			SGT		28/02/1946
WALKER	A	R			CPL		07/04/1955
WALKER	A	W	C		CPL		16/03/1950
WALKER	A				CPL		24/11/1949
WALKER	A				LAC		28/07/1955
WALKER	D				FL		21/03/1946
WALKER	D				CPL		16/03/1950
WALKER	E	A			FL		01/07/1948
WALKER	E				FL		17/01/1952
WALKER	F	R			SGT		16/10/1963
WALKER	G	A			FL		09/08/1945
WALKER	G	C			SQD LDR		30/12/1943
WALKER	G	W	J		FO		17/04/1947
WALKER	G				FLT SGT		03/11/1989
WALKER	G				FLT SGT		03/11/1989
WALKER	H				SGT		01/07/1948
WALKER	J	C			FL	Acting	17/04/1947
WALKER	J	W	H		SQD LDR	Acting	20/09/1945
WALKER	N	O	F	E	SQD LDR	Acting	25/05/1950
WALKER	N	O	F	E	SQD LDR	Acting	13/10/1955
WALKER	N	W			FL		03/06/1959
WALKER	R	C			FL		24/01/1946
WALKER	R	E			FLT SGT		30/09/1948
WALKER	S				SQD LDR		03/04/1990
WALKER	V	S			FL		28/02/1946
WALKER	W	A			FO		06/03/1947
WALKER	W	L	B		FL		15/09/1949
WALKLEY	B	A	J		FLT SGT	Acting	24/11/1949
WALL	A	T			LAC		06/01/1955
WALL	T	C			FO		14/10/1943
WALLACE	C	R			LAC		26/08/1948
WALLACE	P	J			LAC		26/04/1961
WALLACE	R	D			CPL		21/09/1995
WALLACE	T	D			LAC		24/11/1949
WALLACE	T				FL		09/05/1946
WALLACE	W	F			SGT		10/06/1948
WALLACE-PANNELL	P				FL		05/06/1947
WALLACE-WILLIAMS	G	S			FL		13/02/1957
WALLACE-WILLIAMS	G	S			FL		30/12/1964
WALLBUTTON	C	J	A		LAC		16/10/1947
WALLEN	D	S			WING COMM	Acting	18/04/1946
WALLER	G	A			FL		24/01/1946
WALLER	G	S			WING COMM		06/03/1947
WALLINGTON	D	C			FL		08/07/1959
WALLIS	A	G			FL		02/09/1943
WALLIS	A	G			FO		07/02/1952
WALLIS	K	E			WO		30/09/1948
WALLS	G	A			SGT		22/02/1961
WALLS	R				FO		08/11/1945
WALLS	T	K			SQD LDR		30/08/1945
WALMSLEY	P	S	F		SQD LDR	Acting	24/01/1946
WALSH	T				SGT	Temporary	20/08/1958
WALSH	W	F			SGT		20/05/1948
WALSH	W	F			SGT		13/12/1951
WALSH	W	H			SGT		07/07/1949

Source	Service No	Force	Notes	Surname
N1 1950	88646	RAFVR		WALES
YB 1996	n/a	RAuxAF		WALKER
YB 1996	n/a	RAuxAF	1st Clasp	WALKER
YB 1996	n/a	RAuxAF	2nd Clasp	WALKER
N953 1949	61940	RAFVR		WALKER
N197 1946	805359	AAF		WALKER
N277 1955	857542	RAuxAF		WALKER
N257 1950	840511	RAuxAF		WALKER
N1193 1949	817142	RAuxAF		WALKER
N565 1955	858245	RAuxAF		WALKER
N263 1946	144759	RAFVR		WALKER
N257 1950	808397	RAuxAF		WALKER
N527 1948	61603	RAFVR		WALKER
N35 1952	72693	RAFVR	DFC	WALKER
N756 1963	2601733	RAFVR	wef 13/9/62	WALKER
N874 1945	113499	RAFVR		WALKER
N1350 1943	70703	RAFVR		WALKER
N289 1947	197282	RAFVR		WALKER
YB 1996	n/a	RAuxAF		WALKER
YB 2000	n/a	RAuxAF	1st Clasp	WALKER
N527 1948	870912	RAuxAF		WALKER
N289 1947	133913	RAFVR		WALKER
N1036 1945	102075	RAFVR		WALKER
N517 1950	78449	RAFVR	DFC	WALKER
N762 1955	78449	RAFVR	DFC; 1st Clasp	WALKER
N368 1959	112454	RAFVR	DFC; wef 22/2/54	WALKER
N82 1946	146794	RAFVR		WALKER
N805 1948	871594	RAuxAF		WALKER
YB 1996	4251237	RAuxAF		WALKER
N197 1946	120936	RAFVR		WALKER
N187 1947	138040	RAFVR		WALKER
N953 1949	82662	RAFVR		WALKER
N1193 1949	848340	RAuxAF		WALKLEY
N1 1955	840074	RAuxAF		WALL
N1065 1943	121112	RAFVR		WALL
N694 1948	861115	RAuxAF		WALLACE
N339 1961	2681123	RAuxAF	wef 15/1/61	WALLACE
YB 1997	n/a	RAuxAF		WALLACE
N1193 1949	817159	RAuxAF		WALLACE
N415 1946	117935	RAFVR	DFC	WALLACE
N456 1948	808239	RAuxAF		WALLACE
N460 1947	122386	RAFVR		WALLACE-PANNELL
N104 1957	195158	RAFVR	wef 30/12/54	WALLACE-WILLIAMS
MOD 1967	195158	RAFVR	1st Clasp	WALLACE-WILLIAMS
N846 1947	813234	AAF		WALLBUTTON
N358 1946	77347	RAFVR	OBE	WALLEN
N82 1946	149145	RAFVR	DFC	WALLER
N187 1947	72758	RAFVR	OBE	WALLER
N454 1959	104509	RAFVR	wef 16/5/59	WALLINGTON
N918 1943	81061	RAFVR		WALLIS
N103 1952	81061	RAFVR	1st Clasp	WALLIS
N805 1948	755690	RAFVR		WALLIS
N153 1961	2699538	RAuxAF	wef 29/10/60	WALLS
N1232 1945	161019	RAFVR		WALLS
N961 1945	72031	RAFVR		WALLS
N82 1946	66501	RAFVR	DFC	WALMSLEY
N559 1958	808295	AAF	wef 23/4/44	WALSH
N395 1948	815111	RAuxAF		WALSH
N1271 1951	2681503	RAuxAF	1st Clasp	WALSH
N673 1949	865639	RAuxAF		WALSH

Surname	Initials			Rank	Status	Date
WALSH	W	N		FL		21/03/1946
WALSHE	P	W		WO		30/12/1943
WALSHE	T	C		WO		17/01/1952
WALTER	H	W		SQD LDR		10/12/1942
WALTER	L	J		CPL		26/08/1948
WALTERS	B	F		FLT SGT		01/03/1945
WALTERS	G	D	B	FO		13/12/1951
WALTERS	G	G	G	FL		04/10/1945
WALTERS	G			SGT		05/02/1948
WALTERS	H	C		SQD LDR	Acting	08/11/1945
WALTERS	H	J		SQD LDR	Acting	21/06/1945
WALTERS	H	N		FL		05/06/1947
WALTERS	J	R	C	SGT		20/05/1948
WALTERS	W	H	G	FL	Acting	06/03/1947
WALTHAM	S			SGT		07/06/1951
WALTON	A	J	C	SAC		08/06/1995
WALTON	A			FL		16/02/1968
WALTON	F	E		SGT		27/03/1952
WALTON	J			FO		02/09/1943
WALTON	J			FL		15/01/1953
WALTON	P	M	R	SQD LDR		29/04/1954
WALTON	P	R		FLT OFF		28/11/1956
WANDER	W	E		SGT		01/11/1951
WANSTALL	J	E		FLT SGT		27/04/1944
WANSTALL	J	E	H	SGT		17/05/1996
WARBURTON	A	T		FL		01/05/1952
WARBURTON	A	T		FL		06/06/1962
WARBURTON	G	C	G	FL		09/05/1946
WARBURTON	H	W		WO		20/09/1951
WARD	A	B		SGT		05/06/1947
WARD	D	J	N	WO		24/01/1946
WARD	E	W		SQD LDR		06/01/1955
WARD	F			SGT		25/06/1953
WARD	H	W		LAC		15/01/1953
WARD	J	H		FL		24/01/1946
WARD	J	R		FL		10/03/1994
WARD	J			SGT		30/09/1948
WARD	N	E		PO		17/02/1944
WARD	P	C		FL		06/03/1947
WARD	R	G		WO		01/07/1948
WARD	S	E		FL		01/05/1957
WARD	W	J		FL		01/03/1945
WARDELL	M	G		FLT SGT		04/11/1992
WARDEN	W	E		SGT		14/10/1948
WARDLE	J	E		CPL		20/01/1960
WARDLEY	A			LAC		26/01/1950
WARD-SMITH	P			FL		31/05/1945
WARE	F			LAC		15/09/1949
WARE	G			FL		15/01/1958
WARE	H	W		WO		07/07/1949
WAREHAM	M	P		FL		24/01/1946
WAREING	E	J		SGT		28/02/1962
WAREING	P	T		FL		11/07/1946
WAREING	R			SQD LDR		19/10/1960
WAREING	S			FL		20/09/1945
WARING	E			SAC		16/08/1961
WARING	S	A		FO		12/06/1947
WARLOW	R	P		FL		05/10/1944
WARMAN	F	J		CPL	Temporary	16/04/1958

Source	Service No	Force	Notes	Surname
N263 1946	135427	RAFVR		WALSH
N1350 1943	817066	AAF		WALSHE
N35 1952	803528	RAuxAF		WALSHE
N1680 1942	90118	RAFVR	MD,BS	WALTER
N694 1948	842045	RAuxAF		WALTER
N225 1945	804230	AAF		WALTERS
N1271 1951	171511	RAFVR		WALTERS
N1089 1945	129147	RAFVR	AFC	WALTERS
N87 1948	746438	RAFVR		WALTERS
N1232 1945	77467	RAFVR		WALTERS
N671 1945	100612	RAFVR		WALTERS
N460 1947	121955	RAFVR		WALTERS
N395 1948	841671	RAuxAF		WALTERS
N187 1947	121970	RAFVR		WALTERS
N560 1951	749612	RAFVR		WALTHAM
YB 1996	n/a	RAuxAF		WALTON
MOD 1968	3039086	RAFVR		WALTON
N224 1952	863278	RAuxAF		WALTON
N918 1943	131472	RAFVR		WALTON
N28 1953	90633	RAuxAF		WALTON
N329 1954	150601	RAuxAF		WALTON
N833 1956	1523	WRAAF		WALTON
N1126 1951	750212	RAFVR		WANDER
N396 1944	812023	AAF		WANSTALL
YB 1997	n/a	RAuxAF		WANSTALL
N336 1952	91299	RAuxAF		WARBURTON
N435 1962	91299	RAuxAF	MBE; 1st Clasp; wef 25/3/62	WARBURTON
N415 1946	160675	RAFVR		WARBURTON
N963 1951	812156	RAuxAF		WARBURTON
N460 1947	815130	AAF		WARD
N82 1946	742056	RAFVR		WARD
N1 1955	90684	RAuxAF		WARD
N527 1953	870951	RAuxAF		WARD
N28 1953	864189	RAuxAF		WARD
N82 1946	60300	RAFVR		WARD
YB 1996	2631760	RAuxAF		WARD
N805 1948	869365	RAuxAF		WARD
N132 1944	157456	RAFVR		WARD
N187 1947	61236	RAFVR		WARD
N527 1948	770550	RAFVR		WARD
N277 1957	193151	RAuxAF	wef 28/10/56	WARD
N225 1945	130691	RAFVR		WARD
YB 1996	n/a	RAuxAF		WARDELL
N856 1948	847561	RAuxAF		WARDEN
N33 1960	2605677	RAFVR	wef 18/12/59	WARDLE
N94 1950	841321	RAuxAF	BEM	WARDLEY
N573 1945	100039	RAFVR		WARD-SMITH
N953 1949	864221	RAuxAF		WARE
N40 1958	140918	RAuxAF	DFC; wef 6/5/57	WARE
N673 1949	799765	RAFVR		WARE
N82 1946	85269	RAFVR	DFC	WAREHAM
N157 1962	2663781	WRAAF	wef 15/1/62	WAREING
N594 1946	155258	RAFVR	DCM	WAREING
N775 1960	86325	RAFVR	DFC; wef 7/2/44	WAREING
N1036 1945	66497	RAFVR	DFC	WAREING
N624 1961	2650915	RAuxAF	wef 24/5/61	WARING
N478 1947	137084	RAFVR		WARING
N1007 1944	84683	RAFVR		WARLOW
N258 1958	847799	AAF	wef 9/7/44	WARMAN

Surname	Initials			Rank	Status	Date
WARMAN	G	E		WO		07/07/1949
WARMINGER	A	H		FL		17/06/1954
WARN	W	J		FL		08/11/1945
WARNCKEN	B	C		SQD LDR		18/06/2000
WARNER	H	F		FL	Acting	15/01/1958
WARNER-HORNE	H	E		SQD LDR		06/01/1955
WARNES	W	H		FL	Acting	30/09/1948
WARREN	D	J		CPL		05/06/1947
WARREN	J	H		LAC		15/07/1948
WARREN	M	J		CPL		03/07/1952
WARREN	N	L	McN.	SQD OFF		17/04/1947
WARREN	P	C		FL		29/06/1944
WARREN	P	C		FL		01/10/1953
WARREN	R			FLT SGT		05/06/1947
WARRINGTON	D	J		FL		21/10/1959
WASE	R	C		FLT SGT		29/04/1954
WASS	E	A		FL		25/03/1954
WASTIE	E	S		WING COMM	Acting	12/06/1947
WATCHMAN	A	L		SGT		17/04/1947
WATCHORN	W			SGT		16/03/1950
WATERER	G	D		SQD LDR	Acting	05/01/1950
WATERS	D	K		FL		28/02/1946
WATERS	S	M		CPL		10/03/1994
WATERS-WEBB	N	F		SQD LDR	Acting	30/08/1945
WATERWORTH	S	B		SGT		26/04/1951
WATHES	J	W		SQD LDR		26/01/1950
WATKINS	A			CPL		04/09/1947
WATKINS	D	H		WING COMM	Acting	09/08/1945
WATKINS	E	E		FL		31/10/1956
WATKINS	H	W		AC1		06/01/1955
WATKINS	I	R		CPL		15/09/1949
WATKINSON	A	R		PO		10/06/1948
WATKINSON	J	L		SGT		16/03/1950
WATKISS	N	S		SGT		28/06/1984
WATKISS	N	S		SGT		28/06/1994
WATLING	M	I		CPL		24/10/1997
WATLING	R	B		FO		05/01/1950
WATSON	B	H	J	FL		13/12/1945
WATSON	B	J		FL		14/10/1943
WATSON	C	D		SQD LDR	Acting	05/10/1944
WATSON	D	A		SQD LDR	Acting	18/04/1946
WATSON	G			FO		07/01/1943
WATSON	G			SGT		02/08/1951
WATSON	H			SGT		07/08/1947
WATSON	I	E	C	SQD LDR		04/02/1943
WATSON	J	H		FL		04/09/1957
WATSON	J	T		LAC		19/08/1954
WATSON	R	A		FO		25/03/1954
WATSON	R	G		FLT SGT		18/07/1956
WATSON	R			CPL		24/11/1949
WATSON	R			FL		16/04/1958
WATSON	R			FL		09/12/1959
WATSON	T	H		SGT		24/11/1949
WATSON	T	R		FO		30/12/1943
WATSON	V	H	C	FL		12/06/1947
WATSON	W	A	H	SQD LDR	Acting	13/12/1945
WATSON	W			CPL		13/01/1949
WATSON	W			SGT		10/08/1991
WATT	H	B		AC1		10/02/1949
WATT	J			SQD LDR	Acting	20/09/1945

Source	Service No	Force	Notes	Surname
N673 1949	799818	RAFVR		WARMAN
N471 1954	129716	RAFVR		WARMINGER
N1232 1945	135406	RAFVR	DFC	WARN
YB 2001	91454	RAuxAF		WARNCKEN
N40 1958	100672	RAFVR	wef 28/6/44	WARNER
N1 1955	72918	RAFVR		WARNER-HORNE
N805 1948	161951	RAFVR		WARNES
N460 1947	801460	AAF		WARREN
N573 1948	846875	RAuxAF		WARREN
N513 1952	859683	RAuxAF		WARREN
N289 1947	241	WAAF	MBE	WARREN
N652 1944	61059	RAFVR		WARREN
N796 1953	61059	RAFVR	1st Clasp	WARREN
N460 1947	811015	AAF		WARREN
N708 1959	163291	RAFVR	wef 19/9/53	WARRINGTON
N329 1954	2602253	RAFVR		WASE
N249 1954	187849	RAFVR		WASS
N478 1947	72207	RAFVR		WASTIE
N289 1947	743324	RAFVR		WATCHMAN
N257 1950	771072	RAFVR		WATCHORN
N1 1950	77214	RAFVR		WATERER
N197 1946	130414	RAFVR		WATERS
YB 1996	n/a	RAuxAF		WATERS
N961 1945	81653	RAFVR		WATERS-WEBB
N402 1951	816197	RAuxAF		WATERWORTH
N94 1950	73451	RAFVR		WATHES
N740 1947	871700	AAF		WATKINS
N874 1945	90363	AAF	DFC	WATKINS
N761 1956	190496	RAFVR		WATKINS
N1 1955	2650025	RAuxAF		WATKINS
N953 1949	865597	RAuxAF		WATKINS
N456 1948	204308	RAuxAF		WATKINSON
N257 1950	804084	RAuxAF		WATKINSON
YB 1997	n/a	RAuxAF		WATKISS
YB 1997	n/a	RAuxAF	1st Clasp	WATKISS
YB 1997	n/a	RAuxAF		WATLING
N1 1950	162272	RAFVR		WATLING
N1355 1945	79545	RAFVR		WATSON
N1065 1943	89311	RAFVR		WATSON
N1007 1944	82658	RAFVR		WATSON
N358 1946	85019	RAFVR	DFC	WATSON
N3 1943	86663	RAFVR		WATSON
N777 1951	866272	RAuxAF		WATSON
N662 1947	811051	AAF		WATSON
N131 1943	70717	RAFVR		WATSON
N611 1957	87599	RAFVR	wef 18/1/54	WATSON
N672 1954	866572	RAuxAF		WATSON
N249 1954	199108	RAFVR		WATSON
N513 1956	2696019	RAuxAF		WATSON
N1193 1949	817279	RAuxAF		WATSON
N258 1958	129130	RAFVR	wef 7/1/57	WATSON
N843 1959	172490	RAFVR	DFM; wef 2/8/59	WATSON
N1193 1949	747492	RAFVR		WATSON
N1350 1943	146125	RAFVR		WATSON
N478 1947	109995	RAFVR		WATSON
N1355 1945	78268	RAFVR		WATSON
N30 1949	840054	RAuxAF		WATSON
YB 1996	n/a	RAuxAF		WATSON
N123 1949	873183	RAuxAF		WATT
N1036 1945	81928	RAFVR		WATT

Surname	Initials			Rank	Status	Date
WATT	R			SAC		21/01/1959
WATT	R			FL		05/09/1962
WATTHEY	S			CPL		05/09/1962
WATTON	S	M		CPL		18/12/1963
WATTS	A	N		FLT SGT		15/09/1949
WATTS	B	A		SGT		19/10/1960
WATTS	B	W		CPL		26/08/1948
WATTS	C	S		CPL		08/05/1947
WATTS	E	G		SGT		02/07/1958
WATTS	E	H		FL		13/10/1955
WATTS	J	H		SGT		30/09/1948
WATTS	R	H		SAC		03/06/1959
WAUD	K	L		FL		07/06/1945
WAUGH	W	A	O'N	FL		13/02/1957
WAY	J	B		FL		26/11/1958
WAY	W	E		AC2		28/03/1962
WAYMAN	R	D		SQD LDR		30/03/1943
WEALE	D	A		SQD LDR	Acting	04/11/1943
WEALE	D	A		FL		03/07/1957
WEARMOUTH	L	P		FO		09/08/1945
WEATHERELL	H			CPL		07/07/1949
WEATHERLAND	D			FL	Acting	10/08/1950
WEAVER	J	F		SGT		05/06/1947
WEAVER	K	W		CPL		16/10/1947
WEAVER	P	C		FLT OFF	Acting	04/09/1947
WEBB	A	E		LAC		21/10/1959
WEBB	D	A		FL		28/02/1946
WEBB	E	M		SQD LDR		03/12/1942
WEBB	E	N		FLT SGT	Acting	07/04/1955
WEBB	F	G		FO		26/01/1950
WEBB	F	R		FO		22/02/1951
WEBB	F	W		FLT SGT	Acting	10/02/1949
WEBB	G	F	H	SQD LDR	Acting	06/05/1943
WEBB	G	H		FL		26/10/1944
WEBB	G	W	A	SQD LDR		01/07/1948
WEBB	G			FO		26/01/1950
WEBB	H	P		FO		05/06/1947
WEBB	L	E	W	FL		24/01/1946
WEBB	L	E	W	SQD LDR	Acting	25/03/1959
WEBB	P	C		WING COMM	Acting	11/01/1945
WEBB	R	A		CPL		21/02/1952
WEBB	R	J		FL		05/01/1950
WEBB	S	E		FO		05/01/1950
WEBBER	D	H		SAC		12/09/1989
WEBBER	T	H		FL		26/01/1950
WEBDELL	R			FL		06/03/1947
WEBDELL	R			FO		19/09/1956
WEBSTER	E	F		SGT		16/04/1958
WEBSTER	E	R		FO		23/05/1946
WEBSTER	H	G		FL		07/08/1947
WEBSTER	J	C		SGT		01/04/1985
WEBSTER	J	C		FLT SGT		01/04/1995
WEBSTER	W	F		SAC		28/08/1963
WEBSTER-GRINLING	H			FL	Acting	10/06/1948
WEEKES	N	C	F	FL		20/07/2004
WEEKS	R	H		FL		19/09/1956
WEIR	D	R		FL		28/02/1946
WEIR	T	H		AC1		01/05/1957
WELCH	B	E		CPL		16/03/1950

Source	Service No	Force	Notes	Surname
N28 1959	2681037	RAuxAF	wef 17/11/57	WATT
N655 1962	205371	RAFVR	wef 16/9/57	WATT
N655 1962	2690570	RAFVR	wef 28/2/62	WATTHEY
N917 1963	2663815	WRAAF	wef 15/10/63	WATTON
N953 1949	863262	RAuxAF		WATTS
N775 1960	2656731	WRAAF	wef 22/6/60	WATTS
N694 1948	770759	RAFVR		WATTS
N358 1947	845765	AAF		WATTS
N452 1958	2655010	WRAAF	wef 6/5/58	WATTS
N762 1955	154489	RAuxAF		WATTS
N805 1948	804299	RAuxAF		WATTS
N368 1959	2693527	RAuxAF	wef 21/3/59	WATTS
N608 1945	126771	RAFVR	DFC	WAUD
N104 1957	143683	RAuxAF	wef 28/2/52	WAUGH
N807 1958	150161	RAFVR	wef 12/6/58	WAY
N242 1962	2612752	RAFVR	wef 11/9/61	WAY
N1014 1943	70718	RAFVR		WAYMAN
N1144 1943	102159	RAFVR		WEALE
N456 1957	102159	RAFVR	1st Clasp; wef 14/3/49	WEALE
N874 1945	169689	RAFVR		WEARMOUTH
N673 1949	870957	RAuxAF		WEATHERELL
N780 1950	133300	RAFVR		WEATHERLAND
N460 1947	813022	AAF		WEAVER
N846 1947	844913	AAF		WEAVER
N740 1947	2602	WAAF		WEAVER
N708 1959	859565	AAF	wef 22/6/44	WEBB
N197 1946	135467	RAFVR		WEBB
N1638 1942	72149	RAFVR		WEBB
N277 1955	2655680	WRAAF		WEBB
N94 1950	172327	RAFVR		WEBB
N194 1951	137987	RAFVR		WEBB
N123 1949	746323	RAFVR		WEBB
N526 1943	90987	AAF	DFC	WEBB
N1094 1944	63425	RAFVR		WEBB
N527 1948	73473	RAFVR		WEBB
N94 1950	169270	RAFVR		WEBB
N460 1947	154488	RAFVR		WEBB
N82 1946	116996	RAFVR	DFC,DFM	WEBB
N196 1959	116996	RAuxAF	DFC,DFM; 1st Clasp; wef 19/3/57	WEBB
N14 1945	90171	AAF	DFC	WEBB
N134 1952	2689502	RAuxAF		WEBB
N1 1950	121209	RAFVR		WEBB
N1 1950	121971	RAFVR		WEBB
YB 1996	n/a	RAuxAF		WEBBER
N94 1950	105163	RAFVR		WEBBER
N187 1947	124211	RAFVR		WEBDELL
N666 1956	124211	RAF	1st Clasp	WEBDELL
N258 1958	2692004	RAuxAF	wef 26/2/58	WEBSTER
N468 1946	158315	RAFVR		WEBSTER
N662 1947	90931	AAF		WEBSTER
YB 1996	n/a	RAuxAF		WEBSTER
YB 2000	n/a	RAuxAF	1st Clasp	WEBSTER
N621 1963	2684245	RAuxAF	wef 12/4/63	WEBSTER
N456 1948	128078	RAFVR		WEBSTER-GRINLING
LG 2004	02137420M	RAuxAF		WEEKES
N666 1956	146708	RAFVR		WEEKS
N197 1946	157321	RAFVR		WEIR
N277 1957	858247	RAuxAF	wef 5/7/44	WEIR
N257 1950	804448	RAuxAF		WELCH

Surname	Initials			Rank	Status	Date
WELCH	F	E		SGT		23/08/1951
WELCH	F	W		FL		15/07/1948
WELCH	L	W	H	FL		17/02/1944
WELCH	R	H		FL		28/02/1946
WELCH	T	M		FL		28/02/1946
WELDON	A	E	W	SQD LDR		04/11/1948
WELDON	D	R		FL		15/07/1948
WELDON	T	G	M	FL		04/09/1947
WELFORD	G	H	E	FL		26/01/1950
WELFORD	J	W		FO		03/12/1942
WELLER	E	J		LAC		05/01/1950
WELLER	R	L		SGT		23/02/1950
WELLER	W	G		SGT		14/10/1948
WELLER	W	J		SQD LDR	Acting	31/05/1945
WELLS	E	C		FL		15/07/1948
WELLS	F	W	G	FL	Acting	01/10/1953
WELLS	H	W		FL	Acting	08/05/1947
WELLS	J	C	A	SQD LDR		09/07/1995
WELLS	L	A		FLT SGT	Acting	01/12/1955
WELLS	P	H	V	SQD LDR		12/04/1945
WELLWOOD	W	D		FO		15/04/1943
WELSH	G	C		FLT SGT		04/05/1944
WELSH	J			SGT		15/09/1949
WELSHMAN	J	M		SQD LDR		26/04/1956
WENMAN	A	C	G	SQD LDR	Acting	13/12/1945
WENMAN	L	W		FL		04/10/1945
WENT	J	T		FL		29/04/1954
WERTHEIMER	C	R		SAC		25/03/1998
WESCOMBE	R	C		SQD LDR	Acting	21/03/1946
WESLEY	A	R		FL		13/12/1945
WESLEY	W	G		SQD LDR	Acting	28/02/1946
WEST	C	W		FL		28/02/1946
WEST	D			FL		16/03/2004
WEST	J	D		FL		07/02/1962
WEST	L			CPL		13/07/1950
WEST	P	G		FL		11/07/1946
WEST	S	C		FL		07/04/1955
WEST	S	O		SGT		01/02/1975
WEST	S	O		SGT		01/02/1985
WEST	W	A		FO		17/03/1949
WESTAWAY	A	J		WO		27/03/1947
WESTLAKE	G	H		SQD LDR		29/06/1944
WESTON	F			LAC		21/06/1951
WESTON-HAYS	B	M		CPL	Acting	14/11/1962
WESTWOOD	A	H		FO		23/02/1950
WESTWOOD	W	A		SGT		07/07/1949
WETHERALL	E	C		FO		16/04/1958
WETTON	W	H		SQD LDR		03/12/1942
WHARAM	D			FL		18/04/1946
WHARTON	P	D		SQD LDR		06/03/1947
WHEADON	B	E		FL	Acting	30/03/1943
WHEATLEY	G	W		CPL		13/12/1951
WHEATLEY	R	A		FLT SGT		20/05/1948
WHEATON	J	P		SGT		01/11/1992
WHEELER	F	D		SQD LDR	Acting	07/08/1947
WHEELER	H	E		WO		18/03/1948
WHEELER	H	W		SEC OFF		27/03/1947
WHEELER	M	E		WO		01/07/1948

Source	Service No	Force	Notes	Surname
N866 1951	859509	RAuxAF		WELCH
N573 1948	112080	RAFVR		WELCH
N132 1944	61463	RAFVR	DFC	WELCH
N197 1946	89312	RAFVR	AFC	WELCH
N197 1946	133266	RAFVR	AFC	WELCH
N934 1948	90534	RAuxAF	Bart; Sir	WELDON
N573 1948	82752	RAFVR		WELDON
N740 1947	91037	AAF		WELDON
N94 1950	90756	RAuxAF		WELFORD
N1638 1942	89619	RAFVR		WELFORD
N1 1950	841880	RAuxAF		WELLER
N179 1950	812324	RAuxAF		WELLER
N856 1948	819193	RAuxAF	Forfeited under N777 1951 due to GCM verdict; Re-instated 7/12/63.	WELLER
N573 1945	128355	RAFVR	DSO,DFC	WELLER
N573 1948	119513	RAFVR		WELLS
N796 1953	200863	RAuxAF		WELLS
N358 1947	147819	RAFVR		WELLS
YB 1997	5200660	RAuxAF		WELLS
N895 1955	2686025	RAuxAF		WELLS
N381 1945	72098	RAFVR	DSO	WELLS
N454 1943	126486	RAFVR		WELLWOOD
N425 1944	802355	AAF		WELSH
N953 1949	867275	RAuxAF		WELSH
N318 1956	73669	RAFVR		WELSHMAN
N1355 1945	85270	RAFVR		WENMAN
N1089 1945	79174	RAFVR		WENMAN
N329 1954	132463	RAuxAF		WENT
YB 1999	n/a	RAuxAF		WERTHEIMER
N263 1946	111104	RAFVR		WESCOMBE
N1355 1945	81860	RAFVR		WESLEY
N197 1946	115726	RAFVR	DFC	WESLEY
N197 1946	124003	RAFVR		WEST
LG 2004	0300865N	RAuxAF		WEST
N102 1962	2505685	RAFVR	wef 25/10/61	WEST
N675 1950	857531	RAuxAF		WEST
N594 1946	142022	RAFVR		WEST
N277 1955	141775	RAFVR		WEST
YB 1996	n/a	RAuxAF		WEST
YB 1996	n/a	RAuxAF	1st Clasp	WEST
N231 1949	108458	RAFVR		WEST
N250 1947	801462	AAF		WESTAWAY
N652 1944	84019	RAFVR	DFC	WESTLAKE
N621 1951	860477	RAuxAF		WESTON
N837 1962	2664308	WRAFVR	wef 25/9/62	WESTON-HAYS
N179 1950	187525	RAFVR		WESTWOOD
N673 1949	840056	RAuxAF		WESTWOOD
N258 1958	2603438	RAFVR	wef 26/2/56	WETHERALL
N1638 1942	90111	AAF		WETTON
N358 1946	171690	RAFVR		WHARAM
N187 1947	73560	RAFVR		WHARTON
N1014 1943	108264	RAFVR	MBE	WHEADON
N1271 1951	842452	RAuxAF		WHEATLEY
N395 1948	841294	RAuxAF		WHEATLEY
YB 1996	n/a	RAuxAF		WHEATON
N662 1947	72379	RAFVR		WHEELER
N232 1948	744386	RAFVR		WHEELER
N250 1947	2854	WAAF		WHEELER
N527 1948	770005	RAFVR	BEM	WHEELER

Surname	Initials		Rank	Status	Date
WHEELER	N	J	FL		17/12/1942
WHEELER	S	H	LAC		04/11/1948
WHEELER	S	S	GRP CAP	Acting	26/06/1947
WHEELER	W	G	FO		10/06/1948
WHEELEY	P	L	FL		26/08/1948
WHEELOCK	T	P	FL		30/09/1954
WHEELWRIGHT	W	E	PO		07/12/1960
WHERRY	F	L	FO		06/03/1947
WHICHELOW	A	J	SGT		04/09/1947
WHIFFEN	G	K	FL		26/01/1950
WHISSELL	H	B	WING COMM	Acting	20/05/1948
WHITE	A	E	FO		07/06/1945
WHITE	C	R	FO		26/01/1950
WHITE	E	H	FL		08/11/1945
WHITE	G	C	SQD LDR		28/06/1945
WHITE	G	J	LAC		15/09/1949
WHITE	G	P	FL		23/06/1955
WHITE	G	P	FL		28/11/1956
WHITE	J	C	FL		20/04/1944
WHITE	J	S	FO		04/10/1945
WHITE	J	S	SQD LDR	Acting	26/08/1948
WHITE	J		FO		20/01/1960
WHITE	L	A	S	SQD LDR	02/07/1958
WHITE	L	N	SQD LDR		17/04/1947
WHITE	L	W	C	FL	10/06/1948
WHITE	N		LAC		22/02/1951
WHITE	P	D	FL		24/01/1946
WHITE	S	J	T	CPL	15/07/1948
WHITE	T	P	H	SGT	13/01/1949
WHITE	W	R	FL		13/07/1950
WHITE	W		LAC		10/12/1942
WHITEFORD	W	G	FL		24/01/1946
WHITEHEAD	F	E	CPL		15/09/1949
WHITEHEAD	H	M	SQD LDR	Acting	05/01/1950
WHITEHEAD	J	W	W	WING COMM	Acting	05/06/1947
WHITEHEAD	R	O	FL		11/07/1946
WHITEHOUSE	H	D	SGT		10/06/1948
WHITEHOUSE	H		LAC		07/04/1955
WHITEHOUSE	S	A	H	WING COMM	Acting	13/12/1945
WHITEMAN	W	E	FL		06/03/1947
WHITEN	W	H	CPL		06/05/1943
WHITESIDE	A	B	CPL		04/09/1947
WHITESIDE	C	G	J	SQD LDR	14/09/1950
WHITFIELD	F	A	FL	Acting	27/04/1950
WHITFIELD	G		FL		08/11/1945
WHITING	A	A	SGT		05/06/1947
WHITING	T	F	C SGT		20/08/1958
WHITLEY	H	E	FL		24/07/1999
WHITLOCK	P	E	SEC OFF		27/03/1947
WHITMORE	A		LAC		30/10/1957
WHITMORE	H		FLT SGT		04/11/1948
WHITNELL	C	L	FL		08/11/1945
WHITNEY	R	J	SQD LDR		04/07/1998
WHITTAKER	A	R	FL		20/01/1960
WHITTAKER	W		FO		28/02/1946
WHITTARD	G	R	SQD LDR	Acting	28/02/1946
WHITTET	J	A	SQD LDR	Acting	24/01/1946
WHITTINGHAM	C	D	SQD LDR	Acting	24/12/1942
WHITTY	W	H	R	SQD LDR	30/11/1944
WHITWORTH	N		FLT SGT		15/09/1994

Source	Service No	Force	Notes	Surname
N1708 1942	82668	RAFVR	AFC	WHEELER
N934 1948	846163	RAuxAF		WHEELER
N519 1947	91127	AAF		WHEELER
N456 1948	121749	RAFVR		WHEELER
N694 1948	75296	RAFVR		WHEELEY
N799 1954	91134	RAuxAF		WHEELOCK
N933 1960	2650863	RAuxAF	wef 10/10/60	WHEELWRIGHT
N187 1947	202337	RAFVR		WHERRY
N740 1947	842329	AAF		WHICHELOW
N94 1950	81668	RAFVR	Deceased	WHIFFEN
N395 1948	72273	RAFVR		WHISSELL
N608 1945	145471	RAFVR		WHITE
N94 1950	108589	RAFVR		WHITE
N1232 1945	131480	RAFVR		WHITE
N706 1945	90275	AAF		WHITE
N953 1949	747077	RAFVR		WHITE
N473 1955	202551	RAFVR		WHITE
N833 1956	105842	RAFVR		WHITE
N363 1944	61234	RAFVR		WHITE
N1089 1945	128434	RAFVR		WHITE
N694 1948	76260	RAFVR		WHITE
N33 1960	2655046	WRAAF	wef 13/6/59	WHITE
N452 1958	155890	RAuxAF	MB,BCh; wef 4/1/58	WHITE
N289 1947	72208	RAFVR		WHITE
N456 1948	104384	RAFVR		WHITE
N194 1951	868770	RAuxAF		WHITE
N82 1946	86403	RAFVR		WHITE
N573 1948	800515	RAuxAF		WHITE
N30 1949	843412	RAuxAF		WHITE
N675 1950	85717	RAFVR		WHITE
N1680 1942	321738	EX-AAF		WHITE
N82 1946	158651	RAFVR		WHITEFORD
N953 1949	848372	RAuxAF		WHITEHEAD
N1 1950	90749	RAuxAF		WHITEHEAD
N460 1947	73456	RAFVR		WHITEHEAD
N594 1946	118049	RAFVR		WHITEHEAD
N456 1948	848370	RAuxAF		WHITEHOUSE
N277 1955	2677661	RAuxAF		WHITEHOUSE
N1355 1945	88438	RAFVR		WHITEHOUSE
N187 1947	133516	RAFVR		WHITEMAN
N526 1943	805169	AAF		WHITEN
N740 1947	856624	AAF		WHITESIDE
N926 1950	90741	RAuxAF		WHITESIDE
N406 1950	109516	RAFVR		WHITFIELD
N1232 1945	126158	RAFVR	DFM	WHITFIELD
N460 1947	840489	AAF		WHITING
N559 1958	22982854	TA	wef 21/6/44	WHITING
YB 2000	212474	RAuxAF		WHITLEY
N250 1947	2619	WAAF		WHITLOCK
N775 1957	857403	AAF	wef 1/6/44	WHITMORE
N934 1948	848592	RAuxAF		WHITMORE
N1232 1945	116064	RAFVR		WHITNELL
YB 1999	n/a	RAuxAF		WHITNEY
N33 1960	186705	RAFVR	wef 26/1/53	WHITTAKER
N197 1946	170600	RAFVR		WHITTAKER
N197 1946	79742	RAFVR	DFC	WHITTARD
N82 1946	84013	RAFVR	DFC	WHITTET
N1758 1942	70734	RAFVR		WHITTINGHAM
N1225 1944	90288	AAF		WHITTY
YB 1996	n/a	RAuxAF		WHITWORTH

Surname	Initials			Rank	Status	Date
WHORLOW	G	K		FL		21/01/1959
WHORLOW	G	K		FL		16/02/1964
WHYARD	R	D		FL		09/08/1945
WHYTE	A	McL		LAC		13/01/1949
WIARD	C	A		FL		07/08/1947
WICKENDEN	E	C		FL		15/11/1961
WICKHAM	W	N		FL		21/03/1946
WICKINS	C			SGT		22/04/1943
WICKMAN	P	A	G	FL		09/11/1944
WICKS	H	F		SQD LDR	Acting	14/10/1948
WIFFEN	A	J		FL		04/10/1945
WIGGINS	J			FLT SGT		17/03/1949
WIGGLESWORTH	A	W		LAC		02/05/1962
WIGGLESWORTH	R	A		FL	Acting	01/05/1952
WIGHT-BOYCOTT	C	M		WING COMM	Acting	10/08/1944
WIGHTMAN	R	W	F	FL		21/03/1946
WIGMORE	F	G		FO		13/01/1949
WIGMORE	S	E		FL		25/05/1960
WILBERFORCE	R	G		FL		30/03/1943
WILBY	F	J		SGT		23/02/1950
WILBY	R	A		FL		26/11/1953
WILCOCKSON	M	C		SQD OFF	Acting	30/09/1948
WILCOX	A	J		SGT		13/01/1949
WILCOX	F	W		SGT		12/06/1947
WILCOX	W	A		FL		31/05/1945
WILD	E	F		SGT		15/04/1943
WILD	S	W		FL		24/01/1946
WILDE	D	C		SQD LDR	Acting	24/01/1946
WILES	H	B		FLT SGT	Acting	13/10/1955
WILFORD	F	B		FLT SGT	Acting	27/04/1950
WILFORD	S			FL		06/03/1947
WILKES	A			FL		13/12/1945
WILKES	H	E		FO		10/06/1948
WILKIE	C	B		LAC		26/08/1948
WILKIE	D			SQD LDR	Acting	27/03/1947
WILKINS	A	E		WO		13/12/1945
WILKINS	N	J	Mcl.	FO		09/08/1945
WILKINS	R	W		FO		26/01/1950
WILKINSON	A	T		FL		10/08/1950
WILKINSON	A	W		FLT SGT		01/10/1953
WILKINSON	B	D	C	FL		10/05/1945
WILKINSON	E	H		FL		12/04/1951
WILKINSON	F	C		SGT		16/10/1947
WILKINSON	F			SGT		20/05/1948
WILKINSON	G	L		SGT		05/02/1948
WILKINSON	G			LAC		25/06/1953
WILKINSON	H	B		SGT		20/05/1948
WILKINSON	H			WO		28/02/1946
WILKINSON	H			FO		16/01/1957
WILKINSON	J	E		CPL		25/05/1960
WILKINSON	J	G	G	FO		07/01/1943
WILKINSON	K	A		FO		13/12/1945
WILKINSON	P	A		CPL		18/03/1948
WILKS	J	M		CPL		13/01/1949
WILKS	J			FL		30/03/1960
WILLETT	J	B		CPL		13/01/1949
WILLGOSS	E	S		LAC		14/10/1948
WILLIAMS	A	V		LAC		14/01/1954
WILLIAMS	A			WING COMM	Acting	05/01/1950
WILLIAMS	B	A		FL		16/01/1957

Source	Service No	Force	Notes	Surname
N28 1959	137515	RAFVR	wef 16/2/54	WHORLOW
MOD 1964	137515	RAFVR	1st Clasp	WHORLOW
N874 1945	104335	RAFVR	DFC,AFC	WHYARD
N30 1949	841709	RAuxAF		WHYTE
N662 1947	86446	RAFVR		WIARD
N902 1961	151383	RAFVR	wef 25/8/61	WICKENDEN
N263 1946	155487	RAFVR		WICKHAM
N475 1943	800022	AAF		WICKINS
N1150 1944	81338	RAFVR		WICKMAN
N856 1948	85205	RAFVR		WICKS
N1089 1945	84337	RAFVR		WIFFEN
N231 1949	810066	RAuxAF		WIGGINS
N314 1962	2684618	RAuxAF	wef 1/6/58	WIGGLESWORTH
N336 1952	133363	RAuxAF	DFC	WIGGLESWORTH
N812 1944	72005	RAFVR	DSO	WIGHT-BOYCOTT
N263 1946	79165	RAFVR	AFC	WIGHTMAN
N30 1949	106024	RAFVR		WIGMORE
N375 1960	177278	RAFVR	wef 5/9/57	WIGMORE
N1014 1943	70736	RAFVR	DFC	WILBERFORCE
N179 1950	755661	RAFVR		WILBY
N952 1953	145550	RAuxAF		WILBY
N805 1948	1805	WAAF		WILCOCKSON
N30 1949	702078	RAFVR		WILCOX
N478 1947	801424	AAF		WILCOX
N573 1945	126696	RAFVR		WILCOX
N454 1943	805188	AAF		WILD
N82 1946	86682	RAFVR		WILD
N82 1946	83291	RAFVR		WILDE
N762 1955	770357	RAFVR		WILES
N406 1950	760276	RAFVR		WILFORD
N187 1947	86677	RAFVR		WILFORD
N1355 1945	142857	RAFVR		WILKES
N456 1948	123703	RAFVR		WILKES
N694 1948	874124	RAuxAF		WILKIE
N250 1947	62103	RAFVR		WILKIE
N1355 1945	812229	AAF	DFM	WILKINS
N874 1945	147609	RAFVR		WILKINS
N94 1950	134651	RAFVR	Deceased	WILKINS
N780 1950	130316	RAFVR		WILKINSON
N796 1953	2603312	RAFVR		WILKINSON
N485 1945	90023	AAF		WILKINSON
N342 1951	78142	RAFVR		WILKINSON
N846 1947	844044	AAF		WILKINSON
N395 1948	815251	RAuxAF		WILKINSON
N87 1948	869320	RAuxAF		WILKINSON
N527 1953	868553	RAuxAF		WILKINSON
N395 1948	756921	RAFVR		WILKINSON
N197 1946	749428	RAFVR		WILKINSON
N24 1957	2685622	RAuxAF	DFM	WILKINSON
N375 1960	2656364	WRAAF	wef 13/3/60	WILKINSON
N3 1943	61233	RAFVR		WILKINSON
N1355 1945	172142	RAFVR		WILKINSON
N232 1948	869370	RAuxAF		WILKINSON
N30 1949	859623	RAuxAF		WILKS
N229 1960	199557	RAFVR	wef 31/1/60	WILKS
N30 1949	747297	RAFVR		WILLETT
N856 1948	840828	RAuxAF		WILLGOSS
N18 1954	850221	RAuxAF		WILLIAMS
N1 1950	72433	RAFVR		WILLIAMS
N24 1957	142973	RAFVR		WILLIAMS

Surname	Initials				Rank	Status	Date
WILLIAMS	C	E			WO		29/06/1944
WILLIAMS	C	F			FLT SGT		30/09/1948
WILLIAMS	C	T			FO		02/09/1943
WILLIAMS	D	F			FL		07/07/1949
WILLIAMS	E	A			FL	Acting	17/04/1947
WILLIAMS	E	S			WING COMM	Acting	11/05/1944
WILLIAMS	F	A			SQD LDR	Acting	13/12/1945
WILLIAMS	G	H			SQD LDR		27/03/1947
WILLIAMS	G	H			FO		07/07/1949
WILLIAMS	G	M			CPL		23/02/1950
WILLIAMS	G				FO		23/05/1946
WILLIAMS	H	G			SQD LDR	Acting	09/02/1951
WILLIAMS	H	J			SQD LDR		22/04/1943
WILLIAMS	H	T			SGT		01/05/1963
WILLIAMS	H				SGT		14/10/1948
WILLIAMS	H				LAC		27/04/1950
WILLIAMS	J	C			FO		10/06/1948
WILLIAMS	J	F			WO		28/02/1946
WILLIAMS	J	H			WING COMM		11/01/1945
WILLIAMS	J	H	M	B	FO		09/11/1950
WILLIAMS	J				FL		24/01/1946
WILLIAMS	J				CPL		26/01/1950
WILLIAMS	L	C			LAC		05/06/1947
WILLIAMS	L	S			LAC		15/09/1949
WILLIAMS	M				SACW		03/07/1957
WILLIAMS	N	E			FL		06/03/1947
WILLIAMS	O	M			FLT OFF		21/06/1951
WILLIAMS	P	R			CPL		05/02/1948
WILLIAMS	P	R	E		FLT SGT		08/01/1998
WILLIAMS	R	B			SQD LDR		02/03/2001
WILLIAMS	R	C			WING COMM	Acting	13/01/1949
WILLIAMS	R	T	G		FL		16/01/1957
WILLIAMS	S	J			FL		03/06/1959
WILLIAMS	S	R			LAC		07/07/1949
WILLIAMS	T	A			SQD LDR	Acting	27/04/1950
WILLIAMS	T	D			SQD LDR	Acting	13/12/1945
WILLIAMS	T	R			FL		31/10/1956
WILLIAMS	T	W			SQD LDR	Acting	27/03/1947
WILLIAMS	W	D			SQD LDR	Acting	25/01/1945
WILLIAMS	W	H			AC1		24/05/2005
WILLIAMSON	C	M			WO		28/02/1946
WILLIAMSON	C	W			CPL		15/07/1948
WILLIAMSON	J	A			SGT		12/10/1950
WILLIAMSON	L	A			CPL		30/09/1948
WILLIAMSON	S	V			GRP OFF		27/03/1947
WILLINGHAM	A	H			FL		30/10/1957
WILLINGHAM	E	A			CPL	Acting	26/11/1958
WILLIS	C	H			SQD LDR	Acting	30/12/1943
WILLIS	C	H			FO		17/01/1952
WILLIS	C	M			WING OFF	Acting	08/05/1947
WILLIS	F				SGT		14/09/1950
WILLIS	H	R			WO		07/06/1945
WILLIS	L				CPL		10/08/1950
WILLIS	W	H			SGT		26/11/1953
WILLIS	W	H			CPL		21/10/1959
WILLIS-FRANCIS	P				FO		24/07/1952
WILLOUGHBY	J	H			SGT		01/08/1987
WILLOUGHBY	J	H			SGT		01/08/1997
WILLS	C	W			CPL		10/12/1942
WILLS	E	D			FLT SGT		04/09/1947

Source	Service No	Force	Notes	Surname
N652 1944	740534	RAFVR		WILLIAMS
N805 1948	849107	RAuxAF		WILLIAMS
N918 1943	126105	RAFVR		WILLIAMS
N673 1949	72412	RAFVR		WILLIAMS
N289 1947	198911	RAFVR		WILLIAMS
N444 1944	90021	AAF		WILLIAMS
N1355 1945	116661	RAFVR	DFC	WILLIAMS
N250 1947	72971	RAFVR		WILLIAMS
N673 1949	177598	RAFVR		WILLIAMS
N179 1950	755189	RAFVR		WILLIAMS
N468 1946	174917	RAFVR	DFM	WILLIAMS
N145 1951	66041	RAFVR		WILLIAMS
N475 1943	90303	AAF		WILLIAMS
N333 1963	750841	RAF	wef 10/8/44	WILLIAMS
N856 1948	853593	RAuxAF		WILLIAMS
N406 1950	850271	RAuxAF		WILLIAMS
N456 1948	162275	RAFVR		WILLIAMS
N197 1946	741584	RAFVR		WILLIAMS
N14 1945	90144	AAF		WILLIAMS
N1113 1950	123705	RAFVR		WILLIAMS
N82 1946	124667	RAFVR	DFC,AFC	WILLIAMS
N94 1950	850270	RAuxAF		WILLIAMS
N460 1947	744490	RAFVR		WILLIAMS
N953 1949	865635	RAuxAF		WILLIAMS
N456 1957	2657241	WRAAF	wef 9/4/57	WILLIAMS
N187 1947	90934	AAF	DFC	WILLIAMS
N621 1951	2558	WAAF		WILLIAMS
N87 1948	744505	RAuxAF		WILLIAMS
YB 1999	n/a	RAuxAF		WILLIAMS
YB 2001	211366	RAuxAF		WILLIAMS
N30 1949	73481	RAFVR	MC	WILLIAMS
N24 1957	197717	RAuxAF	AFC	WILLIAMS
N368 1959	157899	RAFVR	wef 28/8/53	WILLIAMS
N673 1949	846352	RAuxAF		WILLIAMS
N406 1950	72972	RAFVR	MC	WILLIAMS
N1355 1945	90658	AAF	DFC	WILLIAMS
N761 1956	745088	RAFVR		WILLIAMS
N250 1947	78149	RAFVR		WILLIAMS
N70 1945	78985	RAFVR	DFC	WILLIAMS
LG 2005	751764	RAFVR		WILLIAMS
N197 1946	742545	RAFVR		WILLIAMSON
N573 1948	846461	RAuxAF		WILLIAMSON
N1018 1950	816195	RAuxAF		WILLIAMSON
N805 1948	743167	RAFVR		WILLIAMSON
N250 1947	5	WAAF		WILLIAMSON
N775 1957	201659	RAFVR	wef 24/7/52	WILLINGHAM
N807 1958	2693514	RAuxAF	wef 29/10/58	WILLINGHAM
N1350 1943	73013	RAFVR	AFC	WILLIS
N35 1952	73013	RAFVR	AFC; 1st Clasp	WILLIS
N358 1947	118	WAAF		WILLIS
N926 1950	800026	RAuxAF		WILLIS
N608 1945	741942	RAFVR		WILLIS
N780 1950	750140	RAFVR		WILLIS
N952 1953	2699561	RAFVR		WILLIS
N708 1959	2699561	RAuxAF	1st Clasp; wef 19/12/56	WILLIS
N569 1952	131782	RAuxAF		WILLIS-FRANCIS
YB 1999	n/a	RAuxAF		WILLOUGHBY
YB 1999	n/a	RAuxAF	1st Clasp	WILLOUGHBY
N1680 1942	805166	AAF		WILLS
N740 1947	813135	AAF		WILLS

Surname	Initials			Rank	Status	Date
WILLS	J	F		SQD LDR	Acting	05/10/1944
WILLS	S	B		SQD LDR	Acting	05/01/1950
WILSON	A	D		SAC		08/07/1959
WILSON	A	G		WING COMM	Acting	03/06/1943
WILSON	A	McC		GRP CAP		05/01/1950
WILSON	A	McP		SGT		15/07/1948
WILSON	A	S	A	SGT	Acting	08/07/1959
WILSON	A	W	G	FL		20/08/1958
WILSON	A			CPL		10/02/1949
WILSON	A			SQD LDR		23/02/1950
WILSON	A			CPL		01/10/1989
WILSON	A			CPL		01/10/1999
WILSON	B			FLT SGT		19/01/1994
WILSON	C	B		LAC		05/06/1947
WILSON	C	G	J	SGT		08/05/1947
WILSON	D	G		FL		01/05/1963
WILSON	D	S		SQD LDR		30/12/1943
WILSON	D	W		SAC		03/10/1962
WILSON	E	B		SQD LDR		05/06/1947
WILSON	F			FL		13/12/1945
WILSON	G	J		WO		30/03/1943
WILSON	G	R	A	SQD LDR		11/07/1946
WILSON	G			SGT		20/05/1948
WILSON	G			SGT		01/11/1951
WILSON	G			SGT		21/06/1961
WILSON	H	J		FL		08/07/1959
WILSON	H	P		SGT		10/02/1949
WILSON	H			SQD LDR	Acting	01/07/1948
WILSON	J	A		FLT SGT		10/08/1944
WILSON	J	B		FL		03/10/1946
WILSON	J	K		FL		27/07/1944
WILSON	J	W		SQD LDR		17/04/1947
WILSON	J	W		SGT		16/10/1952
WILSON	J			SGT		13/12/1945
WILSON	J			SQD LDR	Acting	01/11/1951
WILSON	L	G		LAC		01/07/1948
WILSON	L			FL		28/06/1945
WILSON	N	L		SQD LDR	Acting	27/03/1947
WILSON	P	E		FL		04/09/1957
WILSON	P			FL		10/02/1949
WILSON	R	E		WING COMM		15/07/1948
WILSON	R	J	S	SQD LDR		30/09/1954
WILSON	R			FL		18/04/1946
WILSON	S	C	A	FO		05/06/1947
WILSON	T	A	C	LAC		16/03/1950
WILSON	V			FL		23/05/1946
WILSON	W	A		FL		12/02/1953
WILSON	W	B		WING COMM	Acting	14/10/1943
WILSON	W	B		FL		01/11/1951
WILSON	W	C		FL		25/01/1945
WILSON	W	H		FL		09/05/1946
WILSON				CPL		06/11/1997
WILTON-JONES	L	T		FO		21/09/1944
WIMHURST	T	J		SGT		10/02/1949
WINCH	F	E		FO		19/08/1954
WINCOTE	G	L		WO		30/08/1945
WINDER	A	J		FL		20/04/1944
WINDSOR	H	C		WING COMM	Acting	16/10/1947
WINDSOR	M			SEC OFF		18/03/1948
WINFIELD	R	J		FL	Acting	24/01/1946

Source	Service No	Force	Notes	Surname
N1007 1944	80549	RAFVR	DFC	WILLS
N1 1950	83735	RAFVR		WILLS
N454 1959	2685036	RAFVR	wef 20/7/58	WILSON
N609 1943	72114	RAFVR	MC,MRCS,LRCP	WILSON
N1 1950	72973	RAFVR	OBE; Deceased	WILSON
N573 1948	744007	RAFVR		WILSON
N454 1959	2659192	WRAAF	wef 11/11/58	WILSON
N559 1958	193487	RAFVR	wef 21/12/53	WILSON
N123 1949	803444	RAuxAF		WILSON
N179 1950	72347	RAFVR	MB,ChB	WILSON
YB 2000	n/a	RAuxAF		WILSON
YB 2000	n/a	RAuxAF	1st Clasp	WILSON
YB 1997	n/a	RAuxAF		WILSON
N460 1947	858219	AAF		WILSON
N358 1947	812003	AAF		WILSON
N333 1963	2614248	RAFVR	wef 25/8/62	WILSON
N1350 1943	90342	AAFRO		WILSON
N722 1962	2683686	RAuxAF	wef 17/8/62	WILSON
N460 1947	72348	RAFVR		WILSON
N1355 1945	115948	RAFVR		WILSON
N1014 1943	801296	AAF		WILSON
N594 1946	90069	AAF		WILSON
N395 1948	815082	RAuxAF		WILSON
N1126 1951	2684511	RAuxAF	1st Clasp	WILSON
N478 1961	2684511	RAFVR	2nd Clasp; wef 15/4/61	WILSON
N454 1959	56795	RAuxAF	wef 26/5/59	WILSON
N123 1949	853890	RAuxAF		WILSON
N527 1948	136581	RAFVR		WILSON
N812 1944	802255	AAF		WILSON
N837 1946	101509	RAFVR		WILSON
N763 1944	60301	RAFVR	AFC	WILSON
N289 1947	61150	RAFVR		WILSON
N798 1952	799939	RAFVR		WILSON
N1355 1945	802313	AAF		WILSON
N1126 1951	91174	RAuxAF	MC	WILSON
N527 1948	847048	RAuxAF		WILSON
N706 1945	127028	RAFVR		WILSON
N250 1947	110079	RAFVR		WILSON
N611 1957	136176	RAFVR	DFC; wef 21/4/53	WILSON
N123 1949	91192	RAuxAF		WILSON
N573 1948	90941	RAuxAF		WILSON
N799 1954	67546	RAuxAF		WILSON
N358 1946	106190	RAFVR		WILSON
N460 1947	149445	RAFVR		WILSON
N257 1950	808299	RAuxAF		WILSON
N468 1946	179842	RAFVR		WILSON
N115 1953	163213	RAuxAF		WILSON
N1065 1943	70742	RAFVR		WILSON
N1126 1951	70742	RAFVR	1st Clasp	WILSON
N70 1945	130108	RAFVR	AFC	WILSON
N415 1946	171799	RAFVR		WILSON
YB 1999	n/a	RAuxAF		WILSON
N961 1944	142539	RAFVR		WILTON-JONES
N123 1949	812098	RAuxAF		WIMHURST
N672 1954	2682542	RAuxAF		WINCH
N961 1945	742275	RAFVR		WINCOTE
N363 1944	64893	RAFVR		WINDER
N846 1947	72247	RAFVR		WINDSOR
N232 1948	7510	WAAF		WINDSOR
N82 1946	183186	RAFVR		WINFIELD

Surname	Initials			Rank	Status	Date
WINNER	G	L		CPL		16/03/1950
WINNING	T	N	G	FO		17/12/1942
WINSKILL	A	L		WING COMM	Acting	20/04/1944
WINSLAND	S			SGT		28/09/1950
WINSLAND	S			FLT SGT		19/01/1956
WINTER	C	R		SAC		20/10/1994
WINTER	R	A		FL		27/01/1944
WINTERBURN	R	L		FL		04/10/1951
WINTON	G			SAC		30/11/1993
WISE	L	G		FL		10/05/1945
WISEMAN	E	W		FL		21/03/1946
WISHER	R	R		SQD LDR		03/06/1943
WISKEN	G	W		AC1		17/06/1954
WITHELL	J	A		LAC		18/11/1948
WITHERDEN	E	C		SGT		30/09/1948
WITHERS	J	R		CPL	Temporary	14/11/1962
WITTS	A	C		FL		25/06/1953
WODEHOUSE	P	A		FL		30/03/1960
WOLFENDEN	H	R		FL		18/04/1946
WOLFF	L			FLT SGT		15/07/1948
WOLSTENHOLME	K			FL		18/04/1946
WOLSTENHOLME	T			FL		08/11/1945
WOMACK	W	T	A	SQD LDR	Acting	28/02/1946
WOOD	A	W		SQD LDR	Acting	05/10/1944
WOOD	E	H		SQD LDR	Acting	26/08/1948
WOOD	E	H		CPL	Temporary	21/01/1959
WOOD	F	E		FL		23/06/1955
WOOD	F	E		WING COMM	Acting	18/01/1964
WOOD	F	R		FL		01/03/1945
WOOD	G			FLT SGT		24/05/2005
WOOD	H	E		WO		27/07/1944
WOOD	H	E		SQD LDR	Acting	09/11/1950
WOOD	H			FL		17/12/1942
WOOD	J	D		FL		04/02/1943
WOOD	J	M		FL		24/01/1946
WOOD	J			WO		24/12/1942
WOOD	L	H		FL		17/02/1944
WOOD	M			CPL		07/12/1960
WOOD	P	H		SQD LDR		26/06/1947
WOOD	R	B		SGT		01/05/1982
WOOD	R	B		SGT		01/05/1992
WOOD	R			CPL		17/03/1949
WOOD	T	J		LAC		13/01/1949
WOOD	W	D		SGT		26/11/1953
WOOD	W	H		LAC		26/01/1950
WOOD	W	J		WO		09/08/1945
WOOD	W			WO		13/01/1949
WOODALL	P	J		SGT	Temporary	22/10/1958
WOODCOCK	S			LAC		14/10/1948
WOODEN	W	A		FL		14/01/1954
WOODGATE	M	L		FL		18/05/1944
WOODHALL	H	C		CPL		08/05/1947
WOODHEAD	H	C		FL		19/10/1960
WOODHEAD	L			FL		07/04/1955
WOODHEAD	N	E		WING OFF	Acting	06/03/1947
WOODHEAD	N	H		WING COMM		27/04/1944
WOODHEAD	S	J	M	FL		23/11/1997
WOODHOUSE	A	J		FL		21/01/1959
WOODHOUSE	J	M		FL		09/05/1946
WOODIN	J	E		FO		17/04/1947

Source	Service No	Force	Notes	Surname
N257 1950	808192	RAuxAF		WINNER
N1708 1942	116985	RAFVR		WINNING
N363 1944	84702	RAFVR	DFC	WINSKILL
N968 1950	853568	RAuxAF		WINSLAND
N43 1956	2697006	RAuxAF	1st Clasp	WINSLAND
YB 1996	n/a	RAuxAF		WINTER
N59 1944	73015	RAFVR		WINTER
N1019 1951	72227	RAFVR		WINTERBURN
YB 1996	n/a	RAuxAF		WINTON
N485 1945	109010	RAFVR		WISE
N263 1946	127787	RAFVR		WISEMAN
N609 1943	72228	RAFVR		WISHER
N471 1954	844786	RAuxAF		WISKEN
N984 1948	866345	RAuxAF		WITHELL
N805 1948	841006	RAuxAF		WITHERDEN
N837 1962	848353	AAF	wef 30/5/44	WITHERS
N527 1953	151644	RAFVR		WITTS
N229 1960	2606203	RAuxAF	wef 17/2/60	WODEHOUSE
N358 1946	136168	RAFVR		WOLFENDEN
N573 1948	880187	RAuxAF		WOLFF
N358 1946	121908	RAFVR	DFC	WOLSTENHOLME
N1232 1945	158795	RAFVR		WOLSTENHOLME
N197 1946	79379	RAFVR		WOMACK
N1007 1944	106086	RAFVR	DFC,BEM	WOOD
N694 1948	79588	RAFVR		WOOD
N28 1959	812103	AAF	wef 17/1/44	WOOD
N473 1955	85532	RAFVR		WOOD
MOD 1964	85532	RAuxAF	1st Clasp	WOOD
N225 1945	81669	RAFVR		WOOD
LG 2005	817234	RAFVR		WOOD
N763 1944	740542	RAFVR		WOOD
N1113 1950	87364	RAFVR	DFC	WOOD
N1708 1942	70746	RAFVR		WOOD
N131 1943	70747	RAFVR		WOOD
N82 1946	88211	RAFVR		WOOD
N1758 1942	802048	AAF		WOOD
N132 1944	73017	RAFVR		WOOD
N933 1960	2656736	WRAAF	wef 21/9/60	WOOD
N519 1947	73490	RAFVR		WOOD
YB 1996	n/a	RAuxAF		WOOD
YB 1996	n/a	RAuxAF	1st Clasp	WOOD
N231 1949	857614	RAuxAF		WOOD
N30 1949	812155	RAuxAF		WOOD
N952 1953	2696008	RAuxAF		WOOD
N94 1950	750126	RAFVR		WOOD
N874 1945	742244	RAFVR		WOOD
N30 1949	749510	RAFVR		WOOD
N716 1958	813026	AAF	wef 28/12/42	WOODALL
N856 1948	857568	RAuxAF		WOODCOCK
N18 1954	153966	RAFVR		WOODEN
N473 1944	118710	RAFVR		WOODGATE
N358 1947	847538	AAF		WOODHALL
N775 1960	114654	RAFVR	wef 3/11/55	WOODHEAD
N277 1955	80516	RAuxAF		WOODHEAD
N187 1947	570	WAAF		WOODHEAD
N396 1944	70750	RAFVR	DSC,AFC	WOODHEAD
YB 2000	91441	RAuxAF		WOODHEAD
N28 1959	72736	RAFVR	wef 22/5/44	WOODHOUSE
N415 1946	156022	RAFVR		WOODHOUSE
N289 1947	202156	RAFVR		WOODIN

Surname	Initials			Rank	Status	Date
WOODLEY	E	W		FL		04/10/1945
WOODLEY	J	H		CPL	Acting	16/01/1957
WOODLEY	W	L		SQD OFF		05/06/1947
WOODRUFF	C	G		FL		30/09/1954
WOODS	A	H		SQD LDR		30/12/1943
WOODS	A	J		CPL		30/09/1948
WOODS	A			SGT		17/03/1949
WOODS	D	N		FL	Acting	28/10/1943
WOODS	E	H		FL		09/05/1946
WOOLASTON	J			CPL		05/01/1950
WOOLBERT	R	F		SGT		21/03/1946
WOOLCOCK	J	W		WING COMM		31/05/1945
WOOLDRIDGE	J	de L		WING COMM	Acting	12/08/1943
WOOLEY	J	W		FO		17/01/1952
WOOLLATT	J	G		SQD LDR	Acting	08/11/1945
WOOLLEY	A	W		FL		08/11/1945
WOOLNOUGH	A	S		FL		23/05/1946
WOOLNOUGH	H	M		FLT OFF		28/07/1955
WOOLSTON	B			FO		27/04/1944
WOOLVERIDGE	A	M		SGT		26/08/1948
WOOSTER	K	R		CPL		04/09/1947
WOOTTON	J	W		WO		02/08/1951
WOOTTON	J	W		FO		21/10/1959
WORBOYS	A			CPL		24/09/1998
WORBY	J	R		FL		08/07/1959
WORDIE	P			SGT		06/03/1947
WORDSELL	L	V		FL	Acting	10/12/1942
WORDSELL	L	V		SQD LDR		20/09/1951
WORDSELL	L	V		FL		04/09/1957
WORDSWORTH	J	T		WING COMM		20/01/1960
WORMERSLEY	P	A		SQD LDR	Acting	28/02/1946
WORRALL	C			CPL	Temporary	25/05/1960
WORRALL	T	C		FL		18/07/1956
WORTERS	S	J		LAC		27/03/1947
WORTH	D			SQD LDR	Acting	09/05/1946
WORTHINGTON	A	R		SGT		05/02/1948
WORTHINGTON	A	S		SQD LDR		04/10/1945
WORTLEY	C	E		FL		20/05/1948
WRAGG	A	A		SGT		02/09/1998
WRAGG	E	R		SGT		18/03/1948
WRAGG	G			FL		06/03/1947
WREN	E	A		FLT SGT	Acting	24/12/1942
WRIGHT	A	E		WO		24/06/1943
WRIGHT	A	G		FL		16/03/1950
WRIGHT	A	T		FO		27/03/1947
WRIGHT	C	D	C	FL		02/05/1962
WRIGHT	E	J		SGT		24/11/1949
WRIGHT	E	W		FL		21/03/1946
WRIGHT	E			FL		08/02/1945
WRIGHT	F	J		SGT		17/06/1954
WRIGHT	F	W		FL		03/06/1959
WRIGHT	G	A		FL		07/07/1949
WRIGHT	G	A	C	FL		09/12/1959
WRIGHT	G	C		FL		20/04/1944
WRIGHT	G	J	D	CPL		06/01/1955
WRIGHT	G	J		WING COMM	Acting	21/06/1961
WRIGHT	G	L		SQD LDR		27/04/1950
WRIGHT	G	S		CPL		27/02/1963
WRIGHT	G	V		FL		13/01/1949

Source	Service No	Force	Notes	Surname
N1089 1945	147989	RAFVR	DFC	WOODLEY
N24 1957	2601043	RAFVR		WOODLEY
N460 1947	674	WAAF		WOODLEY
N799 1954	151878	RAFVR		WOODRUFF
N1350 1943	90539	AAF		WOODS
N805 1948	842616	RAuxAF		WOODS
N231 1949	852260	RAuxAF		WOODS
N1118 1943	119084	RAFVR		WOODS
N415 1946	162541	RAFVR		WOODS
N1 1950	849218	RAuxAF		WOOLASTON
N263 1946	804192	AAF		WOOLBERT
N573 1945	90307	AAF		WOOLCOCK
N850 1943	84687	RAFVR	DFC,DFM	WOOLDRIDGE
N35 1952	187417	RAFVR		WOOLEY
N1232 1945	82710	RAFVR	DFC	WOOLLATT
N1232 1945	67602	RAFVR		WOOLLEY
N468 1946	109158	RAFVR	DFC	WOOLNOUGH
N565 1955	4146	WRAAF		WOOLNOUGH
N396 1944	133265	RAFVR	DFM	WOOLSTON
N694 1948	753671	RAFVR		WOOLVERIDGE
N740 1947	743743	RAFVR		WOOSTER
N777 1951	740468	RAFVR		WOOTTON
N708 1959	2602003	RAFVR	1st Clasp	WOOTTON
YB 1999	n/a	RAuxAF		WORBOYS
N454 1959	162998	RAuxAF	wef 29/1/54	WORBY
N187 1947	874601	AAF		WORDIE
N1680 1942	84694	RAFVR	DFC	WORDSELL
N963 1951	84694	RAFVR	DFC; 1st Clasp	WORDSELL
N611 1957	84694	RAFVR	DFC; 2nd Clasp; wef 9/10/53	WORDSELL
N33 1960	90465	AAF	wef 28/1/44	WORDSWORTH
N197 1946	90648	AAF		WORMERSLEY
N375 1960	869381	AAF	wef 1/7/44	WORRALL
N513 1956	116825	RAuxAF		WORRALL
N250 1947	844569	AAF		WORTERS
N415 1946	104403	AAF		WORTH
N87 1948	746596	RAFVR		WORTHINGTON
N1089 1945	72080	RAFVR		WORTHINGTON
N395 1948	118204	RAFVR		WORTLEY
YB 1999	n/a	RAuxAF		WRAGG
N232 1948	869298	RAuxAF		WRAGG
N187 1947	155251	RAFVR		WRAGG
N1758 1942	804100	AAF		WREN
N677 1943	814020	AAF		WRIGHT
N257 1950	133312	RAFVR		WRIGHT
N250 1947	190324	RAFVR		WRIGHT
N314 1962	104087	RAFVR	wef 5/3/62	WRIGHT
N1193 1949	842291	RAuxAF		WRIGHT
N263 1946	64870	RAFVR	DFM	WRIGHT
N139 1945	134755	RAFVR	DFC	WRIGHT
N471 1954	2650091	RAuxAF		WRIGHT
N368 1959	91323	RAuxAF	wef 14/11/52	WRIGHT
N673 1949	90792	RAuxAF		WRIGHT
N843 1959	203041	RAFVR	wef 26/9/56	WRIGHT
N363 1944	67031	RAFVR	AFC	WRIGHT
N1 1955	2603656	RAFVR		WRIGHT
N478 1961	86684	RAFVR(T)	wef 29/6/44	WRIGHT
N406 1950	73674	RAFVR	DSM	WRIGHT
N155 1963	2650966	RAuxAF	wef 15/1/63	WRIGHT
N30 1949	90651	RAuxAF		WRIGHT

Surname	Initials			Rank	Status	Date
WRIGHT	H	C		SQD LDR		01/05/1963
WRIGHT	H	T		FL	Acting	28/02/1946
WRIGHT	H	W	B	FL		21/03/1946
WRIGHT	J	A	C	GRP CAP	Acting	10/12/1942
WRIGHT	J	F	M	PO	Acting	17/12/1942
WRIGHT	J	F	M	FL	Acting	19/07/1951
WRIGHT	J	F	M	FL		19/08/1954
WRIGHT	J	W	S	SGT		16/03/1950
WRIGHT	J			SGT		12/04/1945
WRIGHT	J			FL		18/04/1946
WRIGHT	J			SGT		17/04/1947
WRIGHT	J			SGT	Temporary	16/08/1961
WRIGHT	K	E		FLT OFF		13/01/1949
WRIGHT	K	P		SAC		09/07/1994
WRIGHT	K	S		FL		08/11/1945
WRIGHT	L	A		FL		07/04/1955
WRIGHT	R	R		SQD LDR	Acting	13/12/1945
WRIGHT	S	C		FO		18/04/1946
WRIGHT	T	L		FL		18/06/1995
WRIGHT	T			CPL		30/09/1948
WRIGHT	W	A		FO		14/09/1950
WRIGHT	W	A		FL		16/01/1957
WRIGHT	W	F		FL		01/05/1993
WRIGHT	W			CPL		10/06/1948
WRIGHT	W			LAC		17/01/1952
WROE	J			SGT		25/06/1953
WROUGHTON	W	E		SQD LDR	Acting	18/03/1948
WYATT	E	L		SGT		07/06/1951
WYATT	H	D		FL		10/02/1949
WYLES	T	K		FL	Acting	24/01/1946
WYNELL-SUTHERLAND	B	St.J		FL		03/07/1952
WYNN	H	H		FL		08/03/1956
WYNNE	B	P		WO	Acting	15/07/1948
WYNNE	G	J		SGT		01/12/1955
WYNNE	G	W		SGT		15/07/1948
WYNNE-POWELL	G	T		SQD LDR		10/12/1942
WYNNE-POWELL	G	T		WING COMM		19/07/1951
WYTHE	D	H		CPL		28/07/1955
YARDLEY	S	W		SQD LDR		01/07/1948
YARDLEY	S	W		SQD LDR		03/07/1957
YARDY	A	J		LAC		27/03/1947
YATES	F	D		SQD LDR		10/08/1944
YATES	F	L		FLT SGT		15/07/1948
YATES	F	W		SGT		09/11/1944
YATES	R	C		WING COMM		13/01/1949
YAXLEY	G	L		WO		12/06/1990
YEADELL	W	H		FL		08/11/1945
YEATS	G			FL		27/02/1963
YEEND	F	E		AC1		07/04/1955
YEO	R	N		FLT SGT		05/06/1947
YEO	W			SGT	Acting	01/07/1948
YORK	F			LAC		01/12/1955
YORKE	H	R	F	SQD LDR	Acting	26/08/1948
YORKE	R	L		FO		28/02/1946
YOUNG	A	A		SQD LDR	Acting	11/01/1945
YOUNG	A	B		FL		01/03/1945
YOUNG	A	F		FL		19/08/1959
YOUNG	F	R		FL		26/04/1956
YOUNG	F	R		FL		24/12/1964
YOUNG	H			PO	Acting	04/02/1943

Source	*Service No*	*Force*	*Notes*	*Surname*
N333 1963	161768	RAFVR	DFC; wef 21/4/53	WRIGHT
N197 1946	183765	RAFVR		WRIGHT
N263 1946	114576	RAFVR	DFC	WRIGHT
N1680 1942	90261	AAF	AFC,TD,MP	WRIGHT
N1708 1942	117819	RAFVR		WRIGHT
N724 1951	91241	RAuxAF	MBE; 1st Clasp	WRIGHT
N672 1954	91241	RAuxAF	MBE; 2nd Clasp	WRIGHT
N257 1950	770601	RAFVR		WRIGHT
N381 1945	803407	AAF		WRIGHT
N358 1946	81929	RAFVR	DFC,AFC	WRIGHT
N289 1947	880252	WAAF		WRIGHT
N624 1961	2611855	RAFVR	wef 9/6/61	WRIGHT
N30 1949	111	WAAF		WRIGHT
YB 1996	n/a	RAuxAF		WRIGHT
N1232 1945	132335	RAFVR		WRIGHT
N277 1955	161958	RAFVR		WRIGHT
N1355 1945	60514	RAFVR	DFC	WRIGHT
N358 1946	175391	RAFVR		WRIGHT
YB 1996	2635858	RAuxAF		WRIGHT
N805 1948	840050	RAuxAF		WRIGHT
N926 1950	174013	RAFVR		WRIGHT
N24 1957	174013	RAFVR	1st Clasp	WRIGHT
YB 1996	208886	RAuxAF		WRIGHT
N456 1948	808248	RAuxAF		WRIGHT
N35 1952	855914	RAuxAF		WRIGHT
N527 1953	2686626	RAuxAF		WROE
N232 1948	74031	RAFVR		WROUGHTON
N560 1951	842612	RAuxAF		WYATT
N123 1949	78150	RAFVR		WYATT
N82 1946	87440	RAFVR		WYLES
N513 1952	110566	RAFVR		WYNELL-SUTHERLAND
N187 1956	173003	RAFVR		WYNN
N573 1948	701728	RAFVR		WYNNE
N895 1955	807222	RAuxAF		WYNNE
N573 1948	744689	RAFVR		WYNNE
N1680 1942	90195	AAF		WYNNE-POWELL
N724 1951	90195	RAuxAF	DFC; 1st Clasp	WYNNE-POWELL
N565 1955	755411	RAFVR		WYTHE
N527 1948	72274	RAFVR		YARDLEY
N456 1957	72274	RAFVR	1st Clasp; wef 19/5/55	YARDLEY
N250 1947	841377	AAF		YARDY
N812 1944	91053	AAF		YATES
N573 1948	810012	RAuxAF		YATES
N1150 1944	800431	AAF		YATES
N30 1949	90619	RAuxAF		YATES
YB 1997	n/a	RAuxAF		YAXLEY
N1232 1945	116109	RAFVR		YEADELL
N155 1963	176744	RAFVR	DFC; wef 24/5/61	YEATS
N277 1955	846432	RAuxAF		YEEND
N460 1947	811018	AAF		YEO
N527 1948	751998	RAFVR		YEO
N895 1955	855968	RAuxAF		YORK
N694 1948	90923	RAuxAF		YORKE
N197 1946	184385	RAFVR		YORKE
N14 1945	88603	RAFVR		YOUNG
N225 1945	102981	RAFVR		YOUNG
N552 1959	68320	RAFVR	wef 1/4/59	YOUNG
N318 1956	164970	RAFVR		YOUNG
MOD 1966	164970	RAFVR	1st Clasp	YOUNG
N131 1943	132569	RAFVR		YOUNG

Surname	Initials			Rank	Status	Date
YOUNG	I	D		FL		22/09/1955
YOUNG	J	H	D	SQD LDR	Acting	07/08/1947
YOUNG	J	H		SAC		25/03/1959
YOUNG	K	E		SQD LDR	Acting	01/07/1948
YOUNG	M	D		FL		09/11/1944
YOUNG	R	J		FL		21/10/1959
YOUNG	R	N	W	SGT		21/01/1959
YOUNG	T	D		LAC		30/09/1948
YOUNG	W			LAC		31/10/1956
YOUNGMAN	R	H		FO		01/07/1948
YUILL	G	F		SQD LDR		11/11/1943
YULE	J	C		FL		24/07/1963
ZERVOUDAKIS	A			FL		06/06/1999
ZIGMOND	W	L		SQD LDR		25/06/1953
ZIMMERMAN	R	G	D	SGT		09/08/1945
ZORAB	C	P		SQD LDR	Acting	27/04/1950

Source	Service No	Force	Notes	Surname
N708 1955	164715	RAFVR		YOUNG
N662 1947	81733	RAFVR		YOUNG
N196 1959	2691015	RAuxAF	wef 3/7/58	YOUNG
N527 1948	73798	RAFVR		YOUNG
N1150 1944	89589	RAFVR		YOUNG
N708 1959	179589	RAFVR	wef 8/9/52	YOUNG
N28 1959	2696005	RAuxAF	wef 20/8/58	YOUNG
N805 1948	848462	RAuxAF		YOUNG
N761 1956	2684527	RAuxAF		YOUNG
N527 1948	174524	RAFVR	DFC	YOUNGMAN
N1177 1943	70767	RAFVR		YUILL
N525 1963	3129920	RAuxAF	wef 20/10/62	YULE
YB 2000	2636047	RAuxAF		ZERVOUDAKIS
N527 1953	116917	RAFVR		ZIGMOND
N874 1945	741903	RAFVR		ZIMMERMAN
N406 1950	83192	RAFVR		ZORAB

NOMINAL ROLL FOR AUSTRALIA

Surname	Initials			Rank	Status	Date	Source
ADAMS	J	I		WING COMM		29/04/1965	N35
ADDISON	W			WING COMM		12/12/1963	N105
ALEXANDER	J	S		SGT		27/05/1954	N34
AMBROSE	R	H	C	FO		23/11/1950	N70
ANDERSON	A	B		WING COMM	Temporary	15/03/1951	N19
ANDERSON	A	N		LAC		15/03/1951	N19
ANDERSON	F	K		FL		01/10/1964	N81
ANDERSON	R	A		SGT		07/11/1957	N61
ANDREWS	G	W		FL	Temporary	23/11/1950	N70
ANDREWS	K	C		WING COMM	Honorary	13/01/1966	N2
ANTHONY	T	W		FL		01/04/1965	N28
ARNOLD	C	L		FL		04/04/1968	N31
ARNOLD	F	J	G	SQD LDR		29/09/1966	N83
ASKER	R	H		FL	Temporary	23/11/1950	N70
ATHERTON	A	A		SGT		22/01/1953	N3
ATKINSON	K	E		SQD LDR	Temporary	23/11/1950	N70
AULADELL	S	J		SGT		06/09/1951	N66
AVISON	R	L		FLT SGT		03/12/1953	N79
BAILEY	A	J		LAC		30/08/1973	N116
BAILEY	T	E		CPL		27/05/1954	N34
BAKER	S	E		FLT SGT		27/05/1954	N34
BALLANTINE	B	N	C	FO		19/03/1953	N17
BAMFORD	G	M		CPL		27/05/1954	N34
BARBLETT	A	J		FL	Acting	03/02/1966	N9
BARNES	D	G		FLT SGT		04/09/1952	N60
BARRY	L	F		FL		18/03/1965	N20
BARTLETT	K	R		SQD LDR		29/08/1968	N72
BASTICK	J	E		FL	Temporary	06/09/1951	N66
BAXTER	A	T		WO	Temporary	23/11/1950	N70
BAYNE	R	D		FL		12/01/1967	N2
BEAL	J	T		CPL		12/10/1972	N97
BEATON	A			LAC		28/05/1953	N33
BEATTIE	P	R	G	SGT		06/09/1951	N66
BEATTON	J	S	B R	WO		28/05/1953	N33
BEAVAN	D			CPL	Temporary	23/11/1950	N70
BEECROFT	A	J		FO		08/11/1951	N83
BELL	J	N	R	FL		05/08/1965	N65
BENJAMIN	P	J		GRP CAP	Temporary	25/01/1951	N7
BENNETTS	H	R		SGT		31/08/1972	N83
BERRIMAN	W	R		FL		21/06/1951	N42
BINGLE	F	J		WO		13/08/1964	N68
BIRKS	E	J		SGT		03/02/1966	N9
BLECKLY	F	A		SGT	Temporary	23/11/1950	N70
BLIGHT	K	J		FL		19/11/1964	N94
BLOWER	C	W		WO		28/05/1953	N33
BOITTIER	E	J		LAC		07/04/1960	N25
BOULTON	R	J		WO	Temporary	23/11/1950	N70
BOURKE	F	G		FL		29/07/1965	N63
BOURNE	C	A	B	SQD LDR		04/09/1952	N60
BOX	N	E	H	GRP CAP	Temporary	25/01/1951	N7
BRAMMER	W	K		FL		23/09/1971	N90
BRASIER	S	J		WING COMM		01/10/1964	N81
BRAY	K	W		CPL		26/07/1973	N90
BRENNAN	D	B		FL		13/08/1964	N68
BRESSINGTON	E			FLT SGT		22/01/1953	N3
BRILL	W	L		WING COMM		28/07/1955	N35
BRINSLEY	K	G		WING COMM		19/08/1965	N69
BRITT	L			WING COMM		29/07/1965	N63
BROADBENT	J	A		WING COMM	Temporary	23/11/1950	N70
BROOKES	W	D		WING COMM	Temporary	23/11/1950	N70

Service No	Force	Notes	Surname
033119	PAF	DFC,AFC; gd	ADAMS
011344	PAF	MVO,AFC; gd	ADDISON
33709	EX-CAF-R		ALEXANDER
5060	EX-CAF-R		AMBROSE
291187	EX-CAF-R		ANDERSON
32608	EX-CAF-R		ANDERSON
036236	CAF-R	sd; No 21 (City of Melbourne) (Auxiliary) Squadron	ANDERSON
207861	EX-CAF-R		ANDERSON
402932	EX-CAF-R		ANDREWS
021988	RETIRED	gd	ANDREWS
033089	PAF	DFC; sd	ANTHONY
05819	PAF	sd	ARNOLD
025085	CAF-R	medic; Western Australian University Squadron	ARNOLD
255907	EX-CAF-R	N42 31/5/62 - WITHDRAWN IN FAVOUR OF ELIGIBILITY FOR ED	ASKER
8224	EX-CAF-R		ATHERTON
252399	EX-CAF-R		ATKINSON
A31457	PAF		AULADELL
A175	PAF		AVISON
A315629	CAF-R	No 21 (City of Melbourne) (Auxiliary) Squadron	BAILEY
A34256	PAF		BAILEY
9090	EX-CAF-R		BAKER
8610	EX-CAF-R		BALLANTINE
6119	EX-CAF-R		BAMFORD
051451	CAF-R	sd; Western Australian University Squadron	BARBLETT
15961	EX-CAF-R		BARNES
051813	CAF-R	sd; Western Australian University Squadron	BARRY
022184	PAF	gd	BARTLETT
408037	EX-CAF-R		BASTICK
22302	EX-CAF-R		BAXTER
033620	CAF-R	sd; No 25 (City of Perth) (Auxiliary) Squadron	BAYNE
A43188	CAF-R	No 24 (City of Adelaide) (Auxiliary) Squadron	BEAL
8914	EX-CAF-R		BEATON
A31413	PAF		BEATTIE
5733	EX-CAF-R		BEATTON
27507	EX-CAF-R		BEAVAN
30267	EX-CAF-R		BEECROFT
022001	PAF	sd	BELL
251170	EX-CAF-R		BENJAMIN
A54578	CAF-R	No 25 (City of Perth) (Auxiliary) Squadron	BENNETTS
021981	PAF	gd	BERRIMAN
207699	EX-CAF-R		BINGLE
27347	EX-CAF-R		BIRKS
27160	EX-CAF-R		BLECKLY
031569	PAF	tech	BLIGHT
A113	PAF		BLOWER
33626	EX-CAF-R		BOITTIER
20024	EX-CAF-R		BOULTON
022079	PAF	AFC; gd	BOURKE
05805	PAF	gd	BOURNE
251172	EX-CAF-R		BOX
01647	CAF-R	sd; No 23 (City of Brisbane) (Auxiliary) Squadron	BRAMMER
021925	PAF	gd	BRASIER
A16783	CAF-R	No 23 (City of Brisbane) (Auxiliary) Squadron	BRAY
021991	CAF-R	DFC; gd; No 22 (City of Sydney) (Auxiliary) Squadron	BRENNAN
207626	EX-CAF-R		BRESSINGTON
021978	PAF	DSO,DFC; gd	BRILL
04414	PAF	gd	BRINSLEY
05844	PAF	DFC; gd	BRITT
250192	EX-CAF-R	MC	BROADBENT
250299	CAF-R	DSO	BROOKES

Surname	Initials			Rank	Status	Date	Source
BROWN	A	W	E	WO		28/05/1953	N33
BROWN	C	L		SQD LDR	Acting	04/08/1966	N69
BROWN	D	H		WING COMM	Temporary	23/11/1950	N70
BROWN	J	G		WING COMM	Temporary	25/01/1951	N7
BROWNE	L	R		FLT SGT	Temporary	13/08/1964	N68
BROWNE	L	R		FLT SGT		18/08/1966	N72
BROWNING	W	P		CPL	Temporary	15/03/1951	N19
BRUNCKHURST	E	J		WING COMM		22/01/1953	N3
BRYCE	M	J		FL	Acting	08/10/1974	N82
BUCHANON	L	W		SGT	Acting	23/11/1950	N70
BUCKENHAM	J	F	H	SGT		27/07/1972	N67
BUCKHAM	J	H		WING COMM		28/05/1953	N33
BULS	K	L		FO		05/11/1974	N90
BURGE	E	H		SGT		22/01/1953	N3
BURGESS	G	J		SQD LDR		28/09/1967	N84
BURGESS-LLOYD	D	A		SQD LDR		27/10/1966	N90
BURR	R	H	E	FL		28/01/1965	N10
BUSHELL	F	E		FLT SGT		24/04/1952	N31
BUTHERWAY	P			CPL		28/05/1974	N44
BYRNE	J	P	B	CPL		27/05/1954	N34
BYRNE	J	T		FL		28/01/1965	N10
CABRAL	D	G		FO		13/04/1972	N29
CADAN	L	L		FL		10/09/1953	N55
CAHIR	J	T		FL		21/05/1953	N30
CAHIR	T	G		WING COMM		18/08/1966	N72
CAIRNS	J	M		WING COMM		13/08/1964	N68
CALLAGHAN	L	C		WO	Temporary	23/11/1950	N70
CALVERT	G	J		WO		19/11/1964	N94
CALVERT	G	J		WO		17/10/1968	N85
CAMERON	J	G		FL	Temporary	23/11/1950	N70
CAMERON	W	V	D	FLT SGT	Temporary	23/11/1950	N70
CAMPBELL	A	A		FL		13/08/1964	N68
CAMPBELL	A	A		FL		16/06/1966	N48
CAMPBELL	R	D		FO		29/11/1973	N179
CAMPBELL	W	S		FO		31/03/1960	N23
CANT	A	R		FLT SGT		07/04/1971	N39
CAREY	G	D		FL		22/10/1964	N87
CAREY	G	D		FL		23/04/1974	N34
CAREY	L	G		LAC		28/05/1974	N44
CARLIN	R			SQD LDR		01/04/1965	N28
CARR	M	O		SQD LDR		06/09/1951	N66
CARROLL	M	D		FL	Acting	23/06/1966	N53
CARTER	M	J		AC		28/05/1974	N44
CASSELL	H	J		FL		21/08/1969	N71
CHALLIS	C	J		SQD LDR		07/01/1965	N1
CHAMPION	N	A		FO		22/01/1953	N3
CHAPMAN	H	W		FO		23/11/1950	N70
CHAVASSE	K	F	N	SGT		24/04/1952	N31
CHILTON	C	D		SGT	Temporary	23/11/1950	N70
CHISHOLM	J	H		FL	Temporary	23/11/1950	N70
CHIVERS	M	E		WO		31/01/1957	N7
CHOATE	R	S		SQD LDR	Temporary	10/12/1964	N98
CHOATE	R	S		SQD LDR	Temporary	10/12/1964	N98
CHRISTOFIS	L	G		WING COMM		03/02/1966	N9
CLABBURN	G	E	S	FL		22/01/1953	N3
CLARK	A	N		FO		05/08/1965	N65
CLARKE	R	G		WO		27/05/1954	N34
CLARKO	J	G		FL		06/11/1969	N92
COATE	E	F		FL	Temporary	05/04/1951	N23
COATS	D			SGT		04/03/1975	G9

Service No	Force	Notes	Surname
11592	EX-CAF-R		BROWN
042525	CAF-R	sd; No 24 (City of Adelaide) (Auxiliary) Squadron	BROWN
250104	CAF-R	OBE	BROWN
25147	EX-CAF-R		BROWN
A210754	CAF-R	No 22 (City of Sydney) (Auxiliary) Squadron	BROWNE
A210754	CAF-R	1st Clasp; tech; No 22 (City of Sydney) (Auxiliary) Squadron	BROWNE
6131	EX-CAF-R		BROWNING
270293	EX-CAF-R		BRUNCKHURST
015320	AFR	sd	BRYCE
40679	EX-CAF-R		BUCHANON
A16761	CAF-R	No 23 (City of Brisbane) (Auxiliary) Squadron	BUCKENHAM
250297	EX-CAF-R		BUCKHAM
0211978	CAF-R		BULS
A31377	WRAAF		BURGE
018411	CAF-R	medic; No 23 (City of Brisbane) (Auxiliary) Squadron	BURGESS
05802	PAF	gd	BURGESS-LLOYD
04415	PAF	sd	BURR
A2135	PAF		BUSHELL
A53696	CAF-R		BUTHERWAY
8540	EX-CAF-R		BYRNE
035197	PAF	DFC; sd	BYRNE
053670	CAF-R	sd; No 25 (City of Perth) (Auxiliary) Squadron	CABRAL
033097	PAF	gd	CADAN
051850	EX-CAF-R		CAHIR
033147	PAF	sd	CAHIR
031473	PAF	AFC; gd	CAIRNS
300400	EX-CAF-R		CALLAGHAN
A4936	CAF-R	No 24 (City of Adelaide) (Auxiliary) Squadron	CALVERT
A4936	CAF-R	1st Clasp; No 24 (City of Adelaide) (Auxiliary) Squadron	CALVERT
5327	EX-CAF-R		CAMERON
9023	EX-CAF-R		CAMERON
0210022	CAF-R	tech; No 22 (City of Sydney) (Auxiliary) Squadron	CAMPBELL
0210022	CAF-R	1st Clasp; tech; No 22 (City of Sydney) (Auxiliary) Squadron	CAMPBELL
043534	AFR	engin	CAMPBELL
8612	AFR		CAMPBELL
A12685	CAF-R	No 23 (City of Brisbane) (Auxiliary) Squadron	CANT
04780	CAF-R	equip; Adelaide University Squadron	CAREY
04780	AFR	1st Clasp; equip	CAREY
A53695	CAF-R		CAREY
011345	PAF	DFC,AFC; gd	CARLIN
04404	PAF	gd	CARR
054371	CAF-R	sd; Western Australian University Squadron	CARROLL
A53693	CAF-R		CARTER
0314498	CAF-R	sd; No 21 (City of Melbourne) (Auxiliary) Squadron	CASSELL
035437	AFR	DFC; sd	CHALLIS
300128	EX-CAF-R		CHAMPION
207703	EX-CAF-R		CHAPMAN
A1114	PAF		CHAVASSE
23251	EX-CAF-R		CHILTON
265930	EX-CAF-R		CHISHOLM
11001	EX-CAF-R		CHIVERS
051846	CAF-R	OBE; sd; No 25 (City of Perth) (Auxiliary) Squadron	CHOATE
051846	CAF-R	OBE; 1st Clasp; sd; No 25 (City of Perth) (Auxiliary) Squadron	CHOATE
011355	PAF	gd	CHRISTOFIS
205745	EX-CAF-R		CLABBURN
018149	CAF-R	DFC; sd; No 23 (City of Brisbane) (Auxiliary) Squadron	CLARK
12108	EX-CAF-R		CLARKE
054346	CAF-R	sd; No 25 (City of Perth) (Auxiliary) Squadron	CLARKO
254886	EX-CAF-R		COATE
A18648	CAF-R	No 23 (City of Brisbane) (Auxiliary) Squadron	COATS

Surname	Initials			Rank	Status	Date	Source
COBBY	A	H		AIR COMMD	Temporary	23/11/1950	N70
COLEBROOK	H	L		FL		05/08/1954	N47
COLLINS	J	D		FL		08/11/1951	N83
COLLINS	J	E		FL	Temporary	06/09/1951	N66
COLLIS	D	F		CPL		08/06/1967	N48
COLQUHOUN	D	W		WING COMM		06/09/1951	N66
COOK	N	L		CPL		19/06/1952	N43
COOMBES	G	M		SGT	Temporary	23/11/1950	N70
COOMBES	W	M		WING COMM		28/01/1965	N10
COOPER	B			FL		06/09/1951	N66
COOPER	E	W		WING COMM	Temporary	23/11/1950	N70
COOPER	R	D		SQD LDR		21/02/1952	N18
CORMIE	R	E		WING COMM		07/01/1965	N1
COSTELLO	J	L		FO		05/07/1956	N37
COSTIN	W	J		FLT SGT		08/04/1975	G13
COUNSELL	W	D		GRP CAP	Temporary	18/10/1951	N78
CRAIGIE	R	G		FL	Acting	24/07/1969	N62
CRAWFORD	H	W		SGT		28/05/1953	N33
CREIGHTMORE	M	L		WING COMM	Temporary	06/09/1951	N66
CROMBIE	W	D		SQD LDR		17/05/1973	N56
CROSSER	J			FO		12/11/1953	N71
CROWTHER	J	R		FL		23/06/1966	N53
CRUMP	K	S		FL	Acting	29/10/1970	N95
CRUSE	K	A	J	FL		04/02/1965	N12
CUMING	D	R		SQD LDR		06/09/1951	N66
CUNNINGHAM	R	H		WING COMM	Temporary	23/11/1950	N70
CURTIS	O	V		FL	Temporary	25/01/1951	N7
CURWEN	N			CPL		26/07/1973	N90
CURZON-SIGGERS	C	V		SGT		22/01/1953	N3
CUTHBERTSON	G	A		CPL		28/05/1953	N33
DALE	W	A	C	GRP CAP		21/06/1951	N42
DALKIN	R	N		WING COMM		06/09/1951	N66
DALLYWATER	S	W		WING COMM		24/06/1965	N50
DALTON	K	R		FL		07/04/1971	N39
DAVIES	H	R		CPL		30/11/1972	N121
DAVIS	R	B		GRP CAP		21/06/1951	N42
DAYTON	T	W		CPL		15/08/1957	N45
De MEDICI	J	R		FL		28/01/1965	N10
De YOUNG	H	R		FLT SGT		19/11/1964	N94
De YOUNG	H	R		FLT SGT		08/06/1967	N48
DEAN	W	T		FLT SGT	Temporary	23/11/1950	N70
DEAS	M	W		SGT		26/07/1973	N90
DEE	K	M		FL		14/04/1954	N23
DELLAR	E	R	L	SGT	Temporary	23/11/1950	N70
DENT	A	J	F	SQD LDR	Acting	01/07/1971	N66
DEWHURST	J	J		WO		22/01/1953	N3
DIXON	D	E		LAC		13/03/1969	N23
DONNELLAN	F	X		CPL		19/11/1964	N94
DONNELLAN	F	X		CPL		19/11/1964	N94
DOOREY	R	A		FO		13/08/1964	N68
DOOREY	R	A		FO		03/06/1971	N57
DOUDY	C	T		SQD LDR	Temporary	15/03/1951	N19
DOUGHTY	D	J		SQD LDR		21/06/1951	N42
DOUGLAS	J	R		CPL		24/07/1969	N62
DOWLING	K	L		FL	Temporary	23/11/1950	N70
DRUMMOND	L	J		WO		13/08/1964	N68
DRUMMOND	L	J		WO		28/07/1966	N67
DRYBURGH	N	B		FLT SGT	Temporary	23/11/1950	N70
DUFFIELD	A	E		SGT	Temporary	23/11/1950	N70
DUFFY	B	J		SGT		05/08/1965	N65

Service No	Force	Notes	Surname
251134	EX-CAF-R	CBE,DSO,DFC,GM	COBBY
051217	PAF	equip	COLEBROOK
024305	PAF	sd	COLLINS
401204	EX-CAF-R		COLLINS
A211667	CAF-R	sd; No 22 (City of Sydney) (Auxiliary) Squadron	COLLIS
033011	PAF	DFC,AFC; gd	COLQUHOUN
33330	EX-CAF-R		COOK
37039	EX-CAF-R		COOMBES
021987	PAF	gd	COOMBES
02147	PAF	tech	COOPER
260093	CAF-R	AFC	COOPER
021927	AFR	gd	COOPER
03246	PAF	tech	CORMIE
18358	EX-CAF-R		COSTELLO
207654	EX-PAF		COSTIN
251164	EX-CAF-R		COUNSELL
025025	CAF-R	sd; No 22 (City of Sydney) (Auxiliary) Squadron	CRAIGIE
5218	EX-CAF-R		CRAWFORD
291210	EX-CAF-R		CREIGHTMORE
0311304	CAF-R	medic; No 21 (City of Melbourne) (Auxiliary) Squadron	CROMBIE
5795	EX-CAF-R		CROSSER
033085	PAF	gd	CROWTHER
016566	AFR	sd; No 23 (City of Brisbane) (Auxiliary) Squadron	CRUMP
039588	CAF-R	equip; No 21 (City of Melbourne) (Auxiliary) Squadron	CRUSE
033012	PAF	AFC; tech	CUMING
250107	EX-CAF-R		CUNNINGHAM
205714	EX-CAF-R		CURTIS
A119546	CAF-R	No 23 (City of Brisbane) (Auxiliary) Squadron	CURWEN
300282	EX-CAF-R		CURZON-SIGGERS
19032	EX-CAF-R		CUTHBERTSON
021917	PAF	DSO; equip	DALE
021919	PAF	DFC; gd	DALKIN
05824	PAF	gd	DALLYWATER
034743	EX-AFR	gd	DALTON
054192	CAF-R	No 25 (City of Perth) (Auxiliary) Squadron	DAVIES
033026	PAF	medic	DAVIS
30465	EX-CAF-R		DAYTON
11142	EX-CAF-R	gd; DECEASED	De MEDICI
A51113	CAF-R	No 25 (City of Perth) (Auxiliary) Squadron	De YOUNG
A51113	CAF-R	1st Clasp; sd; No 25 (City of Perth) (Auxiliary) Squadron	De YOUNG
26269	EX-CAF-R		DEAN
A16790	CAF-R	No 23 (City of Brisbane) (Auxiliary) Squadron	DEAS
051503	PAF	gd	DEE
12268	EX-CAF-R		DELLAR
0217997	CAF-R	engin; No 22 (City of Sydney) (Auxiliary) Squadron	DENT
207575	EX-CAF-R		DEWHURST
A215946	CAF-R	medic	DIXON
A33444	CAF-R	No 21 (City of Melbourne) (Auxiliary) Squadron	DONNELLAN
A33444	CAF-R	1st Clasp; No 21 (City of Melbourne) (Auxiliary) Squadron	DONNELLAN
022914	CAF-R	tech; No 22 (City of Sydney) (Auxiliary) Squadron	DOOREY
022914	CAF-R	1st Clasp; engin; No 22 (City of Sydney) (Auxiliary) Squadron	DOOREY
403125	EX-CAF-R	DFC	DOUDY
034066	PAF	tech	DOUGHTY
A211778	CAF-R	sd; No 22 (City of Sydney) (Auxiliary) Squadron	DOUGLAS
400554	EX-CAF-R		DOWLING
A210834	CAF-R	No 22 (City of Sydney) (Auxiliary) Squadron	DRUMMOND
A210834	CAF-R	1st clasp; sd; No 22 (City of Sydney) (Auxiliary) Squadron	DRUMMOND
10028	EX-CAF-R		DRYBURGH
11795	EX-CAF-R		DUFFIELD
A16008	CAF-R	No 23 (City of Brisbane) (Auxiliary) Squadron	DUFFY

Surname	Initials			Rank	Status	Date	Source
DUFFY	B	J		SGT		05/08/1965	N65
DUKE	G	A		CPL		15/01/1970	N3
DUNCAN	D	N		LAC		18/02/1971	N18
DUNCAN	J	M		SQD LDR		01/10/1964	N81
DUNCAN	J	M		SQD LDR		01/10/1964	N81
DUNCAN	J	M		SQD LDR		25/01/1973	N8
DUNCAN	W	J		GRP CAP	Temporary	23/11/1950	N70
DUNN	J	H		FL	Temporary	15/03/1951	N19
DURAN	M	E		FL		09/03/1972	N18
DYER	D	J		FL		05/08/1965	N65
DYER	D	J		SQD LDR		30/07/1974	N62
EASTWOOD	E	W		FL	Temporary	21/06/1951	N42
EDWARDS	F	P	J	WO		22/01/1953	N3
EGERTON	T	A		FL		20/04/1967	N33
ELEISON	H	G		SGT		11/09/1969	N74
ELLERTON	R	E		FL		21/06/1951	N42
ELLIOT	N	S		FL		13/08/1964	N68
ELLIOT	N	S		FL		04/04/1968	N31
ELLIOT	W	D		FLT SGT		03/12/1974	N98
ELLIS	G	H		FL		22/01/1953	N3
ELLIS	G	H		FL	Temporary	03/09/1964	N74
ELPHICK	J	R	H	FL		10/09/1953	N55
EMMETT	E	R		FLT SGT	Temporary	23/11/1950	N70
ESSLEMONT	J	P		FL		14/08/1952	N55
EVANS	F			WING COMM	Acting	16/06/1966	N48
EVANS	G	B		FL	Temporary	06/09/1951	N66
FAIRBURN	T	S		WING COMM		18/03/1965	N20
FARQUHAR	J			SGT		06/09/1951	N66
FARRELL	J	H		WO		10/06/1954	N36
FAY	J	H		WING COMM		12/11/1964	N92
FAY	J	H		WING COMM		28/01/1965	N10
FELLOWS	E	J		FL	Honorary	10/09/1953	N55
FIRTH	M	W	D	FO		10/06/1954	N36
FISHER	D	Le	S	FL		08/12/1960	N82
FLETCHER	H	B		SQD LDR	Temporary	23/11/1950	N70
FORMAN	C	C		WING COMM		19/03/1953	N17
FORSTER	W	E	S	FL		28/05/1953	N33
FOSTER	K	I		SQD LDR		12/11/1964	N92
FOSTER	R	J		CPL		06/09/1951	N66
FRANCIS	J	M		SQD LDR		28/07/1955	N35
FRANKS	J	L		WO	Temporary	23/11/1950	N70
FRASER	H	E		SQD LDR	Honorary	23/11/1950	N70
FREDRICKSEN	N	C		WO		22/01/1953	N3
FROST	E	M		SQD LDR	Temporary	23/11/1950	N70
FURZE	M	M		FO		20/05/1965	N41
GALLWEY	C	R		SQD LDR		21/06/1951	N42
GAMBLE	H	A		SQD LDR		01/10/1964	N81
GARDNER	J	J		CPL	Temporary	05/04/1951	N23
GARDNER	W	E		WING COMM		12/08/1954	N49
GARRETT	W	R		GRP CAP	Temporary	06/09/1951	N66
GASTON	T	A		FLT SGT	Temporary	23/11/1950	N70
GAULTON	K	J	F	WO		05/07/1956	N37
GAVIN	H	V		GRP CAP		17/12/1970	N116
GEPP	R	R	C	CPL		06/09/1951	N66
GERBER	J	E		WING COMM		28/01/1965	N10
GIBBONS	J	P		FLT SGT		09/03/1972	N18
GIBSON	H	B		FL	Acting	23/11/1950	N70
GIDDINGS	A	W	J	WO		22/01/1953	N3
GILFEATHER	L	V		LAC		06/09/1951	N66
GILLIS	A	A		SQD LDR		12/11/1964	N92

Service No	Force	Notes	Surname
A16008	CAF-R	1st Clasp; No 23 (City of Brisbane) (Auxiliary) Squadron	DUFFY
A16705	CAF-R	No 23 (City of Brisbane) (Auxiliary) Squadron	DUKE
A16726	CAF-R		DUNCAN
04980	CAF-R	MBE; sd; No 21 (City of Melbourne) (Auxiliary) Squadron	DUNCAN
04980	CAF-R	MBE; 1st Clasp; sd; No 21 (City of Melbourne) (Auxiliary) Squadron	DUNCAN
04980	CAF-R	MBE; 2nd Clasp; sd; No 21 (City of Melbourne) (Auxiliary) Squadron	DUNCAN
270065	CAF-R	OBE	DUNCAN
255313	EX-CAF-R		DUNN
015378	CAF-R	sd; Queensland University Squadron	DURAN
015143	CAF-R	sd; No 23 (City of Brisbane) (Auxiliary) Squadron	DYER
015143	CAF-R	1st Clasp; sd; No 23 (City of Brisbane) (Auxiliary) Squadron	DYER
255787	EX-CAF-R		EASTWOOD
300444	EX-CAF-R		EDWARDS
036036	AFR	DFC; gd	EGERTON
A116353	CAF-R	No 23 (City of Brisbane) (Auxiliary) Squadron	ELEISON
033042	PAF	sd	ELLERTON
05820	CAF-R	AFC; sd; No 22 (City of Sydney) (Auxiliary) Squadron	ELLIOT
05820	CAF-R	AFC; 1st Clasp; sd; No 22 (City of Sydney) (Auxiliary) Squadron	ELLIOT
A25140	CAF-R	No 22 (City of Sydney) (Auxiliary) Squadron	ELLIOT
205448	EX-CAF-R	GM	ELLIS
205448	EX-CAF-R	GM; 1st Clasp	ELLIS
402157	EX-CAF-R		ELPHICK
205681	EX-CAF-R		EMMETT
205718	EX-CAF-R		ESSLEMONT
035444	PAF	sd	EVANS
400791	EX-CAF-R		EVANS
033103	PAF	DFC; gd	FAIRBURN
A123	PAF		FARQUHAR
18108	EX-CAF-R		FARRELL
021926	PAF	equip	FAY
021926	PAF	1st Clasp; equip	FAY
207606	EX-CAF-R		FELLOWS
300377	EX-CAF-R		FIRTH
407480	AFR	gd	FISHER
260181	EX-CAF-R	DFC	FLETCHER
290487	EX-CAF-R		FORMAN
04981	CAF-R	gd; No 24 (City of Adelaide) Squadron	FORSTER
01132	PAF	DFC; gd	FOSTER
A280	PAF		FOSTER
012324	PAF	sd	FRANCIS
24155	EX-CAF-R		FRANKS
1343	CAF-R		FRASER
205679	EX-CAF-R		FREDRICKSEN
407420	EX-CAF-R	DFC	FROST
0315232	PAF	tech	FURZE
011323	PAF	AFC; gd	GALLWEY
0250499	AFR	gd	GAMBLE
8761	EX-CAF-R		GARDNER
250124	EX-CAF-R	DSC	GARDNER
250118	EX-CAF-R	AFC	GARRETT
300375	EX-CAF-R		GASTON
404669	WRAAF		GAULTON
06131	PAF	DFC; sd	GAVIN
A31497	PAF		GEPP
03486	PAF	OBE,AFC; gd	GERBER
A53668	CAF-R	No 25 (City of Perth) (Auxiliary) Squadron	GIBBONS
1356	EX-CAF-R		GIBSON
300417	EX-CAF-R		GIDDINGS
A33068	PAF		GILFEATHER
03462	PAF	equip	GILLIS

Surname	Initials			Rank	Status	Date	Source
GIRARDAU	D	E		FL	Temporary	13/08/1964	N68
GISSING	L	H		WO		10/09/1953	N55
GLANVILE	K	R		SQD LDR		28/07/1966	N67
GLANVILE	K	R		SQD LDR		28/07/1966	N67
GLASSOP	F	A		SGT		02/12/1971	N113
GLASSOP	R	H		WING COMM		03/06/1965	N44
GLEESON	N	G		WO		22/01/1953	N3
GLENN	D	A		WING COMM		13/08/1964	N68
GODFREY	K	H		SQD LDR		28/09/1967	N84
GOLDBERG	R	G		FL	Temporary	23/11/1950	N70
GOLDSMITH	J	M		SGT		24/01/1957	N5
GOLDSWORTHY	R	H		SGT		14/02/1963	N13
GOOCH	J	S		WING COMM		10/12/1964	N98
GOOD	A	T		CPL		10/06/1954	N36
GORDON	J	R		WING COMM	Temporary	22/02/1968	N18
GOW	F	E		WO		10/09/1953	N55
GRADING	R	C		CPL		26/07/1973	N90
GRAHAM	J	E		GRP CAP	Temporary	23/11/1950	N70
GRAHAM	J	E		GRP CAP		01/10/1964	N81
GRAY	J			WING COMM		03/12/1964	N97
GRAY	R	H	S	WING COMM		21/06/1951	N42
GREEN	S	A		WO		10/09/1953	N55
GREENHAM	R	I		WING COMM	Temporary	15/03/1951	N19
GREENTREE	R	A		FL		21/06/1951	N42
GREENWOOD	C	A		GRP CAP	Acting	03/12/1964	N97
GREET	E	E		FL		01/10/1974	N80
GREGG	G	R		WO		13/09/1951	N68
GREGORY	K	W		FL		27/06/1968	N57
GREY	J	W		SQD LDR		01/10/1964	N81
GRIGGS	F	M		GRP CAP		31/08/1972	N83
GRIMLEY	G	J		FL	Acting	10/06/1975	G22
GULLAN	B	W		WO		05/07/1956	N37
GUNDELACH	C	P		SQD LDR		20/05/1965	N41
GUTHRIE	V	D		SQD LDR		13/08/1964	N68
HACKETT	D	L		FL	Temporary	23/11/1950	N70
HAMILTON	H	D		WING COMM		21/06/1951	N42
HAMMOND	H	T		GRP CAP	Temporary	23/11/1950	N70
HANSFORD	B	W		PO		07/01/1965	N1
HARDING	H	R		WING COMM	Temporary	23/11/1950	N70
HARPER	J	H		GRP CAP	Acting	15/03/1951	N19
HART	B	M		SGT		19/06/1952	N43
HARTLEY	R	J		SGT		28/05/1974	N44
HATHERLEY	H	C		FL	Acting	11/08/1960	N56
HAYES	A	P		FL		13/08/1964	N68
HAYES	B	J	F X	SQD LDR		07/01/1965	N1
HAYES	R	B		FLT SGT	Temporary	23/11/1950	N70
HAYES	R	B		FL		13/08/1964	N68
HAYES	R	B		FL		16/06/1966	N48
HEAD	P	R		SGT		03/02/1972	N9
HEANES	R			LAC		19/05/1966	N42
HEAP	E	P		WO		13/08/1964	N68
HEAP	E	P		WO		13/08/1964	N68
HEAVYSIDE	W	T		CPL		28/05/1974	N44
HEGGIE	R	K		SGT	Temporary	23/11/1950	N70
HEMBROW	C	H		GRP CAP	Temporary	15/03/1951	N19
HENRY	B	J		FLT SGT		06/09/1951	N66
HENSLEY	W	J		WING COMM	Temporary	07/09/1972	N84
HERBERT	J	P	J	FO		01/04/1965	N28
HILL	V	J		SQD LDR		01/10/1964	N81
HITCHENS	F	W		FO		12/11/1953	N71

Service No	Force	Notes	Surname
409688	AFR	DFC; gd	GIRARDAU
300371	EX-CAF-R		GISSING
017919	CAF-R	sd; No 23 (City of Brisbane) (Auxiliary) Squadron	GLANVILE
017919	CAF-R	1st Clasp; sd; No 23 (City of Brisbane) (Auxiliary) Squadron	GLANVILE
A56096	CAF-R	No 25 (City of Perth) (Auxiliary) Squadron	GLASSOP
021989	PAF	DFC; gd	GLASSOP
A210246	EX-CAF-R		GLEESON
03412	PAF	gd	GLENN
022017	PAF	gd	GODFREY
407422	CAF-R	DFC	GOLDBERG
27398	EX-CAF-R		GOLDSMITH
205882	EX-CAF-R		GOLDSWORTHY
05826	PAF	AFC; gd	GOOCH
60528	EX-CAF-R		GOOD
250854	AFR	MC; sd	GORDON
207685	EX-CAF-R		GOW
A16779	CAF-R		GRADING
250095	CAF-R		GRAHAM
250095	AFR	1st Clasp; gd	GRAHAM
033081	PAF	sd	GRAY
021923	PAF	DFC; gd	GRAY
205773	EX-CAF-R		GREEN
291188	EX-CAF-R		GREENHAM
021980	PAF	tech	GREENTREE
03110	PAF	OBE,AFC; gd	GREENWOOD
402795	EX-AFR	gd	GREET
A31551	PAF		GREGG
0116817	CAF-R	sd; No 23 (City of Brisbane) (Auxiliary) Squadron	GREGORY
04411	PAF	sd	GREY
035376	PAF	DFC,DFM; gd	GRIGGS
0226235	AFR		GRIMLEY
16143	WRAAF		GULLAN
022018	PAF	sd	GUNDELACH
022015	PAF	AFC; gd	GUTHRIE
407323	EX-CAF-R		HACKETT
033024	PAF	acct	HAMILTON
260180	EX-CAF-R	OBE	HAMMOND
0311085	AFR	sd	HANSFORD
250122	CAF-R		HARDING
251382	EX-CAF-R	AFC	HARPER
A256	PAF		HART
A53691	CAF-R		HARTLEY
407286	AFR	MBE; gd	HATHERLEY
0210999	CAF-R	medic; No 22 (City of Sydney) (Auxiliary) Squadron	HAYES
033086	PAF	DFM; sd	HAYES
8538	EX-CAF-R		HAYES
0210018	CAF-R	1st Clasp; sd; No 22 (City of Sydney) (Auxiliary) Squadron	HAYES
0210018	CAF-R	MBE; 2nd Clasp; sd; No 22 (City of Sydney) (Auxiliary) Squadron	HAYES
A16748	CAF-R	No 23 (City of Brisbane) (Auxiliary) Squadron	HEAD
A4920	AFR	sd	HEANES
A210479	CAF-R	No 22 (City of Sydney) (Auxiliary) Squadron	HEAP
A210479	CAF-R	1st Clasp; No 22 (City of Sydney) (Auxiliary) Squadron	HEAP
A311592	CAF-R		HEAVYSIDE
19091	EX-CAF-R		HEGGIE
1184	EX-CAF-R		HEMBROW
A21912	PAF		HENRY
0211188	AFR	medic	HENSLEY
04904	CAF-R	sd	HERBERT
033618	PAF	DFC,AFC; gd	HILL
0288	PAF	sd	HITCHENS

Surname	Initials			Rank	Status	Date	Source
HODGE	A	R		GRP CAP		22/10/1964	N87
HODGINS	J	R		FL	Temporary	06/04/1966	N31
HODGSON	A	W		SGT		13/08/1964	N68
HOLBERTON	R	W		FL	Temporary	23/11/1950	N70
HOLDSWORTH	M			SQD LDR		30/09/1965	N78
HOLFORD	W	E		CPL		19/03/1953	N17
HOLSTEN	F	D		WO		22/01/1953	N3
HOLTEN	L	J	K	SQD LDR		21/06/1951	N42
HOOPER	N	C		FL		21/06/1951	N42
HOPTON	J	R		FL		19/03/1953	N17
HOSKING	R	A		WING COMM		03/12/1964	N97
HOWES	D	J	F	PO		24/04/1952	N31
HOWIE	A	E	S	WO		03/12/1953	N79
HOWIE	W	H		FLT SGT		24/04/1952	N31
HUBBARD	R	V		FL		16/06/1966	N48
HUBBLE	J	W		AIR COMMD		01/06/1972	N43
HUCKER	R	A		LAC		04/03/1975	G9
HUDSON	N	H		FL		20/05/1965	N41
HUGHES	G			WING COMM	Honorary	01/10/1964	N81
HUMPHREY	M	S		SQD LDR		21/06/1951	N42
HURDITCH	D	D		AIR COMMD		19/08/1971	N81
HURDITCH	D	D		AIR COMMD		19/08/1971	N81
HURFORD	W	T		FL	Acting	29/09/1966	N83
HURLEY	A	G	M	WO	Temporary	23/11/1950	N70
HUTCHINSON	J	E		FL		02/12/1971	N113
HUTCHINSON	K	N		SGT	Acting	23/11/1950	N70
HUTCHINSON	M	P		SGT	Temporary	23/11/1950	N70
HYNES	W	H		FLT SGT		31/01/1957	N7
INGER	F	J		WING COMM		13/08/1964	N68
INKSON	J			WO		27/08/1974	N70
ISAACS	K			WING COMM		07/01/1965	N1
JACKSON	H	H		FL		24/04/1952	N31
JACKSON	R	J		LAC		23/11/1950	N70
JAMES	L	T		FLT SGT	Temporary	23/11/1950	N70
JARVIS	J	R		SGT		22/01/1953	N3
JEWELL	J	B		SQD LDR		07/01/1965	N1
JOHNSON	F	W	S	WING COMM		06/09/1951	N66
JOLLY	A	P		FO		23/11/1950	N70
JONES	A	S		FL		28/05/1953	N33
JONES	E	J		SQD LDR		10/06/1954	N36
JORGENSEN	T	H		FO		09/03/1972	N18
JOSEPH	P	G		FO		05/04/1951	N23
JOY	J	J		WO		24/04/1952	N31
JOYCE	D	T		WO		01/04/1965	N28
JUETT	J	C		SQD LDR		01/10/1964	N81
KEARNEY	J	R		SGT		06/09/1951	N66
KEITH	T	D		CPL		02/09/1971	N86
KELLY	B	A		CPL		08/07/1971	N68
KELLY	F	M		FLT SGT		06/09/1951	N66
KELLY	M	J		SGT		06/09/1951	N66
KEMMIS	V	D		FL		21/06/1951	N42
KING	G	S		WING COMM		13/08/1964	N68
KINNEAR	J	C		SQD LDR	Temporary	25/01/1951	N7
KIRKHOUSE	H	A		FL		21/06/1951	N42
KIRKWOOD	W	W		FL		03/06/1971	N57
KNIGHT	F	F		WING COMM		23/11/1950	N70
KNIGHT	F	F		WING COMM	Temporary	03/09/1964	N74
KNIGHT	H	W		SGT	Temporary	23/11/1950	N70
KNOX	P	V		FLT SGT	Temporary	23/11/1950	N70
KNUSDEN	F	O		WING COMM		18/03/1965	N20

Service No	Force	Notes	Surname
03109	PAF	AFC; gd	HODGE
0210658	AFR	sd	HODGINS
A210508	CAF-R	No 22 (City of Sydney) (Auxiliary) Squadron	HODGSON
404482	EX-CAF-R		HOLBERTON
021992	PAF	AFC; gd	HOLDSWORTH
32264	EX-CAF-R		HOLFORD
300402	EX-CAF-R		HOLSTEN
033018	PAF	equip	HOLTEN
033010	PAF	gd	HOOPER
407175	EX-CAF-R		HOPTON
033102	PAF	DFC,AFC; gd	HOSKING
031544	PAF	equip	HOWES
A5857	PAF		HOWIE
A246	PAF		HOWIE
033118	PAF	equip	HUBBARD
n/a	PAF	CBE,DSO,AFC; gd	HUBBLE
044538	CAF-R	No 24 (City of Adelaide) (Auxiliary) Squadron	HUCKER
022011	PAF	DFC; sd	HUDSON
021974	RETIRED	DFC,AFC; gd	HUGHES
033079	PAF	DFC; gd	HUMPHREY
n/a	PAF	CBE; gd	HURDITCH
n/a	PAF	CBE; 1st Clasp; gd	HURDITCH
036022	AFR	sd	HURFORD
22682	EX-CAF-R		HURLEY
015321	CAF-R	equip; Queensland University Squadron	HUTCHINSON
8718	EX-CAF-R		HUTCHINSON
10887	EX-CAF-R		HUTCHINSON
33892	EX-CAF-R		HYNES
031424	PAF	AFC; gd	INGER
A31506	CAF-R	No 22 (City of Sydney) (Auxiliary) Squadron	INKSON
05861	PAF	AFC; gd	ISAACS
251379	CAF-R		JACKSON
26622	EX-CAF-R		JACKSON
10804	EX-CAF-R		JAMES
300301	EX-CAF-R		JARVIS
03338	PAF	DFC; gd	JEWELL
033023	PAF	acct	JOHNSON
180078	CAF-R		JOLLY
407003	EX-CAF-R		JONES
2663484	EX-CAF-R	AFC	JONES
053512	CAF-R	equip; No 25 (City of Perth) (Auxiliary) Squadron	JORGENSEN
407365	EX-CAF-R		JOSEPH
207568	EX-CAF-R		JOY
A51008	CAF-R		JOYCE
05812	PAF	sd	JUETT
A31488	PAF		KEARNEY
A211911	CAF-R	No 22 (City of Sydney) (Auxiliary) Squadron	KEITH
A314262	CAF-R	No 21 (City of Melbourne) (Auxiliary) Squadron	KELLY
A33007	PAF		KELLY
A224	PAF		KELLY
021985	PAF	gd	KEMMIS
033095	PAF	gd	KING
231367	EX-CAF-R		KINNEAR
02112	PAF	AFC,DFM; tech	KIRKHOUSE
411497	EX-AFR	gd	KIRKWOOD
250022	EX-CAF-R		KNIGHT
250022	EX-CAF-R	1st Clasp	KNIGHT
205776	EX-CAF-R		KNIGHT
9042	EX-CAF-R		KNOX
033094	PAF	gd	KNUSDEN

Surname	Initials		Rank	Status	Date	Source
KOOREY	A	J	WO		12/04/1973	N43
KROLL	L	N	SQD LDR		28/07/1955	N35
KUSCHERT	D W	K	FL		10/12/1964	N98
LAIDLAW	H	B	WO		22/01/1953	N3
LAMBERT	D	B	CPL		23/06/1966	N53
LAMONT	N	A	FO		09/08/1956	N43
LAND	W	A	SQD LDR		07/09/1972	N84
LARDEN	G	H	SGT		13/08/1964	N68
LARDEN	G	H	FLT SGT		29/04/1965	N35
LARGE	W	A	SGT		16/09/1954	N56
LAWRENCE	F	A	SGT		05/08/1965	N65
LAWRENCE	F	G	WO	Acting	23/11/1950	N70
LAWRENCE	R	D	FLT SGT		19/11/1964	N94
LAWRENCE	R	D	FLT SGT		21/03/1968	N26
Le ROSSIGNOL	K	E	SGT	Temporary	23/11/1950	N70
LEONARD	N C	K	WO		29/03/1973	N37
LEWIS	O	H	FL		11/03/1954	N17
LINCOLN	A N	A	SGT		11/08/1960	N56
LINGHAM	L	S	SGT		06/09/1951	N66
LINNEHAN	N	G	CPL		28/05/1974	N44
LISSENDEN	A	E	FL	Temporary	21/02/1952	N18
LITTLEHALES	G	J	SQD LDR		01/10/1964	N81
LOCK	J	F	FLT SGT		10/09/1953	N55
LOVE	A	R	GRP CAP		05/08/1954	N47
LOVELESS	A	T	FL	Temporary	03/09/1964	N74
LUNN	E	S	SGT		17/10/1968	N85
LUPTON	J	S	FL	Acting	28/10/1965	N83
LYNCH	J	J	WING COMM		19/11/1964	N94
LYONS	I	A	FL		03/09/1964	N74
MacBEAN	W	H	FLT SGT		19/03/1953	N17
MACEY	L	E	WO	Temporary	08/11/1951	N83
MacKAY	G	W	SQD LDR		05/08/1954	N47
MacKAY	K C	K	LAC		22/01/1953	N3
MacKERRAS	H	J	SGT	Temporary	24/01/1952	N5
MADELEY	L	R	WO	Acting	23/11/1950	N70
MAGNUS	G	C	FO		08/11/1951	N83
MANNING	N	J	LAC	Temporary	24/01/1952	N5
MARDEN	W	R	LAC		05/11/1974	N90
MARFLEET	D	L	WO		22/01/1953	N3
MARKS	P	B	FL		15/10/1970	N87
MARSH	N	V	SGT	Temporary	23/11/1950	N70
MARSH	W	J	SGT	Temporary	03/04/1952	N27
MARSHALL	H		FO		22/01/1953	N3
MARSHALL	I W	A	FL		13/04/1972	N29
MARSHALL	V	E	GRP CAP	Temporary	27/09/1962	N81
MARTIN	A	K	FL	Temporary	23/11/1950	N70
MARTIN	F	B	FL	Temporary	01/04/1965	N28
MARTYN	C	E	FO		05/04/1951	N23
MASON	E	L	FL		06/09/1951	N66
MATHER	A	E	GRP CAP		29/04/1965	N35
McBROOM	I	C	CPL		05/11/1974	N90
McCARTHY	J	F	SQD LDR		29/03/1973	N37
McCARTHY	J	M	WING COMM	Honorary	10/12/1964	N98
McCARTHY	J	M	LAC		18/02/1971	N18
McCORMACK	J	D	CPL		30/08/1973	N116
McCORMICK	D		CPL		29/11/1973	N179
McCUE	R	L	FL		08/11/1951	N83
McDONALD	J	W	LAC		23/06/1966	N53
McFARLANE	A	B	WING COMM		07/11/1957	N61
McFARLANE	A	B	WING COMM		13/08/1964	N68

Service No	Force	Notes	Surname
433077	EX-CAF-R		KOOREY
011334	PAF	MBE; gd	KROLL
022029	PAF	sd	KUSCHERT
205689	EX-CAF-R		LAIDLAW
A51377	CAF-R	No 25 (City of Perth) (Auxiliary) Squadron	LAMBERT
6539	EX-CAF-R		LAMONT
0211975	CAF-R	medic; No 22 (City of Sydney) (Auxiliary) Squadron	LAND
A210483	CAF-R	No 22 (City of Sydney) (Auxiliary) Squadron	LARDEN
A210483	CAF-R	1st clasp; No 22 (City of Sydney) (Auxiliary) Squadron	LARDEN
6935	EX-CAF-R		LARGE
A15105	CAF-R	No 23 (City of Brisbane) (Auxiliary) Squadron	LAWRENCE
11920	EX-CAF-R		LAWRENCE
A4940	CAF-R	No 24 (City of Adelaide) (Auxiliary) Squadron	LAWRENCE
A4940	CAF-R	1st clasp; sd; No 24 (City of Adelaide) (Auxiliary) Squadron	LAWRENCE
18862	EX-CAF-R		Le ROSSIGNOL
A6143	CAF-R	No 23 (City of Brisbane) (Auxiliary) Squadron	LEONARD
5594	EX-CAF-R		LEWIS
18510	AFR	acct	LINCOLN
A31595	PAF		LINGHAM
A111257	CAF-R		LINNEHAN
253194	EX-CAF-R		LISSENDEN
033111	PAF	gd	LITTLEHALES
207644	EX-CAF-R		LOCK
250185	EX-CAF-R		LOVE
0310534	AFR	gd	LOVELESS
A43241	CAF-R		LUNN
0211608	CAF-R	sd; No 22 (City of Sydney) (Auxiliary) Squadron	LUPTON
021969	PAF	AFC; gd	LYNCH
033072	PAF	MBE; gd	LYONS
19409	EX-CAF-R		MacBEAN
207688	EX-CAF-R		MACEY
262506	EX-CAF-R		MacKAY
A263	PAF		MacKAY
21538	EX-CAF-R		MacKERRAS
31048	EX-CAF-R		MADELEY
205723	EX-CAF-R		MAGNUS
20139	EX-CAF-R		MANNING
A211302	CAF-R	No 22 (City of Sydney) (Auxiliary) Squadron	MARDEN
A35597	EX-CAF-R		MARFLEET
015328	CAF-R	engin; No 23 (City of Brisbane) (Auxiliary) Squadron	MARKS
26543	EX-CAF-R		MARSH
16095	CAF-R		MARSH
205752	EX-CAF-R		MARSHALL
016593	CAF-R	medic; Queensland University Squadron	MARSHALL
250245	CAF-R	tech	MARSHALL
205163	CAF-R	MBE	MARTIN
052939	CAF-R	equip; Western Australian University Squadron	MARTIN
21390	EX-CAF-R		MARTYN
0247	PAF		MASON
021970	PAF	DFC,AFC; gd	MATHER
A53703	CAF-R	No 25 (City of Perth) (Auxiliary) Squadron	McBROOM
0316838	CAF-R	No 21 (City of Melbourne) (Auxiliary) Squadron	McCARTHY
03393	RETIRED	DFC,AFM; tech	McCARTHY
A16723	CAF-R		McCARTHY
A116037	CAF-R	No 23 (City of Brisbane) (Auxiliary) Squadron	McCORMACK
A44174	CAF-R	No 24 (City of Adelaide) (Auxiliary) Squadron	McCORMICK
011331	PAF	gd	McCUE
10175	EX-CAF-R		McDONALD
250207	EX-CAF-R	DFC	McFARLANE
250207	EX-CAF-R	CBE,DFC; 1st Clasp; gd (the first award of a clasp)	McFARLANE

Surname	Initials			Rank	Status	Date	Source
McFARLANE	J	A		FO		25/07/1968	N64
McGOLDRICK	P	G		FO		22/09/1966	N81
McHARDIE	E	D		SQD LDR		01/04/1965	N28
McKENZIE	D			FLT SGT		06/09/1951	N66
McKINNON	A	E	C	CPL		31/08/1972	N83
McLEAN	J	A		FO		23/11/1950	N70
McLEAY	D	B		SQD LDR		16/06/1966	N48
McLEOD	K	D		FL		10/09/1953	N55
McMURRAY	J	M		SGT	Acting	23/11/1950	N70
McNAMARA	N	P		WING COMM		18/03/1965	N20
McPHAIL	N			SQD LDR		29/03/1962	N21
McPHERSON	C	J		SGT	Temporary	23/11/1950	N70
MELVILLE	D	McK		SGT		06/09/1951	N66
MEYERS	N	F		WO		27/05/1954	N34
MILKINS	H	W		SGT	Temporary	15/03/1951	N19
MILLEN	D	R		WO		31/05/1962	N42
MILLER	R	J		FO		23/11/1950	N70
MILLER	S	G		WO		22/01/1953	N3
MITCHELL	R	O		PO		07/04/1971	N39
MOIR	S	J		WING COMM	Temporary	29/05/1952	N38
MONTGOMERY	F	J		SQD LDR		01/10/1964	N81
MOON	B	F	G	FLT SGT	Temporary	21/02/1952	N18
MOON	E	J		FL	Temporary	24/01/1957	N5
MOONEY	T	R		FL		18/02/1971	N18
MOORE	H	F		SQD LDR		21/06/1951	N42
MORRIS	J	W		SGT		18/10/1951	N78
MORRISON	J			SGT		16/09/1954	N56
MORRISON	K	R		CHAP		21/04/1955	N18
MOSS	R			SGT		22/01/1953	N3
MOTLEY	G	A	P	SGT		03/02/1966	N9
MOTTERAM	A	G		WO		21/04/1955	N18
MOYLE	A	F		SGT		02/11/1972	N103
MUIRHEAD	R	M		FL		27/05/1954	N34
MULLER	F	K		SQD LDR		01/07/1965	N53
MULRONEY	N			WING COMM	Temporary	29/05/1952	N38
MUNDT	B	C		LAC		26/07/1973	N90
MURPHY	C	D		WING COMM		18/03/1965	N20
MUSGRAVE	I	A		SQD LDR		02/10/1969	N83
MUSTAR	E	A		WING COMM	Temporary	23/11/1950	N70
MYERS	W	F		FL	Acting	23/11/1950	N70
MYERS	W	F		FO		28/01/1965	N10
NEAVE	F	J		FLT SGT		31/01/1957	N7
NEILSEN	W	J		FL		23/01/1969	N7
NEWBURY	T	W		LAC		29/04/1965	N35
NEWBURY	T	W		LAC		24/06/1965	N50
NEWITT	R	L		CPL		28/09/1967	N84
NEWLANDS	W	A		LAC		20/05/1971	N53
NICHOL	S	J		WING COMM		13/08/1964	N68
NICHOLAS	K	W		FO		07/01/1954	N1
NICHOLLS	J	B		WING COMM	Acting	01/10/1964	N81
NIVEN	H			FL	Honorary	11/02/1954	N8
NOBLE	C	R		FL		08/11/1951	N83
NOLAN	S	T	A	CPL		05/08/1965	N65
NOONAN	M	J		CPL		08/03/1956	N11
NORRIS	H	C		SGT		29/04/1965	N35
O'BRIEN	C	L		FLT SGT		10/09/1953	N55
O'LEARY	J	B		FO		06/08/1970	N65
OLIVE	C	G	G	WING COMM	Acting	05/03/1970	N15
OLORENSHAW	I	R		SQD LDR		28/04/1955	N19
O'NEILL	J	V		FO		13/08/1964	N68

Service No	Force	Notes	Surname
0314272	CAF-R	sd; No 22 (City of Sydney) (Auxiliary) Squadron	McFARLANE
023696	AFR	gd	McGOLDRICK
011430	PAF	equip	McHARDIE
A32998	PAF		McKENZIE
A45602	CAF-R	No 24 (City of Adelaide) (Auxiliary) Squadron	McKINNON
9292	CAF-R		McLEAN
042513	CAF-R	medic; No 24 (City of Adelaide) (Auxiliary) Squadron	McLEAY
031479	PAF	DFC; gd	McLEOD
18802	EX-CAF-R		McMURRAY
011353	PAF	AFC; gd	McNAMARA
022010	PAF	DFC,DFM; gd	McPHAIL
34820	EX-CAF-R		McPHERSON
A314417	PAF		MELVILLE
26438	EX-CAF-R		MEYERS
5921	EX-CAF-R		MILKINS
207595	EX-CAF-R		MILLEN
27897	CAF-R		MILLER
207695	EX-CAF-R		MILLER
043711	CAF-R	sd; No 24 (City of Adelaide) (Auxiliary) Squadron	MITCHELL
261340	EX-CAF-R		MOIR
03384	PAF	AFC; sd	MONTGOMERY
10064	EX-CAF-R		MOON
205502	EX-CAF-R		MOON
016731	CAF-R	sd; No 23 (City of Brisbane) (Auxiliary) Squadron	MOONEY
033083	PAF	gd	MOORE
A259	PAF		MORRIS
A163	PAF		MORRISON
035500	AFR	chap	MORRISON
300177	EX-CAF-R		MOSS
10213	EX-CAF-R		MOTLEY
A33189	WRAAF		MOTTERAM
A32699	CAF-R	No 24 (City of Adelaide) (Auxiliary) Squadron	MOYLE
407127	EX-CAF-R		MUIRHEAD
031492	PAF	AFC; tech	MULLER
260064	EX-CAF-R		MULRONEY
A115539	CAF-R		MUNDT
033188	PAF	DFC,AFC; gd	MURPHY
0115703	CAF-R	medic; No 23 (City of Brisbane) (Auxiliary) Squadron	MUSGRAVE
250062	CAF-R	DFC	MUSTAR
14080	EX-CAF-R		MYERS
14080	EX-CAF-R	1st Clasp; sd	MYERS
207646	EX-CAF-R		NEAVE
022006	PAF	MBE,BEM; equip	NEILSEN
A51336	CAF-R	No 25 (City of Perth) (Auxiliary) Squadron	NEWBURY
A51336	CAF-R	1st Clasp; No 25 (City of Perth) (Auxiliary) Squadron	NEWBURY
A15355	CAF-R	sd; No 23 (City of Brisbane) (Auxiliary) Squadron	NEWITT
A311697	CAF-R		NEWLANDS
03487	PAF	DFC; sd	NICHOL
207729	EX-CAF-R		NICHOLAS
0210990	AFR	DFC; gd	NICHOLLS
036165	EX-CAF-R		NIVEN
033061	PAF	DFC; gd	NOBLE
A51368	CAF-R	No 25 (City of Perth) (Auxiliary) Squadron	NOLAN
A216216	WRAAF		NOONAN
A16019	EX-CAF-R		NORRIS
207711	EX-CAF-R		O'BRIEN
042251	CAF-R	equip; South Australian University Squadron	O'LEARY
017934	AFR	MBE,DFC; sd	OLIVE
04410	EX-CAF-R	DFC	OLORENSHAW
0210013	CAF-R	acct; No 22 (City of Sydney) (Auxiliary) Squadron	O'NEILL

Surname	Initials		Rank	Status	Date	Source
O'NEILL	J V		FO		28/09/1967	N84
O'REILLY	C S		FL		26/07/1973	N90
ORR	R N	W	WO		20/01/1955	N4
ORR	W J		SGT		02/11/1972	N103
O'SACHY	G Y		LAC		11/09/1969	N74
OSBORNE	H N		FLT SGT		14/08/1952	N55
O'SULLIVAN	F L		SGT		03/12/1953	N79
OWSTON	A		FLT SGT		06/09/1951	N66
OXER	G		SQD LDR	Temporary	15/03/1951	N19
OXFORD	J A		LAC		27/08/1974	N70
PAGE	C D		SGT	Temporary	25/01/1951	N7
PAGET	T B		SQD LDR		24/01/1963	N6
PARKER	B		FO		12/10/1972	N97
PARKER	L G		LAC		09/03/1972	N18
PARKINSON	J V		FO		23/11/1950	N70
PATTERSON	G J		FLT SGT		22/01/1953	N3
PAUL	D V		SQD LDR	Acting	13/08/1964	N68
PAUL	D V		SQD LDR	Acting	03/12/1964	N97
PAULL	N W		CPL		27/07/1967	N68
PAYNE	M H		SQD LDR		28/01/1965	N10
PEAREE	L A		WO		24/04/1952	N31
PECK	E N		CPL		30/08/1973	N116
PEEL	N H		FL		18/03/1965	N20
PEGG	J A		WO		15/08/1957	N45
PELLETIER	J W		SQD LDR		03/12/1953	N79
PELLING	L R		LAC		09/05/1957	N25
PELLY	B R		WING COMM	Temporary	15/03/1951	N19
PENNISI	S J		LAC		03/02/1972	N9
PEPPER	A R		FLT SGT		19/11/1964	N94
PHILLIPS	G E		SQD LDR	Temporary	05/04/1951	N23
PICKARD	E T		WING COMM		01/10/1964	N81
PICKERING	H A	H	GRP CAP		01/06/1972	N43
PIGDON	G		FLT SGT		05/08/1965	N65
PIKE	M R		FLT SGT		30/07/1959	N47
PILL	W H		FO		22/01/1953	N3
PINDER	M C		FLT SGT		19/11/1964	N94
PINDER	M C		WO		02/04/1970	N22
PLATT	J T		SGT	Temporary	23/11/1950	N70
PLUNKETT	B E		CPL	Temporary	13/08/1964	N68
POATE	H R	G	GRP CAP	Temporary	15/03/1951	N19
POCKLEY	E V	W	WING COMM	Temporary	29/09/1966	N83
POLLOCK	J E		FL		17/10/1968	N85
PORTCH	C R		FLT SGT		22/01/1953	N3
PORTER	M J		SQD LDR	Acting	23/07/1970	N61
PORTER	R C		FL		15/10/1970	N87
PORTWAY	C		FL	Temporary	06/09/1951	N66
POTTER	L R		WING COMM	Temporary	05/04/1951	N23
PRATT	S J		LAC		09/03/1972	N18
PRESTON	W D	A	FLT SGT		30/07/1974	N62
PRIME	N D		FO		23/11/1950	N70
PRINCE	C A		SGT		01/05/1969	N36
PRITCHARD	F J		SGT		19/05/1955	N22
PYKE	G P		SGT		12/10/1972	N97
RABY	M J		FLT SGT	Temporary	25/01/1951	N7
RACKLEY	L W		FL	Acting	15/12/1966	N104
RAE	K		CPL	Temporary	15/03/1951	N19
RAINBOW	C J		FL	Temporary	01/10/1964	N81
RANSLEY	C P		FL		01/10/1964	N81
REDENBACH	C E		SQD LDR	Temporary	09/04/1959	N23
REDENBACH	C J	J	FL	Acting	01/10/1964	N81

Service No	Force	Notes	Surname
0210013	CAF-R	1st Clasp; sd; No 22 (City of Sydney) (Auxiliary) Squadron	O'NEILL
011154	CAF-R	No 23 (City of Brisbane) (Auxiliary) Squadron	O'REILLY
A31406	WRAAF		ORR
A52635	CAF-R	No 25 (City of Perth) (Auxiliary) Squadron	ORR
A16713	CAF-R	No 23 (City of Brisbane) (Auxiliary) Squadron	O'SACHY
205772	EX-CAF-R		OSBORNE
A210663	EX-CAF-R		O'SULLIVAN
A522	PAF		OWSTON
251201	EX-CAF-R		OXER
A44191	CAF-R	No 24 (City of Adelaide) (Auxiliary) Squadron	OXFORD
10833	EX-CAF-R		PAGE
035806	PAF	sd	PAGET
041232	CAF-R	sd; No 24 (City of Adelaide) (Auxiliary) Squadron	PARKER
A56085	CAF-R	No 25 (City of Perth) (Auxiliary) Squadron	PARKER
411177	CAF-R		PARKINSON
207722	EX-CAF-R		PATTERSON
0210106	CAF-R	DFC; gd; No 22 (City of Sydney) (Auxiliary) Squadron	PAUL
0210106	CAF-R	DFC; 1st Clasp; gd; No 22 (City of Sydney) (Auxiliary) Squadron	PAUL
8094	EX-PAF		PAULL
031468	PAF	AFC; gd	PAYNE
300276	EX-CAF-R		PEAREE
A315626	CAF-R	No 21 (City of Melbourne) (Auxiliary) Squadron	PECK
0156	PAF	AFC; equip	PEEL
207657	EX-CAF-R		PEGG
400483	EX-CAF-R	DFC	PELLETIER
24352	EX-CAF-R		PELLING
260226	EX-CAF-R	OBE	PELLY
A16747	CAF-R	No 23 (City of Brisbane) (Auxiliary) Squadron	PENNISI
A41153	CAF-R	No 24 (City of Adelaide) (Auxiliary) Squadron	PEPPER
1456	EX-CAF-R		PHILLIPS
033123	PAF	gd	PICKARD
033131	PAF	gd	PICKERING
A13328	CAF-R	No 23 (City of Brisbane) (Auxiliary) Squadron	PIGDON
A431	WRAAF		PIKE
035440	PAF	sd	PILL
A4910	CAF-R	No 24 (City of Adelaide) (Auxiliary) Squadron	PINDER
A4910	CAF-R	1st Clasp; No 24 (City of Adelaide) (Auxiliary) Squadron	PINDER
43488	EX-CAF-R		PLATT
A210603	CAF-R	No 22 (City of Sydney) (Auxiliary) Squadron	PLUNKETT
1162	EX-CAF-R	MVO	POATE
3752	AFR	medic	POCKLEY
0211726	CAF-R	No 22 (City of Sydney) (Auxiliary) Squadron	POLLOCK
205769	EX-CAF-R		PORTCH
042503	CAF-R	engin; No 24 (City of Adelaide) (Auxiliary) Squadron	PORTER
061188	CAF-R	sd; Tasmanian University Squadron	PORTER
263753	EX-CAF-R		PORTWAY
250278	EX-CAF-R		POTTER
A56098	CAF-R	No 25 (City of Perth) (Auxiliary) Squadron	PRATT
A44187	CAF-R	No 24 (City of Adelaide) (Auxiliary) Squadron	PRESTON
26629	CAF-R		PRIME
A16714	CAF-R	No 23 (City of Brisbane) (Auxiliary) Squadron	PRINCE
A310599	WRAAF		PRITCHARD
A31955	CAF-R	No 24 (City of Adelaide) (Auxiliary) Squadron	PYKE
18085	EX-CAF-R		RABY
016231	AFR	DFC; gd	RACKLEY
24817	EX-CAF-R		RAE
035200	CAF-R	tech; No 21 (City of Melbourne) (Auxiliary) Squadron	RAINBOW
016126	PAF	sd	RANSLEY
408755	EX-CAF-R		REDENBACH
035673	CAF-R	tech; No 21 (City of Melbourne) (Auxiliary) Squadron	REDENBACH

Surname	Initials			Rank	Status	Date	Source
REID	R	A		LAC		26/07/1973	N90
RICE	J	J		SQD LDR		13/09/1951	N68
RICHARDS	N	D		CPL		03/06/1971	N57
RIGBY	H	A		WING COMM	Temporary	21/06/1951	N42
ROBERTS	A	V	L	SQD LDR		12/08/1954	N49
ROBERTSON	K	V		WING COMM		01/10/1964	N81
ROBEY	F	S		SQD LDR		21/02/1952	N18
ROCKS	W	H		WO		08/06/1967	N48
ROGERS	G	J		SGT		26/06/1958	N36
ROGERS	J	W		CPL		18/03/1965	N20
ROGERS	W	McI		FL		30/09/1965	N78
ROLES	E	S		FLT SGT		19/11/1964	N94
ROPER	A	A		SQD LDR	Temporary	13/08/1964	N68
RUSS	W	H		FO		07/01/1954	N1
RUSSELL	S	W		SQD LDR		19/11/1964	N94
RUTHOF	M	C	A	FLT SGT		01/05/1969	N36
RYLAND	J	P		WING COMM	Temporary	15/03/1951	N19
RYLAND	J	P		GRP CAP	Acting	28/01/1965	N10
SANDER	N	J		LAC		02/11/1972	N103
SANDERS	E	J		SGT		16/04/1970	N29
SCALFS	T	S		WO	Temporary	25/01/1951	N7
SCARONI	G	T		CPL		04/03/1975	G9
SCATES	F	W		FL		30/11/1972	N121
SCHAAF	F	R		WING COMM		22/10/1964	N87
SCHOLES	A	J		WO	Temporary	15/03/1951	N19
SCOTT	T	W		FL		07/01/1965	N1
SCOTT	W	D		FL	Acting	29/04/1965	N35
SCOTT	W	J	M	FL		10/09/1953	N55
SCRIVEN	G	C		FO		23/11/1950	N70
SFORCINA	H	G		SQD LDR		10/12/1964	N98
SHADFORTH	R	J		SQD LDR		03/09/1964	N74
SHANKS	J			FL		01/10/1964	N81
SHAW	L	McG	D	FO		19/05/1966	N42
SHEARN	H	V		SQD LDR		19/08/1965	N69
SHEPHERD	A	W		PO		21/06/1951	N42
SHEPPARD	D	R		SQD LDR		04/09/1952	N60
SIMMS	M	J		CPL		23/06/1966	N53
SIMMS	M	J		SGT		04/03/1975	G9
SINNETT	K	M		CPL	Temporary	25/01/1951	N7
SKINNER	N	F		FO		27/05/1954	N34
SMALLWOOD	B	T		FLT SGT		20/04/1972	N32
SMITH	A	G		SGT		14/08/1952	N55
SMITH	E	J	D	FLT SGT		21/02/1952	N18
SMITH	E			WO		28/05/1953	N33
SMITH	F	H		SGT	Temporary	25/01/1951	N7
SMITH	G	T		CPL		26/06/1958	N36
SMITH	H	T		SQD LDR		28/10/1965	N83
SMITH	J	C		FL		22/01/1953	N3
SMYTH	J	W		CPL		28/06/1973	N75
SNIDE	H	A	W	SGT		06/09/1973	N123
SOLOMON	A	H		SQD LDR		04/09/1952	N60
SPALDING	J	W		FO		23/12/1954	N77
SPOONER	J	W		WO	Temporary	15/03/1951	N19
SPURRITT	R			SGT	Temporary	13/09/1951	N68
STANLEY	R	N		FL	Temporary	16/06/1966	N48
STAR	J	C		SQD LDR		20/05/1965	N41
STAUNTON	J	K		FL	Temporary	14/08/1952	N55
STEELE	A	H		WING COMM		22/10/1964	N87
STEELE	C			WING COMM		03/12/1964	N97
STEVENS	W	L		WO	Temporary	25/01/1951	N7

Service No	Force	Notes	Surname
A16786	CAF-R		REID
035153	CAF-R	sd	RICE
A43720	CAF-R	No 24 (City of Adelaide) (Auxiliary) Squadron	RICHARDS
250063	EX-CAF-R	MC	RIGBY
290294	EX-CAF-R		ROBERTS
033013	PAF	DFC,AFC; gd	ROBERTSON
05787	PAF	gd	ROBEY
A211181	CAF-R	sd; No 22 (City of Sydney) (Auxiliary) Squadron	ROCKS
A210264	EX-CAF-R		ROGERS
A41189	CAF-R	No 24 (City of Adelaide) (Auxiliary) Squadron	ROGERS
041301	CAF-R	sd; Adelaide University Squadron	ROGERS
A4956	CAF-R	No 24 (City of Adelaide) (Auxiliary) Squadron	ROLES
0211015	CAF-R	medic; No 22 (City of Sydney) (Auxiliary) Squadron	ROPER
016120	EX-CAF-R		RUSS
016675	AFR	sd	RUSSELL
A12269	CAF-R	No 23 (City of Brisbane) (Auxiliary) Squadron	RUTHOF
250188	EX-CAF-R	DFC	RYLAND
250188	EX-CAF-R	DFC; 1st Clasp; gd	RYLAND
A16771	CAF-R	No 23 (City of Brisbane) (Auxiliary) Squadron	SANDER
A40177	CAF-R	No 21 (City of Melbourne) (Auxiliary) Squadron	SANDERS
404058	EX-CAF-R		SCALFS
A111264	CAF-R	No 23 (City of Brisbane) (Auxiliary) Squadron	SCARONI
0315332	CAF-R	No 21 (City of Melbourne) (Auxiliary) Squadron	SCATES
03382	PAF	DFC; gd	SCHAAF
19191	EX-CAF-R	DFM	SCHOLES
0210038	AFR	gd	SCOTT
051866	CAF-R	tech; No 25 (City of Perth) (Auxiliary) Squadron	SCOTT
40186	EX-CAF-R		SCOTT
5016	CAF-R		SCRIVEN
036455	AFR	equip	SFORCINA
05829	PAF	gd	SHADFORTH
407991	AFR	gd	SHANKS
016018	AFR	gd	SHAW
05974	PAF	DFC,AFC; gd	SHEARN
011966	PAF	tech	SHEPHERD
031436	PAF	gd	SHEPPARD
A41174	CAF-R	No 24 (City of Adelaide) (Auxiliary) Squadron	SIMMS
A41174	CAF-R	1st Clasp; No 24 (City of Adelaide) (Auxiliary) Squadron	SIMMS
30150	EX-CAF-R		SINNETT
402209	EX-CAF-R		SKINNER
A39268	CAF-R	No 21 (City of Melbourne) (Auxiliary) Squadron	SMALLWOOD
19623	EX-CAF-R		SMITH
A32997	PAF		SMITH
A51653	EX-CAF-R		SMITH
18515	EX-CAF-R		SMITH
34055	EX-CAF-R		SMITH
03488	PAF	tech	SMITH
023145	PAF	tech	SMITH
A115309	CAF-R	sd; No 23 (City of Brisbane) (Auxiliary) Squadron	SMYTH
A16797	CAF-R	No 23 (City of Brisbane) (Auxiliary) Squadron	SNIDE
021929	PAF	acct	SOLOMON
403116	AFR	gd	SPALDING
5984	EX-CAF-R		SPOONER
28614	EX-CAF-R		SPURRITT
403963	CAF-R	gd	STANLEY
03570	PAF	sd	STAR
012714	CAF-R	DFC; gd	STAUNTON
03380	PAF	sd	STEELE
03108	PAF	sd	STEELE
11646	EX-CAF-R		STEVENS

Surname	Initials			Rank	Status	Date	Source
STEWART	N	W		FO		05/08/1954	N47
STOKES	T	V		WING COMM	Temporary	15/03/1951	N19
STRUGNELL	H	E	C	SGT		06/09/1951	N66
SUGDEN	C	J		SQD LDR		01/04/1965	N28
SUMMERVILLE	A	D		WO	Temporary	25/01/1951	N7
SUNDERLAND	J	P		SGT		22/01/1953	N3
SUTTON	R	A		CPL		19/03/1953	N17
SVENSON	G	T		FL	Acting	20/04/1967	N33
SYMMANS	A	J		SGT		24/07/1969	N62
TALBERG	W	H	S	GRP CAP	Honorary	05/08/1965	N65
TARPLEE	W			SGT		14/08/1952	N55
TARRANT	L	H		SGT	Temporary	18/10/1951	N78
TAYLOR	C	W		SGT		20/04/1972	N32
TAYLOR	D	C	F	CPL		18/02/1971	N18
TAYLOR	G			WO		12/11/1953	N71
TENISWOOD	J	C		FL		23/07/1970	N61
TERRY	A	J		WING COMM		25/09/1952	N65
THOMAS	A	E		WO		28/05/1953	N33
THOMAS	E	A		SGT		01/05/1969	N36
THOMAS	F	W		GRP CAP	Temporary	23/11/1950	N70
THOMAS	F	W		GRP CAP		22/10/1964	N87
THOMAS	F	W		GRP CAP		22/10/1964	N87
THOMPSON	G	E	C	FLT SGT	Temporary	25/01/1951	N7
THOMPSON	M	M		WO	Temporary	25/01/1951	N7
THOMPSON	R	W		WO		31/01/1957	N7
THOMPSON	S	G		SQD LDR		05/08/1954	N47
THORNTON	R	J		SGT		22/01/1953	N3
TITCHENER	R	C		FL		07/01/1954	N1
TORPY	K	M		CPL		10/06/1954	N36
TOWERS	I	F		WO		01/09/1955	N41
TOWN	H	A		FL		05/12/1963	N104
TREDENICK	A	F		WO	Temporary	23/11/1950	N70
TREZONA	T	A		CPL		19/11/1964	N94
TRIGGS	A	W	R	FL		07/01/1954	N1
TRIMMER	A	E		CPL		28/04/1955	N19
TROTTER	A	K		SGT		05/11/1974	N90
TROUT	D			WO		12/10/1972	N97
TURNER	J	R		FLT SGT		19/11/1964	N94
TURNER	J	R		FLT SGT		21/05/1970	N41
TURNER	R	A		SGT		27/05/1954	N34
TURNER	S	J		SGT	Temporary	06/09/1951	N66
TURNER	W	J		WO		14/08/1952	N55
UPSTON	C	J		FO		29/09/1966	N83
VAUDREY	G	C		FL	Temporary	23/11/1950	N70
VEALE	K	J		LAC		27/07/1972	N67
VOWLES	E	J	R	FL		06/04/1967	N28
WADE	W	L		FL		29/05/1952	N38
WALFORD	G	A		CPL	Temporary	13/08/1964	N68
WALFORD	G	A		FLT SGT		02/09/1971	N86
WALKER	G	F		WING COMM	Temporary	23/11/1950	N70
WALLER	G	L		SQD LDR		07/01/1965	N1
WALLIKER	F	V		FL		28/05/1953	N33
WALTER	K	J		FLT SGT		05/03/1970	N15
WALTERS	C	H		FL	Temporary	23/11/1950	N70
WALTERS	J	A		FL		11/03/1954	N17
WALTERS	W	R		CPL		19/03/1953	N17
WARD	B	P		FL		23/06/1966	N53
WARD	C	E		SGT		23/09/1971	N90
WATSON	P	M		FO		12/11/1953	N71
WENNERBOM	E			CPL	Temporary	25/01/1951	N7

Service No	Force	Notes	Surname
432098	EX-CAF-R		STEWART
250279	EX-CAF-R	DFC	STOKES
A142	PAF		STRUGNELL
05813	PAF	DFC; gd	SUGDEN
22562	EX-CAF-R		SUMMERVILLE
14045	EX-CAF-R		SUNDERLAND
33224	EX-CAF-R		SUTTON
016097	AFR	sd	SVENSON
A51396	CAF-R	sd; No 25 (City of Perth) (Auxiliary) Squadron	SYMMANS
021918	RETIRED	OBE; gd	TALBERG
32914	EX-CAF-R		TARPLEE
205779	EX-CAF-R		TARRANT
A16759	CAF-R	No 23 (City of Brisbane) (Auxiliary) Squadron	TAYLOR
A16725	CAF-R		TAYLOR
5414	EX-CAF-R		TAYLOR
061187	CAF-R	sd; Tasmanian University Squadron	TENISWOOD
291224	CAF-R		TERRY
A21911	PAF		THOMAS
A13026	CAF-R	No 23 (City of Brisbane) (Auxiliary) Squadron	THOMAS
250097	CAF-R		THOMAS
250097	RETIRED	KBE; Sir; 1st Clasp; gd	THOMAS
250097	RETIRED	KBE; Sir; 2nd Clasp; gd	THOMAS
25070	EX-CAF-R		THOMPSON
300347	EX-CAF-R		THOMPSON
A314417	WRAAF		THOMPSON
0210401	EX-CAF-R		THOMPSON
A2109	PAF		THORNTON
205650	EX-CAF-R		TITCHENER
A148	PAF		TORPY
20154	EX-CAF-R		TOWERS
207662	EX-CAF-R		TOWN
404307	EX-CAF-R		TREDENICK
A52730	CAF-R	No 25 (City of Perth) (Auxiliary) Squadron	TREZONA
460500	EX-CAF-R	MBE,DFC	TRIGGS
A21508	EX-CAF-R		TRIMMER
A211308	CAF-R		TROTTER
A51052	CAF-R	No 25 (City of Perth) (Auxiliary) Squadron	TROUT
A4934	CAF-R	No 24 (City of Adelaide) (Auxiliary) Squadron	TURNER
A4934	CAF-R	1st Clasp; No 24 (City of Adelaide) (Auxiliary) Squadron	TURNER
11385	EX-CAF-R		TURNER
10069	EX-CAF-R		TURNER
300126	EX-CAF-R		TURNER
0211183	AFR	tech; No 22 (City of Sydney) (Auxiliary) Squadron	UPSTON
407919	EX-CAF-R		VAUDREY
A315562	CAF-R	No 21 (City of Melbourne) (Auxiliary) Squadron	VEALE
0114160	CAF-R	equip; Queensland University Squadron	VOWLES
033054	PAF	sd	WADE
A24252	CAF-R	No 22 (City of Sydney) (Auxiliary) Squadron	WALFORD
A24252	CAF-R	1st Clasp; No 22 (City of Sydney) (Auxiliary) Squadron	WALFORD
260312	EX-CAF-R	DFC	WALKER
033087	PAF	gd	WALLER
400033	EX-CAF-R		WALLIKER
A41153	CAF-R	No 24 (City of Adelaide) (Auxiliary) Squadron	WALTER
408055	EX-CAF-R		WALTERS
033067	PAF	equip	WALTERS
A31912	WRAAF		WALTERS
016476	CAF-R	tech; No 23 (City of Brisbane) (Auxiliary) Squadron	WARD
A16618	CAF-R	No 23 (City of Brisbane) (Auxiliary) Squadron	WARD
407921	EX-CAF-R		WATSON
28851	EX-CAF-R		WENNERBOM

Surname	Initials			Rank	Status	Date	Source
WEST	J	G		FL		22/01/1953	N3
WHEELER	P	W		WO	Temporary	25/01/1951	N7
WHITE	F	H		WO		05/08/1965	N65
WILLIAMS	C	G		SQD LDR	Temporary	23/11/1950	N70
WILLIAMS	F	N	T	FL		01/10/1964	N81
WILLIAMS	H			SGT		19/11/1964	N94
WILLIAMS	H			SGT		03/12/1974	N98
WILLIAMS	W	H		FL	Temporary	15/03/1951	N19
WILLIAMSON	L	H		SQD LDR		21/06/1951	N42
WILLMETT	N	J		PO		02/11/1972	N103
WILLOX	A	C		SGT		30/07/1974	N62
WILLS	V	J		FL		04/04/1968	N31
WILSON	D	L		GRP CAP		27/10/1966	N90
WILSON	J	S		WING COMM	Acting	12/11/1964	N92
WILSON-SMITH	W			FLT SGT	Temporary	25/01/1951	N7
WINTERSON	M	R		FO		20/02/1969	N16
WINZAR	C	F		SQD LDR		18/02/1971	N18
WITHAM	M	D		FL		02/09/1965	N72
WOLFE	G	E		WO		15/08/1957	N45
WOOD	S	R	C	SQD LDR	Temporary	18/03/1965	N20
WOODMAN	J	H		FL	Temporary	23/11/1950	N70
WOOLLEY	A	W		SQD LDR	Temporary	23/11/1950	N70
WRAITH	A	L		SQD LDR		19/11/1964	N94
WRIGHT	F	N		GRP CAP	Temporary	23/11/1950	N70
WRIGHT	F	N		AIR COMMD	Honorary	22/10/1964	N87
WRIGHT	R	F	G	WO		18/10/1951	N78
YARROW	L	R		CPL		02/09/1971	N86

Service No	Force	Notes	Surname
205746	EX-CAF-R		WEST
26439	EX-CAF-R		WHEELER
A546	CAF-R	No 23 (City of Brisbane) (Auxiliary) Squadron	WHITE
270311	EX-CAF-R		WILLIAMS
n/a	PAF	equip	WILLIAMS
A51007	CAF-R	No 25 (City of Perth) (Auxiliary) Squadron	WILLIAMS
A51007	CAF-R	1st Clasp; No 25 (City of Perth) (Auxiliary) Squadron	WILLIAMS
27807	EX-CAF-R		WILLIAMS
021972	PAF	gd	WILLIAMSON
011659	CAF-R	sd; No 23 (City of Brisbane) (Auxiliary) Squadron	WILLMETT
A53700	CAF-R	No 25 (City of Perth) (Auxiliary) Squadron	WILLOX
0216820	PAF	equip	WILLS
021924	PAF	DFC; gd	WILSON
022003	PAF	AFC; gd	WILSON
9044	EX-CAF-R		WILSON-SMITH
033580	CAF-R	tech; No 24 (City of Adelaide) (Auxiliary) Squadron	WINTERSON
016745	CAF-R	medic; No 23 (City of Brisbane) (Auxiliary) Squadron	WINZAR
011562	AFR	gd	WITHAM
30227	EX-CAF-R		WOLFE
035909	AFR	DFC; gd	WOOD
264614	EX-CAF-R		WOODMAN
252409	CAF-R		WOOLLEY
033105	PAF	sd	WRAITH
250061	CAF-R	OBE,MVO	WRIGHT
250061	RETIRED	OBE,MVO; 1st Clasp; gd	WRIGHT
22506	EX-CAF-R		WRIGHT
A16738	CAF-R	No 23 (City of Brisbane) (Auxiliary) Squadron	YARROW

NOMINAL ROLL FOR CANADA

Surname	Initials			Rank	Status	Date
ALLEN	J	S		FO		20/07/1945
BARRACLOUGH	T	E		FL		03/08/1945
BEARDMORE	E	W		WING COMM		23/03/1945
BEGG	R	F		WING COMM		26/01/1945
BELL	G	W	A	SGT		17/10/1958
BELL-IRVING	A	D		GRP CAP		06/10/1944
BENCHIER	B	J		SQD LDR		16/11/1945
BIRNIE	W	A		FLT SGT		04/05/1945
BOUGHNER	C	W		FL		12/09/1945
BUNSTAD	R	J		WO1		29/09/1944
CARPENTER	G			FL		23/03/1945
CARVER	A	E		WO1		28/05/1948
COOPER	P	J		PO		24/11/1944
COPLEY	W			FL		03/01/1947
CORBETT	V	B		GRP CAP		15/12/1944
CRABB	H	P		GRP CAP		26/01/1945
CUNNINGHAM	C	D		WO2		10/11/1944
CURTIS	W	A		AVM		24/11/1944
DELHAYE	R	A		AIR COMMD		03/11/1944
DIAMOND	G	G		WING COMM		12/04/1946
DOE	W	A		FL		14/05/1950
DUTEMPLE	G	W		WING COMM		26/01/1945
ELLIS	H	C		WO2		23/02/1945
ELLISON	J	H	K	FO		29/09/1944
ELMS	G	H		GRP CAP		07/06/1946
FOLKINS	G	A		WING COMM		22/12/1944
FOSS	R	H		GRP CAP		29/09/1944
FRASER	J	R		FL		28/07/1950
FRASER	N	R		FO		22/03/1946
GARDNER	E	R		GRP CAP		20/07/1945
GILES	J	L		WO2		20/10/1944
GLEDHILL	J	W		SQD LDR		16/03/1945
GOLDEY	J	L		FLT SGT		06/12/1946
GRIFFITH	J	O		FL		20/10/1944
HALL	K	W		WO1		20/10/1944
HANNA	W	F		GRP CAP		31/08/1945
HARDMAN	A	W		FO		15/12/1944
HARRIS	J	G		WO2		10/11/1944
HENDERSON	J			FLT SGT		27/04/1945
HILL	G	W		FO		05/03/1945
HILLOCK	F	W		WING COMM		25/07/1947
HIVES	A	W		WO2		05/07/1946
HOLLOWAY	C	H		WO2		01/06/1945
HORNELL	H	A		FL		24/07/1953
IRWIN	G	N		AIR COMMD		29/09/1944
IVERSON	W	M		FL		25/07/1947
JACOBI	G	W		WING COMM		24/11/1944
JARDINE	D	T		WO1		07/12/1945
KERR	J			WO1		13/10/1944
LEITCH	G	A		WO1		12/09/1945
LYEL	R	N		FO		01/03/1946
MALCOLMSON	H	G		WING COMM		23/03/1945
MATHEWS	J	M		SGT		04/05/1945
McARDLE	J	S	R	FL		29/09/1944
McCALLUM	H	D		WO1		21/03/1947
McDOWELL	R			FO		13/10/1944
McFADYN	M	D		WING COMM		24/11/1944

Source	Service No	Force	Notes	Surname
	C40315	RCAF		ALLEN
	C16881	RCAF		BARRACLOUGH
	C820	RCAF		BEARDMORE
	C223	RCAF		BEGG
	120074	RCAF		BELL
	C225	RCAF	MC	BELL-IRVING
	C245	RCAF		BENCHIER
	6103A	RCAF		BIRNIE
	C15090	RCAF	BEM	BOUGHNER
	2028A	RCAF		BUNSTAD
	C10937	RCAF		CARPENTER
	10001A	RCAF		CARVER
	J44470	RCAF		COOPER
	C10938	RCAF		COPLEY
	C299	RCAF	DFC	CORBETT
	C305	RCAF		CRABB
	2019A	RCAF		CUNNINGHAM
	C317	RCAF	BCBE,DSC,ED,(OC CB)	CURTIS
	C333	RCAF	DFC	DELHAYE
	C818	RCAF	AFC	DIAMOND
	C20493	RCAF		DOE
	C351	RCAF		DUTEMPLE
	72A	RCAF		ELLIS
	C38539	RCAF		ELLISON
	C826	RCAF		ELMS
	C1013	RCAF	AFC	FOLKINS
	C373	RCAF		FOSS
	120691	RCAF		FRASER
	C50149	RCAF		FRASER
	C380	RCAF		GARDNER
	4067A	RCAF		GILES
	C387	RCAF		GLEDHILL
	122A	RCAF		GOLDEY
	C13010	RCAF		GRIFFITH
	33A	RCAF		HALL
	C409	RCAF	CBE	HANNA
	C5205	RCAF		HARDMAN
	23A	RCAF		HARRIS
	105A	RCAF		HENDERSON
	120371	RCAF		HILL
	C1018	RCAF		HILLOCK
	4058A	RCAF		HIVES
	73A	RCAF		HOLLOWAY
	90131	RCAF		HORNELL
	C450	RCAF	CBE	IRWIN
	19645	RCAF		IVERSON
	C452	RCAF		JACOBI
	2085A	RCAF		JARDINE
	4020A	RCAF		KERR
	64A	RCAF		LEITCH
	C37391	RCAF		LYEL
	C515	RCAF		MALCOLMSON
	6109A	RCAF		MATHEWS
	C4506	RCAF		McARDLE
	95A	RCAF		McCALLUM
	C34377	RCAF		McDOWELL
	C876	RCAF		McFADYN

Surname	Initials			Rank	Status	Date
McFARLANE	W	J		GRP CAP		06/10/1944
McGILL	F	S		AVM		31/08/1945
MONCRIEFF	E	H	G	GRP CAP		25/07/1952
MURRAY	W	A		GRP CAP		06/10/1944
NAIRN	K	G		AVM		27/10/1944
NANTON	E	A		SQD LDR		15/12/1944
ORANGE	G			WO1		23/03/1945
PADWICK	H	W		WING COMM		17/11/1944
PEARCE	W	S		FL		08/02/1946
PENNELLS	A	C		FL		16/11/1945
PICKERING	J	H		WO1		12/10/1945
POLLOCK	W	R		GRP CAP		03/11/1944
RAYMOND	A			AVM		13/10/1944
REFAUSSE	W	H		WO1		12/09/1945
ROGERS	J	E		SQD LDR		29/09/1944
ROSS	W	W	S	GRP CAP		16/03/1945
ROZEN	E			FO		06/07/1945
RUSSELL	A	H		AIR COMMD		29/09/1944
SCOTT	L	S		SQD LDR		15/12/1944
SELLERS	G	H		GRP CAP		29/09/1944
SMITH	F	J	C	FLT SGT		10/11/1944
SOLSKI	J	J		FO		27/10/1944
SPRANGE	S	A		WING COMM		28/09/1945
ST. PIERRE	J	M	W	GRP CAP		16/11/1945
SULLY	J	A		AVM		22/09/1944
TINKER	A	H		SQD LDR		22/09/1944
TUPPING	J	S		WO1		27/04/1945
VADBONCOUER	G			WING COMM		13/10/1944
WALMSLEY	A			WING COMM		20/10/1944
WATTS	A			GRP CAP		15/12/1944
WEAVER	E	A		WING COMM		03/11/1944
WIGGINS	H	J		WO1		24/11/1944
WIGLE	D	H		GRP CAP		04/05/1945
WILSON	A	H		GRP CAP		30/08/1946
WILSON	M			WO2		22/12/1944
WISEMAN	N			FL		19/10/1951
YELLOWLEES	L	A		FL		06/12/1946

Source	Service No	Force	Notes	Surname
	C564	RCAF		McFARLANE
	C565	RCAF	CB	McGILL
	C535	RCAF	OBE,AFC	MONCRIEFF
	C544	RCAF	OBE	MURRAY
	C585	RCAF	CB	NAIRN
	C586	RCAF		NANTON
	2A	RCAF		ORANGE
	C3358	RCAF		PADWICK
	C10525	RCAF		PEARCE
	C28447	RCAF		PENNELLS
	4013A	RCAF		PICKERING
	C616	RCAF		POLLOCK
	C621	RCAF	CBE	RAYMOND
	39A	RCAF		REFAUSSE
	C631	RCAF		ROGERS
	C638	RCAF		ROSS
	J13630	RCAF		ROZEN
	C640	RCAF	CBE	RUSSELL
	C5429	RCAF		SCOTT
	C647	RCAF		SELLERS
	87A	RCAF		SMITH
	J23998	RCAF		SOLSKI
	C811	RCAF		SPRANGE
	C786A	RCAF	AFC	ST. PIERRE
	C686	RCAF	CB,AFC	SULLY
	C5774	RCAF		TINKER
	8041A	RCAF		TUPPING
	C715	RCAF		VADBONCOUER
	C721	RCAF		WALMSLEY
	C723	RCAF	AFC	WATTS
	C725	RCAF		WEAVER
	7A	RCAF		WIGGINS
	C738	RCAF		WIGLE
	C742	RCAF		WILSON
	2046A	RCAF		WILSON
	38786	RCAF		WISEMAN
	C24850	RCAF	BEM	YELLOWLEES

NOMINAL ROLL FOR HONG KONG

Surname	Initials		Rank	Status	Date
ALLEN	J	R	FL		09/12/1966
ANTONIO	G	M	SGT		02/02/1962
ANTONIO	G	M	FLT SGT		19/11/1971
ANTONIO	G	M	WO		13/11/1981
ASHLEY	N	J	FL		28/07/1989
ASPREY	A	P	FL		28/06/1974
ASPREY	A	P	SQD LDR		06/03/1981
ASPREY	A	P	WING COMM		19/08/1988
BELL	G	J	FL		26/09/1958
BELL	G	J	WING COMM		12/02/1965
BERGER	V	E	FL		24/12/1959
BERGER	V	E	FL		16/09/1966
BOTELHO	F	M	FO		20/11/1959
BOTELHO	F	M	FL		05/06/1970
BROOKS	R	W	FL		22/09/1978
BROOKS	R	W	FL		25/01/1985
BUSH	C		SGT		19/03/1965
BUTT	B	Y	FL		25/01/1985
CHAN	K		FLT SGT		16/05/1980
CHAN	K		FO		03/08/1990
CHAN	K		SGT		20/03/1981
CHAN	K		SGT		25/01/1991
CHAN	K		CPL		18/07/1986
CHAN	M	Y	FLT SGT		03/08/1984
CHAN	R	S	CPL		01/10/1965
CHAN	R	S	SGT		18/07/1975
CHAN	R	S	FLT SGT		19/07/1985
CHAN	S		SAC		03/05/1963
CHAN	S		CPL		01/10/1965
CHAN	S		FLT SGT		18/07/1975
CHANG	M		SAC		14/08/1964
CHANG	M		SGT		13/09/1974
CHENG	A	B	FLT SGT		29/07/1983
CHENG	A	B	PO		02/02/1990
CHEONG-LEEN	R	J	FL		14/08/1964
CHEONG-LEEN	R	J	FL		28/06/1974
CHEUNG	D	K	FO		28/04/1961
CHEUNG	D	K	SQD LDR		08/03/1968
CHEUNG	D	K	SQD LDR		13/09/1974
CHEUNG	H	E	SAC		16/09/1960
CHEUNG	K		SGT		24/01/1986
CHEUNG	W		CPL		03/08/1984
CHIU	S		CPL		03/08/1979
CHOI	C		FLT SGT		03/08/1984
CHOW	H		SAC		21/06/1963
CHOW	H		CPL		02/11/1973
CHOW	K		SGT		22/07/1988
CHOW	W		FO		07/08/1992
CHU	K		SGT		18/07/1986
CHU	P		SGT		25/01/1985
CHUNG	C		CPL		28/10/1977
CHUNG	C		SGT		24/07/1987
CHUNG	P		SAC		28/04/1961
CHUNG	P		SGT		04/06/1971
COLLACO	F	J	CPL		03/05/1963
CONNELL	R	M	FL		19/08/1988
CREW	M	E	J	FL		10/05/1974

Source	Service No	Force	Notes	Surname
N3064	18097049	HKAAF		ALLEN
N226	18093593	HKAAF		ANTONIO
N2657	18093593	HKAAF	1st Clasp	ANTONIO
N3363	18093593	HKAAF	2nd Clasp	ANTONIO
N2503	18101258	HKAAF		ASHLEY
N1599	18101045	HKAAF		ASPREY
N726	18101045	HKAAF	1st Clasp	ASPREY
N2697	18101045	HKAAF	2nd Clasp	ASPREY
N1313	18093411	HKAAF		BELL
N370	18093411	HKAAF	1st Clasp	BELL
N1799	18095533	HKAAF		BERGER
N2364	18095533	HKAAF	1st Clasp	BERGER
N1585	18092832	HKAAF		BOTELHO
N1152	18092832	HKAAF	1st Clasp	BOTELHO
N2394	18101093	HKAAF		BROOKS
N269	18101093	HKAAF	1st Clasp	BROOKS
N679	18095964	HKAAF		BUSH
N269	18101201	HKAAF		BUTT
N1464	18101077	HKAAF		CHAN
N2724	18101077	HKAAF	1st Clasp	CHAN
N856	18101084	HKAAF		CHAN
N255	18101084	HKAAF	1st Clasp	CHAN
N2400	18101177	HKAAF		CHAN
N2270	18101137	HKAAF		CHAN
N2340	18095992	HKAAF		CHAN
N1574	18095992	HKAAF	1st Clasp	CHAN
N2307	18095992	HKAAF	2nd Clasp	CHAN
N779	18095276	HKAAF		CHAN
N2340	18095996	HKAAF		CHAN
N1574	18095996	HKAAF	1st Clasp	CHAN
N2222	18095799	HKAAF		CHANG
N2337	18095799	HKAAF	1st Clasp	CHANG
N2375	18101175	HKAAF		CHENG
N362	18101175	HKAAF	1st Clasp	CHENG
N2222	18095786	HKAAF		CHEONG-LEEN
N1600	18095786	HKAAF	1st Clasp	CHEONG-LEEN
N686	18095805	HKAAF		CHEUNG
N450	18095805	HKAAF	AFC; 1st Clasp	CHEUNG
N2338	18095805	HKAAF	AFC; 2nd Clasp	CHEUNG
N1401	18093151	HKAAF		CHEUNG
N283	18101164	HKAAF		CHEUNG
N2270	18101140	HKAAF		CHEUNG
N1960	18101056	HKAAF		CHIU
N2270	18101194	HKAAF		CHOI
N1127	18095488	HKAAF		CHOW
N2851	18095488	HKAAF	1st Clasp	CHOW
N2285	18101205	HKAAF		CHOW
N2760	18101310	HKAAF		CHOW
N2400	18101173	HKAAF		CHU
N269	18101157	HKAAF		CHU
N2593	18101043	HKAAF		CHUNG
N2331	18101043	HKAAF	1st Clasp	CHUNG
N686	18093290	HKAAF		CHUNG
N1175	18093290	HKAAF	1st Clasp	CHUNG
N779	18095416	HKAAF		COLLACO
N2697	18101198	HKAAF		CONNELL
N1164	18100062	HKAAF		CREW

Surname	Initials			Rank	Status	Date
CRUZ	J	E		SGT		10/07/1959
CURTIS	P	J	S	FL		24/06/1977
CURTIS	P	J	S	FL		03/08/1984
Da ROZA	M	A		CPL		10/07/1959
De CARVALHO	A	B		CPL		02/02/1962
De LUZ	C	E		SGT		17/06/1960
De SOUZA	G	L		SAC		22/11/1963
DEWAR	J	G	M	CPL		07/09/1962
ELLIS	S	P	J	WING COMM		02/11/1973
ESTRADA	G	A		FL		18/07/1986
FAN	K			CPL		16/09/1966
FAN	K			SGT		17/09/1976
FERRIS	J	W		FO		26/09/1958
FOO	C			LAC		03/05/1963
FRASER	A	K		FL		05/11/1982
FUNG	F	K		CAD OFF		27/01/1989
GARCIA	M			SAC		21/10/1960
GARDNER	W	H		CPL		23/03/1962
GRAHAM	G			FL		28/04/1961
HA	C			LAC		03/05/1963
HARRIS	W	C		FL		03/08/1990
HAU	C			CPL		27/01/1989
HIGGINSON	G	A		FL		06/08/1976
HO	E	D		FLT SGT		19/03/1993
HO	Y			SGT		28/07/1989
HOLM	J	B		CPL		28/04/1961
HONG	A	K		SNR ATC		06/09/1997
IP	I			CPL		12/07/1963
IP	I			SGT		02/11/1973
IP	W	H		LAC		12/06/1959
IP	W	H		CPL		20/06/1969
ISMAIL	S	A		SAC		28/04/1961
KADER	A	N		SGT		16/09/1960
KAM	C			CPL		14/01/1966
KAM	C			SGT		05/03/1976
KO	T			PO		15/08/1980
KO	T			FL		25/01/1991
KWAN	K			CPL		18/07/1986
KWAN	W			FL		19/07/1985
KWOK	C			FL		18/07/1986
KWOK	K			CPL		13/07/1962
LAI	K			CPL		30/12/1960
LAM	C			SAC		16/08/1963
LAM	C			SGT		02/11/1973
LAM	C			FLT SGT		20/01/1984
LAM	J	Y		FO		06/03/1981
LAM	Y			FL		24/07/1987
LARM	P	K		FO		07/09/1979
LARM	P	K		FL		24/01/1986
LAU	E	V		CPL		02/11/1973
LAU	E	V		FLT SGT		20/01/1984
LAU	E	W		FO		29/07/1983
LAU	E	W		FL		03/08/1990
LAU	F			SAC		12/01/1968
LAU	F			SGT		30/06/1977
LAW	S			SGT		14/10/1966
LAW	S			WO		22/10/1976

Source	Service No	Force	Notes	Surname
N940	18092363	HKAAF		CRUZ
N1413	18101078	HKAAF		CURTIS
N2270	18101078	HKAAF	1st Clasp	CURTIS
N940	18092296	HKAAF		Da ROZA
N226	18093363	HKAAF		De CARVALHO
N917	18093084	HKAAF		De LUZ
N2283	18095618	HKAAF		De SOUZA
N1850	18095126	HKAAF		DEWAR
N2852	18101038	HKAAF		ELLIS
N2400	18101176	HKAAF		ESTRADA
N2364	18096416	HKAAF		FAN
N2022	18096416	HKAAF	1st Clasp	FAN
N1313	18092063	HKAAF		FERRIS
N779	18095292	HKAAF		FOO
N3502	18101089	HKAAF		FRASER
N244	18101215	HKAAF		FUNG
N1579	18093199	HKAAF		GARCIA
N600	18093846	HKAAF		GARDNER
N686	18093301	HKAAF		GRAHAM
N779	18095278	HKAAF		HA
N2724	18101268	HKAAF		HARRIS
N244	18101207	HKAAF		HAU
N1690	18101065	HKAAF		HIGGINSON
N879	18101166	HKAAF		HO
N2503	18101222	HKAAF		HO
N686	18093291	HKAAF		HOLM
N2732	n/a	GFS		HONG
N1291	18095540	HKAAF		IP
N2851	18095540	HKAAF	1st Clasp	IP
N801	18092189	HKAAF		IP
N1169	18092189	HKAAF	1st Clasp	IP
N686	18093292	HKAAF		ISMAIL
N1401	18093183	HKAAF		KADER
N90	18096172	HKAAF		KAM
N504	18096172	HKAAF	1st Clasp	KAM
N2519	18101107	HKAAF		KO
N255	18101107	HKAAF	1st Clasp	KO
N2400	18101183	HKAAF		KWAN
N2307	18101158	HKAAF		KWAN
N2400	18101232	HKAAF		KWOK
N1480	18095078	HKAAF		KWOK
N1991	18093243	HKAAF		LAI
N1553	18095545	HKAAF		LAM
N2851	18095545	HKAAF	1st Clasp	LAM
N195	18095545	HKAAF	2nd Clasp	LAM
N727	18101129	HKAAF		LAM
N2331	18101129	HKAAF	1st Clasp	LAM
N1960	18101105	HKAAF		LARM
N283	18101105	HKAAF	1st Clasp	LARM
N2852	18101009	HKAAF		LAU
N195	18101009	HKAAF	1st Clasp	LAU
N2375	18101182	HKAAF		LAU
N2724	18101182	HKAAF	1st Clasp	LAU
N41	18096586	HKAAF		LAU
N1469	18096586	HKAAF	1st Clasp	LAU
N2612	18096454	HKAAF		LAW
N2326	18096454	HKAAF	BEM; 1st Clasp	LAW

Surname	Initials			Rank	Status	Date
LAW	S			FLT SGT		14/02/1992
LEE	J	N		FL		25/01/1991
LEE	N			SAC		02/02/1962
LEE	P			LAC		28/04/1961
LEE	Q			SAC		14/01/1966
LEE	T			SGT		20/01/1984
LEE	W			SGT		02/10/1991
LEE	W			FLT SGT		14/02/1992
LEUNG	C	C		SGT		08/04/1960
LEUNG	C			SGT		28/07/1989
LEUNG	E	T		FLT SGT		18/07/1986
LEUNG	E	T		PO		19/03/1993
LEUNG	K			FL		18/07/1986
LEUNG	K			FL		19/03/1993
LEUNG	P			CPL		03/05/1963
LEUNG	S			CPL		13/09/1974
LEUNG	S			SGT		25/01/1985
LEUNG	W	Y		CPL		08/07/1977
LEUNG	W	Y		FLT SGT		24/07/1987
LEUNG	W			SGT		07/09/1979
LEUNG	W			FLT SGT	Acting	28/07/1989
LEUNG	Y			CPL		12/09/1980
LEUNG	Y			SGT	Acting	02/02/1990
LI	G	K		CPL		08/04/1960
LI	G	K		FO		15/05/1970
LI	G	K		FLT SGT		10/09/1982
LI	G	K		FLT SGT		28/07/1989
LO	A	W		CPL		03/08/1984
LO	N	S		FL		27/01/1989
LO	T			CPL		25/01/1985
LOPEZ	A	C	De B	FL		15/09/1978
LUJAN	R			SAC		16/09/1960
LUK	K			CPL		05/01/1979
LUK	K			SGT		27/01/1989
MAK	S			FLT SGT		22/07/1988
MANG	N			CPL		08/07/1977
MARSH	N	W	G	FL		26/09/1958
McINTOSH	G	M		SQD LDR		24/01/1986
McINTOSH	G	M		SQD LDR		07/08/1992
MOHAMED	W			FL		01/10/1965
MURRAY	L	U		FO		24/06/1977
MURRAY	L	U		FL		20/01/1984
MURRAY	L	U		SQD LDR		03/08/1990
NG	C			SAC		02/02/1962
NG	C			CPL		20/03/1964
NG	C			FLT SGT		28/06/1974
NG	C			PO		03/08/1984
NG	W			SNR ACM		06/09/1997
OGILVIE	A	C	D	FO		12/03/1982
PARK	S	J	C	FO		20/10/1978
PARK	S	J	C	FL		19/07/1985
PENLINGTON	R	G		FL		01/10/1971
PENLINGTON	R	G		WING COMM		15/09/1978
PERES	R	DA	L	CPL		02/02/1962
PILKINGTON	F	J		FLT SGT		19/07/1985
PILKINGTON	F	J		FL		07/08/1992
POON	C			CPL		19/11/1971

Source	Service No	Force	Notes	Surname
N499	18101308	HKAAF		LAW
N255	18101277	HKAAF		LEE
N226	18093360	HKAAF		LEE
N686	18093317	HKAAF		LEE
N90	18096174	HKAAF		LEE
N195	18101123	HKAAF		LEE
N2939	18101103	HKAAF		LEE
N499	18101103	HKAAF	1st Clasp	LEE
N558	18093020	HKAAF		LEUNG
N2503	18101227	HKAAF		LEUNG
N2400	18101216	HKAAF		LEUNG
N879	18101216	HKAAF	1st Clasp	LEUNG
N2400	18101235	HKAAF		LEUNG
N879	18101235	HKAAF	1st Clasp	LEUNG
N779	18095132	HKAAF		LEUNG
N2339	18101016	HKAAF		LEUNG
N269	18101016	HKAAF	1st Clasp	LEUNG
N1528	18101042	HKAAF		LEUNG
N2331	18101042	HKAAF	1st Clasp	LEUNG
N1960	18101054	HKAAF		LEUNG
N2503	18101054	HKAAF	1st Clasp	LEUNG
N2880	18101061	HKAAF		LEUNG
N362	18101061	HKAAF	1st Clasp	LEUNG
N558	18093025	HKAAF		LI
N996	18093025	HKAAF	1st Clasp	LI
N2893	18101132	HKAAF		LI
N2503	18101132	HKAAF	1st Clasp	LI
N2270	18101135	HKAAF		LO
N244	18101257	HKAAF		LO
N269	18101150	HKAAF		LO
N2324	18101076	HKAAF		LOPEZ
N1401	18093186	HKAAF		LUJAN
N20	18101051	HKAAF		LUK
N244	18101051	HKAAF	1st Clasp	LUK
N2285	18101206	HKAAF		MAK
N1528	18101032	HKAAF		MANG
N1313	18093246	HKAAF		MARSH
N283	18101187	HKAAF		McINTOSH
N2760	18101187	HKAAF	1st Clasp	McINTOSH
N2340	18096075	HKAAF		MOHAMED
N1413	18101069	HKAAF		MURRAY
N195	18101069	HKAAF	1st Clasp	MURRAY
N2724	18101069	HKAAF	2nd Clasp	MURRAY
N226	18093715	HKAAF		NG
N676	18095703	HKAAF		NG
N1600	18095703	HKAAF	1st Clasp	NG
N2270	18095703	HKAAF	2nd Clasp	NG
N2732	n/a	GFS		NG
N822	18101124	HKAAF		OGILVIE
N2671	18101095	HKAAF		PARK
N2307	18101095	HKAAF	1st Clasp	PARK
N2233	18101018	HKAAF		PENLINGTON
N2324	18101018	HKAAF	1st Clasp	PENLINGTON
N226	18093560	HKAAF		PERES
N2307	18101211	HKAAF		PILKINGTON
N2760	18101211	HKAAF	1st Clasp	PILKINGTON
N2657	18097249	HKAAF		POON

Surname	Initials			Rank	Status	Date
RADCLIFFE	P	D		FL		02/02/1990
RANDALL	E	B		CPL		02/02/1962
RAWLINGS	E	H		FL		02/02/1962
RAZACK	K	A		FL		03/08/1984
REMEDIOS	J	O		SAC		23/03/1962
REMEDIOS	P	R		WO		20/11/1959
RIBEIRO	A	J	V	LAC		20/11/1959
RIBEIRO	V	B		FLT SGT		28/04/1961
RITCHIE	D	W		FL		24/06/1977
ROBERTSON	I	R	S	FO		02/04/1965
SALLEH	A	R		CPL		16/09/1960
SCALES	P	O		WING COMM		02/02/1962
SCHREYER	H			LAC		30/12/1960
SHAWCROSS	J	G		FL		04/06/1971
SHAWCROSS	J	G		FL		28/10/1977
SHAWCROSS	J	G		SQD LDR		03/08/1984
SHUN	S			SAC		03/05/1963
SMITH	D	L		FL		28/07/1989
SMITH	G	W		FL		27/01/1989
SMITH	R	P		FL		26/09/1958
SMITH	R	P		SQD LDR		06/03/1964
SMITH	R	P		WING COMM		22/05/1970
SOUZA	L	M		FLT SGT		08/04/1960
SOUZA	L	M		FL		05/02/1970
SPIKINS	B	C		FL		19/03/1993
STRANGE	L	C		FL		26/09/1958
STRANGE	L	C		FL		06/03/1964
STRANGE	L	C		FL		15/05/1970
STYLES	T	P		FL		20/11/1959
SUNG	C			SAC		02/02/1962
SWIRE	A	C		PO		18/08/1961
SZE-TO	H			CPL		14/01/1966
SZE-TO	H			SGT		05/03/1976
TAI	R			FO		17/06/1960
TAI	R			FL		22/04/1966
TAM	P	K		CPL		20/01/1984
TAM	T			SAC		14/06/1963
TAM	Y			SAC		16/09/1966
TAN	K			CPL		03/05/1963
TAN	K			SGT		02/11/1973
TANG	K			SGT		23/03/1962
TANG	R	K	H	FLT SGT		20/11/1959
TANG	R	C		FL		18/07/1986
TARK SINGH	J			CPL		20/01/1984
TONG	G	S		CPL		30/12/1960
TONG	P	K		CPL		10/10/1975
TONG	P	K		SGT		24/01/1986
TONG	S			SAC		10/12/1965
TSAO	K			LAC		28/05/1965
TSE	A	Y		SGT		02/02/1990
TSE	P			CPL		03/05/1963
TSUI	J	T		FLT SGT		10/09/1982
TSUI	J	T		WO		28/07/1989
VAS	R	A		SGT		08/04/1960
WAI	N			SAC		07/09/1962
WAI	N			CPL		11/07/1975
WAN	T			FLT SGT		16/01/1987

Source	Service No	Force	Notes	Surname
N362	18101262	HKAAF		RADCLIFFE
N226	18093792	HKAAF		RANDALL
N226	18093582	HKAAF		RAWLINGS
N2270	18101146	HKAAF		RAZACK
N600	18093829	HKAAF		REMEDIOS
N1585	18092752	HKAAF		REMEDIOS
N1585	18092428	HKAAF		RIBEIRO
N686	18093335	HKAAF		RIBEIRO
N1413	18101068	HKAAF		RITCHIE
N823	18092107	HKAAF		ROBERTSON
N1401	18093138	HKAAF		SALLEH
N226	18093585	HKAAF		SCALES
N1991	18093232	HKAAF		SCHREYER
N1174	18101014	HKAAF		SHAWCROSS
N2592	18101014	HKAAF	1st Clasp	SHAWCROSS
N2270	18101014	HKAAF	2nd Clasp	SHAWCROSS
N779	18095296	HKAAF		SHUN
N2503	18101190	HKAAF		SMITH
N244	18101256	HKAAF		SMITH
N1313	18093056	HKAAF		SMITH
N560	18093056	HKAAF	1st Clasp	SMITH
N1042	18093056	HKAAF	MBE; 2nd Clasp	SMITH
N558	18092964	HKAAF		SOUZA
N314	18092964	HKAAF	1st Clasp	SOUZA
N879	18101317	HKAAF		SPIKINS
N1313	18092241	HKAAF		STRANGE
N560	18092241	HKAAF	1st Clasp	STRANGE
N997	18092241	HKAAF	2nd Clasp	STRANGE
N1585	18092738	HKAAF		STYLES
N226	18093790	HKAAF		SUNG
N1427	18096474	HKAAF		SWIRE
N90	18096173	HKAAF		SZE-TO
N504	18096173	HKAAF	1st Clasp	SZE-TO
N917	18093441	HKAAF		TAI
N1051	18093441	HKAAF	1st Clasp	TAI
N195	18101153	HKAAF		TAM
N1058	18095438	HKAAF		TAM
N2364	18096428	HKAAF		TAM
N779	18095280	HKAAF		TAN
N2851	18095280	HKAAF	1st Clasp	TAN
N600	18093795	HKAAF		TANG
N1585	18092684	HKAAF		TANG
N2400	18101178	HKAAF		TANG
N195	18101126	HKAAF		TARK SINGH
N1991	18093245	HKAAF		TONG
N2214	18101025	HKAAF		TONG
N283	18101025	HKAAF	1st Clasp	TONG
N2843	18096127	HKAAF		TONG
N1289	18096015	HKAAF		TSAO
N362	18101267	HKAAF		TSE
N779	18095340	HKAAF		TSE
N2893	18101163	HKAAF		TSUI
N2503	18101163	HKAAF	1st Clasp	TSUI
N558	18093007	HKAAF		VAS
N1850	18095114	HKAAF		WAI
N1515	18101022	HKAAF		WAI
N162	18101239	HKAAF		WAN

Surname	Initials			Rank	Status	Date
WIGHTMAN	M	A		FL		06/03/1981
WIGHTMAN	M	A		FL		16/01/1987
WONG	B			SAC		30/09/1960
WONG	C			SAC		17/06/1960
WONG	D	C		SGT		24/03/1972
WONG	D	C		FLT SGT		29/01/1982
WONG	C			FLT SGT		16/01/1987
WONG	D	W		FO		28/04/1961
WONG	K			CPL		02/10/1991
WONG	K			FLT SGT		14/02/1992
WONG	K			SGT		18/07/1986
WONG	K			CPL		28/07/1989
WONG	K	T		SGT		28/07/1989
WONG	P			SAC		16/09/1960
WONG	P			SGT		04/06/1971
WONG	P			SGT		03/05/1963
WONG	T			SGT		28/07/1989
YAN	A	K		SGT		04/01/1980
YAN	A	K		FLT SGT		02/02/1990
YAN	C			CPL		16/05/1980
YAU	D	W		CPL		27/01/1989
YEUNG	W	F		CPL		12/06/1959
YEUNG	W	F		CPL		20/06/1969
YICK	P			CPL		17/06/1960
YICK	P			SGT		05/06/1970
YIP	D	P		FL		01/02/1974
YIP	D	P		FL		06/03/1981
YIP	D	P		SQD LDR		03/08/1990
YOUNG	I	A	S	SQD LDR		10/05/1974
YUNG	C			CPL		16/01/1987
ZAMAN	M	J		FO		12/06/1959

Source	Service No	Force	Notes	Surname
N727	18101121	HKAAF		WIGHTMAN
N162	18101121	HKAAF	1st Clasp	WIGHTMAN
N1480	18093139	HKAAF		WONG
N917	18093067	HKAAF		WONG
N717	18097290	HKAAF		WONG
N306	18097290	HKAAF	1st Clasp	WONG
N162	18101184	HKAAF		WONG
N686	18094163	HKWAAF		WONG
N2939	18101102	HKAAF		WONG
N499	18101102	HKAAF	1st Clasp	WONG
N2400	18101172	HKAAF		WONG
N2503	18101223	HKAAF		WONG
N2503	18101305	HKAAF		WONG
N1401	18093141	HKAAF		WONG
N1175	18093141	HKAAF	1st Clasp	WONG
N779	18095415	HKAAF		WONG
N2503	18101243	HKAAF		WONG
N13	18101060	HKAAF		YAN
N362	18101060	HKAAF	1st Clasp	YAN
N1464	18101075	HKAAF		YAN
N244	18101210	HKAAF		YAU
N801	18092192	HKAAF		YEUNG
N1169	18092192	HKAAF	1st Clasp	YEUNG
N917	18093069	HKAAF		YICK
N1152	18093069	HKAAF	1st Clasp	YICK
N255	18101036	HKAAF		YIP
N726	18101036	HKAAF	1st Clasp	YIP
N2724	18101036	HKAAF	2nd Clasp	YIP
N1164	18100034	HKAAF		YOUNG
N162	18101185	HKAAF		YUNG
N801	18092127	HKAAF		ZAMAN

NOMINAL ROLL FOR NEW ZEALAND

Surname	Initials	Rank	Status	Date	Source
ADAMS	J	GSH		08/01/1989	
AKARIRI	B R	GSH		16/03/1997	
ALFORD	A J	FL		18/03/2001	
ALLAN	I E	FL		30/06/1964	
ALLISON	M R	LAC		16/01/1985	
ANDERSON	J L	WO	Acting	08/12/1985	
ANNABELL	G C	GSH		29/11/1999	
ANSCOMBE	H	GSH		01/07/1995	
APPLEGATE	D F A	GSH		21/09/1988	
ARCHIBALD	M K	SQD LDR		n/a	
ARCHIBALD	R J N	SQD LDR		13/05/1954	
ARMSTRONG	P J	CPL	Temporary	05/08/1981	
ATKINS	N J	SQD LDR		31/01/1965	
ATKINS	N J	SQD LDR		15/01/1985	
ATWELL	J H	SQD LDR		21/05/1954	
ATWELL	J H	SQD LDR		21/05/1954	
AVERY	S G	GSH		05/09/1987	
BAILLIE	U B	GSH		28/05/2000	
BAIN	M G	SQD LDR		22/02/1965	
BAIN	M G	SQD LDR		07/11/1975	
BAKER	D H	GSH		28/09/1997	
BAND	J F	FL		21/05/1954	
BARNSTON	D J	FL		16/06/1958	
BEATTIE	J A	CPL	Temporary	09/09/1968	
BEER	C T	WO	Acting	09/10/1987	
BEGG	R J	FL		01/10/1954	
BENNISON	A	FL		06/08/1952	
BENNISON	A	FL		09/04/1959	
BENNISON	A	FL		06/12/1965	
BERRY	D	FL		26/01/2004	
BERRYMAN	C J	SQD LDR		19/10/1960	
BERRYMAN	C J	SQD LDR		19/10/1970	
BISHELL	P L	GSH		11/11/1995	
BLACK	I W	SQD LDR		03/03/1981	
BLACK	I W	SQD LDR		03/03/1991	
BLAKE	D	GSH		16/06/1990	
BLANK	M R	FL		22/08/1954	
BOND	J H	SQD LDR		01/12/1966	
BOND	J H	SQD LDR		01/12/1966	
BOROUGHS	D C	SQD LDR		30/06/1966	
BOYD	N W R	SGT	Temporary	30/10/1956	
BRERETON	R H	FL		17/09/1990	
BRERETON	R H	FL		17/09/2000	
BRILL	R A	SQD LDR		22/11/1987	
BRIZZELL	R E	FL		16/06/1964	
BUCHANAN	J	FL		25/11/1995	
BURGESS	R G A	SQD LDR		26/01/1954	
BURNELL	W H	FL		01/04/1954	
BUTT	W H F	CPL		23/09/2001	
CAIN	M F	LAC		19/07/1983	
CAIRNS	B E	GSH		24/08/1997	
CALDWELL	M P	SGT	Acting	29/06/1981	
CALDWELL	M P	SGT	Acting	29/06/1991	
CAMPBELL	I M	LAC		19/06/1991	
CARROLL	G K	GSH		21/09/1989	
CARTER	G	AIR COMMD		13/06/1955	
CARTER	G	AIR COMMD		13/06/1955	
CASCI	N W	GSH		25/09/1988	
CLARK	A R	GSH		30/01/2001	
CLAUSEN	L B	LAC		19/06/1991	

Service No	Force	Notes	Surname
G644788	RNZAF	sec	ADAMS
K338730	RNZAF	airasst	AKARIRI
A86641	RNZAF	atc	ALFORD
n/a	RNZAF	acc	ALLAN
V134477	RNZAF	musn	ALLISON
D75880	RNZAF	musn	ANDERSON
Y93862	RNZAF		ANNABELL
S14046	RNZAF	airasst	ANSCOMBE
B88873	RNZAF	airasst	APPLEGATE
133329	RNZAF	gd; pilot	ARCHIBALD
130068	RNZAF	gd; pilot	ARCHIBALD
Y134411	RNZAF	musn	ARMSTRONG
J329989	RNZAF	gd; nav	ATKINS
J329989	RNZAF	1st Clasp; gd; nav	ATKINS
130309	RNZAF	gd; pilot	ATWELL
130309	RNZAF	1st Clasp; gd; pilot	ATWELL
W327494	RNZAF	airasst	AVERY
A990426	RNZAF	airasst	BAILLIE
E134117	RNZAF	sec	BAIN
E134117	RNZAF	1st Clasp; sec	BAIN
H92835	RNZAF	sec	BAKER
130141	RNZAF	eng	BAND
130703	RNZAF		BARNSTON
n/a	RNZAF		BEATTIE
L77175	RNZAF	musn	BEER
130754	RNZAF	arm	BEGG
130806	RNZAF	gd; ag / nav	BENNISON
130806	RNZAF	1st Clasp; gd; ag / nav	BENNISON
130806	RNZAF	2nd Clasp; gd; ag / nav	BENNISON
W996817	RNZAF	atc	BERRY
70174	RNZAF	gd; pilot	BERRYMAN
70174	RNZAF	1st Clasp; gd; pilot	BERRYMAN
Y91930	RNZAF	airasst	BISHELL
S134405	RNZAF	atc	BLACK
S134405	RNZAF	1st Clasp; atc	BLACK
Q89852	RNZAF	sup	BLAKE
130092	RNZAF	gd; pilot	BLANK
130310	RNZAF	sig	BOND
130310	RNZAF	1st Clasp; sig	BOND
134139	RNZAF	atc	BOROUGHS
130272	RNZAF		BOYD
Q134619	RNZAF	atc	BRERETON
Q134619	RNZAF	1st Clasp; atc	BRERETON
K74184	RNZAF	gd; nav	BRILL
132697	RNZAF	gd; wo/ag	BRIZZELL
L134767	RNZAF	atc	BUCHANAN
130347	RNZAF	gd; nav	BURGESS
130095	RNZAF	gd; nav	BURNELL
U993204	RNZAF	musn	BUTT
P134448	RNZAF	musn	CAIN
G92788	RNZAF	airasst	CAIRNS
X134410	RNZAF	musn	CALDWELL
X134410	RNZAF	1st Clasp; musn	CALDWELL
T134636	RNZAF	musn	CAMPBELL
A89332	RNZAF	airasst	CARROLL
73638	RNZAF	gd; pilot	CARTER
73638	RNZAF	1st Clasp; gd; pilot	CARTER
C88874	RNZAF	airasst	CASCI
N17538	RNZAF	airasst	CLARK
R134634	RNZAF	musn	CLAUSEN

Surname	Initials			Rank	Status	Date	Source
COLE	E	M		SGT		08/06/1994	
COLLIER	S	J		CPL		26/09/1995	
COLLINS	N	B		SQD LDR		01/04/1954	
COLLIS	R	C		CPL		07/05/1997	
CONLY	R	M		WING COMM		23/09/1964	
CONLY	R	M		WING COMM		23/09/1974	
CONLY	R	M		WING COMM		23/09/1984	
CONLY	R	M		WING COMM		23/09/1994	
COOK	N			GSH		17/09/1989	
COOK	N			GSH		17/09/1999	
COOPER	N	J		FLT SGT		07/11/1963	
COOPER	N	J		FLT SGT		07/11/1963	
COUPER	G	C		FL		21/05/1954	
COUTTS	A	G		WO		16/04/1962	
CRABB	A	L		GSH		24/05/1992	
CRABB	A	L		GSH		24/05/2002	
CREEVEY	J	H		FL		10/10/1961	
CROSSAN	S	H		GSH		07/04/1985	
CROZIER	D			CPL	Temporary	20/05/1975	
CRUMP	M			CPL		12/02/1968	
CRUNDWELL	L	T		FLT SGT		30/10/1956	
CUDBY	N	B		CPL	Temporary	12/01/1969	
CUDBY	N	B		CPL	Temporary	12/01/1979	
CUDBY	N	B		CPL	Temporary	12/01/1989	
DALCOM	B	P	W	FL		29/09/1959	
DANIELL	M	E		SQD LDR		10/10/1962	
DANIELL	M	E		SQD LDR		10/10/1962	
DAVIDSON	J	M		SGT		n/a	
DAVIS	J	R	L	GSH		03/09/1994	
DAY	J	R		SQD LDR		21/05/1954	
De WILLIMOFF	J	J		WING COMM		19/06/1961	
DENNISON	G	N		GSH		29/04/1991	
DENT	D	E		CPL	Temporary	03/08/1991	
DIVE	H	P	B	SQD LDR		01/08/1957	
DIVE	H	P	B	SQD LDR		01/08/1957	
DOWNER	M	J		GSH		12/08/1995	
DURNEY	M	G		CPL	Temporary	04/06/1971	
EADE	V	N		GSH		21/07/1990	
EASTON	J	A		FL		23/06/1971	
EDWARDS	F			SQD LDR		10/10/1961	
EDWARDS	M	J		LAC		08/08/1984	
EDWARDS	M	J		CPL		08/08/1994	
EDWARDS	M	J		CPL		08/08/2004	
ELLIOTT	R	F		WING COMM		21/06/1964	
EMBLING	R	H		FL		21/09/1959	
EMMERSON	A	F		GSH		02/04/1990	
EVERITT	C	N		SGT		20/01/1999	
EVISON	I	W	P	SQD LDR		31/12/1953	
FAMILTON	G	M		FL		17/09/1990	
FAMILTON	G	M		FL		17/09/2000	
FAUSETT	C	J		SGT	Temporary	18/03/1990	
FENWICK	K	E	S	SQD LDR		31/12/1956	
FENWICK	K	E	S	SQD LDR		31/08/1963	
FENWICK	K	E	S	SQD LDR		27/10/1970	
FITZGERALD	L			CPL		01/07/1997	
FOATE	M	F		SQD LDR		n/a	
FORREST	C	J	M	GSH		21/08/1999	
FRANKLAND	A	F		FLT SGT	Temporary	n/a	
FROST	R	J		GSH		20/05/1995	
GAMBLE	E	D		MAST SIG		12/02/1968	

Service No	Force	Notes	Surname
P772244	RNZAF	musn	COLE
G134763	RNZAF	musn	COLLIER
130544	RNZAF	gd; pilot	COLLINS
B134804	RNZAF	musn	COLLIS
P134057	RNZAF	artist	CONLY
P134057	RNZAF	1st Clasp; artist	CONLY
P134057	RNZAF	2nd Clasp; artist	CONLY
P134057	RNZAF	3rd Clasp; artist	CONLY
B538408	RNZAF	sec	COOK
B538408	RNZAF	1st Clasp; sec	COOK
133677	RNZAF	arm	COOPER
133677	RNZAF	1st Clasp; arm	COOPER
130133	RNZAF	gd; pilot	COUPER
132635	RNZAF		COUTTS
P90518	RNZAF	airasst	CRABB
P90518	RNZAF	1st Clasp; airasst	CRABB
132002	RNZAF	gd; wo/ag	CREEVEY
X134433	RNZAF	airasst	CROSSAN
134315	RNZAF	musn	CROZIER
134166	RNZAF	musn	CRUMP
130275	RNZAF		CRUNDWELL
D134208	RNZAF	musn	CUDBY
D134208	RNZAF	1st Clasp; musn	CUDBY
D134208	RNZAF	2nd Clasp; musn	CUDBY
131688	RNZAF	gd; pilot	DALCOM
73479	RNZAF	gd; pilot	DANIELL
73479	RNZAF	1st Clasp; gd; pilot	DANIELL
130049	RNZAF		DAVIDSON
B89494	RNZAF	airasst	DAVIS
130118	RNZAF	MBE,AFC; gd; pilot	DAY
70035	RNZAF	MBE,DFC; gd; pilot	De WILLIMOFF
W16166	RNZAF	airasst	DENNISON
G134648	RNZAF	musn	DENT
130322	RNZAF	gd; pilot	DIVE
130322	RNZAF	1st Clasp; gd; pilot	DIVE
G91707	RNZAF	sec	DOWNER
71865	RNZAF	musn	DURNEY
S89555	RNZAF	airasst	EADE
74238	RNZAF	gd; air sig	EASTON
131433	RNZAF	gd; pilot	EDWARDS
R134473	RNZAF	musn	EDWARDS
R134473	RNZAF	1st Clasp; musn	EDWARDS
R134473	RNZAF	2nd Clasp; musn	EDWARDS
Q133920	RNZAF	med	ELLIOTT
Q132701	RNZAF	gd; wo/ag	EMBLING
E89704	RNZAF	airasst	EMMERSON
D84804	RNZAF	musn	EVERITT
130128	RNZAF	gd; anv	EVISON
M82972	RNZAF	atc	FAMILTON
M82973	RNZAF	1st Clasp; atc	FAMILTON
T134567	RNZAF	musn	FAUSETT
130894	RNZAF	gd; pilot	FENWICK
130894	RNZAF	1st Clasp; gd; pilot	FENWICK
130894	RNZAF	2nd Clasp; gd; pilot	FENWICK
P134816	RNZAF	musn	FITZGERALD
133058	RNZAF	sec	FOATE
A93564	RNZAF	airasst	FORREST
709873	RNZAF		FRANKLAND
C39608	RNZAF	airasst	FROST
77481	RNZAF	sig	GAMBLE

Surname	Initials			Rank	Status	Date	Source
GARMONSWAY	B	P		SGT	Temporary	25/03/1990	
GARRETT	K	D		SGT		10/07/1994	
GARRETT	S	L		LAC		25/03/1996	
GARROD	S	D		LAC		15/01/1986	
GAULT	B	N		FL		09/06/1965	
GAWNE	R	P		MAST ENG		12/02/1968	
GEARD	W	H		LAC		03/03/1968	
GENEFAAS	W	H		SGT	Temporary	26/11/1983	
GIBSON	E	A		WING COMM		14/11/1955	
GIBSON	R	L		GSH		07/04/1988	
GILLIES	R	D		CPL	Temporary	28/01/1976	
GILLIES	R	D		FLT SGT	Temporary	28/01/1986	
GILLIES	R	D		WO		28/01/1996	
GOLDSTONE	B	A		GSH		19/06/1999	
GOODLEY	A	L		FL		02/10/1956	
GOODMAN	D	N		CPL	Temporary	27/01/1987	
GOODMAN	D	N		CPL	Temporary	27/01/1997	
GRACE	J	te	H	SQD LDR		06/05/1954	
GRACE	J	te	H	SQD LDR		06/05/1954	
GRANT	L	J		SGT	Acting	27/01/2003	
GRAY	D	W		FL		16/08/1960	
GREAGER	C	S		FL		13/12/1953	
GUDGEON	R	P		FLT SGT		25/09/1983	
GUINEY	I	E		GSH		01/10/1984	
GUNDERSON	I	H		FL		27/04/1956	
GUNDERSON	I	H		FL		26/12/1962	
HAIG	A	M		LAC		22/05/1982	
HAINES	K	J		GSH		15/01/1983	
HAMAN	T			GSH		25/06/1983	
HAMILTON	A			FO		03/08/1961	
HANIFY	G	L		LAC		20/07/1982	
HANNAFORD	R	J		CPL	Temporary	13/05/1957	
HARDMAN	J	N		CPL		01/10/2001	
HARRIS	B			FL		08/06/1998	
HARTIS	T			FLT SGT		16/06/1958	
HARVEY	G			SQD LDR		03/11/1955	
HEATH	L	T		SGT	Temporary	16/06/1958	
HENTY	P	M	J	CPL		10/03/1981	
HENTY	P	M	J	CPL		10/03/1991	
HENTY	P	M	J	CPL		10/03/2001	
HERRICK	M	P		GSH		28/08/1999	
HEYDER	M	J		CPL		n/a	
HICKLING	R	B		GSH		23/04/1979	
HICKLING	R	B		GSH		23/04/1989	
HICKMAN	B	G		CPL	Temporary	16/03/1992	
HODGSON	K	W		FL		01/12/1955	
HOPE	F	M		CPL	Temporary	23/07/1983	
HORSHAM	R	L		SGT	Temporary	01/05/1965	
HORSHAM	R	L		SGT	Temporary	01/05/1975	
HORSHAM	R	L		SGT	Temporary	01/05/1985	
HOWARD	N	L	A	FL		20/06/1950	
HUNT	O	P	J	GSH		10/08/1997	
HUNT	W	A		FL		15/07/1060	
IVES	M	R		CPL		04/02/1975	
IVES	M	R		FLT SGT		04/02/1985	
IVES	M	R		FLT SGT		04/02/1995	
JACKSON	A	R	C	SQD LDR		31/07/1954	
JAMES	B	K		GSH		24/09/1989	
JAMES	P	H		GSH		18/10/1992	
JOBSON	J			GSH		29/04/1994	

Service No	Force	Notes	Surname
C134575	RNZAF	musn	GARMONSWAY
T134728	RNZAF	musn	GARRETT
B134781	RNZAF	musn	GARRETT
T134498	RNZAF	musn	GARROD
710929	RNZAF	gd; pilot	GAULT
73148	RNZAF	eng	GAWNE
134169	RNZAF		GEARD
H134465	RNZAF	musn	GENEFAAS
NZ1077	RNZAF	OBE; gd; pilot	GIBSON
B88367	RNZAF	airasst	GIBSON
F134325	RNZAF	musn	GILLIES
F134325	RNZAF	1st Clasp; musn	GILLIES
F134325	RNZAF	2nd Clasp; musn	GILLIES
V44294	RNZAF	airasst	GOLDSTONE
130795	RNZAF	gd; pilot	GOODLEY
D134507	RNZAF	musn	GOODMAN
D134507	RNZAF	1st Clasp; musn	GOODMAN
130304	RNZAF	MVO; adm	GRACE
130304	RNZAF	MVO; 1st Clasp; adm	GRACE
S941130	RNZAF	musn	GRANT
70155	RNZAF	gd; wo/ag	GRAY
130368	RNZAF	gd; pilot	GREAGER
A73899	RNZAF	musn	GUDGEON
K640674	RNZAF	airasst	GUINEY
130586	RNZAF	gd; pilot	GUNDERSON
130586	RNZAF	1st Clasp; gd; pilot	GUNDERSON
L134422	RNZAF	musn	HAIG
D85701	RNZAF	mtdrv	HAINES
G85934	RNZAF	airasst	HAMAN
130867	RNZAF	eng	HAMILTON
V134431	RNZAF	musn	HANIFY
130011	RNZAF	eng; arm	HANNAFORD
Q993338	RNZAF	musn	HARDMAN
C134832	RNZAF	atc	HARRIS
130433	RNZAF		HARTIS
130339	RNZAF	DFC; gd; pilot	HARVEY
130023	RNZAF		HEATH
B134367	RNZAF	musn	HENTY
B134367	RNZAF	1st Clasp; musn	HENTY
B134367	RNZAF	2nd Clasp; musn	HENTY
C93796	RNZAF	airasst	HERRICK
130053	RNZAF		HEYDER
U71893	RNZAF	airasst	HICKLING
U71893	RNZAF	1st Clasp; airasst	HICKLING
D134668	RNZAF	musn	HICKMAN
130111	RNZAF	gd; nav	HODGSON
Q134449	RNZAF	musn	HOPE
S708370	RNZAF	musn	HORSHAM
S708370	RNZAF	1st Clasp; musn	HORSHAM
S708370	RNZAF	2nd Clasp; musn	HORSHAM
130759	RNZAF	gd; wo/ag	HOWARD
K211793	RNZAF	airasst	HUNT
S710233	RNZAF	gd; pilot	HUNT
N134309	RNZAF	musn	IVES
N134309	RNZAF	1st Clasp; musn	IVES
N134309	RNZAF	2nd Clasp; musn	IVES
130377	RNZAF	gd; pilot	JACKSON
Y89331	RNZAF	airasst	JAMES
H48238	RNZAF	airasst	JAMES
F91338	RNZAF	airasst	JOBSON

Surname	Initials			Rank	Status	Date	Source
JOHANSEN	P	C		SGT		30/10/1956	
JOHNSON	F	H		FL		15/11/1961	
JOHNSTON	A	P	R	MAST SIG		29/10/1968	
JOHNSTON	A	P	R	MAST SIG		29/10/1978	
JOHNSTONE	W	A		FL		07/02/1999	
JUDD	N	E	K	LAC		22/02/2004	
KARATAU	M	L		GSH		12/03/1990	
KAY	R	M		SQD LDR		20/08/1954	
KENDRICK	J	F		FO		01/03/1957	
KENNARD	N	R		FL		16/09/1981	
KING	A	J		SQD LDR		04/12/1968	
KING	A	J		SQD LDR		04/12/1968	
KIRK	E	B		FL		01/03/1957	
KNOWLES	L	C		LAC		15/01/1986	
KNOWLES	L	C		LAC		15/01/1996	
KOFOED	W	R		WING COMM		01/12/1953	
KOFOED	W	R		WING COMM		02/08/1960	
LASLETT	S	N		FL		27/01/1961	
LASLETT	S	N		FL		29/09/1967	
LAWSON	W	P		FL		13/05/1957	
LEADLEY	E	J		FL		22/04/1969	
LEADLEY	E	J		FL		29/03/1973	
LEE	D	E		GSH		17/03/1996	
LEE	R	D		GSH		03/12/2000	
LEEDEN	A	G		GSH		07/10/1995	
LENNIE	C	M		SQD LDR		31/07/1958	
LICENCE	S			GSH		17/10/1977	
LIGHTFOOT	F	C		FLT SGT		02/03/1955	
LIGHTFOOT	R	W		FL		15/02/1952	
LILL	J	V		SQD LDR		02/02/1960	
LINKHORN	D	R		FL		20/10/1982	
LITTLE	R	F		SQD LDR		01/08/1985	
LITTLE	R	F		SQD LDR		01/08/1995	
LLOYD	I	S	J	LAC		01/10/1954	
LONDON	R	A		GSH		01/09/1980	
LONDON	R	A		GSH		01/09/1990	
LUTTRELL	J	L		FL		10/03/2003	
LYNSKEY	D	P		FL		01/02/1953	
MACK	P	H		FL		23/03/1994	
MACK	P	H		FL		23/03/2004	
MacKAY	A	M		SGT	Temporary	12/08/1982	
MacKAY	G	W	R	LAC		29/09/1997	
MacNAB	T	L		SGT	Temporary	30/10/1956	
MacPHERSON	J	L		GSH		18/09/1988	
MADILL	S	J		SQD LDR		15/05/1955	
MALCOLM	D	J	C	SGT	Temporary	13/05/1957	
MALCOLM	M	R		SGT	Temporary	11/10/1983	
MALCOLM	M	R		SGT	Temporary	11/10/1993	
MALING	J	R		WING COMM		16/07/1954	
MALZARD	C	J		FL		01/02/1954	
MANAHI	T	H		GSH		01/02/1998	
MANDER	D	B		CPL	Temporary	30/05/1974	
MANDER	D	B		CPL	Temporary	30/05/1984	
MARTIN	C	H		GSH		01/04/1984	
MASON	A	F	W	FO		12/02/1968	
MASON	J	G		CPL		16/04/1962	
MATEKUARE	T			GSH		28/03/1999	
MATI	T			GSH		05/08/1995	
MATTHEWS	E	R		GSH		07/01/2001	
McALLUM	M			CPL		01/09/2000	

Service No	Force	Notes	Surname
130279	RNZAF		JOHANSEN
132417	RNZAF	gd; wo/ag	JOHNSON
133061	RNZAF		JOHNSTON
133061	RNZAF	1st Clasp	JOHNSTON
A134849	RNZAF	atc	JOHNSTONE
E997077	RNZAF	musn	JUDD
Y37374	RNZAF	airasst	KARATAU
130311	RNZAF	sig	KAY
130124	RNZAF	gd; pilot	KENDRICK
P134080	RNZAF	gd; pilot	KENNARD
72262	RNZAF	gd; pilot	KING
72262	RNZAF	1st Clasp; gd; pilot	KING
130080	RNZAF	gd; pilot	KIRK
R134496	RNZAF	musn	KNOWLES
R134496	RNZAF	1st Clasp; musn	KNOWLES
130117	RNZAF	DSO,DFC; gd; pilot	KOFOED
130117	RNZAF	DSO,DFC; 1st Clasp; gd; pilot	KOFOED
B133953	RNZAF	gd; pilot	LASLETT
B133953	RNZAF	1st Clasp; gd; pilot	LASLETT
130112	RNZAF	gd; pilot	LAWSON
H133706	RNZAF	gd; air sig	LEADLEY
H133706	RNZAF	1st Clasp; gd; air sig	LEADLEY
E76870	RNZAF	airasst	LEE
J991285	RNZAF	airasst	LEE
T35345	RNZAF	airasst	LEEDEN
130674	RNZAF	equ	LENNIE
R36033	RNZAF	airasst	LICENCE
130620	RNZAF		LIGHTFOOT
130047	RNZAF	eng	LIGHTFOOT
70310	RNZAF	equ	LILL
R330364	RNZAF	eng	LINKHORN
G134487	RNZAF	atc	LITTLE
G134487	RNZAF	1st Clasp; atc	LITTLE
130018	RNZAF		LLOYD
J75195	RNZAF	airasst	LONDON
J75195	RNZAF	1st Clasp; airasst	LONDON
J756271	RNZAF	atc	LUTTRELL
130562	RNZAF	gd; pilot	LYNSKEY
D134714	RNZAF	atc	MACK
D134714	RNZAF	1st Clasp; atc	MACK
R134427	RNZAF	musn	MacKAY
S134819	RNZAF	musn	MacKAY
130055	RNZAF		MacNAB
G79034	RNZAF	airasst	MacPHERSON
130330	RNZAF	OBE,DFC; gd; pilot	MADILL
130013	RNZAF	arm; fitter	MALCOLM
G134464	RNZAF	musn	MALCOLM
G134464	RNZAF	1st Clasp; musn	MALCOLM
130083	RNZAF	AFC; gd; pilot	MALING
130771	RNZAF	arm	MALZARD
W92871	RNZAF	airasst	MANAHI
J134305	RNZAF	musn	MANDER
J134305	RNZAF	1st Clasp; musn	MANDER
C86666	RNZAF	airasst	MARTIN
133038	RNZAF		MASON
130170	RNZAF	eng; fitter	MASON
B40389	RNZAF	sup	MATEKUARE
R91831	RNZAF	airasst	MATI
E991373	RNZAF	sec	MATTHEWS
H991031	RNZAF	musn	McALLUM

Surname	Initials			Rank	Status	Date	Source
McBREARTY	C	J		GSH		01/08/1987	
McCABE	E	D		SQD LDR		28/02/1954	
McCABE	E	D		SQD LDR		28/02/1954	
McCABE	E	D		SQD LDR		13/06/1958	
McDONALD	R	A		WING COMM		20/08/1954	
McDONALD	R	A		WING COMM		29/10/1954	
McDONNELL	A	J		SGT		29/08/1993	
McDONNELL	A	J		SGT		29/08/1993	
McDONNELL	W			CPL		01/11/1995	
McFADDEN	J	H		SQD LDR		15/05/1954	
McLEAN	R	J		GSH		09/02/1997	
McMILLAN	A	C		CPL		10/11/2002	
McMINN	P	A		GSH		20/10/1988	
McNEIL	P	D		FL		09/02/1955	
McPHAIL	M	J		FL		19/09/1965	
McWILLIAMS	R	H		FO		03/05/1961	
MEATCHEM	S	R		SQD LDR		n/a	
MEATCHEM	S	R		SQD LDR		n/a	
MEGGET	A			CPL		08/04/1986	
MEGGET	A			CPL		08/04/1986	
MELVIN	K	E	W	FL		11/06/1959	
MELVIN	K	E	W	FL		11/06/1969	
MEWETT	W	I		GSH		21/01/1990	
MILLETT	D	E		MAST ENG		12/02/1968	
MILLETT	D	E		MAST ENG		12/02/1978	
MITCHELL	D	S		GSH		02/12/1995	
MITCHELL	T			FL		01/08/1985	
MITCHELL	T			FL		01/08/1995	
MONCRIEFF	M	S		GSH		19/02/2000	
MOONEY	L	E		SGT		05/03/1994	
MORAN	F	D	M	SQD LDR		30/11/1956	
MORAN	F	D	M	SQD LDR		30/07/1963	
MORAN	F	D	M	SQD LDR		27/10/1970	
MORLEY	T	B		FL		21/12/1959	
MORLEY	T	B		FL		21/12/1969	
MORRISON	N			WING COMM		10/06/1960	
MORTON	P	J		GSH		11/09/1999	
MOSELEY	K	R		SGT	Temporary	27/06/1987	
MOSELEY	K	R		SGT	Temporary	27/06/1997	
MOSSLEY	J	H		FL		n/a	
MURPHY	A			GSH		11/06/1983	
MURPHY	A			GSH		11/06/1993	
NEWPORT	T	C		GSH		19/09/1993	
NEWPORT	T	C		GSH		19/09/2003	
NICHOLAS	E	W		GSH		06/10/1996	
NICHOLAS	T	R		SGT	Temporary	30/07/1969	
NICHOLAS	T	R		SGT	Temporary	30/07/1979	
NICHOLAS	T	R		SGT	Temporary	30/07/1989	
NIELSEN	W	E		CPL		15/05/2003	
OLDFIELD	J	A		WING COMM		30/09/1957	
OLLERENSHAW	K			CPL		23/01/1968	
OLLERENSHAW	K			CPL		23/01/1978	
OLLERENSHAW	K			CPL		23/01/1988	
OMBLER	E	W		SGT	Temporary	30/10/1956	
OWENS	M	J		GSH		30/04/1989	
OXFORD	H	L		SGT	Temporary	09/09/1961	
OXFORD	H	L		SGT	Temporary	09/09/1971	
PALLISER	E	S		SQD LDR		01/01/1954	
PALLISER	E	S		SQD LDR		01/01/1964	
PALMER	I	P		SQD LDR		08/10/1962	

Service No	Force	Notes	Surname
U88292	RNZAF	mtdrv	McBREARTY
130910	RNZAF	gd; pilot	McCABE
130910	RNZAF	1st Clasp; gd; pilot	McCABE
130910	RNZAF	OBE; 2nd Clasp; gd; pilot	McCABE
130563	RNZAF	gd; pilot	McDONALD
130563	RNZAF	1st Clasp; gd; pilot	McDONALD
K134421	RNZAF	musn	McDONNELL
K134421	RNZAF	1st Clasp; musn	McDONNELL
M134768	RNZAF	musn	McDONNELL
130305	RNZAF	MBE; acc	McFADDEN
B92484	RNZAF	airasst	McLEAN
K995104	RNZAF	musn	McMILLAN
Y88986	RNZAF	sup	McMINN
130681	RNZAF	eng	McNEIL
888664	RNZAF	gd; pilot	McPHAIL
711958	RNZAF	gd; pilot	McWILLIAMS
NZ1151	RNZAF	gd; pilot	MEATCHEM
NZ1151	RNZAF	1st Clasp; gd; pilot	MEATCHEM
X134502	RNZAF	musn	MEGGET
X134502	RNZAF	1st Clasp; musn	MEGGET
329637	RNZAF	gd; pilot	MELVIN
329637	RNZAF	1st Clasp; gd; pilot	MELVIN
D89358	RNZAF	airasst	MEWETT
133763	RNZAF		MILLETT
133763	RNZAF	1st Clasp	MILLETT
F83840	RNZAF	airasst	MITCHELL
J134489	RNZAF	atc	MITCHELL
J134489	RNZAF	1st Clasp; atc	MITCHELL
Q93969	RNZAF	airasst	MONCRIEFF
H675494	RNZAF	musn	MOONEY
130811	RNZAF	gd; pilot	MORAN
130811	RNZAF	OBE; 1st Clasp; gd; pilot	MORAN
130811	RNZAF	OBE; 2nd Clasp; gd; pilot	MORAN
133255	RNZAF	gd; pilot	MORLEY
133255	RNZAF	1st Clasp; gd; pilot	MORLEY
130337	RNZAF		MORRISON
Y93563	RNZAF	airasst	MORTON
K134444	RNZAF	musn	MOSELEY
K134444	RNZAF	1st Clasp; musn	MOSELEY
No Number	RNZAF	DFM; gd; pilot	MOSSLEY
L85915	RNZAF	airasst	MURPHY
L85915	RNZAF	1st Clasp; airasst	MURPHY
R91049	RNZAF	airasst	NEWPORT
R91049	RNZAF	1st Clasp; airasst	NEWPORT
K92331	RNZAF	sec	NICHOLAS
J134213	RNZAF	musn	NICHOLAS
J134213	RNZAF	1st Clasp; musn	NICHOLAS
J134213	RNZAF	2nd Clasp; musn	NICHOLAS
H744494	RNZAF	musn	NIELSEN
130100	RNZAF	DFC; gd; pilot	OLDFIELD
K134168	RNZAF	musn	OLLERENSHAW
K134168	RNZAF	1st Clasp; musn	OLLERENSHAW
K134168	RNZAF	2nd Clasp; musn	OLLERENSHAW
130220	RNZAF		OMBLER
S89279	RNZAF	airasst	OWENS
133615	RNZAF	musn	OXFORD
133615	RNZAF	1st Clasp; musn	OXFORD
130144	RNZAF	acc	PALLISER
130144	RNZAF	1st Clasp; acc	PALLISER
73485	RNZAF	gd; pilot	PALMER

Surname	Initials			Rank	Status	Date	Source
PALMER	I	P		SQD LDR		08/10/1972	
PALMER	R	A		SGT	Temporary	16/06/1958	
PARKER	C	L		MAST SIG		16/04/1962	
PARKER	J	R	T	GSH		14/08/1999	
PARNELL	D			GSH		05/07/1986	
PASCOE	R	T		LAC		19/07/1993	
PEARSON	F	L		SQD LDR		15/04/1955	
PEARSON	G	E		GSH		01/12/1996	
PEARSON	L	I		GSH		09/08/1982	
PEARSON	L	I		GSH		09/08/1992	
PECK	A	J		GSH		01/05/1999	
PENMAN	W			SQD LDR		01/06/1987	
PENMAN	W			SQD LDR		01/06/1997	
PETERS	I	M		CPL		05/03/1960	
PETERS	I	M		SGT		05/03/1970	
PETERS	I	M		SGT		05/03/1980	
PHILLIPS	D	G		LAC		06/09/2003	
PIERCE	C			GSH		11/09/1982	
PIKE	T	R		WING COMM		24/09/1954	
PILCHER	E	A		FL		12/12/1961	
PILKINGTON	K	A		CPL		09/08/1998	
PILKINGTON	S			SGT	Temporary	10/04/1980	
PILKINGTON	S			SGT	Temporary	10/04/1990	
PLENDERLEITH	S	M		GSH		21/07/1990	
POMEROY	M	S		CPL		29/10/1984	
POMEROY	M	S		SGT		29/10/1994	
POWELL	M	P		GSH		15/12/1988	
POWELL	R	A		SQD LDR		25/07/1979	
POWELL	R	A		SQD LDR		25/07/1989	
PRICE	K	E		CPL	Temporary	12/08/1982	
PRING	G	N		GSH		14/01/1975	
PRING	G	N		GSH		14/01/1985	
PRING	G	N		GSH		14/01/1995	
QUINN	J			GSH		02/10/1988	
REANEY	W	H	P	GSH		04/08/1990	
REBER	E	J		GSH		01/06/1987	
REED	D	B		FL		12/12/1968	
REMIHANA	D	J		GSH		06/08/2001	
REYNISH	D	I		FLT SGT	Acting	02/07/1994	
REYNOLDS	S	J		CPL		30/04/1980	
REYNOLDS	S	J		CPL		30/04/1990	
RHODES	K	J		GSH		17/08/1997	
RIACH	G	B		SGT		01/05/1984	
RIACH	G	B		SGT		01/05/1994	
RIACH	J	H		LAC		18/07/1995	
RICH	G	T		GSH		26/02/1983	
RICHARDS	J	D		GSH		12/05/1996	
RICHARDS	S	W		FO		06/09/1954	
ROBERTS	I	J		SQD LDR		14/10/1991	
ROBERTS	I	J		SQD LDR		14/10/2001	
ROBERTSON	D	B		GSH		15/05/1988	
ROBERTSON	K	G		SGT		01/04/2004	
ROSALIND	M			SGT		11/10/1983	
ROSALIND	M			SGT		11/10/1993	
ROSE	M	J		FL		23/04/1951	
RUSHWORTH	L	T		SGT	Temporary	16/06/1958	
SADD	B	R		GSH		15/03/1992	
SADGROVE	D	R		FL		25/05/1961	
SAUNDERS	G	J		GSH		18/08/1990	
SAWYER	E	R		GSH		22/09/1991	

Service No	Force	Notes	Surname
73485	RNZAF	1st Clasp; gd; pilot	PALMER
130026	RNZAF		PALMER
133477	RNZAF		PARKER
S93442	RNZAF	sup	PARKER
S87876	RNZAF	airasst	PARNELL
S134704	RNZAF	musn	PASCOE
130684	RNZAF	eng	PEARSON
L92447	RNZAF	airasst	PEARSON
K90767	RNZAF	airasst	PEARSON
K90767	RNZAF	1st Clasp; airasst	PEARSON
M93552	RNZAF	airasst	PECK
M134515	RNZAF	atc	PENMAN
M134515	RNZAF	1st Clasp; atc	PENMAN
V134385	RNZAF	musn	PETERS
V134385	RNZAF	1st Clasp; musn	PETERS
V134385	RNZAF	2nd Clasp; musn	PETERS
U996033	RNZAF	musn	PHILLIPS
Q775045	RNZAF	airasst	PIERCE
130398	RNZAF		PIKE
132432	RNZAF	gd; wo/ag	PILCHER
R134841	RNZAF	musn	PILKINGTON
F134578	RNZAF	musn	PILKINGTON
F134578	RNZAF	1st Clasp; musn	PILKINGTON
U89557	RNZAF	airasst	PLENDERLEITH
B134735	RNZAF	musn	POMEROY
B134735	RNZAF	1st Clasp; musn	POMEROY
B89218	RNZAF	sec	POWELL
L78831	RNZAF	atc	POWELL
L78831	RNZAF	1st Clasp; atc	POWELL
S134428	RNZAF	musn	PRICE
X91676	RNZAF	airasst	PRING
X91676	RNZAF	1st Clasp; airasst	PRING
X91676	RNZAF	2nd Clasp; airasst	PRING
E88876	RNZAF	sup	QUINN
T89556	RNZAF	airasst	REANEY
T516045	RNZAF	airasst	REBER
783606	RNZAF	gd; pilot	REED
K990642	RNZAF	airasst	REMIHANA
B81237	RNZAF	musn	REYNISH
Y134388	RNZAF	musn	REYNOLDS
Y134388	RNZAF	1st Clasp; musn	REYNOLDS
N209519	RNZAF	airasst	RHODES
G134717	RNZAF	musn	RIACH
G134717	RNZAF	1st Clasp; musn	RIACH
X834898	RNZAF	musn	RIACH
Y85674	RNZAF	airasst	RICH
B92162	RNZAF	sup	RICHARDS
130401	RNZAF	atc	RICHARDS
Q80560	RNZAF	atc	ROBERTS
Q80560	RNZAF	1st Clasp; atc	ROBERTS
Y88733	RNZAF	airasst	ROBERTSON
D997352	RNZAF	musn	ROBERTSON
G134464	RNZAF	musn	ROSALIND
G134464	RNZAF	1st Clasp; musn	ROSALIND
133100	RNZAF	equ	ROSE
130028	RNZAF		RUSHWORTH
K13119	RNZAF	airasst	SADD
131763	RNZAF	gd; pilot	SADGROVE
E89566	RNZAF	airasst	SAUNDERS
V90340	RNZAF	airasst	SAWYER

Surname	Initials	Rank	Status	Date	Source
SCOTT	D B	FL		13/09/1965	
SCOTT	D L	LAC		04/02/1986	
SCOTT	D L	LAC		04/02/1996	
SCOTT	F L	WO		07/03/1960	
SEABROOK	W G	FL		13/09/1957	
SELWYN	B	WO		16/06/1995	
SEMPLE	R W	LAC		01/11/1995	
SHARP	G S	SQD LDR		16/09/1954	
SHAW	J	SQD LDR		03/06/1954	
SINCLAIR	D B	FL		16/04/1962	
SINCLAIR	R J	WING COMM		07/12/1954	
SINCLAIR	R M	FL		01/04/1970	
SKILLING	H H	SQD LDR		29/04/1956	
SKILLING	H H	SQD LDR		30/04/1961	
SKILLING	H H	SQD LDR		29/04/1970	
SMILLIE	G F	SGT		01/08/1990	
SMITH	E R	WO	Acting	25/02/1999	
SMITH	E	GSH		15/10/1989	
SMITH	J C	GSH		20/01/1989	
SMITH	P J	GSH		21/03/1998	
SMITH	S W	LAC		15/05/2003	
SPENCER	J B	SQD LDR		23/05/1959	
SPENCER	J B	SQD LDR		16/06/1966	
SPENCER	M R	SGT	Temporary	04/08/1991	
STANTIALL	S J H	CPL	Temporary	30/06/1987	
STARK	D M	FL		18/12/1966	
STEVENSON	F S	FL		04/10/1955	
STEVENSON	F S	FL		04/06/1962	
STEWART	R J	CPL	Temporary	07/11/1975	
STIRLING	T C	GSH		05/06/2000	
STOKES	W R	FL		01/10/1961	
STROTHER	A D	FL		16/01/1964	
STRUGNELL	E M	LAC		01/08/1988	
SWEARS	D C	FL		09/02/1968	
SYKES	D F	SQD LDR		08/06/1968	
TAYLOR	G D	FL		07/06/1993	
TAYLOR	L F P	WING COMM		16/07/1954	
TAYLOR	L F P	WING COMM		16/07/1954	
TAYLOR	N L	CPL		08/12/1997	
TAYLOR	S J	SQD LDR		28/10/1992	
TE MOANANUI	H	GSH		17/12/1997	
THOMPSON	G	SQD LDR		03/06/1988	
THOMPSON	L J	AIR COMMD		14/01/1960	
THORPE	A J	FLT SGT		03/10/1998	
TODD	G M	LAC		19/04/1973	
TOMPKINS	A	FL		01/05/1964	
TONKS	M S	LAC		16/01/1974	
TONKS	M S	LAC		16/01/1984	
TOPLIS	S P	CPL		15/10/1997	
TREMAYNE	T W	SQD LDR		21/05/1954	
TROTT	I C	GSH		17/04/1999	
UNDERHILL	E J	LAC		14/04/1969	
USMAR	F J	FL		03/12/1989	
VERCOE	G R	CPL		18/06/1995	
VILE	D D	GSH		28/04/1996	
WAGTENDONK	W J C	FL		14/11/1993	
WAGTENDONK	W J C	FL		14/11/2003	
WAKENSHAW	J D M	GSH		18/04/1987	
WALLACE	G	GSH		30/05/1987	
WARDILL	K E	WING COMM		09/10/1973	

Service No	Force	Notes	Surname
897414	RNZAF	gd; pilot	SCOTT
W134501	RNZAF	musn	SCOTT
W134501	RNZAF	1st Clasp; musn	SCOTT
130639	RNZAF		SCOTT
130673	RNZAF	sec	SEABROOK
B78339	RNZAF	musn	SELWYN
E85794	RNZAF	musn	SEMPLE
130332	RNZAF	DSO; gd; pilot	SHARP
130732	RNZAF	DFC; gd; pilot	SHAW
133104	RNZAF	gd; wo/ag	SINCLAIR
131857	RNZAF	gd; pilot	SINCLAIR
551842	RNZAF	gd; pilot	SINCLAIR
130575	RNZAF	gd; pilot	SKILLING
130575	RNZAF	1st Clasp; gd; pilot	SKILLING
130575	RNZAF	2nd Clasp; gd; pilot	SKILLING
B134597	RNZAF	musn	SMILLIE
Y79625	RNZAF	musn	SMITH
R73178	RNZAF	airasst	SMITH
L85432	RNZAF	airasst	SMITH
C744466	RNZAF	airasst	SMITH
F995675	RNZAF	musn	SMITH
133708	RNZAF	MBE; gd; pilot	SPENCER
133708	RNZAF	MBE; 1st Clasp; gd; pilot	SPENCER
W134639	RNZAF	musn	SPENCER
W134524	RNZAF	musn	STANTIALL
70432	RNZAF	eng	STARK
C130665	RNZAF	gd; nav	STEVENSON
C130665	RNZAF	1st Clasp; gd; nav	STEVENSON
690362	RNZAF	musn	STEWART
V555768	RNZAF	airasst	STIRLING
132949	RNZAF	eng	STOKES
131762	RNZAF	eng	STROTHER
E134531	RNZAF	musn	STRUGNELL
C592712	RNZAF	gd; nav	SWEARS
M134193	RNZAF	AFC,DFM; gd; pilot	SYKES
C84826	RNZAF	atc	TAYLOR
130127	RNZAF	gd; pilot	TAYLOR
130127	RNZAF	1st Clasp; gd; pilot	TAYLOR
Y134825	RNZAF	musn	TAYLOR
J102404	RNZAF	leg	TAYLOR
B93036	RNZAF	airasst	TE MOANANUI
D715694	RNZAF	MBE; gd; pilot	THOMPSON
Y920114	RNZAF	med	THOMPSON
X134847	RNZAF	musn	THORPE
X134295	RNZAF	musn	TODD
130255	RNZAF	gd; nav	TOMPKINS
D134300	RNZAF	musn	TONKS
D134300	RNZAF	1st Clasp; musn	TONKS
X134824	RNZAF	musn	TOPLIS
130138	RNZAF	MBE; eng	TREMAYNE
K604633	RNZAF	airasst	TROTT
676307	RNZAF	musn	UNDERHILL
M134561	RNZAF	atc	USMAR
P134632	RNZAF	musn	VERCOE
Y91102	RNZAF	airasst	VILE
X134709	RNZAF	atc	WAGTENDONK
X134709	RNZAF	1st Clasp; atc	WAGTENDONK
S88428	RNZAF	airasst	WAKENSHAW
H88281	RNZAF	airasst	WALLACE
C134299	RNZAF	med	WARDILL

Surname	Initials			Rank	Status	Date	Source
WARDILL	K	E		WING COMM		08/10/1983	
WARREN	A	M		CPL	Temporary	15/04/1955	
WASS	R	L	W	SGT	Acting	03/06/1987	
WASS	R	L	W	SGT	Acting	03/06/1997	
WATERMAN	A	R		CPL	Temporary	27/08/1979	
WATERMAN	A	R		CPL	Temporary	27/08/1989	
WATERMAN	R	W		FLT SGT	Acting	27/08/1979	
WATERMAN	R	W		FLT SGT	Acting	27/08/1989	
WATTS	J	W	P	SQD LDR		01/01/1954	
WATTS	J	W	P	SQD LDR		01/01/1954	
WEBB	T	C		LAC		22/11/1968	
WEBBER	R	M	H	FL		20/11/1961	
WESTON	P	E		FL		01/07/1955	
WHEELER	S	J		CPL		01/07/1990	
WHEELER	S	J		SGT		01/07/1990	
WHITE	A	N		WING COMM		15/05/1973	
WHITE	A	N		WING COMM		15/05/1983	
WHITE	B	J		SGT		15/07/1996	
WHITE	G	M		FL		20/11/1959	
WHITE	L	E	M	LAC		15/07/1996	
WHITE	M	M		FL		01/07/1966	
WHITEMAN	N	M		GSH		19/06/1984	
WHITING	D	A		CPL		07/06/1994	
WHITTLE	R	V		FL		25/05/1959	
WILDEY	W	E		FL		20/11/1956	
WILLIAMS	D	A		FO		29/08/1954	
WILLIAMS	D	A		FO		30/04/1971	
WILLIAMS	W	E	V	CPL		n/a	
WILSON	A	C		GSH		09/10/1988	
WILSON	D			GSH		01/05/1999	
WILSON	K	A		GSH		18/11/1995	
WILSON	S	T		GSH		13/07/1977	
WILSON	S	T		GSH		13/07/1987	
WINSHIP	R	K		FLT SGT		04/11/1965	
WINSHIP	R	K		FLT SGT		04/11/1975	
WINSHIP	R	K		FLT SGT		04/11/1985	
WINTER	R	F		LAC		13/07/1970	
WISE	J	P		CPL		29/06/1992	
WISE	R	G		SGT		03/09/1983	
WISE	R	G		SGT		03/09/1993	
WOOD	R	W		GSH		12/10/1997	
WRIGHT	A	W		CPL		31/01/1990	
WRIGHT	A	W		CPL		31/01/2000	

Service No	Force	Notes	Surname
C134299	RNZAF	1st Clasp; med	WARDILL
130204	RNZAF		WARREN
U213159	RNZAF	musn	WASS
U213159	RNZAF	1st Clasp; musn	WASS
K134375	RNZAF	musn	WATERMAN
K134375	RNZAF	1st Clasp; musn	WATERMAN
C72866	RNZAF	musn	WATERMAN
C72866	RNZAF	1st Clasp; musn	WATERMAN
130031	RNZAF	sec	WATTS
130031	RNZAF	1st Clasp; sec	WATTS
717015	RNZAF	musn	WEBB
133213	RNZAF	gd; wo/ag	WEBBER
130079	RNZAF	gd; pilot	WESTON
V134592	RNZAF	musn	WHEELER
V134592	RNZAF	1st Clasp; musn	WHEELER
Q134242	RNZAF	med	WHITE
Q134242	RNZAF	1st Clasp; med	WHITE
G134786	RNZAF	musn	WHITE
72500	RNZAF	gd; pilot	WHITE
F134785	RNZAF	musn	WHITE
72252	RNZAF	gd; pilot	WHITE
J86718	RNZAF	airasst	WHITEMAN
G78275	RNZAF	musn	WHITING
132804	RNZAF	gd; wo/ag	WHITTLE
G130761	RNZAF	gd; nav	WILDEY
130600	RNZAF	gd; pilot	WILLIAMS
130600	RNZAF	1st Clasp; gd; pilot	WILLIAMS
130065	RNZAF		WILLIAMS
K331554	RNZAF	airasst	WILSON
N93553	RNZAF	airasst	WILSON
X91906	RNZAF	airasst	WILSON
BR71292	RNZAF	airasst	WILSON
BR71292	RNZAF	1st Clasp; airasst	WILSON
Y134135	RNZAF	musn	WINSHIP
Y134135	RNZAF	1st Clasp; musn	WINSHIP
Y134135	RNZAF	2nd Clasp; musn	WINSHIP
134229	RNZAF	musn	WINTER
Q134679	RNZAF	musn	WISE
V825788	RNZAF	musn	WISE
V825788	RNZAF	1st Clasp; musn	WISE
D337574	RNZAF	airasst	WOOD
S134566	RNZAF	musn	WRIGHT
S134566	RNZAF	1st Clasp; musn	WRIGHT

STATISTICS

As with any statistical exercise, the source data can be presented in a variety of ways and to the lowest level of detail. However, rather than include statistics just for the sake of it, the following few tables have been included to highlight those aspects of the volume and spread of awards that may be of interest to the medal collector or researcher.

Statistics for the years 1942 to 2005 inclusive (to the publication of this book) are provided in four key areas:
- Total awards (split by medal and clasps) for each country,
- Annual awards (split by medal and clasps) for each country,
- Awards by Air Force rank for each country,
- Awards by service for each country.

A look at the tables does provoke some interesting ideas and provide possible pointers for anyone considering the addition of an AEA to their medal collection, including:
- the comparatively small numbers of medals issued over the years to certain Air Force trades and ranks,
- the high and low volume of medals issued during certain decades and even individual years,
- the very few medals (and no clasps) issued in total by the Royal Canadian Air Force.

It should also be noted that awards to the Royal Australian, Royal Canadian and Royal Hong Kong Auxiliary Air Forces have ceased, but that a few awards are still made annually to the British and New Zealand air forces, the former despite the fact that the AEA was officially discontinued in 1999.

By way of a high-level summary, from the nominal rolls, a total of 10,605 British and Commonwealth awards have been made between 1942 and 2005 (accepting that this number is most certainly an underestimate induced by the lack of official rolls for the entire period), comprising:
- 10,042 awards of medals,
- 508 awards of first clasps,
- 52 awards of second clasps,
- 3 awards of third clasps.

No fourth clasp has been awarded to date.

Of the total awards (medals and clasps), the ratio between officers and other ranks is approximately 2:1 (65% officers and 35% other ranks).

Table of total awards of the Air Efficiency Award 1942-2005 (split by medal and clasps) for each country

	GB	Aus	Can	HK	NZ	TOTAL
Medals	8757	648	94	185	358	10042
1st Clasp	322	36	0	63	87	508
2nd Clasp	21	3	0	11	17	52
3rd Clasp	2	0	0	0	1	3
TOTAL	9102	687	94	259	463	10605

Table of annual awards of the Air Efficiency Award 1942-2005 (split by
medal and clasps) for each country

Year	GB				Aus				Can			
	Award	1st Clasp	2nd Clasp	3rd Clasp	Award	1st Clasp	2nd Clasp	3rd Clasp	Award	1st Clasp	2nd Clasp	3rd Clasp
1942	155											
1943	371											
1944	447	1							42			
1945	758								32			
1946	745								9			
1947	1208								4			
1948	1092								1			
1949	517											
1950	642				76				2			
1951	251	41	2	1	102				1			
1952	199	36			31				1			
1953	219	20			66				1			
1954	220	14	1		33							
1955	219	17	1		10							
1956	161	16	1		5							
1957	200	26	2		12							
1958	169	17	4		2				1			
1959	227	19		1	2							
1960	227	12	1		5							
1961	112	15	2									
1962	72	9	1		3							
1963	72	15	1		4							
1964	4	10			76	12	1					
1965	5	4			59	6						
1966	6	5			29	4	1					
1967	1				10	2						
1968	4	1			8	3						
1969					14							
1970	2				11	2						
1971					20	3						
1972					30							
1973	1				18		1					
1974					17	3						
1975	1				5	1						
1976												
1977												
1978	2											
1979	1											
1980		2										
1981	1											
1982	1											

Year	HK				NZ				Total
	Award	1st Clasp	2nd Clasp	3rd Clasp	Award	1st Clasp	2nd Clasp	3rd Clasp	
1942									2097
1943									2314
1944									2434
1945									2735
1946									2700
1947									3159
1948									3041
1949									2466
1950					1				2671
1951					1				2350
1952					2				2221
1953					4				2263
1954					29	6			2257
1955					11	1			2214
1956					11				2150
1957					8	1			2206
1958	5				6		1		2163
1959	11				7	2			2228
1960	18				7	1			2231
1961	9				12	1			2112
1962	15				6	3			2071
1963	14				1	3			2073
1964	3	2			6	1			2079
1965	7	1			6	1	1		2055
1966	7	2			4	2			2026
1967						1			1981
1968	1	1			14	1			2001
1969		2			5	1			1991
1970		4	2		2	2	3		1998
1971	3	3			2	2			2004
1972	1					1			2004
1973	2	4			3	1			2003
1974	5	3	1		2	1			2006
1975	2	2			4	2	1		1993
1976	1	4			1				1982
1977	6	2			2				1987
1978	3	1				3			1987
1979	4				4	2			1990
1980	5				3		1		1991
1981	3	2	1		5				1993
1982	4	1			7				1995

Year	GB				Aus				Can			
	Award	1st Clasp	2nd Clasp	3rd Clasp	Award	1st Clasp	2nd Clasp	3rd Clasp	Award	1st Clasp	2nd Clasp	3rd Clasp
1983		1										
1984	4											
1985	5	1										
1986	3											
1987	5											
1988	3	2										
1989	16	3										
1990	19	3	2									
1991	7	1										
1992	21	1										
1993	46		1									
1994	41	4										
1995	55	4										
1996	33	5										
1997	62	3										
1998	54	3	2									
1999	20	5										
2000	12	4										
2001	8	1										
2002	5											
2003	4											
2004	12	1										
2005	10											
No Date												
Total	8757	322	21	2	648	36	3	0	94	0	0	0

Notes:
a) GB statistics include awards subsequently Forfeited
b) NZ awards noted under 'No date' include two pre 1950 awards

Year	HK				NZ				Total
	Award	1st Clasp	2nd Clasp	3rd Clasp	Award	1st Clasp	2nd Clasp	3rd Clasp	
1983	2				10	2			1998
1984	8	3	3		6	2	1		2011
1985	5	3	1		6	4	1		2011
1986	11	2			5	2			2009
1987	3	4			12	1			2012
1988	3		1		11		1		2009
1989	12	4			8	4	2		2038
1990	3	5	2		15	4			2043
1991	3	2			7	3			2014
1992	2	4			6	1			2027
1993	2	2			6	5			2055
1994					8	3		1	2051
1995					15	1	2		2072
1996					8	2	1		2045
1997	2				13	4			2081
1998					5				2062
1999					13	1			2038
2000					5	3			2024
2001					6	1	1		2018
2002					1	1			2009
2003					5	2			2014
2004					3	1	1		2022
2005									2015
No Date					8	1			9
Total	185	63	11	0	358	87	17	1	10605

Table of awards of the Air Efficiency Award 1942-2005 by air force rank for each country

Rank	Abbreviation	GB	Aus	Can	HK	NZ	TOTAL
Air Chief Commandant	AIR CHIEF COMM	1					1
Air Commandant	AIR COMMT	1					1
Air Commodore	AIR COMMD	6	5	3		3	17
Air Vice Marshall	AVM	1		5			6
Aircraftsman	AC		1				1
Aircraftsman 1st Class	AC1	44					44
Aircraftsman 2nd Class	AC2	21					21
Cadet Officer	CAD OFF				1		1
Captain	CAPT	2					2
Chief Technician	CHIEF TECH	2					2
Colour Sergeant	C SGT	1					1
Corporal	CPL	968	56		52	59	1135
Engineer 1A	ENG 1A	1					1
Flight Lieutenant	FL	2715	141	14	59	82	3011
Flight Officer	FLT OFF	120					120
Flight Sergeant	FLT SGT	456	54	4	29	17	560
Flying Officer	FO	719	47	9	16	7	798
General Service Hand	GSH					103	103
Group Captain	GRP CAP	59	28	17			104
Group Officer	GRP OFF	10					10
Gunner	GNR	1					1
Junior Technician	JNR TECH	5					5
Lance Corporal	L CPL	1					1
Leading Aircraftman	LAC	543	29		7	30	609
Leading Aircraftwoman	LACW	1					1
Master Engineer	MST ENG					3	3
Master Gunner	MST GNR	1					1
Master Navigator	MST NAV	1					1
Master Pilot	MST PIL	1					1
Master Signaller	MST SIG	1				4	5
Navigator	NAV	1					1
Pilot 1	PILOT 1	2					2
Pilot 2	PILOT 2	1					1
Pilot Officer	PO	77	5	1	5		88
Principal Chaplain	CHAP		1				1
Reverend	REV	3					3
Section Officer	SEC OFF	19					19
Senior Air Traffic Controller	SNR ATC				1		1
Senior Aircraftman	SAC	161			24		185
Senior Aircraftwoman	SACW	10					10
Senior Aircrewman	SNR ACM				1		1
Senior Tech	SNR TECH	3					3
Sergeant	SGT	1016	87	2	44	52	1201
Signaller I	SIG 1	2					2
Squadron Leader	SQD LDR	1192	88	6	10	74	1370
Squadron Officer	SQD OFF	56					56
Warrant Officer	WO	331	68		4	7	410
Warrant Officer Class 1	WO1			12			12
Warrant Officer Class 2	WO2			7			7
Wing Commander	WING COMM	523	77	14	6	22	642
Wing Officer	WING OFF	23					23
TOTAL		**9102**	**687**	**94**	**259**	**463**	**10605**

Table of awards of the Air Efficiency Award 1942-2005 by service for each country

GB Service	Acronym	TOTAL
Army Emergency Reserve	AER	1
Auxiliary Air Force	AAF	976
Auxiliary Air Force Reserve of Officers	AAFRO	6
Ex - Auxiliary Air Force	EX-AAF	9
Ex - Auxiliary Air Force Reserve of Officers	EX-AAFRO	1
Ex - Royal Air Force Volunteer Reserve	EX-RAFVR	3
Royal Air Force	RAF	39
Royal Air Force Reserve	RAFR	1
Royal Air Force Reserve of Officers	RAFRO	2
Royal Air Force Volunteer Reserve	RAFVR	4991
Royal Air Force Volunteer Reserve (Training)	RAFVR(T)	9
Royal Australian Air Force	RAAF	1
Royal Auxiliary Air Force	RAuxAF	2697
Territorial Army	TA	4
Womens Auxiliary Air Force	WAAF	255
Womens Royal Air Force Volunteer Reserve	WRAFVR	14
Womens Royal Auxiliary Air Force	WRAAF	93
TOTAL		**9102**

Australia Service	Acronym	TOTAL
Air Force Reserve	AFR	38
Citizens Air Force Reserve	CAF-R	219
Ex-AFR	EX-AFR	3
Ex-CAF-R	EX-CAF-R	242
Ex-PAF	EX-PAF	2
Peoples Air Force	PAF	166
Retired	RETIRED	7
Womens Royal Auxiliary Air Force	WRAAF	10
TOTAL		**687**

Canada Service	Acronym	TOTAL
Royal Canadian Air Force	RCAF	94
TOTAL		**94**

Hong Kong Service	Acronym	TOTAL
Government Flying Service	GSF	2
Hong Kong Auxiliary Air Force	HKAAF	256
Hong Kong Womens Auxiliary Air Force	HKWAAF	1
TOTAL		**259**

New Zealand Service	Acronym	TOTAL
Royal New Zealand Air Force	RNZAF	463
TOTAL		**463**

REFERENCES

Congdon, P., *Behind The Hangar Doors*, 1985, Woodhall Spa, Lincs., Sonik Books.

Dickson, A., Gp Capt, RAFVR, *The Royal Air Force Volunteer Reserve – Memories*, 1997, London, Ministry of Defence.

Hanson, C.M., *By Such Deeds – Honours and Awards in the Royal New Zealand Air Force 1923-1999*, 2001, Christchurch, NZ, Volplane Press.

Jefford, C.G. (Ed), *Royal Air Force Reserve and Auxiliary Forces*, 2003, RAF Historical Society Symposium, RAF Museum, Hendon.

Source documents included:

The *Air Force Lists* and *Air Force Retired Lists*, London, HMSO/TSO.

The *London Gazette*, London, HMSO/TSO.

Yearbook of the Royal Auxiliary Air Force, from 1996 to 2003.

At the National Archives, Kew, files in the Public Record Office collection, particularly in AIR 2/6892 (Decorations, Medals, Honours and Awards (Code B,30): Air Efficiency Award: suggestions for design), and T333/44 (Air Efficiency Award: South Africa).

Flight Lieutenant Henry Frank Grubb, AE, RAFVR

A Londoner, Grubb and his identical twin brother Ernest also had near identical air force careers, both serving during the Battle of Britain as pilots with 219 Squadron, a Blenheim night fighter unit with which he achieved one confirmed 'kill'. His service period was comparatively short, 1938 to 1945, but he earned the AEA within the ten-year qualifying period because of his combination of air-crew and wartime service.

His medal entitlement is: 1939-45 Star with Battle of Britain clasp, Air Crew Europe Star, Defence Medal, 1939-45 War Medal, Air Efficiency Award.

Grubb was awarded the AEA under AMO N425 dated 4 May 1944.

(Author's collection)

Squadron Leader Robert Austin Kings, AE, RAF

Kings was a Hurricane pilot in 238 Squadron who baled out twice during the Battle of Britain – once after being shot down and once as a result of a collision. He remained with the squadron when it embarked on HMS *Victorius* for the Middle East. A 1938 RAFVR entrant, he ended his career in 1964.

His medal entitlement is: 1939-45 Star with BATTLE OF BRITAIN clasp, Air Crew Europe Star, Africa Star with NORTH AFRICA 1942–1943 clasp, Defence Medal, 1939-45 War Medal, Air Efficiency Award.

Kings was awarded the AEA under AMO N485 dated 10 May 1945.

(Author's collection)

Group Captain Joseph Shaw Kennedy, DFC*, AE, RAF

A Boston and Blenheim pilot with 226 Squadron, Kennedy earned his DFC in 1941 for low-level attacks on enemy shipping, and an immediate Bar for laying an effective smokescreen for the Dieppe raid in August 1942. He had also previously led the first raid on Europe by American bomber crews, on 4 July 1942. Kennedy continued his service as a post-war regular, and served as Air Attaché in Bucharest in the early 1950s, eventually retiring from the service in 1959.

His medal entitlement is: DFC and Bar, 1939-45 Star, Atlantic Star with AIR CREW EUROPE clasp, Italy Star, Defence Medal, 1939-45 War Medal, Air Efficiency Award.

Kennedy was awarded the AEA under AMO N187 dated 6 March 1947.

(Author's collection)

Squadron Leader Charlton 'Wag' Haw, DFC, DFM, AE, RAF

A pre-war RAFVR Sergeant pilot, Haw was called up in September 1939, and flew Hurricanes with 504 Auxiliary Squadron during the Battle of Britain, and later with 81 Squadron in Russia, for which he was decorated with Russia's highest award for gallantry, the Order of Lenin. He finished the war with four confirmed kills and retired from the RAF in 1951.

His medal entitlement is: DFC, DFM, 1939-45 Star with BATTLE OF BRITAIN clasp, Air Crew Europe Star with ATLANTIC clasp, Defence Medal, 1939-45 War Medal, Air Efficiency Award, Order of Lenin (Russia).

Haw was awarded the AEA under AMO N1355 dated 13 December 1945.

(Medals held in a private collection)

Flying Officer Cecil Frederick Rawnsley, DSO, DFC, DFM*, AE, RAuxAF

One of the finest examples of an aircrew non-pilot AEA group, Rawnsley earned his awards as partner to John 'Cats Eyes' Cunningham, the pair becoming the best known night-fighter partnership. Rawnsley served from 1936 through to 1948, the post-war years in the Secretarial Branch of the Royal Auxiliary Air Force.

His medal entitlement is: DSO, DFC, DFM and Bar, 1939-45 Star with BATTLE OF BRITAIN clasp, Air Crew Europe Star, Defence Medal, 1939-45 War Medal, Air Efficiency Award, DFC (US).

Rawnsley was awarded the AEA under AMO N1177 dated 11 November 1943.

(Medals held in a private collection)

Wing Commander Donald Ernest Kingaby, DSO, AFC, DFM**, AE, RAF

Kingaby's medal group is, of course, quite special, as it contains the unique award of two Bars to the DFM. His RAF career has been well chronicled, as has his tally of enemy aircraft destroyed, the majority of which were achieved whilst serving with 92 Squadron as a Spitfire pilot. His service career spanned the period 1939 to 1958.

His medal entitlement is: DSO, AFC, DFM and Two Bars, 1939-45 Star with BATTLE OF BRITAIN clasp, Air Crew Europe Star with FRANCE AND GERMANY clasp, Defence Medal, 1939-45 War Medal, GSM 1918 with clasp PALESTINE 1945-48, Air Efficiency Award, DFC (US), *Croix De Guerre avec Palme* (Belgium).

Kingaby was awarded the AEA under AMO N197 dated 28 February 1946.

(Medals held in a private collection)

Wing Commander Edgar Bernard Richard Lockwood, DSO, MBE, AE**, RAuxAF

A Liberator pilot with, and later Commanding Officer of, 614 Squadron, Lockwood received his MBE as an immediate award in 1942 for rescuing a crew-man from a burning Wellington that had crashed at Elgin Aerodrome. His DSO followed in 1945 for his pin-point bombing in support of the 8th Army advance in Italy. His group is notable as it contains one of only 52 instances of two Bars to the AEA, earned through his lengthy service that only ended in 1963.

His medal entitlement is: DSO, MBE, 1939-45 Star, Air Crew Europe Star, Italy Star, Defence Medal, 1939-45 War Medal, GSM 1918 with clasp PALESTINE 1945-48, Air Efficiency Award and Two Bars.

Lockwood was awarded the AEA under AMO N1758 dated 24 December 1942, and his Bars/Clasps under AMOs N913 dated 6 September 1951 and N318 dated 26 April 1956.

(Author's collection)

Warrant Officer Brian Davies, RAuxAF

A departure from the more traditional Second World War groups containing AEAs – a superb modern trio to a photographic intelligence specialist, comprising: General Service Medal 1962 with clasps DHOFAR and AIR OPERATIONS IRAQ, RAF Long Service and Good Conduct Medal, Air Efficiency Award.

The group represents a superb 'double long service' regular and reserve RAF commitment spanning some 39 years in total, with the medals earned as follows: RAF LS&GC in 1974, the GSM for Dhofar in 1976, the AEA in 1993 and the clasp for Air Operations over/against Iraq in 1998.

Davies's AEA was awarded on 26 February 1993.

(Medals held in a private collection)